Chemicals and Fuels from Biomass *via* Fischer–Tropsch Synthesis

A Route to Sustainability

Catalysis Series

Editor-in-chief:

Justin S. J. Hargreaves, *University of Glasgow, UK*

Series editors:

Harry Bitter, *Wageningen University, The Netherlands*

Jose Rodriguez, *Brookhaven National Laboratory, USA*

Titles in the series:

1: Carbons and Carbon Supported Catalysts in Hydroprocessing
2: Chiral Sulfur Ligands: Asymmetric Catalysis
3: Recent Developments in Asymmetric Organocatalysis
4: Catalysis in the Refining of Fischer–Tropsch Syncrude
5: Organocatalytic Enantioselective Conjugate Addition Reactions: A Powerful Tool for the Stereocontrolled Synthesis of Complex Molecules
6: N-Heterocyclic Carbenes: From Laboratory Curiosities to Efficient Synthetic Tools
7: *P*-Stereogenic Ligands in Enantioselective Catalysis
8: Chemistry of the Morita–Baylis–Hillman Reaction
9: Proton-Coupled Electron Transfer: A Carrefour of Chemical Reactivity Traditions
10: Asymmetric Domino Reactions
11: C-H and C-X Bond Functionalization: Transition Metal Mediation
12: Metal Organic Frameworks as Heterogeneous Catalysts
13: Environmental Catalysis Over Gold-Based Materials
14: Computational Catalysis
15: Catalysis in Ionic Liquids: From Catalyst Synthesis to Application
16: Economic Synthesis of Heterocycles: Zinc, Iron, Copper, Cobalt, Manganese and Nickel Catalysts
17: Metal Nanoparticles for Catalysis: Advances and Applications
18: Heterogeneous Gold Catalysts and Catalysis
19: Conjugated Linoleic Acids and Conjugated Vegetable Oils
20: Enantioselective Multicatalysed Tandem Reactions
21: New Trends in Cross-Coupling: Theory and Applications
22: Atomically-Precise Methods for Synthesis of Solid Catalysts
23: Nanostructured Carbon Materials for Catalysis
24: Heterocycles from Double-Functionalized Arenes: Transition Metal Catalyzed Coupling Reactions
25: Asymmetric Functionalization of C–H Bonds
26: Enantioselective Nickel-catalysed Transformations
27: N-Heterocyclic Carbenes: From Laboratory Curiosities to Efficient Synthetic Tools, 2nd edition
28: Zeolites in Catalysis: Properties and Applications
29: Biocatalysis: An Industrial Perspective
30: Dienamine Catalysis for Organic Synthesis

31: Metal-free Functionalized Carbons in Catalysis: Synthesis, Characterization and Applications
32: Modern Biocatalysis: Advances Towards Synthetic Biological Systems
33: NO_x Trap Catalysts and Technologies: Fundamentals and Industrial Applications
34: Alternative Catalytic Materials: Carbides, Nitrides, Phosphides and Amorphous Boron Alloys
35: Enantioselective Cobalt-catalysed Transformations
36: Noncovalent Interactions in Catalysis
37: Carbon Nanomaterials in Hydrogenation Catalysis
38: Nanoparticle Design and Characterization for Catalytic Applications in Sustainable Chemistry
39: Catalytic Aerobic Oxidations
40: Catalysis with Earth-abundant Elements
41: Vanadium Catalysis
42: Organocatalytic Dynamic Kinetic Resolution
43: Asymmetric Autocatalysis: The Soai Reaction
44: Chemicals and Fuels from Biomass *via* Fischer–Tropsch Synthesis: A Route to Sustainability

How to obtain future titles on publication:
A standing order plan is available for this series. A standing order will bring delivery of each new volume immediately on publication.

For further information please contact:
Book Sales Department, Royal Society of Chemistry, Thomas Graham House, Science Park, Milton Road, Cambridge, CB4 0WF, UK
Telephone: +44 (0)1223 420066, Fax: +44 (0)1223 420247
Email: booksales@rsc.org
Visit our website at www.rsc.org/books

Chemicals and Fuels from Biomass via *Fischer–Tropsch Synthesis*

A Route to Sustainability

Edited by

Joshua Gorimbo
University of South Africa, South Africa
Email: joshuagorimbo@gmail.com

Xinying Liu
University of South Africa, South Africa
Email: liux@unisa.ac.za

Yali Yao
University of South Africa, South Africa
Email: Yaoy@unisa.ac.za

and

Diane Hildebrandt
University of South Africa, South Africa
Email: hilded@unisa.ac.za

ROYAL SOCIETY
OF **CHEMISTRY**

Catalysis Series No. 44

Print ISBN: 978-1-83916-393-7
PDF ISBN: 978-1-83916-782-9
EPUB ISBN: 978-1-83916-783-6
Print ISSN: 1757-6725
Electronic ISSN: 1757-6733

A catalogue record for this book is available from the British Library

The Royal Society of Chemistry is a charity, registered in England and Wales, Number 207890, and a company incorporated in England by Royal Charter (Registered No. RC000524), registered office: Burlington House, Piccadilly, London W1J 0BA, UK, Telephone: +44 (0) 20 7437 8656.

For further information see our website at www.rsc.org

Printed in the United Kingdom by CPI Group (UK) Ltd, Croydon, CR0 4YY, UK

Preface

This book aims to fill the gaps in the field of converting biomass to specialty chemicals and fuels and to communicate the most interesting and relevant thinking. The utilization of biomass as a feedstock is a potential route to decreasing global reliance on fossil fuels, which will help to reduce CO_2 emissions and thereby contribute to replacing the existing fossil-based value chains with renewable and sustainable technologies. In this way, the book suggests an approach to help address the looming global warming crisis that the world is facing.

Fischer–Tropsch technology is gaining interest as a low-carbon route to producing liquid fuels and specialty chemicals from biomass and other nonconventional resources, such as waste. This book reviews the issues, solutions, novel ideas and technologies that could support the transition to a low-carbon future, making it a significant addition to contemporary literature on the subject of alternative fuels. The book serves as a reference that summarizes state-of-the-art technology and outlines the knowledge and technology development gaps. We believe it makes a contribution to addressing climate change by educating, promoting research and highlighting the latest developments in the waste-to-energy sector. We hope that this will help to bring together the research community and the private sector and to influence government policies and decision making.

To this end, this book:

1. Provides an up-to-date overview of specialty chemicals and fuels that can be produced from biomass *via* the Fischer–Tropsch synthesis pathway.
2. Focuses on the key challenges that impede the use of biomass as a feedstock for manufacturing chemicals and fuels.

Catalysis Series No. 44
Chemicals and Fuels from Biomass *via* Fischer–Tropsch Synthesis: A Route to Sustainability
Edited by Joshua Gorimbo, Xinying Liu, Yali Yao and Diane Hildebrandt

Published by the Royal Society of Chemistry, www.rsc.org

3. Integrates technological and business development challenges to highlight the key technological developments that are necessary to commercialize biofuels on a global scale.
4. Looks at how to tailor the biomass to hydrocarbons pathways in order to produce a specific group of chemicals, with the ultimate aim of reducing downstream processing.
5. Considers new developments, including process synthesis, catalysts and reactors.

A book such as this could not have appeared without the sustained help of many people. Experts from around the world contributed chapters in their fields of expertise and reviewed the material. Thank you to all the contributing authors, the reviewers who participated in checking the scientific accuracy and the relevance of the material, and Juliet Gillies for editing the book.

Joshua Gorimbo
Yali Yao
X. Liu
Diane Hildebrandt

Foreword

Climate change is upon us. The evidence that this is being caused or at least exacerbated by the increase in carbon dioxide levels in the atmosphere is overwhelming. Only the most recalcitrant naysayers deny it.

How can we reduce our carbon dioxide emissions? It is not easy, as its sources are many and varied. In fact, to do this, we need to look at not just the industries that produce it but the systems that are responsible for its production. Thus there are many sources, though most come from the modern world's need for energy and, to a lesser extent, chemicals. It was the use of energy that spurred the initial Industrial Revolution and our subsequent development into the society we are today. To get the energy we needed for this development, we essentially became dependent on the combustion of fossil fuels. What we have been doing is turning the carbon stored in these materials over millions of years into the greenhouse gas, carbon dioxide, which is accumulating in the atmosphere. For instance, carbon capture and storage as currently mooted, and which is much in vogue, is not really a long-term sustainable solution. ("Delay the evil day!")

We need the energy required to run society as we know it without using fossil fuels. Thus we need new sources of energy that do not produce new sources of carbon dioxide. This has been recognized by most forward-thinking people and industries; hence the proliferation of photovoltaic and wind-generated electricity installations. But there is a problem: energy consumption and its use are not necessarily in phase, and so storage mechanisms are needed. Many of the solutions currently offered are not realistic and can create further problems.

Another form of solar collection that is currently not being adequately utilized as an energy source is photosynthesis. Here, the energy is actually stored as biomass rather than being directly available, *e.g.* as electricity. It is estimated that 140 billion tons of biomass waste are produced annually.

Catalysis Series No. 44
Chemicals and Fuels from Biomass *via* Fischer–Tropsch Synthesis: A Route to Sustainability
Edited by Joshua Gorimbo, Xinying Liu, Yali Yao and Diane Hildebrandt

Published by the Royal Society of Chemistry, www.rsc.org

The biomass is not in a form that can easily and conveniently be used an energy source in our modern environment. What can we do to make its use and storage more convenient?

About 34 billion barrels of crude oil are needed annually, and, for instance, it is estimated we could make about 100 billion barrels of synthetic crude from biomass waste using gasification and Fischer–Tropsch (FT) synthesis. Of course, the problem is when we use the synthetic crude oil as an energy source, we release carbon dioxide back into the air. However, if it is left to rot, it produces methane, which is a much more potent greenhouse gas. In any event, as long as we keep growing plants that use up the carbon dioxide we are producing, we are not increasing the carbon dioxide in the atmosphere.

It might be argued that we can store electricity in batteries or convert it to hydrogen. However, the energy density of these storage mechanisms is so low relative to liquid fuel that this represents a major problem, particularly for moving vehicles such as cars, trucks, ships and aircraft. Thus the need for liquid fuel, although diminishing, is likely to remain. The other advantage of liquid fuel for transport is the possibility to go where there is limited infrastructure, such as on farms and in rural areas.

Furthermore, liquid fuel is a very convenient energy storage method for when electricity supply cannot meet demand. This is because generators can be brought online easily and quickly. All of this suggests that while the demand for liquid fuels may decrease, it will continue to be needed in the future. The challenge then seems to be how to turn biomass into liquid fuel efficiently and sustainably.

Clearly what is needed are new ideas, and the people from IDEAS (Institute for the Development of Energy for African Sustainability) have the experience and the ideas to make a difference! They have been developing fundamental methods to look at systems (in particular chemical plants) to make them more efficient – and hence more environmentally sustainable. Over and above this, they have been working on FT – both the theoretical and the practical sides – for many years and have developed international reputations in all these areas. They are thus in a strong position to put forward new ideas and methods to make liquid fuel materials and chemicals sustainably from biomass. Clearly these ideas (IDEAS) are vital for achieving the sustainability that we should all be trying to achieve.

David Glasser

D. Glasser

Editor Biographies

Dr Joshua Gorimbo is a Senior Researcher at the Institute for Development of Energy for African Sustainability (IDEAS), a research unit at the University of South Africa (UNISA). He received his PhD degree from the University of the Witwatersrand, South Africa, in 2017, and then pursued postdoctoral training at UNISA (2018–2020). His research interests are catalysis and Fischer–Tropsch synthesis (FTS).

Diane Hildebrandt is a Distinguished Professor for Future Energy at Wits. She is the Director of the International Joint Research Laboratory of New Energy at Hebei University of Science and Technology and the Director of Keqiao Green Energy Materials Joint Laboratory, Zhijiang College of Zhejiang University of Technology. Both of these laboratories are situated in China. Diane Hildebrandt obtained her BSc, MSc and PhD from the Wits. She has authored or coauthored over 210 scientific papers, including an invited paper in *Science*, and has supervised over 100 postgraduate students. Her research area is the design of energy-efficient processes, with a view to reducing carbon dioxide emissions. She is particularly interested in how she can apply the results of her research to improving the lives of those who do not have access to energy and clean water. This involves working with multidisciplinary teams to understand the social, economic and technological barriers to the adoption of new technologies. Her approach to research is both academic and entrepreneurial.

Prof. Xinying Liu is the head of the catalysis research group at IDEAS. He received his PhD from Nankai University, Tianjin, China, in 1999. His field of study was carbon dioxide utilization. Prof. Liu is an expert in the field of catalytic and nano inorganic materials for energy applications, especially in relation to FTS. He also has 20 years of industrial experience in the carbonaceous material indirect liquefaction process (XTL process), in pilot-scale XTL facility design and operation, and small- and medium-scale

Catalysis Series No. 44
Chemicals and Fuels from Biomass *via* Fischer–Tropsch Synthesis: A Route to Sustainability
Edited by Joshua Gorimbo, Xinying Liu, Yali Yao and Diane Hildebrandt

Published by the Royal Society of Chemistry, www.rsc.org

XTL plant design (process modeling, conceptual and basic engineering). He has expanded his research into various other renewable energy solutions that are suitable for off-grid and rural applications, especially biomass hybrid solutions with concentrated solar power. He is the codirector of the International Joint Research Laboratory of New Energy at the Hebei University of Science and Technology.

Prof. Yali Yao is a professor at the Institute for the Development of Energy for African Sustainability (IDEAS) at UNISA. She received her PhD degree in Chemical Engineering from the University of the Witwatersrand (Wits), South Africa (2011). She continued her research work at Wits in a post-doctoral fellowship position from 2012 to 2013. She began her employment career as a Senior Researcher at UNISA in May 2013. She was promoted to the position of Associate Professor at IDEAS, a research institute at UNISA, in October 2017, and then to full professor in February 2022. She has completed more than 15 years of experimental work in the field of producing alternative fuels and value-added chemicals *via* FTS (XTL/FTS), in both laboratory and industrial projects. Her research areas include alternative energy solutions, crude oil purification and C_1 chemistry.

Contents

Catalysis Series No. 44
Chemicals and Fuels from Biomass *via* Fischer–Tropsch Synthesis: A Route to Sustainability
Edited by Joshua Gorimbo, Xinying Liu, Yali Yao and Diane Hildebrandt

Published by the Royal Society of Chemistry, www.rsc.org

CHAPTER 1

Process Synthesis of BTL, LCA (Scale, Coproduction)

OLUSOLA O. JAMES[a] AND SUDIP MAITY*[b]

[a] Department of Chemistry & Industrial Chemistry, Faculty of Pure & Applied Sciences, Kwara State University, Malete, P.M.B. 1530, Ilorin, Nigeria, olusola.james@kwasu
edu.ng; [b] CSIR-Central Institute of Mining and Fuel Research (Digwadih), PO: FRI, Dhanbad–828108, India
*Emails: sudip_maity@yahoo.com; sudipmaity@cimfr.nic.in

1.1 Introduction

The history of energy use is closely linked to the history of human civilisation. Energy use per capita is a direct measure of living standards and prosperity.[1] Advances in older civilisations and in present modern society are characterised by increasing energy consumption.[2] Since the agrarian era, when wood was the main fuel for industrial economy and later when coal became the preferred fuel for powering steam engines, energy use grew exponentially to enable a higher standard of living and large-scale industrialisation. This was followed by petroleum becoming the dominant global fuel, after the invention of diesel and internal combustion engines. These engines run on liquid fuel that is obtained cheaply from petroleum.

Liquid-fuel-powered engines were poised to displace solid-fuel- (coal-) powered steam engines in the rapid industrialisation drive in the early 1900s. Moreover, petroleum is more portable than coal and is used as a feedstock for a plethora of chemicals and products. However, petroleum deposits are unevenly distributed across the globe,[3] and perhaps the geographical scarcity of petroleum motivated

Catalysis Series No. 44
Chemicals and Fuels from Biomass *via* Fischer–Tropsch Synthesis: A Route to Sustainability
Edited by Joshua Gorimbo, Xinying Liu, Yali Yao and Diane Hildebrandt

Published by the Royal Society of Chemistry, www.rsc.org

early attempts at liquid fuel production from coal.[4] Friedrich Bergius invented the direct coal liquefaction (DCL) process in 1913. Although the impact of DCL was not apparent during World War I, the future role of liquid fuels in military logistics was not missed. This spurred greater interests in coal liquefaction after World War I. Extensive investigations into coal liquefaction culminated in industrial-scale DCL development and the invention of indirect coal liquefaction (ICL) by Franz Fischer and Hans Tropsch in 1923. The two coal liquefaction processes were developed and operated at industrial scale to meet part of Germany's liquid fuel needs during World War II. South Africa was also able to meet her liquid fuel needs during the Apartheid era using coal liquefaction technology.

After World War II, the economic and political importance of petroleum grew rapidly.[5,6] The long-term security of petroleum supplies moving from producing countries to consuming countries could not be guaranteed, with periods of smooth relations interspersed with interludes of disruption of petroleum supply.[5,7–10] Hence, competence in coal liquefaction technologies was generally considered of strategic importance by petroleum-consuming countries. Instances of disruption of petroleum supply reawakened interest in coal liquefaction technologies, but whenever supply was restored, commercial interest in coal liquefaction technologies waned. However, memories of hardship suffered during the disruption interludes remained a motivation, for strategic reasons, for sustained research interest in continuous improvement and preserving competence in the technology.

The DCL process involves subjecting a slurry of dry coal powder and a hydrogen-donor solvent to high temperature and pressure hydrogenation in the presence of a catalyst. It may be considered a top-down approach to coal liquefaction, since it entails the depolymerisation of large molecules into smaller molecules. The type of coal, the nature of the solvent, the reaction conditions and the catalyst used are important variables that determine the liquid fuel yield. Low-grade soft coal – usually lignite – is the most widely used feedstock for DCL. The hydrogen-donor solvent is usually tetralin – a hydrocarbon that can transfer hydrogen to coal. The dehydrogenated donor solvent has low enthalpy of hydrogenation that allows for its regeneration at a high hydrogen pressure in the presence of a catalyst such as MoS_2. This dehydrogenation and regeneration cycle of the donor solvent is crucial to the liquefaction process. Hence, research and developments on DCL are focussed on designing donor solvents and efficient catalyst for their regeneration, as well as on process optimisation to ensure a maximum liquid fuel yield.

The DCL liquid fuel production scheme is shown in Figure 1.1. The resulting DCL crude is made up of wide spectrum aromatic compounds and cannot be used directly in engines. It has to undergo an extensive refining process such as hydrocracking, hydrotreating, distillation, solvent extraction, blending *etc.* before it is suitable for use in engines.

ICL involves a two-stage process: syngas production, followed by syngas-to-hydrocarbon conversion. The syngas-to-hydrocarbon process is called Fischer–Tropsch Synthesis (FTS). The syngas production stage is basically a coal gasification and cleaning process, while FTS is essentially high-temperature

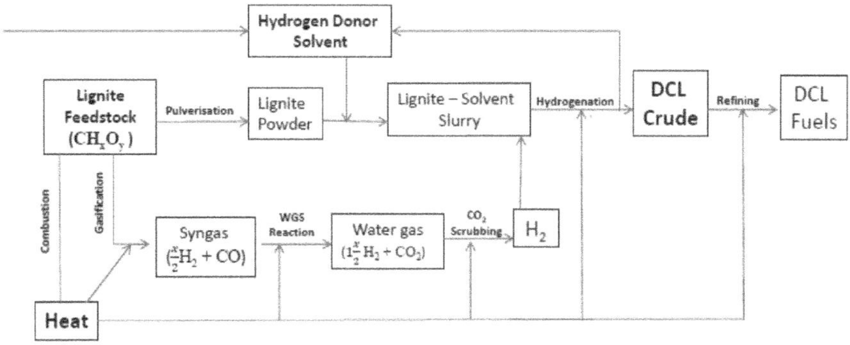

Figure 1.1 Liquid fuel production scheme *via* DCL.

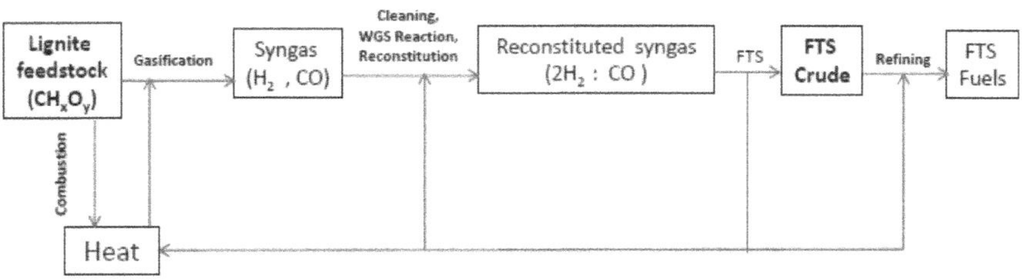

Figure 1.2 Liquid fuel production scheme *via* ICL.

and high-pressure carbon monoxide hydrogenation over a catalyst to produce a mixture of hydrocarbons. In contrast to DCL, with this process, large molecules (liquid hydrocarbon molecules) are built or formed from small molecules (CO and H_2); hence FTS may be considered a bottom-up approach to liquid fuel production. The syngas composition reaction conditions and catalyst are key variables that influence the product delivered by FTS. The ICL liquid fuel production scheme is shown in Figure 1.2. FTS crude – also known as syncrude – usually consists of a wide spectrum of linear hydrocarbon chains. The syncrude must be separated into hydrocarbon fractions, and further refining processes may be necessary before use, *e.g.* hydrocracking, reforming and alkylation.

Although some studies have suggested that DCL is more energy efficient than ICL,[11,12] the two processes essentially entail transforming coal (*e.g.* Turow brown coal, $CH_{1.1}O_{0.3}$)[13] into liquid hydrocarbons (typically of H/C ratio ~2.0) that are suitable for use in engines. A comparison of Figures 1.1 and 1.2 reveals that syngas production is common to the two processes, and it may be argued that DCL appears to be a more complicated process than ICL. DCL is a reductive pyrolysis/hydrogenolysis process that converts coal to liquids (–CH_2–) directly.[14] For a hypothetical coal feedstock – represented as CH_xO_y, – the DCL reactions can be represented by eqn (1.1) and (1.2):

$$CH_xO_y + \frac{1-y}{2}O_2 \xrightarrow{O_2\text{ gasification}} \frac{x}{2}H_2 + CO + H_2O \xrightarrow{\text{WGS}} 1\frac{x}{2}H_2 + CO_2 \quad (1.1)$$

$$CH_xO_y + \frac{2+y-x}{2}H_2 \xrightarrow{\text{hydrogenation}} (-CH_2-) + yH_2O \quad (1.2)$$

Discussions on DCL are often silent about the source or method of hydrogen production for DCL, which is probably through gasification of a portion of the feedstock, as shown in Figure 1.1. The donor solvent intermediate in the hydrogenation process, its regeneration and recycling make it difficult to monitor the DCL fuel yield or carbon utilisation efficiency.

Gasification is a widely discussed feature of ICL, since it is the first step (degradation of the feedstock into syngas) before reassembling the syngas into fuel molecules *via* FTS. Two simple theoretical gasification scenarios can be considered: oxygen and steam gasification. With a coal feedstock, CH_xO_y, the corresponding reactions in the ICL process may be represented as indicated in eqn (1.3) and (1.4):

$$CH_xO_y + \frac{1-y}{2}O_2 \xrightarrow{O_2\text{ gasification}} \frac{x}{2}H_2 + CO \xrightarrow{\text{FTS}} \frac{x}{4}(-CH_2-) + \frac{x}{4}H_2O \quad (1.3)$$

$$\begin{aligned} CH_xO_y + 1 - y\,H_2O &\xrightarrow{H_2O\text{ gasification}} \frac{2+x-2y}{2}H_2 + CO \\ &\xrightarrow{\text{FTS}} \frac{2+x-2y}{4}(-CH_2-) + \frac{2+x-2y}{4}H_2O \end{aligned} \quad (1.4)$$

Eqn (1.3) shows that in oxygen gasification, hydrogen yield depends on the hydrogen content of the feedstock. Eqn (1.4) shows that both the hydrogen and oxygen content of the feedstock influences the hydrogen yield in steam gasification. ICL has the advantage of prerequisite cleaning that removes the sulphur content of the syngas before FTS, so that the resulting syncrude-derived fuel produces very low sulphur emissions. For a given coal sample, $CH_{1.1}O_{0.3}$, the theoretical hydrocarbon yields *via* oxygen gasification and steam gasification are 27.5% and 62.5%, respectively, provided that the energy required for the gasification is supplied from other means than burning a portion of the feedstock. If part of the feedstock is to be burned to supply the energy required for gasification and the other process steps in ICL (including crude refining), the final FTS fuel yield will drop drastically.[15] The low fuel yield of the coal liquefaction process (DCL or ICL) is of little consideration during a time of crisis, such as a war or embargo. However, in noncrisis times, the carbon footprint is a significant consideration, especially in light of global warming and climate change.[16]

Table 1.1 shows a comparison of the fuel properties of coal and biomass samples.[13] Coal and biomass have s similar empirical formula, so biomass can effectively replace coal in either of the liquefaction processes.[17] The resulting biomass liquefaction process is known as the biomass-to-liquid (BTL) process. Since biomass is a product of photosynthesis and has a low sulphur content, BTL biofuel products may be considered carbon neutral and more environment friendly than coal.[18] However, replacing lignite with biomass as the feedstock (Figure 1.2) has little or no effect on the constraint (indicated in eqn (1.3) and (1.4)) on syncrude or fuel yield. A higher fuel yield is only achievable with an external energy and hydrogen input.

Table 1.1 Chemical composition and proximate analysis results of lignite, sunflower hulls and sunflower hulls pellets. Adapted from ref. 13 with permission from Elsevier, Copyright 2018.

	Coal	Biomass	
Property	Turow brown coal	Sunflower hulls	Sunflower hulls pellets
Composition			
C	67.50	50.26	44.31
H	5.98	5.98	5.95
O	24.59	42.23	48.86
N	0.62	1.28	0.77
S	1.31	0.25	0.11
Empirical formula (CH_xO_y)	$CH_{1.1}O_{0.3}$	$CH_{1.4}O_{0.6}$	$CH_{1.6}O_{0.8}$

In the following sections, we discuss the direct biomass liquefaction process and the indirect biomass liquefaction process (DBL and IBL, respectively). We elaborate on IBL: biomass gasification, the FTS mechanism, catalysts and operation modes. Along the way, we detail the limitations of the processes and the strategies used to enhance the BTL biofuel yield.

1.2 Biomass Gasification

Analogous to coal liquefaction, biomass liquefaction can be achieved *via* direct or indirect methods. Gasification is crucial to both of these methods. Gasification is an old practice, and gasifier designs have evolved over the past five centuries. Excellent accounts of gasification technologies can be found in the literature,[19–21] and this section focusses on key highlights relevant to biomass gasification.

Important factors for successful biomass gasification operations include:

- Composition of feedstock (biomass)
- Handling and feeding system
- Gasifying agent
- Choice of gasifier
- Gas clean-up system
- Ash or solid residue removal system

Generally, biomass may be represented by the empirical formula $CH_{1.4}O_{0.6}$ (see Table 1.1). The composition of biomass and coal are similar, so coal gasification technology could be adapted for biomass gasification. In rare cases in which the composition of the biomass feed is significantly different from that of coal, a special gasification system may be warranted. The main objective of biomass gasification is to transform the biomass feed into water gas (for the DBL application) or hydrogen-rich syngas (for the IBL application). It involves a sequence of overlapping processes: drying, pyrolysis, oxidation and heterogeneous and homogeneous reactions (see Figure 1.3[22]). The nature of the gasifying agent influences the energy requirement and the

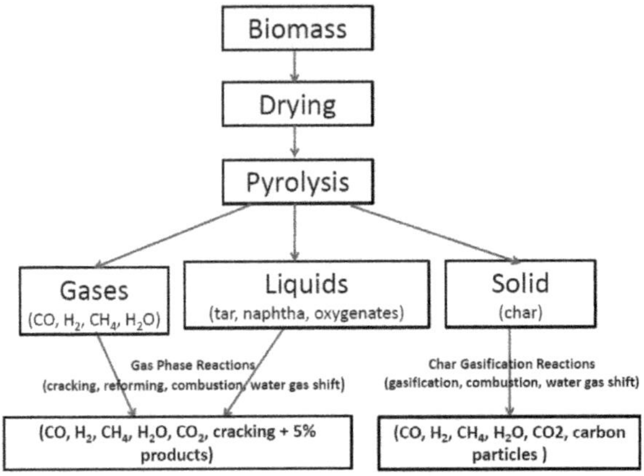

Figure 1.3 Block diagrams of unit processes biomass gasification. Adapted from ref. 22 with permission from the Royal Society of Chemistry.

Table 1.2 Merits and demerits of common gasifying agents for biomass gasification. Adapted from ref. 23, https://iopscience.iop.org/article/10.1088/1757-899X/863/1/012028, under the terms of the CC BY 3.0 license https://creativecommons.org/licenses/by/3.0/.

Gasifying agent	Merits	Demerits
Air	Simple equipment, low cost	High N_2 content in syngas, low heating value
O_2	Good syngas quality	High cost of air separation
H_2O	High H_2 content, high heating value	Indirect or external heat supply, high tar content
CO_2	High heating value, high CO content	Indirect or external heat supply, low H_2 content
Pyrolytic	Syngas with medium heating value	Low system efficiency

H_2/CO ratio of the product syngas. In the absence of a gasifying agent, *i.e.* pyrolysis, gas formation efficiency is very low. The merits and challenges of using common gasifying agents are highlighted in Table 1.2.[23] Qualitatively, the product syngas composition is the same, but quantitatively it is different for the different gasifying agents. None of the gasifying agents satisfy all desirable objectives for biomass gasification for BTL. Among the gasifying agents, oxygen or steam or a mixture of the two are most suited and are commonly used for BTL applications. However, oxygen and steam gasification pose challenges in terms of the high cost related to the required air separation facility and high energy cost.

Contrary to the simple representations of gasification reactions provided in Section 1.1, in practice, a cascade of reactions take place inside a gasifier (see Table 1.3[24]): tandem homogeneous, heterogeneous, exothermic and endothermic processes take place inside a gasifier. The heterogeneous reactions – except carbon (char) hydrogenation – are critical to obtaining

Table 1.3 Main chemical reactions involved in the biomass gasification process. Adapted from ref. 24 with permission from Elsevier, Copyright 2019.

S/N	Reaction	Name	ΔH (kJ mol^{-1})
	Drying and pyrolysis		
1	Biomass → char + tar + H_2O + light gases	Pyrolysis and devolatilisation	>0
	Heterogeneous reactions		
	Char oxidation		
2	$C+\frac{1}{2}O_2 \rightarrow CO$	Partial oxidation	− 111
3	$C+O_2 \rightarrow CO_2$	Complete oxidation	− 394
	Char gasification		
4	$C+H_2O \rightarrow H_2+CO$	Steam gasification	+ 173
5	$C+CO_2 \rightleftharpoons 2CO$	Boudouard reaction	+ 131
6	$C+2H_2 \rightleftharpoons CH_4$	Hydrogenation gasification	− 75
	Homogeneous reactions		
	Gas reactions		
7	$CO+\frac{1}{2}O_2 \rightarrow CO_2$	CO oxidation	− 283
8	$H_2+\frac{1}{2}O_2 \rightarrow H_2O$	H_2 oxidation	− 242
9	$CH_4+2O_2 \rightarrow CO_2+2H_2O$	CH_4 oxidation	− 283
10	$CO+H_2O \rightleftharpoons CO_2+H_2$	WGS	− 41
11	$CO+3H_2 \rightleftharpoons CH_4+H_2O$	Methanation	− 206
	Tar conversion reactions		
12	$C_mH_n+\frac{m}{2}O_2CO+\frac{n}{2}H_2$	Partial oxidation	
13	$C_mH_n+mH_2O \rightarrow \frac{2m+n}{2}H_2+mCO_2$	Steam reforming	Highly
14	$C_mH_n+mCO_2 \rightarrow 2mCO+\frac{n}{2}H_2$	Dry reforming	Endothermic
15	$C_mH_n \rightarrow \frac{n}{4}CH_4+\frac{4m-n}{4}C$	Thermal cracking	

syngas. But the homogeneous gas reactions – except the water-gas-shift (WGS) reaction – are detrimental to syngas selectivity. Overall, syngas formation is endothermic, therefore energy is needed to run the gasification process. A high temperature is favourable with endothermic reactions as it produces syngas with a low H_2/CO ratio and a low tar content. Tar poisons the FTS catalyst, so a high tar content is detrimental to syngas utility, especially in FTS applications. Ultimately, syngas selectivity depends on an intricate optimisation of the parallel competing reactions.

Besides the gasifying agent, the selection of an appropriate gasifier is crucial to achieving the gasification objective. Based on the mode of contact between the biomass and gasifying agent, gasifiers are classified into the following types: fixed bed; fluidised bed; and entrained-flow bed. Each class is further subdivided according to the flow regime of gasifying agents (see Figure 1.4). Some gasifier types are shown in Figure 1.5.[25]

The gasification process can be classified into an autothermal or allothermal process, according to the mode of energy supply. The influence of

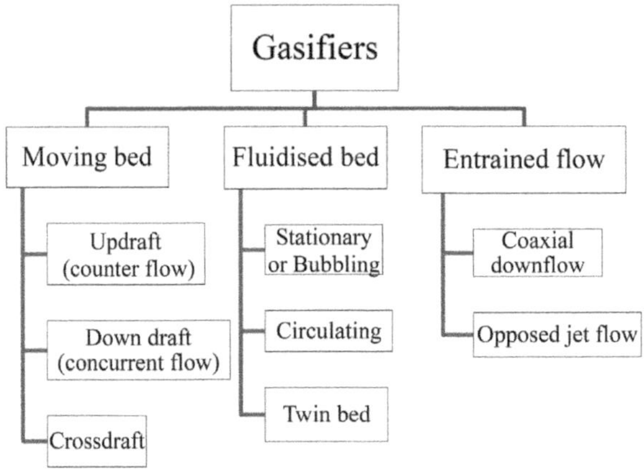

Figure 1.4 Classification of gasifiers based on mode of contact between biomass and gasifying agent.

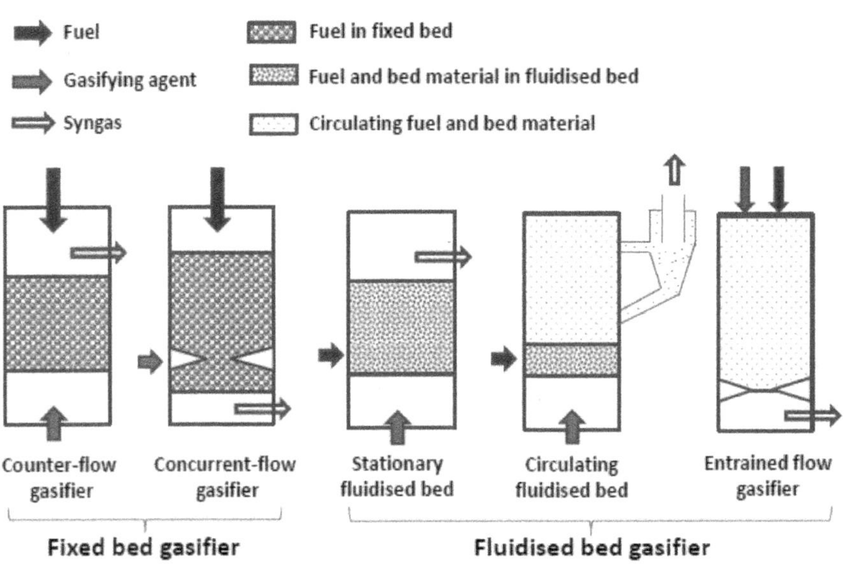

Figure 1.5 Illustration of some gasifier types based on contact mode between fuel (biomass) and gasifying agents. Adapted from ref. 25 with permission from John Wiley & Sons, Copyright © 2013 Wiley Periodicals, Inc.

gasification mode and gasifying agent on syngas composition is summarised in Figure 1.6.[24] In autothermal gasification, the energy is supplied internally through partial combustion reactions inside the gasifier. The advantages of autothermal gasification are that it is simple and easy process to operate under pressurised conditions, and direct heating of the reactants promotes efficient energy utilisation. If air is used as the gasifying agent, the resulting syngas will contain a high amount of nitrogen. Nitrogen-rich syngas will lower the kinetics of the FTS or necessitate the use of an expensive high-pressure

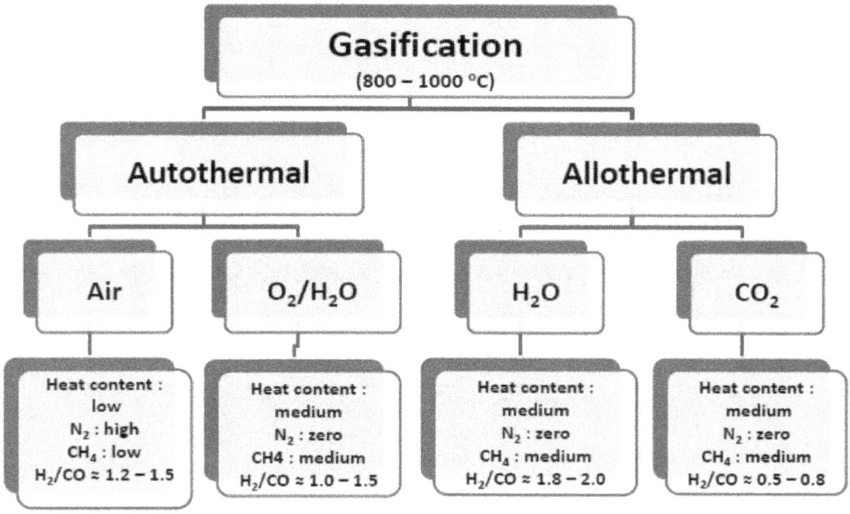

Figure 1.6 Influence of gasification agent and gasification mode on syngas gas composition. Adapted from ref. 24 with permission from Elsevier, Copyright 2019.

system to compensate for nitrogen dilution of the syngas, so pure oxygen and steam are preferred for biomass gasification for BTL applications.

Based on eqn (1.3) and (1.4), pure oxygen gasification is expected to yield syngas of $H_2/CO = 0.7$; steam gasification should yield syngas of $H_2/CO = 1.1$; oxy-steam gasification should yield $0.7 < H_2/CO < 1.1$. The H_2/CO ratios indicated in Figure 1.6 are a reflection of the extent of the carbon loss reactions, such as the WGS reaction, complete oxidation and unconsumed carbon as char. It also explains the report by Dasappa and co-workers, who obtained hydrogen-rich syngas from oxy-steam gasification of biomass using an autothermal mode downdraft gasifier.[26,27] The steam : biomass ratio (SBR) and air/oxygen equivalent ratio (ER) influence the H_2/CO ratio. Syngas with $H_2/CO \sim 2$ that is suitable for FTS was obtained at $ER \approx 0.23$ and $SBR \approx 1.5$. A high ER and SBR favour the WGS reaction, which can lead to pure H_2 after pressure swing adsorption (PSA) and CO_2 scrubbing. This implies that more than half of the carbons in the biomass feedstock will not make it into the second stage (FTS) of the liquefaction process. The oxy-steam autothermal downdraft gasifier has the advantages of being simple and easy to operate and of being flexible to the H_2/CO ratio; however, the need for an air separation facility adds to the energy demand and the cost of the process.

In allothermal gasification, heat production and gasification are separated. This is achieved by connecting a combustion reactor and a gasifier using heat exchangers or bed materials. Allothermal gasifiers are essentially a dual reactor system that produces two gas streams: the combustion exhaust and the syngas (see Figure 1.7[28]). Separation of the combustion exhaust from the syngas product allows for the production of an N_2-free syngas without having to install an air separation facility.

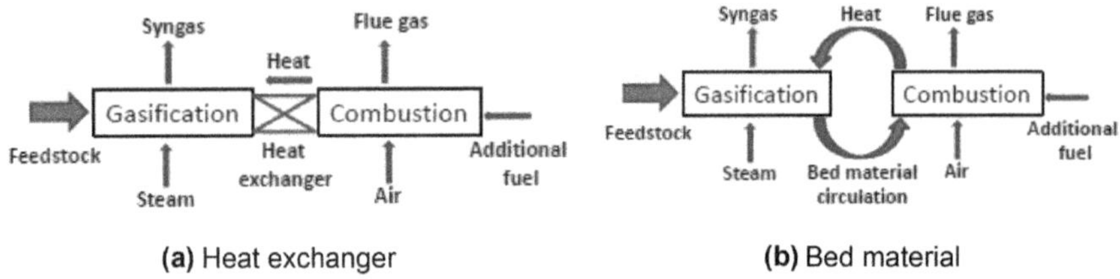

Figure 1.7 Schematics of mode of energy transfer in allothermal gasification. Adapted from ref. 28, https://doi.org/10.3390/app10010002, under the terms of the CC BY 4.0 license, https://creativecommons.org/licenses/by/4.0/.

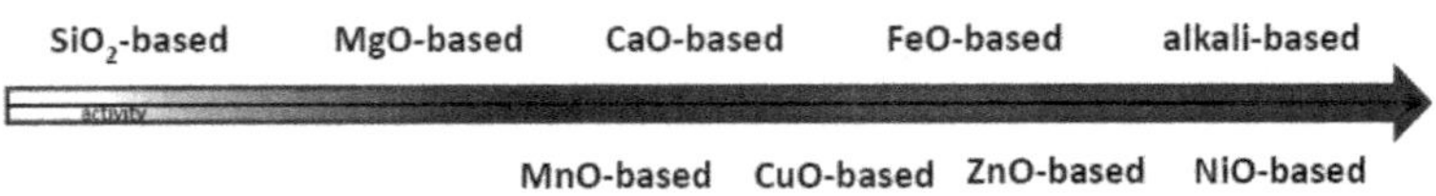

Figure 1.8 Schematic of the activity spectrum for a range of oxide-based bed materials. Adapted from ref. 29 with permission from Jelena Marinkovic, Copyright © Jelena Marinkovic 2016.

The choice and mode of heat exchange between the two reactors dictate efficiency, operational flexibility and the economy of the process. Heat transfer from a combustor to a gasifier may be achieved *via* a sandwich or an envelope design, with the wall of the gasifier serving as the heat exchanger.[29] This allows efficient heat transfer from the combustor to the gasifier but prevents the exchange of material between them. Using circulating bed material can permit the exchange of solid materials between the combustor and the gasifier but prevent mixing of the respective gas streams. The bed materials serve as the carrier of heat from the combustor to the gasifier. This requirement narrows the choice of reactor type to the fluidised bed. Unconsumed char particles from the gasifier can flow to the combustor, where they are combusted to produce heat for the gasifier. The dual fluidised bed (DFB) gasifier coupling through the heat exchanger enjoyed wider interest because of the possibility of recycling the biomass feed between the combustor and the gasifier, thus creating an in-built self-reinforcing and self-regulating system.

Since the bed material serves as the heat carrier, the choice of bed material is critical to achieving the gasification objective. The bed material is an additional process variable involved in controlling the composition of the syngas produced.[29] The bed material is exposed to different chemical environments and attrition as it shuttles between the reactors. Resistance to attrition, sintering and fragmentation is a desirable mechanical property in a bed material, as are low cost, availability, recyclability and disposability. The chemical activity of potential bed materials vary widely (see Figure 1.8[29]), and a chemical property could be advantageous or problematic in the gasification process. For example, SiO_2-sand is commonly used as the bed material in fluidised bed systems because of its attrition resistance and it is inert to on the gas streams. However, at an operating temperature of

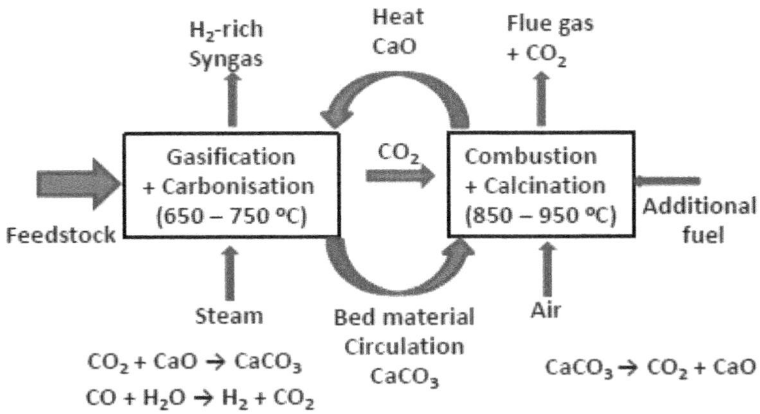

Figure 1.9 Illustration of sorption enhanced gasification effect of limestone in a DFB gasifier. Adapted from ref. 30, https://doi.org/10.1016/j.energy.2019.02.025, under the terms of the CC BY 4.0 license, https://creativecommons.org/licenses/by/4.0/.

650–900 °C, SiO_2-sand can react with the ash by-product from the gasifier and cause agglomeration.

Taking SiO_2-sand as a reference inert bed material, the intrinsic catalytic activity of other bed materials influence the composition of the gas produced by the gasification process. For example, Fe-based and Ni-based bed materials are the oxidation and reduction catalysts, respectively. Fe-based bed material gives CO_2-rich syngas, while Ni-based bed material gives CH_4-rich syngas. Although limestone has poor attrition resistance, its CO_2 sorption property is advantageous for obtaining H_2-rich syngas. Limestone decomposes into quicklime and CO_2 in the combustor but reabsorbs CO_2 in the gasifier. This produces a net effect of transfer of CO_2 from the gasifier to the combustor (see Figure 1.9[30]). The CO_2 is then expelled along with combustor exhaust. In addition, limestone promotes the WGS reaction by this *in situ* removal of CO_2 from the gasifier. The WGS reaction is mildly exothermic and is favoured at a lower temperature. This explains the observed inverse relationship of the H_2/CO ratio with the gasification temperature when using limestone bed material in a DFB gasifier (see Figure 1.9[30]). The advantage of N_2-free syngas production without the use of pure oxygen or the need to install an air purification system has made DFB gasification very attractive for hydrogen and biomethane production. Also, the feed recycling between the combustor and the gasifier allows for complete flexibility with the feed type utilised.

Combinations of bubbling and circulating fluidised bed units can give four variants of combustor-gasifier dual-fluidised bed gasification systems: bubbling–bubbling, bubbling–circulating, circulating–bubbling and circulating–circulating. But the circulating bed (combustor)–bubbling bed (gasifier) appears to be superior in terms of feed/fuel utilisation and the heating value of the syngas output.[31] This gasifier has become synonymous with dual-fluidised bed (DFB) biomass gasification. Several pilot and industrial

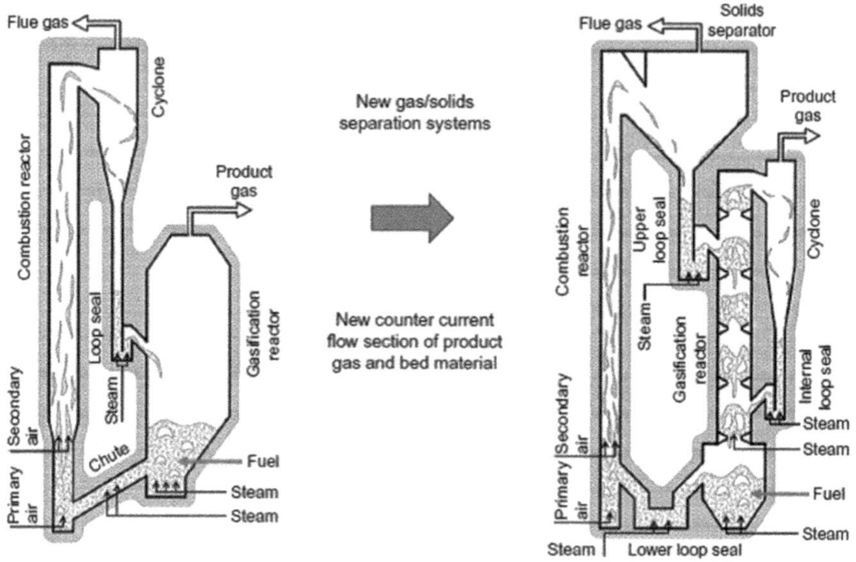

Figure 1.10 Classical (left) *vs.* advanced design (right) DFB gasifier for steam gasification of biomass. Reproduced from ref. 24 with permission from Elsevier, Copyright 2019.

biomass gasification plants that are based on the circulating bed (combustor)-bubbling bed (gasifier) DFB have been built and operated.

To enhance fuel flexibility, transitioning from the classical circulating bed (combustor)-bubbling bed (gasifier) to the circulating bed (combustor)-circulating bed (gasifier) is being investigated.[32–34] The new features added to the classical DFB and the anticipated improvements in performance include (see Figure 1.10):[24]

- Replacing the bubbling bed (gasifier) with a circulation bed allows counterflow contact between the bed material from the combustor and the syngas from the gasifier, which leads to longer contact time for tar conversion and WGS reactions.
- Gently sloped cyclones for loops to minimise the attrition of limestone for sorption enhanced gasification.

Although the development of the DFB gasifier is a significant milestone in biomass gasification, the fundamental deficiency remains with using biomass as the feedstock for FTS biofuel: about 50% of the carbon in the feedstock may end up as CO_2 during syngas production. The amount of carbon in the feedstock that will make it to the final hydrocarbon fuel depends on the efficiency of the FTS and refining processes.

The next section deals with the application of H_2-rich syngas from DFB steam gasification and with strategies for improving carbon utilisation of biomass feedstock.

1.2.1 Direct Biomass Liquefaction

DBL is analogous to DCL, and the chemical basis of DBL is similar to that of DCL. The DBL reactions of a hypothetical biomass feedstock, CH_xO_y, may be represented by eqn (1.5) and (1.6) as DCL:

$$CH_xO_y + \frac{1-y}{2}O_2 + H_2O \xrightarrow{\text{Gasification}} 1\frac{x}{2}H_2 + CO_2 \tag{1.5}$$

$$CH_xO_y + \frac{2+y-x}{2}H_2 \xrightarrow{\text{hydrogenation}} (\text{–}CH_2\text{–}) + yH_2O \tag{1.6}$$

The technical feasibility of carrying out eqn (1.5) through CO_2 sorption-enhanced DFB gasification was discussed in the previous section; however, there are technical challenges with carrying out eqn (1.6). Like DCL, DBL is a reductive pyrolysis process that involves subjecting biomass-solvent slurry to high temperature and H_2 pressure. During the petroleum supply crisis of 1970s, many alternative means of obtaining liquid fuel were explored, including the DBL process.[35] Based on the past success of DCL in Germany (1930–1940s) and South Africa (1960–1980s) and more recently the commercial DCL operation in Shenhua China (since 2009), it is anticipated that DBL will have a lower CO_2 footprint than DCL. However, DBL has not enjoyed the same level of success as DCL, and although several DBL projects were initiated and operated on a pilot scale,[36] none of the projects progressed to the commercial scale.

Although lignite and biomass have a similar empirical formula, they have a markedly different structure.[37] Lignite has a fairly monolithic simple structure compared to biomass, which is a composite of cellulose, hemicellulose and lignin. While there are established hydrogen donor solvents (*e.g.* tetralin) that can solvate and facilitate the hydrogenolysis and hydrogenation of lignite, it is very challenging to find a solvent/compound that can simultaneously solvate the three components of biomass, also the solvent/compound need to possess reversible hydrogenation/dehydrogenation property. The poor success of direct adoption of the DCL process for DBL is traceable to the differences in the structure of the feedstock used. Pretreatment of the biomass feedstock into intermediates that are similar to lignite can improve the results achieved with DBL. This approach has been explored, albeit under a different process nomenclature – bio-oil formation and its upgrade.[38] Consequently Figure 1.1 is revised to include process steps to make DBL feasible (Figure 1.11). Points of CO_2 discharges in the scheme suggest that about three-quarters of the carbons in the feedstock may not be made into the final liquid fuel. An alternate approach to pretreatment of biomass is fractionation followed by conversion of each fraction (cellulose, hemicellulose and lignin) into niche liquid fuels.

Since the objective of transitioning from coal to biomass liquefaction is to achieve neutral or negative CO_2 emissions, considering the energy cost/CO_2 emissions associated with biomass harvesting, transportation, processing

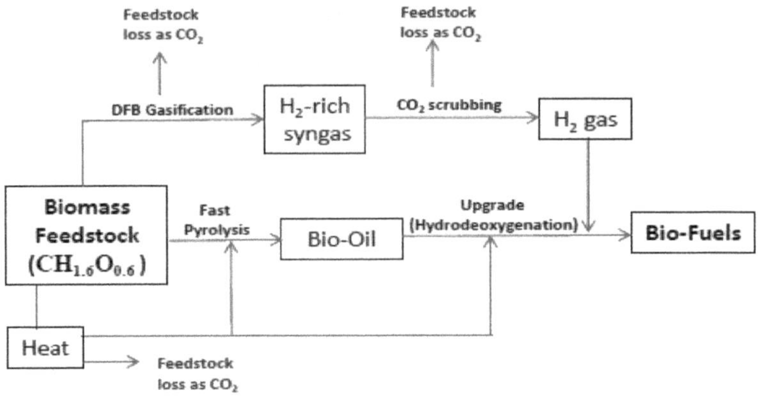

Figure 1.11 Revised scheme for DBL.

and cultivation time, the carbon efficiency or biofuel yield of the current liquefaction processes is very low.[39] The main CO_2 discharge points are hydrogen and heat supply; hence sustainable complementary hydrogen and heat inputs for biomass liquefaction would improve the carbon efficiency.

1.2.2 Complementary Hydrogen and Heat Inputs for Biomass Gasifiers

Assuming the gasification and FTS reactions in eqn (1.3) are stoichiometric as written, only a fraction of the carbon in the feedstock can be converted into liquid fuel. This is because the syngas from oxygen gasification of coal or biomass is hydrogen deficient with respect to the hydrocarbon formation reaction. WGS and CO_2 scrubbing are often used to reconstitute the syngas into $H_2/CO \sim 2.1$ that is suitable for FTS reactions or methanol synthesis. Similarly, the net reaction leading to DBL – eqn (1.7) (the sum of eqn (1.5) and (1.6)) – shows that about 50% of the carbon in the starting feedstock may end up as CO_2.

$$
\begin{aligned}
& CH_xO_y + \frac{1-y}{2}O_2 \frac{2+y-x}{4}H_2 \\
& \xrightarrow{\text{hydrogenation}} \left(-\frac{1}{2}CH_2-\right) + \frac{1}{2}CO_2 + yH_2O
\end{aligned}
\tag{1.7}
$$

The carbon loss steps from hydrogen production *via* gasification. However, alternate feedstocks for the hydrogen production step will help to save the feedstock and enhance the carbon efficiency and liquid fuel yield. Methane and water are important hydrogen-rich feedstock and use of the H_2 feed obtained from methane or water to replace the gasification step in DBL or to augment the H_2-deficiency of bio-syngas for FTS reactions is a viable strategy to improve the carbon efficiency of biomass liquefaction processes.

1.2.2.1 Gas-biomass-to-liquid

Methane is a hydrogen-rich molecule that has the lowest CO_2 emission per unit mass among the fossil fuels. With the increased availability of low-cost natural gas (NG) due to new discoveries, the availability of technology for accessing finds from difficult terrains and the effort to discourage flaring of associated gases, gas-to-liquid (GTL) projects became popular during petroleum supply crisis. With methane as the main constituent, NG is rich in hydrogen, but GTL is not CO_2 emission neutral. Biosyngas is H_2 deficient, so the gas-biomass-to-liquid (GBTL) process – which is a hybrid of GTL and BTL – has the potential to improve the liquid fuel yield of BTL and also lower the CO_2 footprint of GTL.[40,41] Combustion of NG also produces more energy per unit mass than biomass.

GBTL can take different blends and forms; for example, hydrogen from steam reforming of methane can replace biomass gasification, or NG combustion can supply the energy required for fast pyrolysis and subsequent upgrade of the resulting bio-oil. A recent study showed that the economic and environmental performance of GBTL fuels depends on NG/biomass ratio blends.[42,43] GBTL enhances the economic feasibility of the biomass liquefaction process, but a compromise is required to achieve environmental sustainability. A maximum of 28% NG/biomass blend is considered optimum for meeting the CO_2 emission targets for renewable fuel emission standard.[44]

1.2.2.2 Power-biomass-to-liquid

Water is a carbon-free source of hydrogen. Splitting water into hydrogen and oxygen can be achieved using either thermal or electrical energy inputs. Electrical energy input for hydrogen production is the focus of this section, while thermal water splitting will be discussed in the next section. The goal of the power-biomass-to-liquid (PBTL) process is the integration of biomass liquefaction processes with water electrolysis using carbon-free electricity (hydro, wind, solar, nuclear, *etc.*).[45] This is similar to other related power-to-fuel[46] ideas, where the focus is co-electrolysis of CO_2 and H_2O to syngas and the production of gaseous fuels (methane,[47–49] dimethylether[50] or methanol[51]). The similarity is in the use of renewable electricity, but PBTL is different because its target product is liquid fuel.

The process of integrating power (electricity) and BTL depends on the type of biomass liquefaction process (DBL or IBL). The integration can take the form of replacing hydrogen production – *via* biomass gasification – with hydrogen supply from electrolysis (see Figure 1.12). Oxygen from the electrolysis process may be released into the atmosphere or sold as a by-product to improve the process economy.

With IBL, a higher level of integration can be envisaged, as proposed in Figure 1.13. Oxygen from electrolysis then becomes a feed for oxy-steam gasification. The water by-product from FTS and the heat recovered from syngas cooling and FTS could be used to generate steam for the oxy-steam gasification process. The syngas produced by the oxy-gasification could be reconstituted, so that the hydrogen added is sufficient for stoichiometric conversion of

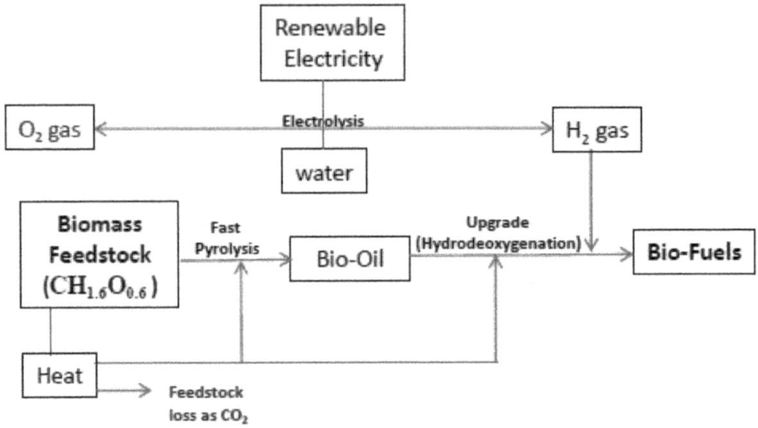

Figure 1.12 Proposed scheme for integration of electricity with DBL.

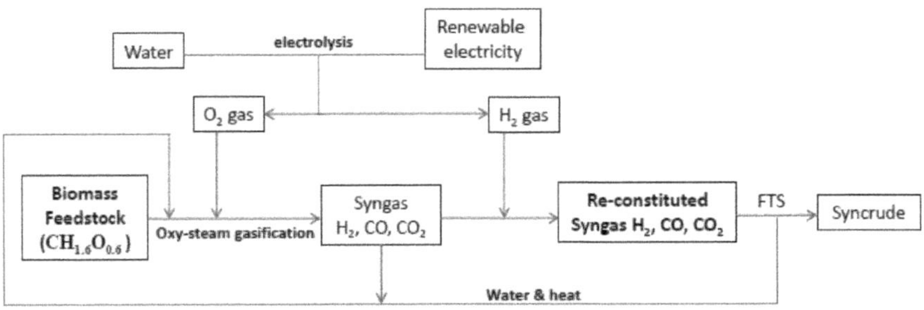

Figure 1.13 Proposed scheme for integration of alkaline electrolysis with oxy-steam biomass gasification.

CO + CO_2 into hydrocarbon fuel. WGS and CO_2 scrubbing are eliminated; as a result, carbon efficiency and liquid fuel yield are greatly enhanced.

Water electrolysis technologies can be classified into alkaline, polymer membrane and solid oxide electrolysis. Polymer electrolyte membrane (PEM) electrolysis gives a higher purity H_2 and O_2; the noble metal electrodes and an expensive polymer membrane used in PEM make it more expensive than alkaline water electrolysis. But alkaline water electrolysis is a mature technology that is commonly used by industry. It has the advantage of low cost due to the use of non-noble metal electrodes that are stable in alkaline media. Purity of H_2 and O_2 obtained from alkaline electrolysis is suitable and recommended for the proposed PBTL shown in Figure 1.13.

Most alkaline and PEM electrolysers usually operate at $<100\ ^{\circ}C$; however, operating at a temperature of $>100\ ^{\circ}C$ can lower the electrical energy required for water splitting (see Figure 1.14[52]). High-temperature operation of solid oxide electrolysis can permit co-electrolysis of H_2O/CO_2 mixtures into syngas.[53,54] These features can be harnessed to modified the integration scheme, as shown in Figure 1.15. Here oxygen gasification is selected to obtain a H_2-lean

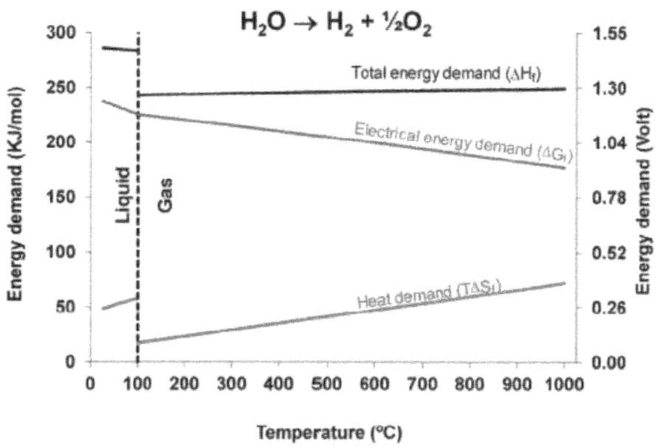

Figure 1.14 Thermodynamics for H_2O electrolysis at atmospheric pressure. Reproduced from ref. 52 with permission from Elsevier, Copyright 2017.

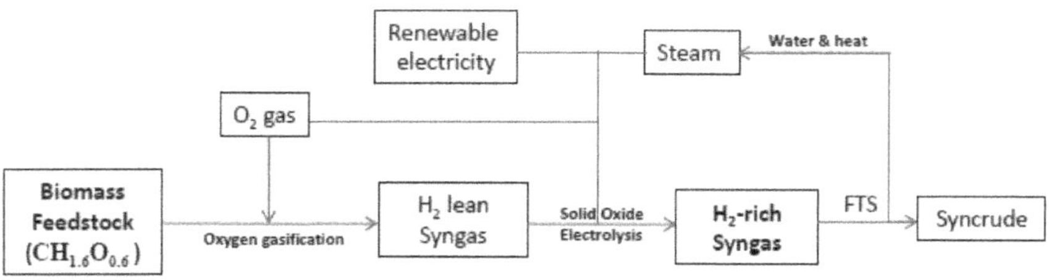

Figure 1.15 Proposed scheme for integration of solid oxide electrolysis with oxygen biomass gasification.

syngas using entrained-flow gasifiers. The hot syngas from the gasifier is fed into the solid oxide electrolyser for co-electrolysis of H_2-rich syngas.

The proposed PBTL above are based on low renewable electricity.[55] The cost of solar and wind electricity has decreased steadily over the past few decades, and a further decrease in price is expected in the future, as the technologies become more mature and competitive. However, sunshine and wind are intermittent in nature, and battery technologies for use in grid scale storage systems are still in their infancy; thus the proposed PBTL offers a means of storing excess renewable electricity as a portable liquid fuel.

1.2.3 Solar-thermal-biomass-to-liquid

A significant fraction of a given biomass feedstock is sacrificed to supply heat and hydrogen in biomass liquefaction processes. Solar thermal is carbon-free and can therefore be harnessed to improve the carbon efficiency of biomass liquefaction processes.[56] Integration of the solar thermal process with the biomass-to-liquid process is termed the solar-thermal-biomass-to-liquid (STBTL) process. Solar thermal energy can be exploited as a source of heat energy or to generate hydrogen for the liquefaction process.[57,58] Both processes require a high temperature: 600–1200 °C. Normal solar insolation on the surface of our planet is between 200 and 400 W m^{-2}, which corresponds to a black body temperature of 18–35 °C. Thus normal solar insolation needs to be concentrated about 500–1000 times[59] before it is useful in thermochemical processes.

Improvement of carbon efficiency has been demonstrated in solar-assisted biomass gasification.[60] Different solar-powered gasifier configurations have been investigated, such as updraft and downdraft.[61,62] Direct and indirect solar heat supply modes to the gasifier have also been reported. It has been shown that the use of solar thermal energy can reduce CO_2 emissions from the gasification process; however, simulated concentrated solar radiators were used in these studies.[63] Very few studies in this field have involved the use of real concentrated solar power for biomass gasification.[64–68] Table 1.4 shows the performance parameters of some typical solar collector installations.[69] The solar thermal installations occupy large areas, and integration with gasifiers would be very challenging. Moreover, most concentrated solar thermal installations are cited in arid areas where biomass is scarce. Nevertheless, electricity produced from concentrated solar thermal is renewable, and it can be transmitted to locations where there is a sustainable supply of biomass for PBTL.

We had identified the factors responsible for low carbon efficiency of biomass liquefaction processes and also highlighted strategies to reduce carbon loss in BTL. The cardinal focus of these strategies is to salvage biomass fractions that would have hitherto been sacrificed to supply heat or hydrogen for the process. Solar energy is the most abundant and sustainable renewable carbon-free resource that can be used to improve the liquid fuel yield from biomass liquefaction processes. *Ab initio*, biomass (and by extension, fossil fuels) are products of solar energy, so the strategies essentially consist of fossilisation of biomass with more solar energy, *i.e.* forcing more solar energy into biomass, thereby increasing its energy density. Next we discuss syngas cleaning, before taking a detailed look at the second part of the IBL process, *i.e.* FTS.

1.2.4 Syngas Cleaning

As indicated in Table 1.1, biomass generally has a low sulphur content (<0.25%) but a significant nitrogen content (~1.28%). The resulting nitrogen and sulphur compounds (H_2S, CS_2, thiophenes, thiols, HCN, NH_3, *etc.*) in the gasification product will not only poison the FTS catalyst but are also potential air pollutants. Hence removal of these gases from the syngas is an important step in the IBL process. Also, tar is a common undesired constituent in virtually

Table 1.4 Performance parameter of typical solar collectors. Adapted from ref. 69 with permission from Elsevier, Copyright 2019.

Solar collector	Concentration ratio	Operating temp (°C)
Parabolic trough	30–100	400
Linear Fresnel	30	300 – 400
Tower collector	500–5000	400–600
Disc collector	1000–10 000	550–750

all biomass-derived syngas irrespective of the gasifying agents but tar density will depend on the gasification technology employed.[70] Tar comprises a wide spectrum of condensable organic compounds, which vary from small oxygenate molecules to large acyclic, cyclic, aromatic and polycyclic aromatic hydrocarbon molecules. The conditions inside the biomass gasifier promote the formation of hundreds or thousands of different tar molecules.[70] Condensation of tar initiates fouling, which can result in malfunctioning or plugged equipment. Other important impurities in biomass-derived syngas are chlorine, ash and char particulates.

A number of strategies are being investigated for cleaning biomass-derived syngas. These strategies can be grouped according to the exit temperature of the syngas from the clean-up device: hot ($T > 300$ °C), warm ($300 > T > 100$ °C) and cold ($T < 100$ °C) gas cleaning strategies. The cold strategy – also known as the wet method – involves the use of wet scrubbers or bubblers.[71] It uses relatively mature techniques such as absorption or adsorption. Wet methods are generally effective but suffer from energy loss and wastewater generation. The hot cleaning strategy has the advantage of less energy loss because it does not include the process of cooling and reheating the gas stream. However, many of the techniques being used are still under development, and technical challenges are involved in handling gas at a high temperature.[72–81] The warm cleaning strategy is a hybrid of the hot and cold methods.

Here, we focus on practical methods for bio-syngas cleaning for FTS applications. In general, the logical sequence of cleaning steps for producing syngas from biomass gasifiers is as follows: particulate removal; removal of organic (tar) impurities; removal of inorganic impurities; removal of volatile (alkali/heavy) metals. The most challenging of the impurities is tar, and it attracts more attention from researchers in terms of cleaning biomass-produced syngas. The oil-based gas washing process (OLGA), developed by the Energy Research Centre of the Netherlands (ECN),[82] focussed on removing tar but also encompassed the removal of other categories of impurities. It is based on the control of dew point (the temperature at which tar starts to condense). OLGA as applied to biomass-derived syngas for FT applications is shown in Figure 1.16.[83]

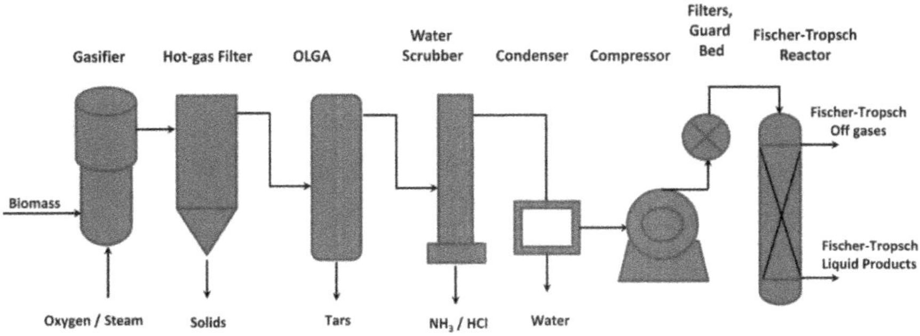

Figure 1.16 OLGA-based biomass-derived syngas cleaning protocol.[83]

Hot syngas from the gasifier is first passed over a solid filter, where ash and particulate impurities are removed. The syngas is then channelled into an oil scrubber or bubbler (the OLGA unit), where lipophilic tar molecules are removed. The exit syngas is then passed through water to remove NH_3, HCl and other inorganic impurities. After the three scrubbing steps are completed, the syngas is considered suitable for FTS. The order of the scrubbing step is critical to ensuring a smooth gas cleaning operation. Biochar is a by-product of biomass gasification that is often employed as the solid filter, while vegetable oil is the most common oil scrubber or bubbler. Bio-oil, biodiesel and surfactant solutions have also been investigated for tar scrubbing.

The sulphur content in biomass is usually very low, and for most practical purposes, the cleaning step used to remove sulphur impurities in biomass-derived syngas is often unnecessary with FT applications. If the sulphur content in the biomass feedstock is high, the sulphur removal step becomes necessary. This can be accomplished using the established conventional acid gas wet scrubbing techniques, which removes acidic gas such as H_2S, SO_2, NO_x and CO_2. This step may be beneficial as part of the gas-conditioning step that reduces the CO_2 load of the commonly CO_2-rich and/or H_2-deficient biomass-derived syngas. Where a low-cost H_2 feed from water electrolysis is available, the rest of the gas-conditioning step involves augmenting the Ribblett ratio $H_2/(2CO+3CO_2)$ to unity.[84]

1.3 Fischer–Tropsch Synthesis

In Section 1.2.2, biomass gasification that leads to syngas production and syngas cleaning was examined. The carbon loss associated with the process was noted, and ways to enhance carbon efficiency were discussed. In this section, the second stage of the IBL process, *i.e.* FTS, is examined. It involves the conversion of cleaned bio-syngas to hydrocarbon products.

Hydrocarbon formation is normally obtained at a temperature of 180–350 °C and 10–30 bar pressure, depending on the type of catalyst and reactor employed. More than one reaction is possible at FTS reaction conditions; Table 1.5 shows a list of likely reactions in an FTS reactor.[85] Generally, product selectivity follows the order of alkanes > alkenes > alcohols, while the full product spectrum is a function of the catalyst type and reaction conditions.

For purposes of simplicity, all the reactions in Table 1.5 are approximated as:

$$2nH_2 + nCO \rightarrow -(CH_2)_n- + nH_2O = -165\ \text{kJ mol}^{-1} \tag{1.8}$$

As pointed out in Section 1.1.2, biosyngas normally contains a significant amount of CO_2, so taking the CO_2 content into consideration, the stoichiometrically balanced syngas feed should have a Ribblett ratio $H_2/(2CO+3CO_2)=1$.[86,87] Catalysts with low WGS activity require a stringent syngas composition of $H_2/CO \approx 2$. However, active WGS catalysts can form hydrocarbon with $H_2/CO < 2$, especially at high-temperature mode, where WGS proceeds to equilibrium. For instance, the net reaction for hydrocarbon

Table 1.5 Fischer–Tropsch synthesis reactions. Adapted from ref. 85 with permission from Elsevier, Copyright 2010.

Main reactions	
1. Methane	$CO + 3H_2 \rightarrow CH_4 + H_2O$
2. Paraffins	$(2n + 1)H_2 + nCO \rightarrow C_nH_{2n} + 2 + nH_2O$
3. Olefins	$2nH_2 + nCO \rightarrow C_nH_{2n} + nH_2O$
4. WGS	$CO + H_2O \rightarrow CO_2 + H_2$
Side reactions	
5. Alcohols	$2nH_2 + nCO \rightarrow C_nH_{2n+1}O + nH_2O$
6. Boudouard reaction	$2CO \rightarrow C + CO_2$
Catalyst modifications	
7. Catalyst oxidation/reduction	a. $M_xO_y + yH_2 \rightarrow yH_2O + xM$
	b. $M_xO_y + yCO \rightarrow yCO_2 + xM$
8. Bulk carbide formation	$yC + xM \rightarrow M_xC_y$

formation on active WGS catalyst using syngas with $H_2/CO = 1$ can be written as per eqn (1.9):

$$2CO + H_2 \rightarrow \text{-}CH_2\text{-} + CO_2 \quad (1.9)$$

Half the CO is ejected as CO_2, which results in a maximum carbon efficiency of 50%, assuming stoichiometric conversion is achieved. For co-conversion of CO and CO_2 to hydrocarbon, $H_2/(2CO + 3CO_2) = 1$ is required. The reverse WGS (rWGS) reaction of CO_2 to CO may be a prerequisite for subsequent hydrogenation to hydrocarbon; however, alternate CO_2 hydrogenation to hydrocarbon *via* the methanol intermediate has also been recognised.[88–90]

The main FTS reaction conditions are temperature, catalyst type, H_2/CO ratio and reactor type. There are two modes of operations in commercial FTS, which are based on temperature:

- High-temperature Fischer–Tropsch (HTFT) mode: the reactor temperature is between 300 and 350 °C. Gasoline and linear low-molecular-mass olefins are selectively produced with an Fe-based catalyst, and significant amounts of oxygenates are also produced.
- Low-temperature Fischer–Tropsch (LTFT) mode: the reactor temperature is usually between 200 and 250 °C, and either Fe- or Co-based catalysts are used. This mode gives products with a high selectivity for paraffins and high-molecular-mass linear waxes.

The dominant design consideration in terms of a Fischer–Tropsch reactor is removing the large exothermic heat of reaction.[91] The choice of reactor type for a commercial operation is largely dictated by the desired product selectivity. Prominent FTS reactor types are:

- Circulating fluidised bed (CFB) reactor,
- Fixed fluidised bed (FFB) reactor, also known as Sasol advanced synthol (SAS) reactor,

- Multitubular fixed-bed (MTFB) reactor, and
- Slurry phase reactor.

1.3.1 Rate Law of the FTS Reaction

An accurate kinetic description is important for the evaluation of activity and comparison of catalysts. It is generally accepted that the specific activity of FT active metals follows the order of Ru > Ni > Co > Fe. Only iron and cobalt are used in commercially FTS operations. Although FTS is generally considered to be a polymerisation reaction in nature, polymerisation kinetic expressions[92–94] are not directly applicable because monomer species are formed *in situ* on the catalyst surface in FTS, and more than one potential monomer is possible. Thus FTS kinetics focus on the initiation step and monomer formation. The rate law of such heterogeneous gas phase reactions are usually derived using the Langmuir–Hinshelwood or Eley–Rideal mechanisms, and the Langmuir–Hinshelwood mechanism has been shown to give a more accurate description of the monomer formation than the Eley–Rideal mechanism.[95,96]

1.3.2 Mechanism of FTS

Although CO activation is the rate-limiting step in monomer formation, there is no agreement on the monomer species that undergoes repeated additions in order to give hydrocarbons. According to Krische *et al.*,[97,98] organometallic intermediates that arise transiently in the course of catalytic hydrogenation of CO are intercepted and re-routed to C–C coupling. Potential intermediate candidates include HCO*, HCOH*, CH* and CH_2^*. These intermediates correspond to the proposed propagation mechanisms CO insertion, enol, alkylidene and alkenyl, respectively. The monomer formation and propagation mechanisms are illustrated in Figure 1.17.[99]

1.3.3 FTS Product Distribution

FTS products fall in a spectrum of hydrocarbons ranging from C_1 to C_{50}. A typical gas chromatogram of syncrude is characterised by a notable degree of order with respect to the carbon number of the molecules.[100–102] The mole fraction of the components in each carbon number fraction declines exponentially with the carbon number, and this behaviour is typical of polymerisation reactions. This is what informed the description of FTS as a surface polymerisation reaction at the earlier stage of its development. The polymerisation ideal conceived as FTS proceeds *via* stepwise addition of a monomer carbon species and one termination constant for all products.[103]

Instead of a tabular list or pictorial presentation of the products, a simple and unambiguous model known as Anderson–Schultz–Flory (ASF) distribution is used to provide a simpler comparison of the selectivity of catalysts

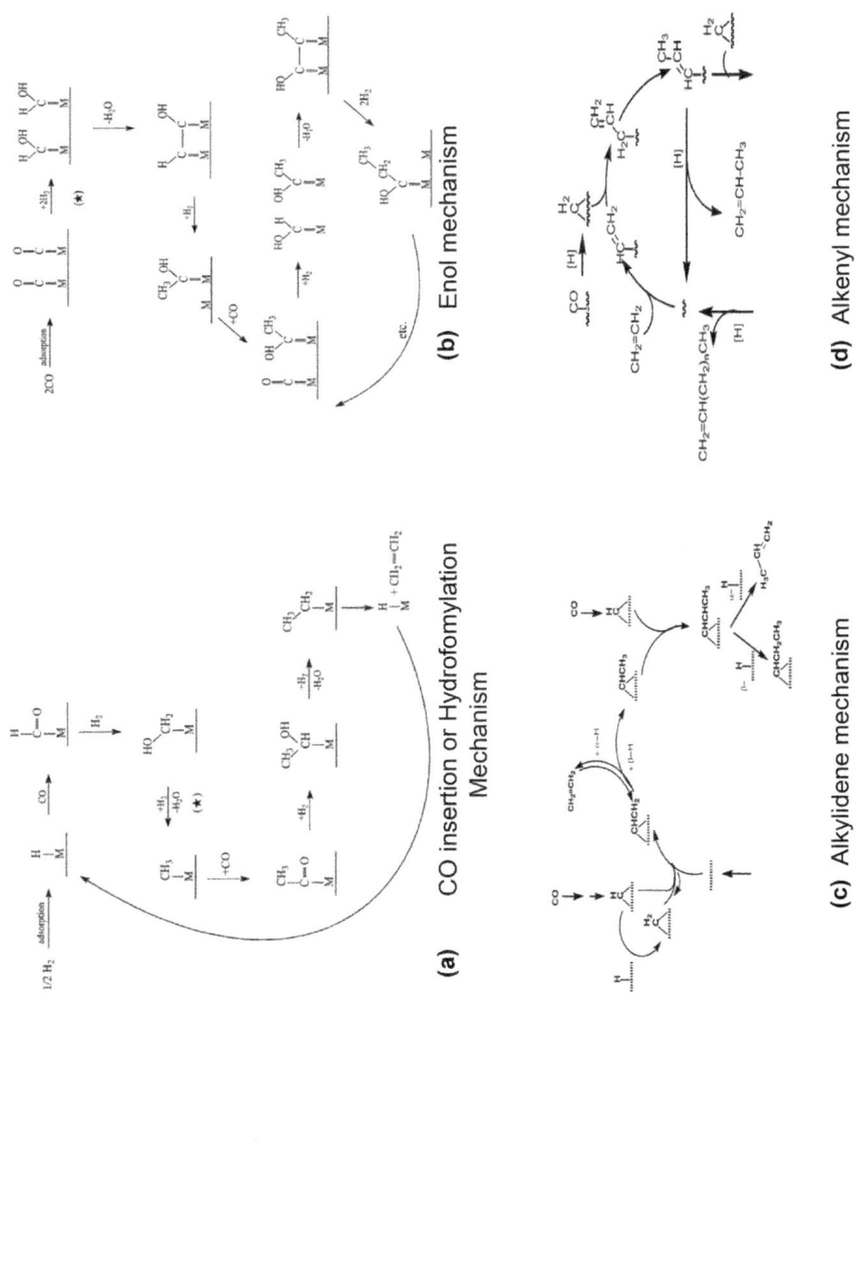

Figure 1.17 Proposed FTS mechanisms. Reproduced from ref. 99 with permission from the Royal Society of Chemistry.

at different reaction conditions. ASF is an empirical statistical distribution. It is described in eqn (1.10).

$$\log\frac{w_n}{n} = n\log\alpha + \log\frac{(1-\alpha)^2}{\alpha} \tag{1.10}$$

A semilogarithm plot of the product mole fractions as a function of the carbon number should give a straight line with a slope, α, the chain growth probability. The ASF distribution is based on the assumption that the chain growth probability is constant and that there is stepwise addition of monomer- leads to saturated hydrocarbons as terminal products. But product distribution of FTS catalysts usually deviate from the ASF distribution law (see Figure 1.18). The beauty of ASF is that, for most practical purposes, is its one model parameter, α, that provides a broad overview of the product spectrum, and it is sufficient for describing syncrude. ASF provides information that guides the selection of the refining technology needed to obtain the target fuel from the syncrude.

1.4 Refining Syncrude

Syncrude (short form of synthetic crude oil) is so-called because like conventional crude oil it is made up of mixtures of hydrocarbon molecules as the main constituent and oxygenates as the minor constituents. Syncrude that has not been refined or upgraded cannot be used as fuel to run engines. The practice in the petroleum refineries in dealing with crude oil is to sort out the constituents based on the boiling point range for different engine types. The next steps are specific processes such as reforming, cracking, extraction blending *etc.*, which are applied in order to meet the requirements for engine fuel. However, refining syncrude demands a holistic approach, and there are very few dedicated syncrude refineries around the globe, consequently, there are very few experts in this area. Arno de Klerk has made a significant contribution to the literature on syncrude refining, based on his experience at Sasol (South Africa).[104] Some of the highlights from his contribution are dealt with here.

Since FTS is a bottom-up process, the target fuel or chemical products should be in view right from syngas production so as to tailor the FTS mode and subsequent refining processes toward maximising the desired products.[104] Furthermore, from a carbon efficiency and process cost point of view, it is desirable to obtain the target fuel using as few process steps as possible. The product distribution of some commercial Fischer–Tropsch technologies is shown in Figure 1.19.[104] It shows that the HTFT mode delivers more straight-run naphtha (gasoline-range carbon number) fractions and lower distillate (diesel-range carbon number) fractions – and *vice versa* for the LTFT mode.[104] Hence, having a target fuel in view from the start, allows for choice of appropriate FTS mode and catalyst selection to be made.

Another important aspect of syncrude refining is to take cognisance of the differences between syncrude and petroleum. The blank adoption of

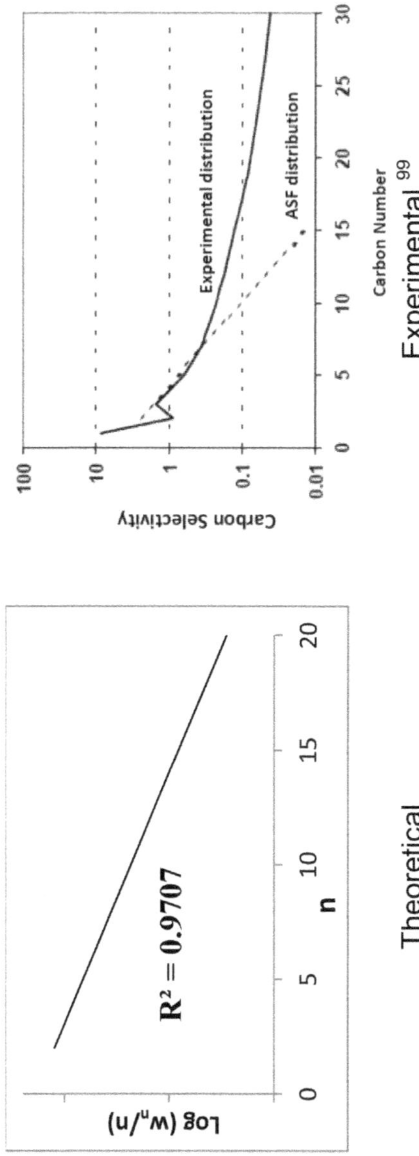

Figure 1.18 Theoretical ASF *vs.* Experimental FTS product distributions.

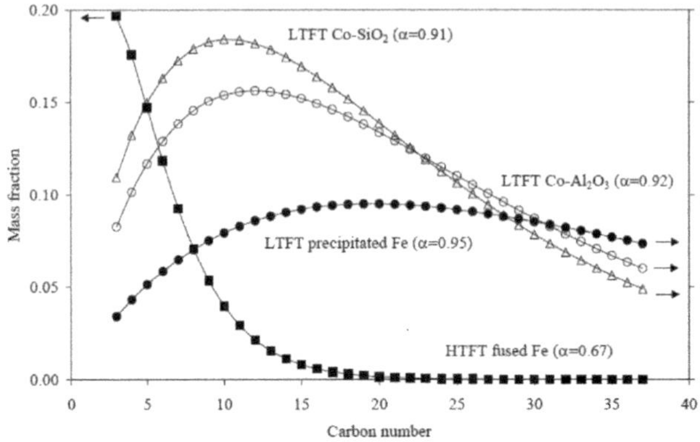

Figure 1.19 Carbon number distribution of some commercial Fischer–Tropsch technologies. Reproduced from ref. 104 with permission from Prof. Arno de Klerk.

established petroleum refining technologies for use with syncrude may prove unwise. Detailed attention should be given to the molecular differences between petroleum and syncrude when selecting refining technologies for syncrude. Although the first diesel engine ran on vegetable oil, development and improvement of the diesel engine (and, by extension, petrol and jet engines) were tailored to use petroleum-derived fuels. Generally, syncrude naphtha and distillate are poor in aromatics and isoparaffins; HTFT naphtha and distillate are rich in olefins, while those of LTFT are rich in *n*-paraffins. Because of the higher olefin content, syncrude fractions (especially HTFT) are more reactive than petroleum fractions.[105] Hence mild catalyst activity and reaction conditions are more appropriate for refining such syncrude. HTFT naphtha and distillate have a higher oxygenates content than LTFTs; in fact, the water by-product of FTS contains some short-chain oxygenates. The aqueous (oxygenates) fraction of syncrudes is about 7% for HTFT and 3% for LTFT. These aqueous products are also FTS products that have to be refined separately from the syncrude. The naphtha fractions in syncrude do not meet the specifications for gasoline fuel. Upgrading the paraffins rich in LTFT naphtha to comply with the specifications for gasoline fuel requires hydroisomerisation of the paraffins and olefins to isoparaffins and then partial aromatisation. This can be accomplished using a Pt/MOR catalyst at 250–280 °C. For the olefin-rich naphthas, the upgrading may lead to greater aromatisation.[105]

HTFT syncrude has a higher naphtha fraction than LTFT; thus it makes sense to build the HTFT syncrude refining process around naphtha and select the processes for converting naphtha to the specified gasoline fuel, as previously discussed. Other fractions of HTFT include distillate, gases and aqueous products. The gas fraction is rich in lower olefins, which can be separated into individual olefins and sold as valuable products (monomers), but this may require expensive cryogenic operations. The olefin mixtures may be oligomerised and hydrogenated into gasoline-range isoparaffins, while a fraction of the olefins may be aromatised into a BTX (benzene, toluene and xylene) mixture. Alkylation of BTX with olefins can give alkyl benzenes that are suitable for use as blending feed (gasoline, jet fuel or diesel), or they can be sold as a high-value chemical for the surfactant industry. The distillate fraction can be used directly or with mild hydroisomerisation as jet fuel or diesel fuel. However, the distillate fraction of HTFT syncrude is barely 7%, and the aqueous fraction may be fractionated into its major constituents (alcohol and carbonyls). The alcohol fraction can further be separated into individual alcohols, which can then be sold as valuable products. The carbonyls can be hydrogenated to deliver products that are suitable for use as gasoline fuel.[106]

Table 1.6 presents a comparison of the fuel properties of Co-LTFT and Fe-HTFT distillates with the European Union's specifications for diesel fuel (EN590:2004).[107] The most critical properties in syncrude distillate utility are density, cetane number and cloud point. As shown in Table 1.6, the Fe-HTFT distillate meets the specifications; however, the distillate fraction of Fe-HTFT syncrude is barely 7%. The Co-LTFT mode, where the distillate yield is higher, does not meet the requirements in terms of density and cloud point. EN590:2004 details the required properties of petroleum-derived diesel, which generally contain fairly equal proportions of *n*-paraffins, cyclic paraffins and aromatics. In this regard, the Fe-HTFT distillate is a close mimic in terms of paraffins, isoparaffins and aromatics. The Co-LTFT distillates comprise mainly *n*-paraffins, which accounts for the lower density and higher cloud point.

A portion of the paraffins can be isomerised into iso-/cyclic paraffins using the low-energy penalty hydroisomerisation step over Pt/ZSM-12 catalyst. This improves the cold flow property, but there is little impact on density. Multistep and high-energy processes are required to upgrade the density at a compromised cetane number. Thus the cetane number and density appear mutually exclusive in terms of diesel fuel. So with respect to low-energy penalty-refining of syncrude to on-specification diesel fuel from syncrude, the low-yield Fe-HTFT distillate meets all three specifications: the high-yield

Table 1.6 Selected fuel properties: Diesel specifications in EN590:2004 *vs.* straight run distillates from Co-LTFT and Fe-HTFT. Adapted from ref. 107 with permission from American Chemical Society, Copyright 2009.

Fuel property	EN590:2004	Co-LTFT	Fe-HTFT
Density at 15 °C ($kg\,m^{-3}$)	820–845	772	844
Cetane number	51 min	80	53
Cloud point (°C)	$-10<x<-34$	0	−11
Viscosity at 40 °C (cSt)	2.0–4.5	2.11	3.62
Lubricity HFRR wear scar (μm)	≤460		251
Aromatic content (mass%)		<0.1	34
T95 boiling point (°C)	≤360	291	339
Fatty acid methyl esters (vol%)	≤5		
Sulphur content ($\mu g\,g^{-1}$)	≤10		

Co-LTFT distillate does not meet the requirements in terms of density. This conundrum of reconciling yield and density is what De Klerk described as the density-cetane-yield triangle.[107]

In tropical countries, where the cold flow requirement for diesel fuel is not stringent, hydroisomerisation of the straight-run distillate becomes less important. Also, where a minimum density of diesel is a requirement, the straight-run LTFT distillate can be used directly in diesel engines, so that the focus is then only on maximising yield. However, when density and cloud point requirements are stringent, hydroisomerisation of the straight-run distillate becomes necessary to improve the cold flow property. Blending or co-refining is the least demanding way to improve density with little compromise in terms of the cetane number. Blending or co-refining involves adding or co-processing a denser feed with the less dense LTFT distillate. Since we have to be conscious of the CO_2 emission footprint of the entire biomass liquefaction process, the complementary feed must be from renewable sources. Feedstock types that satisfy this criterion are bio-oil and biodiesel/vegetable oil.

Bio-oil is crude obtained *via* DBL, which is rich in aromatics and oxygenates. Aromatics are usually denser than paraffins, while oxygenates are beneficial for improving the lubricity of diesel fuel. Hence blending the LTFT distillate with refined bio-oil or co-refining (hydrodeoxygenation) bio-oil with syncrude can help to overcome the density-cetane-yield triangle problem.

Biodiesel/vegetable oils are renewable feedstock that is denser than the LTFT distillate. High viscosity of vegetable oils makes them less suitable for use directly in diesel engines, but they become suitable once transesterified into biodiesel. Most biodiesel products are denser than the LTFT distillate and have a high cetane number; so, blending the LTFT distillate with biodiesel will improve the density without compromising the cetane number. Hydrocracking is another means to lower viscosity of vegetable oil. Vegetable oils such as soybean, sunflower and camelina oils, which are rich in polyunsaturated fatty acids, produce hydrocracked products that are rich in aromatics. Therefore co-hydrocracking over Pt/Al_2O_3/SAPO-11 of LTFT wax or vegetable oil, or co-hydroisomerisation of the LTFT distillate with vegetable oil will deliver a product with a sufficient aromatics content to boost the density of the resulting fuel.[107]

1.5 Modified FTS and Tandem/Cascade Syngas Conversions

Irrespective of the starting feedstock, all indirect liquefaction processes, *i.e.* feed-to-liquid (XTL), can be broken down into three distinct processes[108] (see Figure 1.20). From process cost point of view, BTL, particularly where the energy density of the biomass feedstock is low, it is desirable to reduce the number of distinct categories of processes, so as to minimise the energy expenditure and CO_2 emission of the overall BTL process. The most feasible way to achieve this objective is to merge the latter two processes into one, *i.e.* achieve direct conversion of syngas to the specified fuel or chemicals.

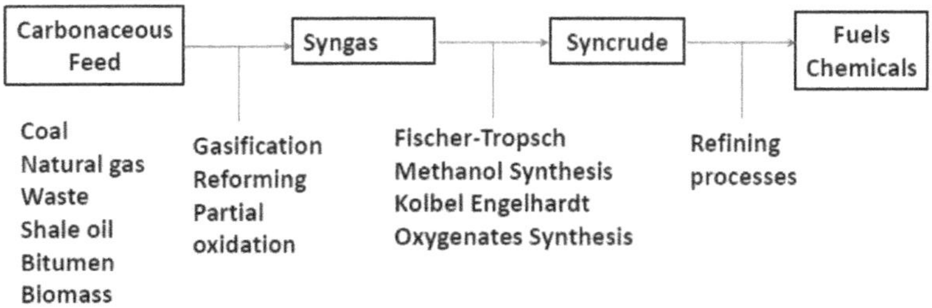

Figure 1.20 Overview of indirect liquefaction processes – XTL.

Traditionally FTS is a polymerisation-type reaction, albeit the monomer formation takes place on the catalyst surface, along with chain propagation. Whether the active sites for the sequence of reactions are the same or different is an open question. The FTS-produced syncrude is a broad spectrum of mainly linear hydrocarbons and some minor oxygenates. However, only a very narrow range within the broad spectrum of products fits the requirements for usage as fuels or chemicals. The product distribution is defined by chain growth probability, α.[100,109,110] Desired product range may be enhanced through catalyst formulation (use of structural and electronic promoters of an active FT metal), but selectivity to a useful product range (*e.g.* C_2–C_4 olefins, C_5–C_{11} gasoline, C_8–C_{16} jet fuel and C_{10}–C_{20} diesel) is still limited.[111] Hence refining (separation and even further conversion processes) becomes necessary to maximise the desired products obtained.

Although syncrude refining is said to constitute 22% of the overall cost of XTL processes, merging the syncrude refining process with the FTS or direct syngas-to-fuels or chemical process would be a huge boost to the process economy.[112] To this end, much effort has been made to overcome the ASF product distribution limitation. A rational approach could be to modify FTS by introducing chain scissoring to curtail the unbridled chain growth or the departure from the polymerisation-type reaction mechanism. These two approaches are examined in the next sections.

1.5.1 Modified FTS

Modified FTS approaches involve controlling product chain length by introducing a chain scissoring catalyst. Bifunctional catalysts that consist of FTS and C–C cleavage (hydrocracking) active sites are used to break the ASF product distribution. Although some measure of chain length control is obtainable with the use of promoters (for example, copper), a promoted iron-based FT catalyst usually displays high distillate selectivity and low wax selectivity.[113,114] This is due to the increase in chain terminations as a result of high hydrogenation activity engendered by copper promotion. However, in the bifunctional catalyst, product selectivity is achieved not by manipulation of chain growth probability, α value, but by scissoring the chain to the desired length by coupling high α-value FT catalysts with zeolites.

The FT catalyst gives the ASF product spectrum defined by α-value, while the zeolite imposed a confining effect or C–C scissoring (hydrocracking) to redefine the product spectrum. Zeolites are framework solids with a varied but well defined pore size and structure. Depending on the acidity, pore size and structure of the zeolite, product selectivity can be tilted toward gasoline, jet fuel or diesel fuel chain length ranges.[115] Another advantage of the zeolite is that it can catalyse other desirable refining reactions, such as hydroisomerisation, oligomerisation, aromatisation, dehydration and etherification. Thus by careful choice of the FT catalyst and zeolite, a one-step syngas process to produce the specified fuel can be achieved.

Spatial arrangement of the FT catalyst and the zeolite catalyst is critical to the performance of the hybrid catalyst. Four spatial arrangements of FT and zeolite catalysts have been reported:[116]

- Mixing the FT catalyst with the zeolite
- Using two catalyst beds
- A core–shell catalyst design
- Using a zeolite-supported FT active metal

The proximity of the FTS and the acid sites of the zeolite is crucial to the efficiency of the tandem reactions in obtaining liquid fuel.[117] Among the spatial arrangements, the core–shell catalyst design and zeolite-supported FT active metal offer the greatest chance of ensuring close proximity of the active sites. The zeolite-supported FT active metal is simply obtained by loading the FT active metal onto the zeolite using the impregnation or ion exchange method, while the core–shell catalyst consists of a conventional FT catalyst as the core, surrounded by zeolite as the shell.

A high gasoline range selectivity (C_5–C_{11}) with an isoparaffin/*n*-paraffin ratio has been reported for FT core–zeolite shell hybrid catalysts.[118] However, core–shell structure bifunctional catalysts are difficult to produce using the hydrothermal method. Growing the zeolite shell around the FT catalyst pellets is a challenge because the strong alkaline condition that is usually employed may be detrimental to the FT core catalyst.[119] Alumina and silica commonly used as the support for the FT active metal, including copper, which is a ubiquitous reduction promoter with iron-based FT catalysts – are prone to dissolution in strong alkaline media. For a silica-supported FT core, *e.g.* Co/SiO_2, a prior silicalite-1 coating placed before growing the zeolite shell gave a catalyst with good performance. Adding the $Fe_{50}Al_{50}$ alloy to the zeolite growth recipe has been demonstrated as a facile one-pot method to FT core–zeolite shell hybrid catalyst.[120] The strong alkaline media for growing the zeolite leaches the aluminium out of the alloy, and the zeolite that formed around the iron or iron oxide is left behind.

Recently, a synthetic method for achieving encapsulation of the metal oxide nanoparticles within the zeolite was demonstrated.[121] Encapsulation is achieved by adding a metal nitrate solution containing the chelating ligand (*N*-[3-(trimethoxysilyl)propyl]ethylenediamine; TPE) to the zeolite recipe.

The ligand prevents precipitation of the metal cations in the alkaline zeolite recipe. The trimethoxysilyl moiety of the ligand forms siloxane linkages with the nucleating zeolite precursors, thereby incorporating the metal oxide precursor into the crystallising frameworks. Oxidative treatment of the crystalline framework removes the ligand species and forms metal-oxide-encapsulated zeolite.

Conventional hydrocrackers are normally operated in the temperature range 350–450 °C.[122] Incorporation of a (de)hydrogenation metal such as Pt enhances catalyst activity and stability. Hydrocracking of FT wax can take place at a lower temperature, depending on the strength of the zeolite acid sites. The formation of cracked products is preceded by an isomerisation step; hence hydrocracking and hydroisomerisation can take place on the same catalyst but at a lower temperature: 250–280 °C. Oligomerisation of lower olefins can also readily take place on zeolite with the LTFT operation mode. Hence hydrocracking, hydroisomerisation and oligomerisation can take place under both LTFT and HTFT conditions.

Typically, the HTFT product spectrum is characterised by high naphtha selectivity and high olefin and oxygenate selectivity. At HTFT temperatures, dehydration of oxygenates, isomerisation of naphtha, oligomerisation and aromatisation of lower olefins readily take place on the zeolite. However, strong acidic zeolite, such as H-ZSM-5 is prone to deactivation, due to coke forming when using this operation mode. Hence weaker acidic zeolite is more suited to achieving tandem FTS and refining reactions under the HTFT mode. With the LTFT mode, hydroisomerisation and oligomerisation are more favourable than hydrocracking over an H-ZSM-5 catalyst, as hydroisomerisation of olefins to iso-olefins discourages chain growth on the catalyst surface. Large linear hydrocarbon chains are also reactive to hydrocracking over H-ZSM-5 at LTFT temperatures. This may account for why most reports on syngas-to-fuel conversions that involve FT-zeolite bifunctional catalysts operated in the LTFT mode display selectivity toward the gasoline-range rather than the diesel-range hydrocarbons.

Direct syngas-to-jet fuel and diesel fuel processing over an FT-zeolite has been less successful, compared to the direct syngas-to-gasoline fuel process.[123] The reactions required on zeolites to increase the distillate selectivity are oligomerisation and hydrocracking. Weaker-acid-strength and lower-acid-site-density zeolite is more suitable for oligomerisation, selectivity and hydroisomerisation, which discourage chain growth, may be favoured over hydrocracking. Use of hierarchical zeolites have been explored to enhance distillate selectivity.[124,125] Hierarchical zeolites combine the advantages of microporous zeolites (*e.g.* acidity functions, molecular sieve features and high temperature stability) and mesopores for easier mass transfer of larger molecules. Syngas conversion over a FT-mesozeolite delivered promising distillate selectivity.[126–128] The acid sites of mesoporous zeolite were treated with an alkali to minimise the unwanted acid-catalysed reactions. Reagents used for synthesis of mesozeolite are expensive, making this catalyst prohibitive in large-scale operations. Spatial separation of the FT and the hydroisomerisation active sites can minimise early shortening of the product chain; thus a LTFT mode that uses a two-bed catalyst arrangement is considered a more practical and plausible approach to direct syngas-to-diesel or syngas-to-jet fuel process.

1.5.2 Tandem/Cascade Syngas Conversion

Syngas conversion aimed at breaking ASF product distribution is achieved by departing from the polymerisation mechanism: the quest for a non-ASF product distribution is approached from monomer formation step. The polymerisation mechanism is avoided by using non-FTS catalysts to effect different pathways of CO activation and monomer formation. For example, FTS and methanol synthesis present the two extremes of syngas reactions. FTS gives an ASF product distribution, while methanol synthesis gives simple oxygenates as the major product. In principle, between these two extremes there is a continuum of possibilities of tuning the product selectivity through catalyst design and varying the reaction conditions. In practice, the catalyst design for such selective syngas conversion has been less successful until recently. The successful catalysts have features that resemble some earlier known syngas conversions, such as methanol synthesis, higher alcohol synthesis and isosynthesis. Since methanol synthesis is a well developed and mature process, and, because methanol can be readily converted into hydrocarbons and chemicals, it is also an attractive and plausible route for non-ASF product selectivity *via* one-pot tandem syngas conversion to fuels and chemicals. Some of the tandem syngas conversion methods are examined in the following section.

1.5.2.1 Direct Syngas-to-dimethyl Ether

Dimethyl ether (DME) is a gas at room temperature, which can be liquefied at a low pressure; hence it is considered an environmentally friendly refrigerant and aerosol propellant alternative to Freon and R-134.[129,130] DME is an intermediate to a plethora of chemicals, such as methylacetate, aromatics, acetic acid, light olefins and aromatics.[130] DME has a high cetane number; hence it has attracted attention in recent years for use as a diesel fuel, and pilot tests of DME-powered buses have been carried out in China. It is clean burning and sulphur free, and it gives lower NO_x and particulate emissions.[131,132] DME is the smallest of the ethers and is usually obtained by dehydration of methanol. Conventional DME involves a two-stage process: methanol synthesis and methanol dehydration. Commercial synthesis of methanol from syngas is usually carried out using a $CuO–ZnO–Al_2O_3$ catalyst, while methanol dehydration can be achieved using a variety of acid catalysts (H_2SO_4, H_3PO_4, γ-alumina, zeolites, polyoxometallate *etc.*).[133,134]

The reactions involved in syngas conversion to DME are provided in Table 1.7.[135] Maximum equilibrium production of methanol is attained at an initial $H_2/CO = 2$ and DME of $H_2/CO = 1$. Biomass-derived syngas usually has an $H_2/(CO + CO_2) \sim 1$; thus biomass-derived syngas is more suitable for DME synthesis than methanol or FTS.

Table 1.7 Reactions for syngas conversion to DME. Adapted from ref. 135 with permission from Springer Nature, Copyright 2000.

Methanol synthesis	$2CO + 2H_2 \leftrightarrow 2CH_3OH$	$\Delta H^\circ = -90.4\ kJ\ mol^{-1}$
	$CO_2 + 3H_2 \leftrightarrow CH_3OH + H_2O$	$\Delta H^\circ = -49.4\ kJ\ mol^{-1}$
DME formation	$2CH_3OH \leftrightarrow CH_3OCH_3 + H_2O$	$\Delta H^\circ = -23.04\ kJ\ mol^{-1}$
WGS	$CO + H_2O \leftrightarrow CO_2 + H_2$	$\Delta H^\circ = -41.0\ kJ\ mol^{-1}$
Overall	$3CO + 3H_2 \leftrightarrow CH_3OCH_3 + CO_2$	$\Delta H^\circ = -258.312\ kJ\ mol^{-1}$

Direct or one-pot syngas-to-DME processing requires integration of the two catalysts. As indicated in the previous section, there are four possible spatial arrangements for the two catalysts and the proximity of the methanol synthesis and dehydrate sites that are crucial to ensuring efficiency of the tandem reactions to DME. However, the catalyst used in methanol synthesis presents an interesting situation. The established methanol synthesis catalyst is CuO–ZnO–Al_2O_3 with a Cu/Zn/Al molar ratio of about 4/5/1. CuO is the active site precursor; ZnO is an electronic promoter, while Al_2O_3 is considered a textural promoter. However, Al_2O_3 is also a methanol dehydration catalyst; so increasing the proportion of Al_2O_3 in the catalyst formulation should give a bifunctional DME catalyst. This was the strategy adopted by Shizuoka University, albeit using the sol–gel method to develop a high-DME-selective catalyst (Cu–Zn(36–4 wt%)/Al_2O_3).[137]

There have been reports regarding stability issues when using γ-Al_2O_3 as a dehydration catalyst because of its hydrophilicity. This issue only occurs when using syngas of $H_2/CO > 1$, when the overall reaction is $2CO + 3H_2 \leftrightarrow CH_3OCH_3 + H_2O$, and the by-product is water. However, silica-supported polyoxometalates or H-ZSM-5 with stronger acid sites than γ-Al_2O_3 are good candidates for methanol dehydration to DME.[136] Integration of these acid catalysts with a methanol synthesis catalyst is more demanding, and fine-tuning the CuO–ZnO–Al_2O_3 composition is still preferable, based on cost and simplicity of process. Most commercial DME synthesis technologies use a syngas of $H_2/CO = 1$, a fixed bed or slurry reactor at 220–290 °C and pressure of 1.4–6 MPa.[137]

1.5.2.2 *Direct Syngas-to-olefins*

Light olefins (C_2–C_4) are important bulk chemicals that are traditionally produced by cracking petroleum naphtha.[138] In recent times, conversion of syngas to light olefins has attracted attention as a viable alternative.[139–143] Direct syngas-to-olefins require tandem catalytic processes. On the basis of intermediates, three syngas-to-olefin processes have been proposed: the methanol-to-olefin (MTO) route, non-FT routes and the FT route. MTO is a well researched process that has proved successful in commercial operations.[144–146] The intermediates involved in the non-FT routes are methanol/DME, carbonylation and ketene.[115] The non-FT and FT routes are briefly examined.

1.5.2.2.1 Non-FT Routes to Direct Syngas-to-olefins. The current commercial syngas-to-olefin process is a two-stage process: methanol synthesis and MTO. There are existing commercial technologies for each of these processes, and integration of the two can lower the energy cost of the overall process.[147–149] However, technical issues must be addressed in order to achieve integration. In principle, in an integrated process, the MTO process should drive the preceding process, according to eqn (1.11):

$$2CO + 2H_2 \leftrightarrow 2CH_3OH \leftrightarrow \tfrac{1}{2}C_2H_4 + H_2O \tag{1.11}$$

The current commercial MTO process operates at ≥400 °C and uses SAPO-34 and SSZ-13 zeolites; the current methanol synthesis process operates at 250–280 °C.[150,151] A combination of low-temperature methanol synthesis catalysts ($CuO–ZnO–Al_2O_3$, $CuO–ZnO–ZrO_2$, $CuO–ZnO–Cr_2O_3$, Pd/ZnO) and the MTO catalyst SAPO-34 gave methane and paraffins, as the catalyst is too hydrogenating at the MTO temperature of 400 °C. However, excluding the hydrogenating metal (Cu) from the catalyst formulation led to selective direct syngas to olefins. Using such catalyst recipes that are without hydrogenation metal were reported about the same time by Wang *et al.* ($ZnZrO_x$/SAPO-34)[151] and Bao *et al.* ($ZrCrO_x$/SAPO-34).[152] Bao *et al.* reportedly obtained 80% $C_2^=$–$C_4^=$ and 94% C_2–C_4 selectivity at a CO conversion of 17%. Wang *et al.* reported 74% C_2–C_4 olefin selectivity with 11% CO conversion at 673 K. In the two catalysts, the CO activation site is attributed to partial reduction of the mixed metal oxides. It is interesting to note that a similar CO activation site was invoked in old and familiar syngas-to-i-C_4 hydrocarbon conversion (isosynthesis) over sulphated or doped ZrO_2 catalysts.[153,154]

In most isosynthesis studies, zinc is not investigated, and the reported isobutene selectivity is very low. But Wu *et al.* recently attributed Zn doping to isobutene selectivity in a $Zn_1Zr_{300}O_z$ catalyst.[155] This shows the critical role of Zn to olefin selectivity in the catalysts. In a follow-up report, Wang and co-workers replaced SAPO-34 with SSZ-13, with the resulting Zn–ZrO_2/SSZ-13 catalyst showing 87% and 77% C_2–C_4 olefin selectivity at 10% and 29% CO conversion, respectively, at 400 °C.[156] Olefin formation on a Zn–ZrO_2/zeolite catalyst has been attributed to the methanol-DME intermediate. CO and H_2 activation to methanol or DME takes place on the Zn–ZrO_2, while the C–C bond formation takes place on the zeolite acid site *via* the CO carbonylation and hydrocarbon-pool mechanism. Here again, the proximity of the active sites and the strength of the acid sites is crucial to olefin selectivity.[115]

In another report, Bao and co-workers expanded the metal oxide-zeolite catalyst design, by replacing $ZnCrO_x$ with MnO, and investigated MnO_2/SAPO-34 for syngas-to-olefin conversion.[157] An optimal performance of 10.1% CO conversion with a $C_2 - C_4$ olefin selectivity of 78.9% was reported at an oxide/zeolite ratio of 2/1. Recently, Su *et al.* replaced SAPO-34 with similar-pore-sized but less acidic, low-Si AlPO-18. The resulting bifunctional catalyst, $ZnCrO_x$/AlPO-18, displayed C_2–C_4 olefin selectivity of 45.0% (86.7% conversion, CO_2 free) and a high olefin/paraffin ratio of 29.9 at CO conversion of 25.2% at 4.0 MPa and 390 °C.[158] In a demonstration of the effect of zeolite pore topography on C–C coupling and olefin selectivity, Bao and co-workers reported 73% ethylene selectivity at 26% CO conversion, using $ZnCrO_x$-mordenite (MOR).[157] This is more than double the maximum ethylene selectivity that is possible based on the ASF product distribution law.

Using highly sensitive synchrotron-based vacuum ultraviolet photoionisation mass spectrometry (SVUV-PIMS), Bao and co-workers detected ketene (CH_2CO) species in the gas phase after the interaction of syngas with $ZnCrO_x$.[157] They explained that the ketene was converted to C_2H_4 and C_3H_6 over the SAPO-34. The hydrocarbon pool mechanism was invoked for ketene-to-olefins transformation, but no explanation was provided for the syngas-to-ketene mechanism. With respect to the MnO_2/SAPO-34 catalyst, it was stated that two CO molecules dissociate over adjacent oxygen vacancy sites that are generated by partial reduction of MnO_x forming a surface C and carbonate species.[157] The carbonate is discharged as CO_2, thereby conserving H_2 consumption for the removal of surface oxygen for syngas H_2/CO ratio <2. Overall, separation of the activation or monomer formation from CO and C–C coupling sites gives greater control of olefin selectivity.

1.5.2.2.2 FT Routes to Direct Syngas-to-olefins. This route is known as Fischer–Tropsch-to-olefins (FTO). Among the FTS active metals (Ru, Co, Fe and Ni), Fe-based FT catalysts have the least hydrogenation activity and display selectivity to α olefins.[159,160] In commercial FTS, Co-catalysts (Co or Fe based) contain noble metal reduction promoters, such as Pt, Ru, Pd, Ir, Cu, *etc.*, and these metals possess high hydrogenating activity, promote reduction and enhance the hydrogenating activity of the FTS active metal *via* a hydrogen spillover effect. Other common components of commercial FTS catalysts are oxides of metals such as Zn, Mn, Zr, Ti, Cr, Sn, Mg, Ca, K and Na. These oxides are textural promoters that increase the dispersion of active metals and sometimes exact an electronic effect through strong metal-support interaction (SMSI).[161–163] An alkali metal (especially K) is usually added to modulate hydrogenation activity of active FTS metal. Overall, commercial FTS catalysts are designed to ensure high hydrogenation activity to maximise C_{5+} selectivity and deliver a high α-value.

Contrarily, FTO catalysts should have a low α-value and low hydrogenating activity.[164] Hence, the first step in tuning a FTS catalyst to a FTO catalyst is to remove the reduction promoter, and it has already been shown that CO activation can take place on ZrO_2, ZnO, CrO_x and MnO_x. The second step is to enhance the status of the metal oxide from a mere textural promoter to a co-partner with the FTS active metal. Third, the alkali is retained to modulate the hydrogenation activity. These three steps can explain the documented higher olefin selectivity displayed by bulk and supported FeMnK catalysts.[165,166] Similar olefin selectivity was reported on FeK promoted with the reducible metal oxides: V, Nb, Pb, Sn and Bi.[167–170] Combined (Na, K) and S promotion of Fe-based FT catalysts has also been reported as displaying high selectivity to lower olefins.[171–174]

Recently, an FT-zeolite core–shell catalyst concept was implemented as the FTO catalyst. Compared to a simple mixture catalyst (Fe-FT/SAPO-34) and a bare Fe–FT catalyst, the new Fe–FT core–zeolite shell Fe@SAPO-34 catalyst displayed a higher lower olefin selectivity of 52.6% and an O/P ratio of 6.4.[175] The nonzeolite core–shell catalyst design has also been explored for the FTO route to olefins.[176] The Fe_3O_4@MnO_2 core–shell is reported to show an exceptionally high lower olefin yield and a high lower olefin selectivity at high CO conversion.[176] The catalyst exhibited 79.60% total alkene selectivity and 64.95% for C_{4+} alkene selectivity at 75% CO conversion. The performance of the catalyst was attributed to electron transfer from the MnO_2 shell to the Fe_3O_4 core, which promotes dissociative adsorption of CO molecules on Fe_3O_4. Additionally, reciprocal H atom spillover from Fe_3O_4 onto MnO_2 enhances C–C coupling, weakens the hydrogenation activity of the catalyst and improves olefin selectivity. Co-based FTS catalysts are generally known for their high wax and paraffin selectivity and have also been tamed as FTO catalysis. Supported $CoMnO_x$ displayed a high lower olefins selectivity, and the selectivity to lower olefins increased when promoted with alkali (Na, K) ion.[177] The lower olefin selectivity of Co-based catalysts has been attributed to the Co_2C phase.[178,179] The widely agreed FT activity of the Co-based catalyst is in the Co^0 phase. However, in $CoMnO_x$ mixed oxide, in addition to the Co^0 phase, the presence of MnO_x promotes Co_2C formation during catalyst activation.[180,181]

1.5.2.3 *Direct Syngas-to-aromatics*

BTX are bulk aromatic chemicals used as the primary feedstock in the production of a wide variety of intermediates, fine chemicals and materials such as phenolic resins, Kevlar, polystyrene, Bakelite *etc.* Current industrial production of BTX involves catalytic reforming of naphtha and, more recently, using shale gas *via* cracking process. With a projected growing demand for BTX, quest to secure the supply of BTX *via* direct syngas-to-aromatic conversion is attracting increasing research attention across the globe.[138,182] Aromatics are not primary FTS products, while little aromatics are often obtained in the Fe-HTFT condition.[183] They are formed from aromatisation of lower olefins. Aromatisation of methanol/DME has also been investigated extensively. Thus lower olefins and methanol can be used as intermediates to aromatics in a direct syngas-to-aromatics *via* a tandem one-pot conversion process.[115] The tandem catalytic processes for syngas-to-aromatics *via* olefin and methanol/DME intermediates are briefly explained in the next sections.

1.5.2.3.1 Olefins Route for the Direct Syngas-to-aromatics Process. Catalyst designs for direct syngas-to-olefin processes eliminate the presence of strong acid sites because they promote olefin-to-aromatics conversion, which decreases olefin selectivity. But in direct syngas-to-aromatics processes, a deliberate effort is made to incorporate strong acid sites into a FTO catalyst to engender a direct syngas-to-aromatics process.

H-ZSM-5 is a prominent aromatisation zeolite catalyst. H-ZSM-5 and its cation- (Zn-, Ga-, Mo-, La- *etc.*) exchanged forms have been extensively studied for aromatisation of lower olefins and other substrates, such as methanol, methane and lower alkanes.[184–187] Any of the FTO catalysts indicated in Section 1.5.2.2 can be integrated with H-ZSM-5 or its cation exchanged form and qualifies as a potential combination for tandem syngas-to-aromatics.[188–191] As discussed, the spatial arrangement of the active sites is critical to the performance of the catalyst combinations in a tandem syngas conversion process. The single-bed catalyst mixture is the simplest arrangement, and core–shell design can also be adopted. However, since aromatisation catalysts are easily prone to deactivation due to coke formation, a two-catalyst bed arrangement is recommended because it allows for flexible adjustment of reaction conditions and independent regeneration and recycling of the catalyst beds.

1.5.2.3.2 Methanol/DME Route for a Direct Syngas-to-aromatics Process. The methanol/DME intermediate is another plausible route for direct formation of aromatics from syngas. Methanol synthesis is an established technology, and direct syngas-to-DME was discussed in Section 1.5.1. Methanol and DME aromatisations over H-ZSM-5 and related cation-exchanged forms are well documented in the literature.[192,193] Therefore, integration of Cu-based methanol/DME synthesis catalysts and H-ZSM-5 should make the direct formation of aromatics from syngas possible. However, this integration faces a temperature mismatch. Aromatisation over H-ZSM-5 usually takes place at a temperature of $\geq$400 °C, while methanol/DME synthesis over a Cu–Zn–Al oxide catalyst is typically carried out at 250–280 °C. The use of the two-catalyst-beds arrangement, which allows for operating the catalyst beds at different temperatures, may help to circumvent the temperature mismatch. But in principle, operating two catalysts at different temperatures is easy to achieve at the laboratory scale but proves very challenging with commercial-scale operations. Thus an isothermal catalyst bed reactor is preferable for large-scale operations.

As discussed in Section 1.5.1, one of the syngas-to-olefins explained involved a methanol/DME intermediate. The reaction is a high-temperature methanol formation process; hence it is more suited than the Cu–Zn–Al-oxide catalyst for tandem conversion with H-ZSM-5. Therefore, replacing the smaller-pore-size and weaker-acidic SAPO-34 with H-ZSM-5 tunes the product selectivity from olefins to aromatics. Thus in a one-pot syngas conversion process, aromatics selectivity of 80% at a CO conversion of 20% was reportedly achieved using a ZnO–ZrO_2/H-ZSM-5 catalyst.[139] The catalyst was integrated by simple dispersion of Zn-doped ZrO_2 nanoparticles dispersed on zeolite H-ZSM-5, and no catalyst deactivation was observed in 1000 h of operation. As previously suggested, methanol and DME are major intermediates with Zn-doped ZrO_2 and are subsequently transformed on H-ZSM-5 into aromatics *via* olefins.[194–196] The easy propensity for coke formation has been a long-standing challenge with aromatisation over H-ZSM-5. The high selectivity and stability obtained with ZnO–ZrO_2/H-ZSM-5 are attributed to the self-promotion mechanism of CO in the selective formation of aromatics. It was explained that CO facilitates the removal of hydrogen species formed on H-ZSM-5 in the process of dehydrogenative aromatisation of olefins. The overall reaction is an elegant tandem process and a new frontier in syngas conversion chemistry.

1.6 Conclusion

Attention is increasingly being given to the use of biomass for the production of fuel and chemicals. For purposes of the security of liquid fuel supply and because of environmental concerns, it is becoming increasingly imperative to develop technologies for transforming biomass into liquid fuel and chemicals. Due to its similarities with coal, established coal liquefaction processes are being adopted and modified into biomass liquefaction processes. These can be classified into two broad categories: direct and IBL processes. Generally, biomass is a composite of three large molecules: cellulose, hemicellulose and lignin. The liquefaction process basically involves transforming the biomass composite into small hydrocarbon molecules that can be used in engines (petrol, jet and diesel). The direct liquefaction process is a top-down approach to transforming biomass into liquid fuel that involves biomass pyrolysis and refining/upgrading processes. It is inflexible in terms of obtainable products. Contrarily, IBL can be considered a bottom-up approach to fuel production from biomass. It involves biomass gasification (syngas production) and FTS (syncrude production), followed by refining (syncrude upgrade to fuels) FTS leads to a wide but predictable product spectrum, it makes IBL more versatile and flexible. Both approaches have challenges in terms of low carbon utilisation.

To address this challenge, renewable energy inputs (electricity, hydrogen and solar thermal) are being advocated and encouraged. To this end, some complementary variant biomass liquefaction processes have been proposed, including GBTL, PBTL and solar-biomass-to-liquid (SBTL). These inputs complement biomass deficiencies and enhance carbon efficiency, thereby reducing CO_2 emissions from the liquefaction process. Another approach to addressing the challenge of low-carbon-efficiency biomass liquefaction processes is the integration of FTS with syncrude refining processes, *i.e.* the development of direct syngas conversions to a target fuel or chemical.

This task goes to the heart of FTS and requires a thorough understanding of the chemistry involved in the process. Conventional FTS is a polymerisation-type reaction, with the statistical product distribution generally described by ASF distribution law. All the reaction steps (monomer formation/initial chain propagation and chain termination) take place on the same active site. In order to control product formation, efforts have been made to truncate the chain growth step or to delineate the reaction steps onto separate active sites. This often involves replacing the conventional one-catalyst FTS with two- or more-catalyst FTS. The integration of the multicatalyst system for syngas conversion is not without some challenges. The tasks constitute emerging frontiers in syngas conversion chemistry.

Abbreviations

ASF	Anderson–Schultz–Flory
BTL	Biomass-to-liquid
BTX	Benzene, toluene and xylene
CFB	Circulating fluidised bed
DBL	Direct biomass liquefaction
DCL	Direct coal liquefaction
DFB	Dual fluidised bed
DME	Dimethyl ether (DME)
ECN	Energy Research Centre of the Netherlands
ER	Equivalent ratio
FFB	Fixed fluidised bed
FTO	Fischer–Tropsch to olefins
FTS	Fischer–Tropsch synthesis
GBTL	Gas-biomass-to-liquid
GTL	Gas-to-liquid
HTFT	High-temperature Fischer–Tropsch
IBL	Indirect biomass liquefaction
ICL	Indirect coal liquefaction
LTFT	Low-temperature Fischer–Tropsch
MTFB	Multitubular fixed bed
MTO	Methanol-to-olefin
NG	Natural gas
OLGA	Dutch acronym for the oil-based gas washing process
PBTL	Power-biomass-to-liquid
PEM	Polymer electrolyte membrane
PSA	Pressure swing adsorption
RFS	Renewable fuel standard
rWGS	Reverse water-gas-shift
SAS	Sasol advanced synthol
SBR	Steam-to-biomass ratio
SMSI	Strong metal-support interaction
STBTL	Solar-thermal-biomass-to-liquid
SVUV-PIMS	Synchrotron-based vacuum ultraviolet photoionisation mass spectrometry
WGS	Water-gas-shift
XTL	Feed-to-liquid

Acknowledgements

S. Maity is thankful to the director of his Institute for allowing the book chapter to be published. O. O. James thanks KWASU management for the conducive work environment that enabled this contribution.

References

1. G. Lalude, *Glob. J. Human-Social Sci. F Polit. Sci.*, 2015.
2. P. Stevens, *Extractive Industries: The Management of Resources as a Driver of Sustainable Development – The Role of Oil and Gas in the Economic Development of the Global Economy*, Oxford University Press, 2018.
3. M. Balcilar, Z. Abidin and Y. Arslanturk, *Energy Econ.*, 2010, **32**, 1398–1410.
4. A. C. Philippe Le Billon, *Ann. Assoc. Am. Geogr.*, 2009, 832–844.
5. J. Noguera-santaella, *Econ. Modell.*, 2016, **52**, 301–309.
6. C. Su, M. Qin, R. Tao and N. Moldovan, *Def. Peace Econ.*, 2019, 1–17.
7. G. Luciani, *Armed Conflicts and Security of Oil and Gas Supplies*, 2015.
8. G. Escribano and J. Valdés, *Geopolitics*, 2016, 1–26.
9. A. Månsson, *Energy Res. Soc. Sci.*, 2014, **4**, 106–116.
10. M. Ratner, *Middle East and North Africa Unrest: Implications for Oil and Natural Gas Markets*, 2011.
11. D. Gao, D. Gao, C. Ye, X. Ren and Y. Zhang, *Fuel Process. Technol.*, 2018, **169**, 42–49.
12. R. H. Williams and E. D. Larson, *Energy Sustainable Dev.*, 2003, 7, 103–129.
13. K. Pstrowska, *J. Energy Inst.*, 2018, **91**, 668–675.
14. S. Vasireddy, B. Morreale, A. Cugini and J. J. Spivey, *Energy Environ. Sci.*, 2011, **4**, 311–345.
15. D. Unruh, K. Pabst, G. Schaub and T. Kit, *Energy Fuels*, 2010, 2634–2641.
16. H. Chandan and E. S. Rubin, *Fuel*, 2013, **103**, 805–813.
17. S. Michailos, D. Parker and C. Webb, *Chem. Eng. Res. Des.*, 2017, **118**, 206–2014.
18. E. Rozzi, F. D. Minuto and A. Lanzini, *Energies*, 2020, **13**, 1–96.
19. V. S. Sikarwar, M. Zhao, P. Clough, J. Yao, X. Zhong, M. Z. Memon, N. Shah, E. J. Anthony and P. S. Fennell, *Energy Environ. Sci.*, 2016, **9**, 2939–2977.
20. X. Luo, T. Wu, K. Shi, M. Song and Y. Rao, in *Gasification for Low-grade Feedstock*, ed. Y. Yun, IntechOpen, 2018, ch. 1, pp. 1–17.
21. A. Molino, S. Chianese and D. Musmarra, *J. Energy Chem.*, 2016, **25**, 10–25.
22. S. K. Sansaniwal, K. Pal, M. A. Rosen and S. K. Tyagi, *Renew. Sust. Energ. Rev.*, 2017, **72**, 363–384.
23. F. N. Sidek, N. A. F. Abdul Samad and S. Saleh, *IOP Conf. Ser.: Mater. Sci. Eng.*, 2020, **863**, 012028.
24. H. Hofbauer and M. Materazzi, in *Substitute Natural Gas from Waste*, ed. M. Materazzi and P. U. Foscolo, Elsevier, 2019, ch. 7, pp. 105–160.
25. R. Rauch, H. Hofbauer, R. Rauch, J. Hrbek and H. Hofbauer, *Wiley Interdiscip. Rev.: Energy Environ.*, 2014, **3**, 343–362.
26. K. Sandeep and S. Dasappa, *Int. J. Energy Res.*, 2014, **38**, 174–188.
27. S. D. Arvind Gupta, *26th European Biomass Conference and Exhibition*, Copenhagen, Denmark HYDROGEN, 2018, pp. 14–17.
28. Y. P. Chyou, D. M. Chang, P. C. Chen, H. Y. Chien, K. T. Wu and R. Y. Chein, *Appl. Sci.*, 2020, **10**, 1–8.
29. J. Marinkovic, PhD thesis, Chalmers University of Technology, 2015.

30. A. M. Mauerhofer, J. C. Schmid, F. Benedikt, J. Fuchs, S. Müller and H. Hofbauer, *Energy*, 2019, **173**, 1256–1272.
31. G. Xu, T. Murakami, T. Suda, Y. Matsuzawa and H. Tani, *Ind. Eng. Chem. Res.*, 2006, **45**, 2281–2286.
32. G. P. Plant, F. Benedikt, J. C. Schmid, J. Fuchs, A. M. Mauerhofer, H. Hofbauer, J. C. Schmid, J. Fuchs, A. M. Mauerhofer and G. P. Plant, *Energy*, 2018, **164**, 329–343.
33. J. C. Schmid, C. Pfeifer, J. C. Schmid, C. Pfeifer, H. Kitzler, T. Pröll and H. Hofbauer, *Proceedings - International Conference on Polygeneration Strategies (ICPS)*, Vienna, Austria, 2011, pp. 1–10.
34. J. C. Schmid, PhD thesis, Vienna University of Technology, 2014.
35. S. Bensaid, R. Conti and D. Fino, *Fuel*, 2012, **94**, 324–332.
36. M. Oevermann, B. Wilmes and N. Zobel, *Chem. Eng. Technol.*, 2008, 667–677.
37. R. M. Davidson, *Coal Sci.*, 1973, **I**, 83–159.
38. M. R. Haverly, PhD thesis, Iowa State University, 2016.
39. J. D. Sterman, L. Siegel and J. N. Rooney-varga, *Environ. Res. Lett.*, 2018, **13**, 1–10.
40. M. M. Wright, N. Seifkar, W. H. Green and Y. Roman-leshkov, *Environ. Sci. Technol.*, 2015, **49**, 8183–8192.
41. G. Liu, R. H. Williams, E. D. Larson and T. G. Kreutz, *Energy Procedia*, 2013, **4**, 1989–1996.
42. A. H. Sahir, Y. Zhang, E. C. D. Tan, L. Tao and B. Analysis, *Biofuels, Bioprod. Biorefin*, 2019, 1–15.
43. M. Rafati, L. Wang, D. C. Dayton, K. Schimmel, V. Kabadi and A. Shahbazi, *Energy Convers. Manage.*, 2017, **133**, 153–166.
44. Y. Zhang, A. H. Sahir, E. C. D. Tan, M. J. Biddy, L. Tao, M. S. Talmadge and R. Davis, *Green Chem.*, 2018, **20**, 5358–5373.
45. S. Müller, P. Groß, R. Rauch, R. Zweiler, C. Aichernig and M. Fuchs, *Biomass Convers. Biorefin.*, 2018, **8**, 275–282.
46. M. Ostadi, E. Rytter and M. Hillestad, *Biomass Bioenergy*, 2019, **127**, 105282.
47. L. Wang, M. Rao, S. Diethelm, T. Lin, H. Zhang and A. Hagen, *Appl. Energy*, 2019, **250**, 1432–1445.
48. M. Götz, J. Lefebvre, J. Lefebvre, M. Friedemann, F. Graf, S. Bajohr, R. Reimert and T. Kolb, *Renew. Energy*, 2016, **85**, 1371–1390.
49. R. Anghilante, C. Müller, M. Schmid, D. Colomar, F. Ortloff, R. Spörl, A. Brisse and F. Graf, *Energy Convers. Manage.*, 2019, **183**, 462–473.
50. M. Pozzo, A. Lanzini and M. Santarelli, *Fuel*, 2015, **145**, 39–49.
51. H. Zhang, L. Wang, J. Van, F. Mar and U. Desideri, *Energies*, 2019, **12**, 1–15.
52. F. M. Sapountzi, J. M. Gracia, C. J. K. Weststrate, H. O. A. Fredriksson and J. W. H. Niemantsverdriet, *Prog. Energy Combust. Sci.*, 2017, **58**, 1–35.
53. Y. Zheng, J. Wang, B. Yu, W. Zhang, J. Chen, J. Qiao and J. Zhang, *Chem. Soc. Rev.*, 2017, **46**, 1427–1463.
54. H. Zhang, L. Wang, F. Maréchal and U. Desideri, *Energy Procedia*, 2019, **158**, 4548–4553.

55. J. B. Hansen, *Faraday Discuss.*, 2015, **182**, 9–48.
56. S. H. Ansari and X. Liu, *AIP Conf. Proc.*, 2018, 130001.
57. B. Coelho, A. C. Oliveira and A. Mendes, *Energy Environ. Sci.*, 2010, **3**, 1398–1405.
58. C. N. R. Rao and S. Dey, *Proc. Natl. Acad. Sci. U. S. A.*, 2017, **114**, 13385–13393.
59. E. G. Hertwich and X. Zhang, *Environ. Sci. Technol.*, 2009, **43**, 4207–4212.
60. A. Nzihou, G. Flamant, B. Stanmore, A. Nzihou, G. Flamant, B. Stanmore, A. Nzihou, G. Flamant and B. Stanmore, *Energy*, 2012, **42**, 121–131.
61. V. Pozzobon, S. Salvador and J. J. Bézian, *Fuel*, 2018, **214**, 300–313.
62. J. Soares, A. C. Oliveira, S. Dieckmann, D. Krüger and F. Orioli, *Int. J. Low-Carbon Technol.*, 2018, **13**, 380–387.
63. I. Janajreh, S. Syed, R. Qudaih and I. Talab, *Int. J. Therm. Environ. Eng. Sol.*, 2010, **2**, 81–90.
64. J. K. Maund and D. M. Earp, *Fuels from biomass by conversion in molten salts*, 1988.
65. B. J. Hathaway and J. H. Davidson, *Sol. Energy*, 2017, **142**, 224–230.
66. B. J. Hathaway and J. H. Davidson, *Energy Fuels*, 2020, **34**, 1811–1821.
67. J. H. D. B. J. Hathaway and D. B. Kittelson, *Energy Procedia*, 2014, **49**, 1950–1959.
68. F. Manenti and Z. Ravaghi-Ardebili, *Energy*, 2013, **55**, 89–97.
69. L. Evangelisti, R. De Lieto Vollaro and F. Asdrubali, *Renewable Sustainable Energy Rev.*, 2019, **114**, 109318.
70. M. Israelsson, PhD thesis, Chalmers University of Technology, 2014.
71. S. Thapa, N. Indrawan, P. R. Bhoi, A. Kumar and R. L. Huhnke, *Energy*, 2019, **175**, 402–409.
72. P. R. Bhoi, R. L. Huhnke, A. Kumar, M. E. Payton, K. N. Patil and J. R. Whiteley, *Fuel Process. Technol.*, 2015, **133**, 97–104.
73. P. Nicolaou, PhD thesis, Delft University of Technology, 2016.
74. P. R. Bhoi, R. L. Huhnke, A. Kumar, K. N. Patil and J. R. Whiteley, *Fuel Process. Technol.*, 2015, **134**, 243–250.
75. M. Balas, M. Lisy, Z. Skala and J. Pospisil, *Adv. Environ. Sci., Dev. Chem.*, 2014, 195–201.
76. S. Nakamura, S. Kitano and K. Yoshikawa, *Appl. Energy*, 2016, **170**, 186–192.
77. T. Tarnpradab, S. Unyaphan and F. Takahashi, *Biofuels*, 2016, **7269**, 1–12.
78. S. Unyaphan, T. Tarnpradab, F. Takahashi and K. Yoshikawa, *Appl. Energy*, 2017, **205**, 802–812.
79. M. Balas, M. Lisy, Z. Skala and J. Pospisil, *WSEAS Trans. Heat Mass Transf.*, 2014, **9**, 195–204.
80. S. Lotfi, W. Ma, K. Austin and A. Kumar, *Fuel Process. Technol. J.*, 2019, **193**, 197–203.
81. S. Unyaphan, T. Tarnpradab and F. Takahashi, *Energy Procedia*, 2017, **105**, 406–412.

82. L. P. L. M. Rabou, R. W. R. Zwart, B. J. Vreugdenhil and L. Bos, *Energy Fuels*, 2009, **23**, 6189–6198.
83. H. Boerrigter, D. J. Slort, H. Bodenstaff, H. Den Uil, H. Boerrigter, R. Van Ree and H. J. Veringa, *Gas Cleaning for Integrated Biomass Gasification (BG) and Fischer–Tropsch (FT) Systems Draft approved by Shell Global Solutions International*, 2004.
84. G. S. M. Iglesias and R. Edzang, *Catal. Today*, 2013, **215**, 194–200.
85. O. O. James, A. M. Mesubi, T. C. Ako and S. Maity, *Fuel Process. Technol.*, 2010, **91**, 136–144.
86. Y. Yao, D. Hildebrandt, D. Glasser and X. Liu, *Ind. Eng. Chem. Res.*, 2010, **49**, 11061–11066.
87. Y. Yao, X. Liu, D. Hildebrandt and D. Glasser, *Ind. Eng. Chem. Res.*, 2011, **50**, 11002–11012.
88. M. Fujlwara, R. Kieffer, H. Ando, Q. Xu and Y. Souma, *Appl. Catal., A*, 1997, **154**, 87–101.
89. B. Rongxian, T. Yisheng and H. Yizhuo, *Fuel Process. Technol.*, 2004, **86**, 293–301.
90. X. Ni, Y. Tan, Y. Han and N. Tsubaki, *Catal. Commun.*, 2007, **8**, 1711–1714.
91. B. H. Davis, *Top. Catal.*, 2005, **32**, 143–168.
92. H. Lynggaard, A. Andreasen, C. Stegelmann and P. Stoltze, *Prog. Surf. Sci.*, 2004, **77**, 71–137.
93. E. Van Steen and H. Schulz, *Appl. Catal., A*, 1999, **186**, 309–320.
94. Y. N. Wang, W. P. Ma, Y. J. Lu, J. Yang, Y. Y. Xu, H. W. Xiang, Y. W. Li, Y. L. Zhao and B. J. Zhang, *Fuel*, 2003, **82**, 195–213.
95. F. G. Botes, *Ind. Eng. Chem. Res.*, 2009, **48**, 1859–1865.
96. F. G. Botes, *Energy Fuels*, 2007, **21**, 1379–1389.
97. M. Y. Ngai, J. R. Kong and M. J. Krische, *J. Org. Chem.*, 2007, **72**, 1063–1072.
98. H. Y. Jang and M. J. Krische, *Acc. Chem. Res.*, 2004, **37**, 653–661.
99. O. O. James, B. Chowdhury, M. A. Mesubi and S. Maity, *RSC Adv.*, 2012, **2**, 7347–7366.
100. A. Y. Krylova, *Solid Fuel Chem.*, 2014, **48**, 22–35.
101. N. O. Egiebor, K. R. Ungar and B. W. Wojciechowski, *Can. J. Chem. Eng.*, 1984, **62**, 425–430.
102. J. Gorimbo, A. Muleja, X. Liu and D. Hildebrandt, *Int. J. Ind. Chem.*, 2018, **9**, 317–333.
103. G. P. van der Laan, PhD thesis, University of Groningen, 1999.
104. A. De Klerk, PhD thesis, University of Pretoria, 2008.
105. A. de Klerk, *Fischer–Tropsch Refining*, Wiley-VCH Verlag & Co. KGaA, Weinheim, Germany, 2011.
106. A. de Klerk, *Energy Environ. Sci.*, 2011, **4**, 1177–1205.
107. A. De Klerk, *Energy Fuels*, 2009, **23**, 4593–4604.
108. A. de Klerk and D. L. King, *Synthetic Liquids Production and Refining*, 2011, ACS Symposium Series, vol. 1084.
109. J. Chen and C. Yang, *ACS Omega*, 2019, **4**, 22237–22244.

110. W. D. Shafer, M. K. Gnanamani, U. M. Graham, J. Yang, C. M. Masuku, G. Jacobs and B. H. Davis, *Catalysts*, 2019, **9**, 3–57.
111. H. Schulz, *Catal. Today*, 2013, **214**, 140–151.
112. R. Zennaro, in *Greener Fischer–Tropsch Processes for Fuels and Feedstocks*, ed. A. de K. Peter and M. Maitlis, Wiley-VCH Verlag & Co. KGaA, 2013, pp. 149–169.
113. O. O. James, B. Chowdhury, A. Auroux and S. Maity, *Appl. Energy*, 2013, **107**, 377–383.
114. S. Li, S. Krishnamoorthy, A. Li, G. D. Meitzner and E. Iglesia, *J. Catal.*, 2002, **206**, 202–217.
115. W. Zhou, K. Cheng, J. Kang and Q. Zhang, *Chem. Soc. Rev.*, 2020, **48**, 3193–3228.
116. O. O. James, A. M. Mesubi, T. C. Ako and S. Maity, *Fuel Process. Technol.*, 2010, **91**, 136–144.
117. P. Lu, J. Sun, D. Shen, R. Yang, C. Xing, C. Lu, N. Tsubaki and S. Shan, *Appl. Energy*, 2018, **209**, 1–7.
118. J. He, B. Xu, Y. Yoneyama, N. Nishiyama and N. Tsubaki, *Chem. Lett.*, 2005, **34**, 148–149.
119. D. Farrusseng and A. Tuel, *New J. Chem.*, 2016, **40**, 3933–3949.
120. B. Sun, G. Yu, J. Lin, K. Xu, Y. Pei, S. Yan, M. Qiao, K. Fan, X. Zhang and B. Zong, *Catal. Sci. Technol.*, 2012, **2**, 1625–1629.
121. T. Otto, S. I. Zones and E. Iglesia, *Microporous Mesoporous Mater.*, 2018, **270**, 10–23.
122. P. Robinson, in *Practical Advances in Petroleum Processing*, ed. P. R. Robinson and C.S. Hsu, Springer, New York, NY, 2006, pp. 177–218.
123. J. Li, Y. He, L. Tan, P. Zhang, X. Peng, A. Oruganti, G. Yang, H. Abe, Y. Wang and N. Tsubaki, *Nat. Catal.*, 2018, **1**, 787–793.
124. R. Güttel and T. Turek, *Energy Technol.*, 2016, **4**, 44–54.
125. C. Xing, G. Yang, M. Wu, R. Yang, L. Tan, P. Zhu, Q. Wei, J. Li, J. Mao, Y. Yoneyama and N. Tsubaki, *Fuel*, 2015, **148**, 48–57.
126. R. Bai, Y. Song, Y. Li and J. Yu, *Trends Chem.*, 2019, **1**, 601–611.
127. L. Emdadi, Y. Wu, G. Zhu, C. C. Chang, W. Fan, T. Pham, R. F. Lobo and D. Liu, *Chem. Mater.*, 2014, **26**, 1345–1355.
128. T. Prasomsri, W. Jiao, S. Z. Weng and J. Garcia Martinez, *Chem. Commun.*, 2015, **51**, 8900–8911.
129. A. A. Zharov, S. A. Garanov, M. S. Talyzin and N. A. Kovalchuk, *AIP Conf. Proc.*, 2019, **2171**, 120009, 1–6.
130. B. Ayyalusamy, *Aust. J. Basic Appl. Sci.*, 2017, **9**, 297–306.
131. E. M. Chapman, S. Bhide, J. Stefanik, H. Glunt, A. L. Boehman, A. Homan and D. Klinikowski, in *Alternative Fuels and Advanced Combustion Techniques as Sustainable Solutions for Internal Combustion Engines*, ed. A. P. Singh, D. Kumar and A. K. Agarwal, Springer Nature, Singapore, 2021, pp. 11–36.
132. K. Koponen and N. Nylund, *Fuel and Technology Alternatives for Buses: Overall Energy Efficiency and Emission*, 2012.

133. U. Mondal and G. D. Yadav, *J. CO2 Util.*, 2019, **32**, 299–320.
134. J. Sun, G. Yang, Y. Yoneyama and N. Tsubaki, *ACS Catal.*, 2014, **4**, 3346–3356.
135. I. A. Kurzina, S. I. Reshetnikov, N. I. Karakchieva and L. N. Kurina, *Chem. Eng. J.*, 2017, **329**, 135–141.
136. W. Shen, K. Jun, H. Choi and K. Lee, *Korean J. Chem. Eng.*, 2000, **17**, 210–216.
137. K. Takeishi, *Biofuels*, 2010, **1**, 217–226.
138. A. M. Niziolek and O. Onel, *AIChE J.*, 2016, **62**, 1531–1556.
139. K. Cheng, W. Zhou, Y. Pan, W. Wen, K. Cheng, W. Zhou, J. Kang, S. He, S. Shi, Q. Zhang and Y. Pan, *Chem*, 2017, **3**, 334–347.
140. S. Kasipandi and J. W. Bae, *Adv. Mater.*, 2019, **31**, 1803390.
141. D. Miao, Y. Ding, T. Yu, J. Li, X. Pan and X. Bao, *ACS Catal*, 2020, **10**, 7389–7397.
142. S. Wei, Y. Shi, L. Wang, Zhang and Y. Wang, *ChemCatChem*, 2019, **11**, 1681–1688.
143. J. Yang, X. Pan, F. Jiao, J. Li and X. Bao, *Chem. Commun.*, 2017, **53**, 11146–11149.
144. M. R. Gogate, *Pet. Sci. Technol.*, 2019, **37**, 559–565.
145. M. Yang, D. Fan, Y. Wei, P. Tian and Z. Liu, *Adv. Mater.*, 2019, **31**, 1–15.
146. P. Tian, Y. Wei, M. Ye and Z. Liu, *ACS Catal.*, 2015, **5**, 1922–1938.
147. V. M. Lebarbier, R. A. Dagle, L. Kovarik, J. A. Lizarazo-Adarme, D. L. King and D. R. Palo, *Catal.: Sci. Technol.*, 2012, **2**, 2116–2127.
148. Q. I. Roode-gutzmer, D. Kaiser and M. Bertau, *ChemBioEng Rev.*, 2019, 209–236.
149. C. Du, P. Lu and N. Tsubaki, *ACS Omega*, 2020, **5**, 49–56.
150. M. R. Gogate, *Pet. Sci. Technol.*, 2019, **37**, 559–565.
151. P. Tian, Y. Wei, M. Ye and Z. Liu, *ACS Catal.*, 2015, **5**, 1922–1938.
152. J. Xiao, Y. Pan and Z. Zhou, *Science*, 2016, **351**, 1065–1068.
153. W. Khaodee, N. Tangchupong, B. Jongsomjit and N. Laosiripojana, *J. Ind. Eng. Chem.*, 2010, **16**, 411–418.
154. Y. T. Shah and A. J. Perrotta, *Ind. Eng. Chem. Prod. Res. Dev*, 1976, **15**, 123–131.
155. X. Wu, M. Tan, S. Tian, F. Song, Q. Ma and Y. He, *Fuel*, 2019, **243**, 34–40.
156. X. Liu, W. Zhou, Y. Yang, K. Cheng, J. Kang, L. Zhang, G. Zhang, X. Min, Q. Zhang and Y. Wang, *Chem. Sci.*, 2018, **9**, 4708–4718.
157. Y. Zhu, P. Northwest, Y. Han and Z. Liu, *ACS Catal.*, 2017, 7, 2800–2804.
158. J. Su, H. Zhou, S. Liu, C. Wang, W. Jiao, Y. Wang, C. Liu, Y. Ye, L. Zhang, Y. Zhao, H. Liu, D. Wang, W. Yang, Z. Xie and M. He, *Nat. Commun.*, 2019, **10**, 1–8.
159. A. Masudi, N. W. C. Jusoh and O. Muraza, *Catal. Sci. Technol.*, 2020, **10**, 1582–1596.
160. H. M. T. Galvis and K. P. De Jong, *ACS Catal.*, 2013, **3**, 2130–2149.
161. S. J. Tauster, *Acc. Chem. Res.*, 1987, **20**, 389–394.

162. C. H. Mejía, T. W. van Deelen and K. P. de Jong, *Nat. Commun.*, 2018, **9**, 4459, 1–8.
163. R. P. Mogorosi, N. Fischer, M. Claeys and E. Van Steen, *J. Catal.*, 2012, **289**, 140–150.
164. O. O. James, B. Chowdhury, M. Adediran and S. Maity, *RSC Adv.*, 2012, **2**, 7347–7366.
165. J.-B. Li, H.-F. Ma, H.-T. Zhang, Q.-W. Sun, W.-Y. Ying and D.-Y. Fang, *React. Kinet., Mech. Catal.*, 2014, **112**, 409–423.
166. L. Jiang-Bing, M. A. Hong-Fang, Z. Hai-Tao, S. Qi-Wen, Y. Wei-Yong and F. Ding-Ye, *Acta Phys. -Chim. Sin*, 2014, **30**, 1932–1940.
167. S. Ali, N. Asmawati, M. Zabidi and D. Subbarao, *J. Fuel Chem. Technol.*, 2012, **40**, 48–53.
168. M. Saglam, *Ind. Eng. Chem. Res.*, 1989, **1989**, 150–154.
169. B. Gu, V. V. Ordomsky, M. Bahri, O. Ersen, P. A. Chernavskii, D. Filimonov and A. Y. Khodakov, *Appl. Catal., B*, 2018, **234**, 153–166.
170. B. Gu, D. V. Peron, A. J. Barrios, M. Bahri, O. Ersen, M. Vorokhta, B. Šmíd, D. Banerjee, M. Virginie, E. Marceau, R. Wojcieszak, V. V. Ordomsky and A. Y. Khodakov, *Chem. Sci.*, 2020, **11**, 6167–6182.
171. H. M. T. Galvis, A. C. J. Koeken, J. H. Bitter, T. Davidian, M. Ruitenbeek, A. I. Dugulan and K. P. De Jong, *J. Catal.*, 2013, **303**, 22–30.
172. S. Yang, S. Lee, C. Kang, J. Han and K. Jun, *RSC Adv.*, 2019, **9**, 14176–14187.
173. P. P. Paalanen, S. H. Van Vreeswijk and B. M. Weckhuysen, *ACS Catal.*, 2020, **10**, 9837–9855.
174. F. Jiang, M. Zhang, B. Liu, Y. Xu and X. Liu, *Catal. Sci. Technol.*, 2017, **7**, 1245–1265.
175. T. Qiu, L. Wang, S. Lv, B. Sun, Y. Zhang, Z. Liu, W. Yang and J. Li, *Fuel*, 2017, **203**, 811–816.
176. Z. Di, T. Zhao, X. Feng and M. Luo, *Catal. Lett.*, 2019, **149**, 441–448.
177. J. Xie, P. P. Paalanen, T. W. van Deelen, B. M. Weckhuysen, M. J. Louwerse and K. P. de Jong, *Nat. Commun.*, 2019, **10**, 167, 1–10.
178. M. Athariboroujeny, A. Raub, V. Iablokov, S. Chenakin, L. Kovarik and N. Kruse, *ACS Catal.*, 2019, **9**, 5603–5612, Research.
179. X. Wang, T. Lin, J. Li, F. Yu, D. Lv, X. Qi and H. Wang, *RSC Adv.*, 2019, **9**, 4131–4139.
180. F. Sun, R. Yang, Z. Xia, Y. Yang, Z. Zhao and S. Gu, *J. Energy Chem.*, 2020, **42**, 227–232.
181. B. Liu, W. Li, Y. Xu, Q. Lin, F. Jiang and X. Liu, *ACS Catal.*, 2019, **9**, 7073–7089.
182. W. Song, Y. Hou, Z. Chen, D. Cai, W. Qian, Y. Hou, Z. Chen, D. Cai and W. Qian, *Chem. Eng. Sci.*, 2020, **212**, 115328.
183. D. A. Makeeva, L. A. Kulikov and M. I. Afokin, *Russ. J. Appl. Chem.*, 2020, **93**, 933–953.
184. E. A. Uslamin, H. Saito, Y. Sekine, E. J. M. Hensen and N. Kosinov, *Catal. Today*, 2021, **369**, 184–192.
185. S. Wei, Y. Xu, Z. Jin and X. Zhu, *Front. Chem. Sci. Eng.*, 2020, **14**, 783–792.

186. S. R. Kanitkar and J. J. Spivey, *Nat. Gas Process. Midstream Downstream*, 2018, 379–401.
187. Y. Ni, A. Sun, X. Wu, J. Hu, T. Li and G. Li, *Chin. J. Chem. Eng.*, 2011, **19**, 439–445.
188. B. Zhao, P. Zhai, Q. Zhang, W. Fan, D. Ma, B. Zhao, P. Zhai, P. Wang, J. Li, T. Li, M. Peng, M. Zhao and G. Hu, *Chem*, 2017, **3**, 323–333.
189. M. Li, M. A. Nawaz, G. Song, W. Q. Zaman and D. Liu, *Ind. Eng. Chem. Res.*, 2020, **59**, 9043–9054.
190. T. Wang, Y. Xu, C. Shi, F. Jiang, B. Liu and X. Liu, *Catal. Sci. Technol.*, 2019, **9**, 3933–3946.
191. Y. Xu, J. Wang, G. Ma, J. Bai, Y. Du and M. Ding, *Appl. Catal., A*, 2020, **598**, 117589.
192. T. Omojola and A. J. O. Malley, *Catal.: Sci. Technol.*, 2020, **10**, 4305–4320.
193. M. Conte, J. A. Lopez-sanchez, Q. He, D. J. Morgan, Y. Ryabenkova, J. K. Bartley, A. F. Carley, S. H. Taylor, C. J. Kiely, K. Khalid and G. J. Hutchings, *Catal. Sci. Technol.*, 2012, **2**, 105–112.
194. R. Brosius and M. Claeys, *Chem*, 2017, **3**, 202–204.
195. Y. Fu, Y. Ni, W. Zhu and Z. Liu, *J. Catal. J.*, 2020, **383**, 97–102.
196. S. Wang, Y. Fang, Z. Huang, H. Xu and W. Shen, *Catalysts*, 2020, **10**, 262.

CHAPTER 2

A High-level Analysis of the Environmental Sustainability of Biomass-derived Fuels

JAMES ALISTAIR FOX

Institute for the Development of Energy for African Sustainability, College of Science, Engineering and Technology, University of South Africa, 28 Pioneer Avenue, Johannesburg, South Africa
Email: foxj@unisa.ac.za

2.1 Introduction

2.1.1 The Return to Biomass: A Brief History of Fuel Sources

Prior to the industrial revolution in the 18th century, the principal energy source was either biomass or the muscular power of animals. The start of the industrial revolution saw coal become the dominant energy source through the 19th and early 20th centuries.[1] Economists often credit coal with driving the industrial revolution.[2,3] The industrial revolution saw a high demand for iron and steel, for which coal was an ideal coking ingredient. It was also cheap and plentiful, albeit dirty and challenging to mine.

As the 20th century progressed, coal began to lose its dominance in the energy market to crude oil and natural gas. The liquid form of oil and the gaseous phase of natural gas made them far more versatile than coal. They have far greater energy density and are easier to extract and transport.

Catalysis Series No. 44
Chemicals and Fuels from Biomass *via* Fischer–Tropsch Synthesis: A Route to Sustainability
Edited by Joshua Gorimbo, Xinying Liu, Yali Yao and Diane Hildebrandt

Published by the Royal Society of Chemistry, www.rsc.org

The "energy density" of a fuel refers to how much energy can be extracted from a given volume of fuel. The higher the energy density, the less physical space is needed to store fuel of a given energy content.

The advent of the nuclear age in the 20th century also saw nuclear energy begin to play a role in the energy market. Nuclear energy has extraordinary energy density, but events of the 20th century also showed just how dangerous nuclear energy can be.

The advent of worldwide conflict in the early 20th century saw restrictions in the availability of preferred liquid and gas fuel sources. Liquid fuels were still preferred over reverting back to coal entirely, and so efforts were undertaken to convert coal reserves into synthetically derived oil. This resulted in the development of the Fischer–Tropsch process, with the purpose of converting coal reserves into far more versatile liquid fuels.

In recent times there has been a growing awareness among the general public that the fuels being used over the last three centuries have been having a detrimental effect on the environment. The public perception has been focused on the emission of greenhouse gases, most notably carbon dioxide, through the combustion of fossil fuels. Fossil fuels are formed from carbon dioxide captured through photosynthesis that occurred in biomass that existed millions of years ago. It takes millions of years to form naturally, and the rate at which greenhouse gases are being produced from the combustion of fossil fuels and released into the atmosphere far exceeds the rate at which these greenhouse gases are being removed from the atmosphere by current natural processes. There is concern that this accumulation in the levels of greenhouse gases will result in the Earth's surface temperature rising to levels that are harmful to life.

Many climate scientists have warned that a temperature increase of 1.5 °C or more in the Earth's atmosphere will result in irreversible damage to the environment. Some scientists believe that this can still be prevented if immediate action is taken to curb greenhouse gas emissions, while others believe that it is already too late and efforts should focus on mitigating the damage.[4] The solution is nevertheless the same: reliance on fossil fuels needs to change, and emissions need to decrease.

A potential solution to this problem may lie in the past, in going back to the use of biomass as a dominant fuel source. This could be viewed as "coming full circle" with fuels; we began using biomass as the primary energy source and we are now looking at returning to it, as illustrated by Figure 2.1.

While biomass still emits greenhouse gases when burned, the biomass itself captures carbon dioxide from the atmosphere and can be replenished at a far greater rate than fossil fuels. This is why biomass is referred to as a renewable fuel.

If the rate of carbon emissions from the utilisation of biomass can be made to match the rate at which carbon dioxide is captured by newly grown biomass, the fuel becomes carbon neutral. However, biomass does suffer from some similar problems to coal. Biomass is a solid fuel and therefore

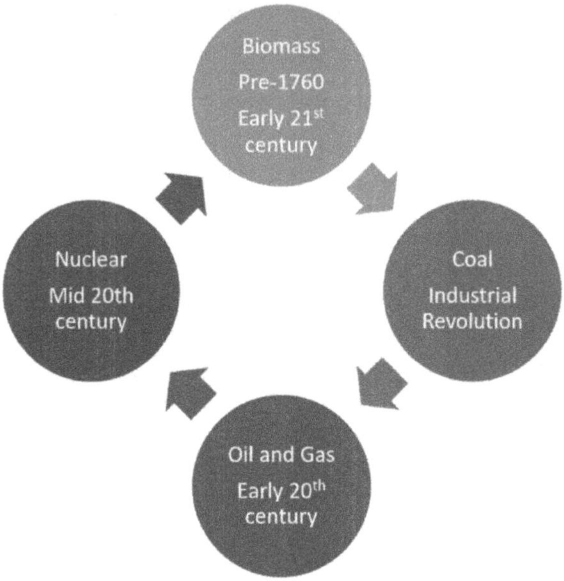

Figure 2.1 Coming full circle, the return to biomass as the primary energy source.

Table 2.1 Energy density of selected fuels.

	Calorific value (MJ kg^{-1})	Bulk density (kg m^{-3})	Energy density (GJ m^{-3})
Biomass	16	250	4
Bituminous coal	18	1683	30.294
Crude oil	44	890	39.16
Natural gas (STP)	50	0.9	0.045

difficult to handle. It also has an energy density lower than that of coal, making it difficult and expensive to utilise.

Table 2.1 shows the thermodynamic data for some fuels. The data have been taken from various sources and averaged to give a general idea of the differences between the fuels. Clearly, biomass has an extremely low energy density. Vast volumes of biomass are required to match the same amount of energy from other fuels. The natural gas data can be misleading as the bulk density for gases is highly dependent on temperature and pressure. Table 2.1 reports the density at standard conditions. With compression, the density of natural gas can be increased substantially.

It seems clear that using biomass directly, as a transportation fuel, is not going to be practical. Fuel storage in vehicles would need to be some 10 times larger than current fuel tanks, just to match the energy output. If biomass is instead used as a feedstock for liquid fuel synthesis, the bulky biomass could be converted into a fuel that is easier to store. This would take advantage of both the renewable nature of biomass and the versatility of liquid fuels.

2.1.2 Biomass and the Goal of Carbon Neutrality

In the previous section, the idea of carbon neutrality was introduced in a manner that implied this to be a desirable state. What is meant by this idea of carbon neutrality? First, consider that the majority of energy is produced by means of combustion. A carbon-based fuel is reacted with oxygen to form combustion products (carbon dioxide and often water as well). A simplified schematic of this is shown in Figure 2.2.

Any system that overall releases carbon into the surroundings is termed "carbon positive". This terminology mostly applies to the release of carbon dioxide. However, a system that releases carbon in any form can also be called carbon positive.

A "carbon-neutral" system is then any system that has zero carbon emissions. Achieving this neutrality is the goal of using biomass as a fuel source. Biomass can be burned to harness its energy content, and the resulting carbon dioxide (and water) can then be converted back into biomass by photosynthesis.

Since a carbon-positive is any system that releases carbon, and carbon-neutral is any system that has zero carbon emissions, it would logically follow that a "carbon-negative" system is any system that removes carbon from the surroundings. It is a system that absorbs carbon dioxide. In Figure 2.3, photosynthesis meets the definition for a carbon-negative system. It absorbs carbon dioxide and produces biomass. The combination of a carbon-positive system with a carbon-negative system, makes a carbon-neutral system, as demonstrated by Figure 2.4.

While carbon-negative systems do exist, the law of conservation of mass makes it such that the entire planet can never be carbon negative – short of venting carbon dioxide into space. Carbon neutral is the best that can be achieved, and biomass has the potential to do this.

2.1.3 Biomass and Biofuels

While biomass has the potential to achieve carbon neutrality, it has also been established that it has a very low energy density in comparison to liquid

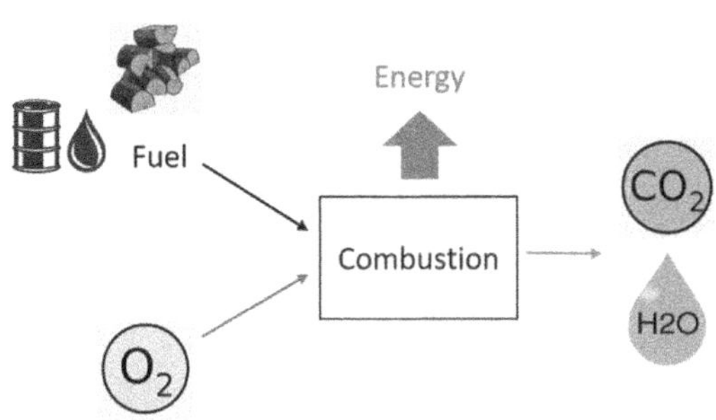

Figure 2.2 Carbon-positive fuel combustion.

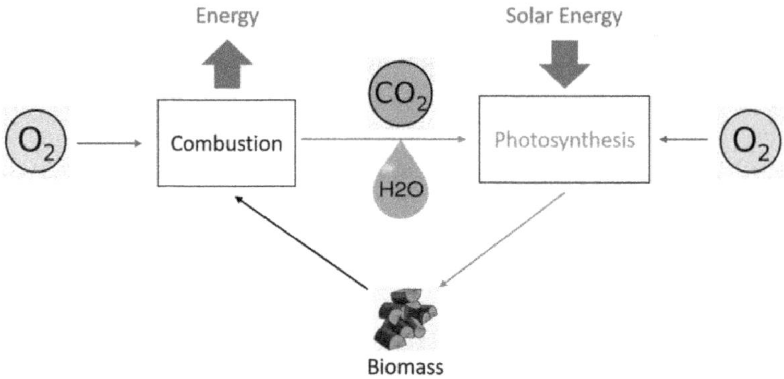

Figure 2.3 Carbon neutral: the goal of biomass as a fuel source.

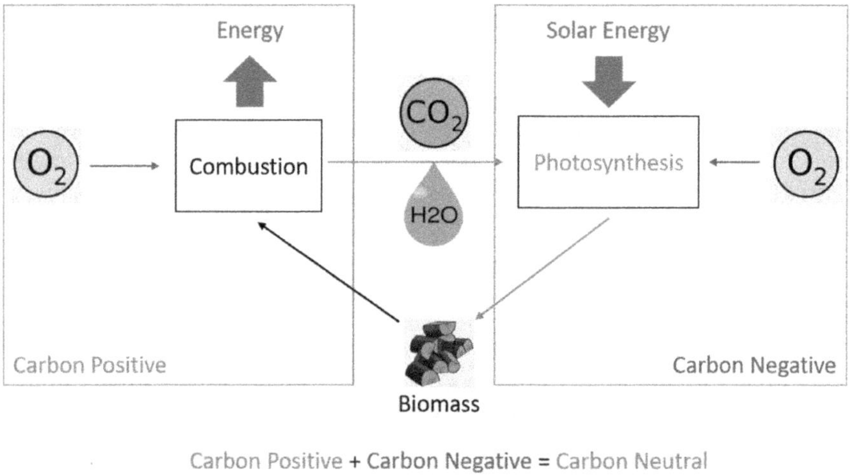

Figure 2.4 Carbon neutrality achieved through combined systems.

fuels. Hauling around massive stores of biomass to use as fuel is simply not a practical solution to the environmental impact problem. The energy density of the biomass needs to be increased. This can be achieved by converting the biomass into an intermediate fuel rather than using the biomass directly. The downside is that for every intermediate step that is introduced, additional inefficiencies are introduced, and the goal of carbon neutrality becomes more difficult to reach.

All biomass-derived fuels are called "biofuels", this is irrespective of the processing done to produce the fuel. Biofuels are currently produced primarily through methods involving bacterial fermentation or thermal processing.

Converting biomass into ethanol is the most widely known of the fermentation techniques,[5] which has been practiced by humans for thousands of years. The original purpose was to produce alcoholic beverages, but the ethanol can be purified for use as a fuel. Fermentation can also produce methane, in the form of biogas[6,7] and butanol.[8,9] The downside of fermentation is that it tends to be slow, and the alcohol product is highly diluted with water. This slow rate and the need for extensive purification can lead to a mismatch in production rates. At steady state, the rate of the overall process matches the rates of the individual processes; thus the rate at which fuels are burned for energy, the rate at which new alcohols can be produced and the rate at which new biomass can be grown must all be the same. In a situation where fuels are burned at a vastly higher rate than they can be regenerated (which is the current global situation), carbon neutrality can never be achieved.

The production rates of biological processes are usually quite low and to address this problem with low production rates, thermal processes are used. The use of significantly higher temperature allows the biomass to be converted much faster, even if this does come with its own problems. The most straightforward of the thermal treatment processes is pyrolysis.[10,11] Heat is applied directly to the biomass, which breaks the chemical bonds of the biomass itself. This results in the production of a variety of hydrocarbon oils, which then have potential to be used as a fuel. Pyrolysis has the advantage of producing products at a faster rate than biological processes, but it still has a weakness with dilution. Pyrolysis products have a wide and difficult-to-predict distribution of products, not all of which are appropriate for use as a fuel and which can result in substantial wastage and negatively impact the sustainability of the process.

Processes using gasification[12–14] can produce improved quality biofuel. In gasification, the biomass is partially oxidised to produce synthesis gas (often referred to as "syngas"). Synthesis gas is primarily composed of hydrogen and carbon monoxide; although it may also contain water, carbon dioxide and light hydrocarbons (such as methane). The synthesis gas can then be used as a fuel directly,[15,16] but conversion into liquid hydrocarbons is also widely practiced.[17,18] The advantage of using syngas is that both the composition and the quality of the syngas are far more predictable than the products produced by pyrolysis. This predictability is desirable since chemical processes can convert the syngas to products that can be used as fuels with properties that can be tuned and optimised with greater ease.

The purpose of this chapter and of this whole book is to focus on this thermal conversion of biomass into liquid fuels. This chapter presents a number of strategies for the thermal conversion of biomass, with particular attention given to the analysis of the quantity of carbon dioxide emissions from the process, to quantify whether the goal of carbon neutrality is achievable. This analysis is at the conceptual level, with many of the systems being highly idealised and simplified. If the goal of carbon neutrality cannot be achieved at the conceptual level, it will never be achievable when the ideal constraints are removed and real constraints, such as equipment inefficiencies, are introduced.

2.2 Thermal Conversion of Biomass

The thermal conversion of biomass can be summarised in two primary steps, as shown by the block flow diagram in Figure 2.5. The two main steps are gasification into synthesis gas, followed by the synthesis of hydrocarbon fuels. To begin quantifying the amount of carbon dioxide emitted from this process, it is necessary to develop equations that represent the steps in the process.

2.2.1 Biomass Gasification

Biomass is a catch-all term that collectively refers to plant-based matter. Such a collective term does little to assist with modelling biomass conversion in a mathematical sense. Biomass is comprised of complex polymer chains including cellulose, hemicellulose and lignin, with cellulose forming the largest proportion,[19,20] but the chemical composition of biomass varies widely within plant species, growing conditions, climate and location, to name just a few. One way to handle this variability is to use an approximation, the monomer of cellulose being a convenient analogue. The chemical composition of the monomer for cellulose can be approximated as

$$C_6H_{10}O_5 \tag{2.1}$$

While this is not a perfect analogue of biomass, it should prove accurate enough for the purposes of the analysis in this chapter. Using this equation for biomass, it is now possible to write equations that describe the conversion of biomass into syngas.

Conversion into syngas is an oxidation reaction or, more precisely, partial oxidation. As such, an oxidising agent is need to facilitate the reaction. The two most readily available oxidising reactants are oxygen (both as pure oxygen and as air) and water (in the form of steam). A possible third oxidising agent for gasification is carbon dioxide. This third reactant has not been used industrially, but there are clear advantages in the use of carbon dioxide as a reactant, especially where the goal is to eliminate the overall emissions of carbon dioxide.

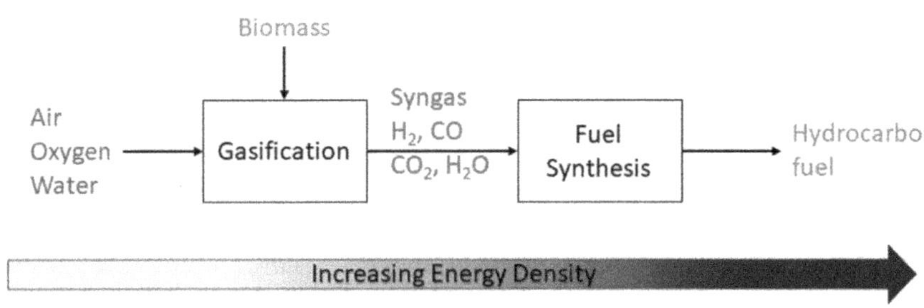

Figure 2.5 Block flow diagram showing the increase of energy density through thermal biomass conversion.

It is best to visualise the equations shown in Table 2.2 as material balances rather than as reaction equations, as a number of reactions may occur during gasification that combine to give the overall material balance shown in Table 2.2. These three balances can be linearly combined with one another in different propositions to create combined oxidation material balances. For example, the gasification of biomass with equal proportions of oxygen and water would be given by the sum of the material balances referred to as "partial oxidation" and "steam reforming". Overall, such a combined system begins to resemble autothermal reforming with a material balance given by

$$C_6H_{10}O_5 + \frac{1}{4}O_2 + \frac{1}{2}H_2O \rightarrow 5\frac{1}{2}H_2 + 6CO \tag{2.2}$$

Similarly, the equations in Table 2.2 can themselves be a combination of different reactions. For example, biomass can be anaerobically digested to produce biogas following the material balance:

$$C_6H_{10}O_5 + H_2O \rightarrow 3CH_4 + 3CO_2 \tag{2.3}$$

If the resulting biomass is then subjected to methane dry reforming, to produce the syngas, the material balance for this step is

$$CH_4 + CO_2 \rightarrow 2CO + 2H_2 \tag{2.4}$$

If there is no net production of methane, the overall material balance effectively becomes (eqn (2.3) + 3×eqn (2.4)), which results in the dry reforming material balance shown in Table 2.2.

Mathematically, there are six types of molecules ($C_6H_{10}O_5$, CO, H_2, O_2, CO_2, H_2O) and three atomic species (C, H, O), all of which must be conserved. The system must be defined by three independent material balances, which are the material balances shown in Table 2.2. Any gasification process can be described by a combination of these three independent material balances. There are an infinite number of combinations of the three independent material balances, and thus attempting to find and analyse every possible gasification material balance is not possible. However, by analysing each of the three independent material balances, one could quickly identify whether some combination of these would be beneficial and thus give a more desirable material balance for the gasification process.

Table 2.2 Reaction or material balance equations and enthalpies for the three oxidising (gasification) reactions.

System	Oxidising reactant	Gasification equation	$\Delta H^0_{reaction}$ (kJ mol^{-1})
Partial oxidation	Oxygen	$C_6H_{10}O_5 + \frac{1}{2}O_2 \rightarrow 5H_2 + 6CO$	367
Steam reforming	Water	$C_6H_{10}O_5 + H_2O \rightarrow 6H_2 + 6CO$	652.83
Dry reforming	Carbon dioxide	$C_6H_{10}O_5 + CO_2 \rightarrow 5H_2 + 7CO$	650

Further examination of Table 2.2 shows that all the gasification material balances for biomass gasification are endotherms. Thus, in order for these reactions to proceed at all, this energy needs to be supplied from somewhere, and how this energy could be supplied is one of the main topics in this chapter. Furthermore, any combination of the three independent material balances for gasification, with the constraint that 1 mol s^{-1} of $C_6H_{10}O_5$, is consumed, is endothermic, and thus any and all biomass gasification processes face the same problem of how to supply the energy required for gasification. It is also noteworthy that biomass gasification tends to be very lean in hydrogen. At best, the proportions of hydrogen and carbon monoxide are equal, but carbon monoxide is usually in higher proportion. This is not significant right now but should be kept in mind when considering the liquid fuel production section where the preferred feed to the synthesis section usually is a 2/1 ratio of hydrogen to carbon monoxide.

2.2.2 Synthesis of Liquid Fuels

2.2.2.1 *Liquid Methanol*

While it is possible to use syngas as fuel directly, the focus of this book is on the synthesis of liquid fuels. There are two main routes for producing liquid fuels from syngas: methanol synthesis and Fischer–Tropsch synthesis.

Methanol synthesis is the more straightforward of the two processes as the reaction is highly selective to methanol. The material balance for the methanol synthesis process is

$$CO + 2H_2 \rightarrow CH_3OH \tag{2.5}$$

This route for the synthesis of methanol is widely used industrially, where the syngas is reacted using metal oxide catalysts at conditions of around 230 °C and pressures of between 50 and 100 bars.[21] The reaction is exothermic, and the low-temperature heat produced from the reaction can be used for process steam. We note that the feed to process has a 2/1 ratio of hydrogen to carbon monoxide, which differs from the ratio produced by gasification. An additional process is required to change the H_2/CO ratio so that it is suitable for synthesis; this is discussed later in this chapter. Methanol is a compelling replacement for fossil-derived fuels. While there are mechanical challenges to converting petroleum-fuelled internal combustion engines into methanol fuelled, much of the current research indicates that there are no safety- or emissions-related problems with using methanol instead of petroleum.[22,23] The reluctance to move away from petrol engines is due not to the technical issues of methanol being used in combustion engines but primarily to the safety and social issues.

2.2.2.2 *Fischer–Tropsch Synthesis – Synthetic Crude*

One way to overcome this reluctance towards adopting methanol fuel is to instead use syngas to produce a synthetic oil that has similar properties to the crude oil currently being drawn from oil reserves. This can be achieved by using Fischer–Tropsch synthesis (FTS), which has an overall material balance given by

$$CO + 2H_2 \rightarrow \frac{1}{2}(C_2H_4)_n + H_2O \tag{2.6}$$

The mechanisms involved in Fischer–Tropsch (FT) synthesis are complex and have been extensively researched over the last 70 years.[24–26] It is hard to do justice to such a long history of research, but, to provide a very brief summary, FT synthesis does not produce a single product – unlike methanol synthesis, which is a highly selective reaction. Rather, FT synthesis produces a wide array of hydrocarbons, of the form $(C_2H_4)_n$ and $(C_{2n}H_{2n+2})$ where the value of the variable n ranges from 1 to "very large". The result is a mixture of hydrocarbons not unlike that seen in crude oil. In fact, FT synthesis products are often referred to as "synthetic crude". This synthetic crude can be subjected to similar refining processes that are used for crude oil, resulting in a petroleum fuel that be used in current engines without modification.

Depending on the desired product range and catalyst used, the reaction operates between 180 °C up to 350 °C. The reaction is exothermic, and the heat rejected by the reaction is typically utilised for process steam.

The distribution of hydrocarbons in the synthetic crude is described by using a product distribution parameter that is typically defined using the Andersen–Schulz–Flory (ASF) law,[27–29] where the product distribution is defined using the symbol α. The product distribution itself is affected by many of the process operating parameters, especially the type of catalyst and the reaction temperature. In summary, the product distribution in FT synthesis (α) can be defined as a ratio:

$$\alpha = \frac{C_{x+1}}{C_x} \tag{2.7}$$

where:

- α is the product distribution, varying between 0 and 1,
- C_x is the number of carbon atoms in the alkane hydrocarbon, and
- C_{x+1} is the number carbon atoms in the next alkane.

The distribution of products from FT varies from methane to long waxes, and smaller values of α (<0.6) favours the production of lighter products, while larger values of α (>0.9) favours the production of oils and waxes. A higher value for α is preferred for the production of liquid fuels. A synthetic crude with a higher value for α has a higher proportion of longer chain hydrocarbons, which are the liquid products needed for the production of liquid fuels.

These intricacies of FT synthesis are well beyond the intended scope of this chapter. The main idea to take away from this brief discussion is that a higher α is better for producing liquid fuels. As an example, consider the following:

$$\alpha = 0.9 \tag{2.8}$$

$$C_{x+1} = 0.9C_x \tag{2.9}$$

C_1 is methane; if methane is assigned a value of 1, the C_2 ethane would have a value of

$$C_2 = 0.9C_1 = 0.9(1) = 0.9 \tag{2.10}$$

Then C_3 (propane) would have a value of

$$C_3 = 0.9C_2 = 0.9(0.9) = 0.81 \tag{2.11}$$

This calculation can be repeated up to any length of alkane hydrocarbon.

Table 2.3 shows a summary of the calculation. Note that the product distribution is described by a series that converges to zero, and thus very high carbon numbers have less effect on the material balance since their proportion in the product becomes smaller as the chain of the hydrocarbon product becomes longer.

The total number of moles in the product can be determined by summing the terms in the distribution; the actual mole fraction of each alkane in the FT product is then simply the number of moles of the alkane divided by the total number of moles in the product. If we continue using Table 2.3, where the iterations were terminated at C_{50}, we can determine the mole fraction of the various components in the product as shown in Table 2.4.

The overall material balance for FT synthesis must not only take into account the product distribution shown in Table 2.4 but must also include the water produced in the reaction. The material balance equation can be solved as follows:

$$\begin{aligned} a\mathrm{CO} + b\mathrm{H_2} \rightarrow\ & 0.1005\mathrm{CH_4} + 0.0905\mathrm{C_2H_6} + 0.0814\mathrm{C_3H_8} + 0.0733\mathrm{C_4H_{10}} + \cdots \\ & + 0.0256\mathrm{C_{14}H_{30}} + \cdots + 0.0006\mathrm{C_{50}H_{102}} + c\mathrm{H_2O} \end{aligned} \tag{2.12}$$

Table 2.3 Fischer–Tropsch product distribution for an ASF distribution of 0.9.

Alpha Carbon number	0.9 Alkane	Distribution
C1	Methane	1.00
C2	Ethane	0.90
C3	Propane	0.81
C4	Butane	0.73
...		
...		
C14	Tetradecane	0.25
...		
...		
C50	Pentacontane	0.0057

Table 2.4 Mole fraction of alkanes produced in FT synthesis for an ASF distribution of 0.9.

Alpha	0.9		
Carbon number	Alkane	Distribution	Mole fraction
C1	Methane	1.00	0.1005
C2	Ethane	0.90	0.0905
C3	Propane	0.81	0.0814
C4	Butane	0.73	0.0733
...			
...			
C14	Tetradecane	0.25	0.0256
...			
...			
C50	Pentacontane	0.0057	0.0006
Sum		9.9484622	1

where:

- a is the stoichiometric coefficient for carbon monoxide,
- b is the stoichiometric coefficient for hydrogen, and
- c is the stoichiometric coefficient for water.

The coefficient a corresponds to the moles of C in the feed and must therefore be equal to the sum of all the carbon present in alkane product, Table 2.4, giving

$$a = \sum_{x=1}^{50} C_x = 9.74 \tag{2.13}$$

All the oxygen in the carbon monoxide must be converted to water, where the ratio of CO in the feed to water in the product is 1/1, giving

$$c = a = 9.74 \tag{2.14}$$

Finally, the coefficient b corresponds to the amount of H_2 in the feed, and thus $2b$ must be equal to the sum of all the hydrogen (H) in the alkane product as well as the hydrogen (H) in the water product, or

$$b = \frac{1}{2}\sum_{x=1}^{50} \mathrm{H}_{2x+2} + a = 20.48 \tag{2.15}$$

Thus the overall material balance for an FT synthesis process producing alkanes with a product distribution described by $\alpha = 0.9$ is

$$9.74\mathrm{CO} + 20.48\mathrm{H_2} \rightarrow 0.1005\mathrm{CH_4} + 0.0905\mathrm{C_2H_6} + 0.0814\mathrm{C_3H_8} + 0.0733\mathrm{C_4H_{10}} + \cdots + 0.0256\mathrm{C_{14}H_{30}} + \cdots + 0.0006\mathrm{C_{50}H_{102}} + 9.74\mathrm{H_2O} \tag{2.16}$$

This material balance is still quite unwieldy for most applications, and it can be simplified by grouping products with similar properties. FT synthesis produces products in three main groups, namely gases, liquids and waxes, and the overall material balance for the process can be simplified to

$$CO + 2H_2 \rightarrow \{gas\} + \{liquid\} + \{wax\} + \{water\} \tag{2.17}$$

The FT synthesis products that are gases can be added together, and the group can be characterised by a representative component. A group of products is referred to as a "cut", such as a gas cut or a liquid cut. Which alkanes fall into which cut is also up to the judgement of the person doing the calculation. As an example, consider a gas cut of C_1 to C_7 (methane to heptane), a liquid cut of C_8 to C_{19} and a wax cut of C_{20} to C_{50}. The gas cut is the sum of the components C_1 to C_7 and is represented by C_3. Similarly, the liquid cut is represented by C_{14}, and the wax cut by C_{30}, giving an overall material balance of

$$10.32CO + 21.64H_2 \rightarrow 0.52C_3H_8 + 0.35C_{14}H_{30} + 0.13C_{30}H_{62} + 10.32H_2O \tag{2.18}$$

This equation has been presented using only two decimal places, resulting in a small discrepancy in the carbon balance. This is purely due to the number of significant figures; this error vanishes when more significant figures are used. While the gas and wax cuts do have a wide range of applications, the liquid cut is typically the product that is of the greatest interest for fuel production. In this analysis, we thus target the production of the liquid cut for use as a liquid fuel.

This analysis has included only the saturated hydrocarbons, *i.e.* paraffins. The unsaturated hydrocarbons, olefins, were omitted from the calculations for the sake of keeping the discussion relatively concise. However, the material balances shown in this section can be expanded to include these olefins. The proportion of olefins to paraffins is expressed as a ratio that depends on the same operating parameters as the previous ASF product distribution (α).

Both methanol synthesis and FT synthesis prefer a hydrogen-rich syngas, but biomass gasification produces a carbon-dioxide-rich syngas. Optimal use of syngas in the synthesis of liquid fuels requires the composition of syngas to be adjusted.

2.2.2.3 Synthesis Gas Refining

Both methanol synthesis and FT synthesis prefer a hydrogen-rich synthesis gas, with a preferred ratio of 2/1 of hydrogen to carbon monoxide. The previous section showed that biomass gasification is only capable of producing at best syngas with equal proportions of hydrogen and carbon monoxide. This mismatched ratio is a problem. From an environmental viewpoint, carbon monoxide emissions are worse than carbon dioxide emissions since biomass cannot be grown from carbon monoxide, and thus the excess carbon monoxide must be utilised and cannot be rejected to the environment.

The composition of the syngas can be adjusted by using the water-gas-shift reaction:

$$CO + H_2O \rightarrow H_2 + CO_2 \tag{2.19}$$

Water-gas shift allows the excess carbon monoxide to be reacted with water to produce hydrogen, thereby enriching the hydrogen composition in the syngas. The downside is that this results in the co-production of carbon dioxide, the very thing that all this work has been trying to avoid. The heat requirements of the water-gas-shift reaction are significantly lower than that of gasification and thus are not considered at this initial high level of analysis.

The goal of overall carbon neutrality appears to be becoming more difficult and unlikely, but more analysis is required before such a judgement can really be made.

2.3 Power and Biomass

2.3.1 Testing the Feasibility of Biomass-to-liquids

An important result of Table 2.2 is that all of the primary gasification routes for biomass are endotherms. Thus gasification needs a source of high-temperature energy. Typically, this energy is supplied by combustion, usually of some of the feed, which then produces carbon dioxide. The requirements for a high-temperature energy source for gasification present a conundrum: the whole point of biomass gasification is to produce a green source of energy, but utilising this green energy requires an energy source of its own.

For biomass to have any potential as a green fuel source, the energy consumed by gasification must be at least less than the energy produced from the combustion of the final liquid fuel. Failing such a test means there is actually an energy deficit, which must be covered by an outside energy source and which further escalates the production of emissions.

2.3.1.1 Energy Supply through Methanol Combustion

A simple calculation can be performed to check whether biomass systems are viable at all: if the product, for example methanol, was produced as the biomass-derived liquid fuel and was then burned to produce energy, is the energy from combustion of the product greater than the energy required for gasification? If not then the production of that liquid product from biomass cannot proceed without the addition of more energy. Methanol combustion is given by

$$CH_3OH + \frac{3}{2}O_2 \rightarrow CO_2 + 2H_2O \tag{2.20}$$

The higher heating value for methanol combustion (liquid water as a combustion product) is 727 $kJ\,mol^{-1}$. Recovering the latent heat of water is difficult to do practically, so the lower heating value would give a better indication of the energy production, 638 $kJ\,mol^{-1}$.

To properly assess whether this is enough energy, all the material balances need to be combined in the correct proportions. This includes gasification (from Table 2.2), water-gas shift and methanol synthesis.

As an example, consider the gasification of biomass using partial oxidation. Starting from a basis of 1 mol s^{-1} of biomass, 0.5 mol s^{-1} of oxygen is needed for the reaction, the reaction produces 5 mol s^{-1} of hydrogen and 6 mol s^{-1} of carbon monoxide. This requires 367 kW of energy. A 2/1 hydrogen-to-carbon monoxide ratio is required for methanol synthesis; the water-gas-shift reaction can be used to adjust the ratio. The amount of hydrogen and carbon monoxide that needs to be reacted with the water gas shift reaction can be calculated using material balances:

$$H_2 : CO = \frac{N^0_{H_2} + N^{wgs}_{H_2}}{N^0_{CO} - N^{wgs}_{CO}} \tag{2.21}$$

$$2 = \frac{5 + N^{wgs}_{H_2}}{6 - N^{wgs}_{CO}} \tag{2.22}$$

where:

- $H_2 : CO$ is the desired syngas ratio, a value of 2;
- $N^0_{H_2}$ is the amount of hydrogen produced by gasification, for this example: 5 mol s^{-1};
- $N^{wgs}_{H_2}$ is the amount of hydrogen produced by the water gas shift reaction;
- N^0_{CO} is the amount of carbon monoxide produced by gasification, for this example: 6 mol s^{-1}; and
- N^{wgs}_{CO} is the amount of carbon monoxide consumed by the water-gas-shift reaction.

Since all the components of the water gas shift reaction have a stoichiometric coefficient of 1, the amount of hydrogen produced is equal to the amount of carbon monoxide consumed.

$$N^{wgs}_{H_2} = N^{wgs}_{CO} \tag{2.23}$$

Thus

$$N^{wgs}_{H_2} = N^{wgs}_{CO} = 2\frac{1}{3}\,\text{mol s}^{-1} \tag{2.24}$$

This results in 7.33 mol s^{-1} of hydrogen (and 3.67 mol s^{-1} of carbon monoxide) being available for methanol production. This results in 1.84 mol s^{-1} of methanol. This methanol can then be burned to produce 1171 kW of energy.

As flowsheet for this case study is presented in Figure 2.6.

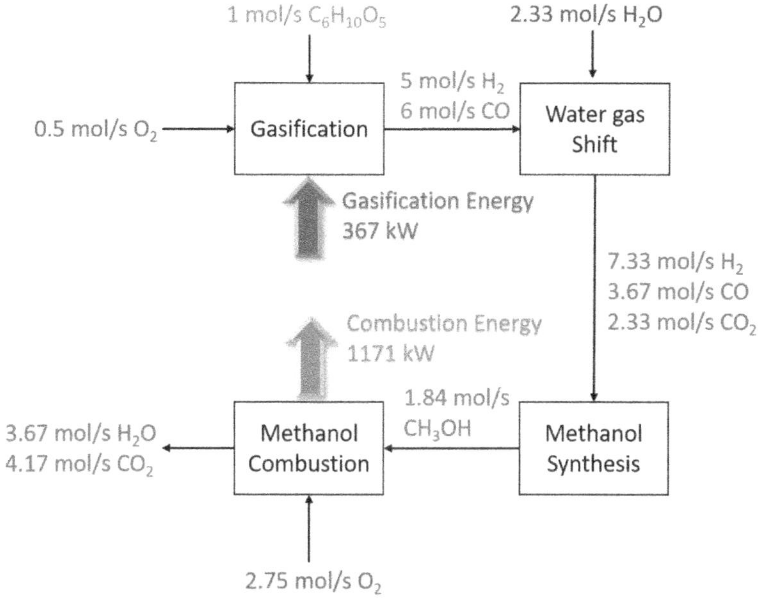

Figure 2.6 Example of energy-based calculation for biomass-to-methanol viability.

Table 2.5 Gasification energy requirements met by burning a portion of the methanol product.

System	Methanol produced ($mol\ s^{-1}$)	Methanol burned ($mol\ s^{-1}$)	CO_2 emissions ($mol\ s^{-1}$)
Partial oxidation	1.26	0.58	2.9
Steam reforming	0.98	0.98	3.02
Dry reforming	0.98	0.98	3.02

The results of the calculation and the summary in Figure 2.6 show that energy produced from methanol combustion is indeed greater than the energy needed in gasification. Even if the combustion is less than 100% efficient, there is still sufficient energy to make the entire system a net producer of energy.

It is also not necessary to combust all the methanol for the sake of gasification. Rather, just enough can be burned to satisfy the gasification requirements, and the rest of the methanol can be used for other applications.

Table 2.5 shows that steam reforming and dry reforming are essentially equal. This is because their respective gasification energy requirements are very similar. Dry reforming actually produces the most carbon dioxide from water-gas shift and methanol combustion (a total of 5 $mol\ s^{-1}$), but the dry reforming itself uses 1 $mol\ s^{-1}$ of carbon dioxide for gasification. This brings the total emissions to the same as that of steam reforming, implying there is not much advantage to dry reforming despite the initial advantage of using carbon dioxide as a reactant. Partial oxidation has the best performance, producing the most methanol and having the lowest carbon dioxide emissions.

Biomass-derived methanol is a feasible for increasing the energy density of the biomass; but for every 1 mol s^{-1} of biomass that is converted to methanol, around 3 mol s^{-1} of carbon dioxide are released. How does FT synthesis compare?

2.3.1.2 *Energy Supply through FTS Fuel Combustion*

In the case where FT synthesis is being used instead of methanol, there is a choice of which hydrocarbon product to burn for energy. The simplest choice is to burn the gas cut and avoid burning the liquid cut entirely. The combustion of the gas cut is given by

$$C_3H_8 + 5O_2 \rightarrow 3CO_2 + 4H_2O \tag{2.25}$$

The lower heating value for the combustion of the gas cut is 2043 kJ mol^{-1}. If the liquid cut is burned, it would have a lower heating value of 9471 kJ mol^{-1}.

What Table 2.6 does not show is that the partial oxidation system requires the combustion of all the gas cut to supply the energy required in gasification. For the two reforming systems, insufficient gas cut is being produced to cover the energy required by gasification. This requires the combustion of all the gas cut and a portion of the liquid cut. The production rate of fuel in these FT systems appears to be very low, but these results are being presented on a molar basis. Consider an alternative perspective, what is the energy potential of the fuels if they were all completely burned.

The result of Table 2.7 is that despite FT synthesis producing smaller quantities of fuel, the energy contained in those fuels is greater.

Overall, biomass-derived methanol and biomass-derived FT liquids perform very similarly, and they are viable processes from an energy viewpoint. However,

Table 2.6 Gasification energy requirements met by burning a portion of the FT synthesis product.

System	Liquid cut produced (mol s^{-1})	Fuel combustion (mol s^{-1})	CO_2 emissions (mol s^{-1})
Partial oxidation	0.12	0.18	3
Steam reforming	0.1	0.22	3.1
Dry reforming	0.1	0.22	3.1

Table 2.7 Comparing the potential power derived from using methanol as a fuel of the FT liquid cut.

System	Total methanol combustion (kW)	Total FT liquid cut combustion (kW)
Partial oxidation	916	1136
Steam reforming	712	947
Dry reforming	712	947

all of these systems emit carbon dioxide, making the goal of carbon neutrality seem out of reach. Will a different energy source perform any better?

2.3.2 Supplying Energy for Gasification: Biomass Combustion

It does not make a great deal of sense to burn a product, like the methanol or FT synthesis products, as an energy source for gasification. The process goes through all the trouble of producing a liquid fuel only to burn that same fuel? It would be more practical to burn a portion of the biomass feed as a source of gasification energy. The energy for gasification is typically supplied by combusting some of the feed biomass. The direct combustion of biomass is given by

$$C_6H_{10}O_5 + 6O_2 \rightarrow 5H_2O + 6CO_2 \tag{2.26}$$

The lower heating value for this direct combustion is 2540 kJ mol^{-1}, which is significantly higher than the energy released by combusting methanol. However, do not forget that the whole point of all this work is to increase the energy density of biomass. In this example, the energy density is less of a problem because the direct combustion is happening at the production site (rather than in a car engine, for instance).

The calculations of the previous section can be repeated, this time diverting just enough biomass for combustion to supply the energy for gasification. We can thus take a given amount of biomass, send a fraction of the feed to the gasification process and use the remainder of the feed for combustion to supply heat to the gasification reactions so that the combined or overall process is adiabatic. This gives the lower limit of the amount of biogas that needs to be combusted, as in practice more heat would need to be supplied due to inefficiencies, heat losses *etc.* in the process.

We will go through the calculations for the partial oxidation system, and similar calculations can be done for the other gasification process in Table 2.2. Gasification of 1 mol s^{-1} of biomass by partial oxidation requires some 367 kW of energy (Table 2.2); the amount of biomass that needs to be burned to meet this requirement is given by a simple ratio:

$$N_{\text{combustion}} = \frac{\Delta H^0_{\text{partial oxidation}}}{\Delta H^0_{\text{biomass combustion}}} = \frac{367}{2540} = 0.14 \text{ mol s}^{-1} \tag{2.27}$$

To meet the gasification energy requirements, 0.14 mol s^{-1} of biomass needs to be burned. This means the entire process requires a feed of 1.14 mol s^{-1} of biomass, 1 mol s^{-1} for the gasification and 0.14 mol s^{-1} for the combustion. This also requires 1.37 mol s^{-1} of oxygen, 0.5 mol s^{-1} for the gasification and 0.87 mol s^{-1} for the combustion.

The products of this now adiabatic gasification are 5 mol s^{-1} of hydrogen, 6 mol s^{-1} of carbon monoxide, 0.87 mol s^{-1} of carbon dioxide and 0.72 mol s^{-1} of water.

Next the water-gas-shift reaction is used to adjust the hydrogen/carbon monoxide ratio to be 2/1 as required by methanol synthesis. Using a material balance, we can see that

$$H_2 : CO = \frac{N_{H_2}^0 + N_{H_2}^{wgs}}{N_{CO}^0 - N_{CO}^{wgs}} \tag{2.28}$$

$$2 = \frac{5 + N_{H_2}^{wgs}}{6 - N_{CO}^{wgs}} \tag{2.29}$$

$$N_{H_2}^{wgs} = N_{CO}^{wgs} = 2\frac{1}{3} \text{ mol s}^{-1} \tag{2.30}$$

where:

- $H_2 : CO$ is the desired syngas ratio, a value of 2;
- $N_{H_2}^0$ is the amount of hydrogen produced by gasification, for this example: 5 mol s^{-1};
- $N_{H_2}^{wgs}$ is the amount of hydrogen produced by the water-gas-shift reaction;
- N_{CO}^0 is the amount of carbon monoxide produced by gasification, for this example: 6 mol s^{-1}; and
- N_{CO}^{wgs} is the amount of carbon monoxide consumed by the water-gas-shift reaction.

This water-gas-shift reaction requires 2.33 mol s^{-1} of water, but 0.72 mol s^{-1} are already present from the gasification. So 1.61 mol s^{-1} of water are needed to make up the difference.

The gas leaving the water-gas shift is composed of 7.33 mol s^{-1} of hydrogen (5 mol s^{-1} from gasification, plus 2.33 mol s^{-1} produced from water-gas shift), 3.67 mol s^{-1} of carbon monoxide and 3.2 mol s^{-1} of carbon dioxide.

Methanol synthesis then produces 1.84 mol s^{-1} of methanol.

A flow sheet summarising these calculations is presented in Figure 2.7.

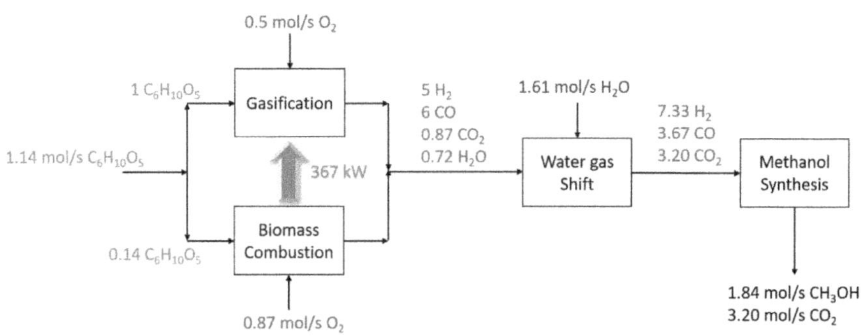

Figure 2.7 Summary of biomass-to-methanol system using biomass combustion to supply the energy for gasification.

Table 2.8 Gasification energy requirements met by combustion of biomass.

System	Methanol produced (mol s^{-1})	Biomass burned (mol s^{-1})	CO_2 emissions (mol s^{-1})	Moles CO_2 emitted per mole of CH_3OH produced	Moles CH_3OH produce per mole of $C_6H_{10}O_5$ feed
Partial	oxidation	1.84	0.14	3.2	1.74
1.61					
Steam	reforming	2	0.26	3.54	1.77
1.59					
Dry	reforming	2	0.26	3.54	1.77
1.59					

The effects of using biomass combustion can readily be observed from the numbers given in Figure 2.7. It can be seen that the minimum carbon dioxide emissions for the process are 1.74 moles of carbon dioxide per mole of methanol produced. A very similar result is expected for the other gasification systems, the results for those systems are presented in Table 2.8.

The results of Table 2.8 show that steam reforming and dry reforming are identical in their carbon dioxide emissions per mol of methanol produced. Compared to partial oxidation, the two reforming systems produce more methanol products but also produce more carbon dioxide. The reforming systems also do not use the biomass feed as effectively as partial oxidation, but the difference in methanol produced per unit of biomass feed is nearly negligible. As we are using the metric of carbon dioxide emissions per mol of methanol in this analysis to compare processes, reforming and dry reforming are nearly identical in terms of this metric. From this point forward, steam reforming and dry reforming consider them as the same process in terms of energy demands, and we will call the system and its associated material balance the reforming process.

This analysis has been highly idealised, with reactions proceeding to completion and energy being transferred with perfect efficiency. These figures represent a best-case scenario; under more realistic conditions the numbers presented here can only get worse – less fuel produced and higher carbon dioxide emissions.

There are two primary sources of carbon emissions in these systems. The first is from the water-gas-shift reaction. This reaction is being used to enrich the hydrogen content of the syngas in order to meet the requirements of the synthesis reactions. If the syngas ratio is simply ignored, there is unreacted carbon monoxide that needs to be emitted from the system or burned as a fuel and subsequently emitted as carbon dioxide; these are both still carbon dioxide emissions. The second source of carbon emissions is from the combustion of carbonaceous fuels for energy supply. Combustion is very problematic for achieving carbon neutrality, as this inevitably contributes to carbon dioxide emissions. The most effective way of reducing carbon emissions therefore requires the elimination of combustion.

2.3.3 Reducing Carbon Emissions: Renewable Energy Supply for Gasification

The only real alternative to combustion as an energy source for gasification reactions is through the use of renewable energy sources. The energy needs to be supplied as heat at high temperature (around 1000 °C). It has already been demonstrated that the combustion of carbon-based fuels, even biomass-derived ones, produces emissions. Renewable energy represents one of the few ways to decarbonise the energy sector. However, adoption of renewable energy technology has been fairly slow, especially by large-scale industries.

Market competitiveness appears to be the main driver behind the hesitancy in adopting renewable energy.[30–32] Renewable energy is perceived to be expensive compared to supplying energy by combustion. In addition, one of the main technical hurdles faced by renewable energy is the intermittent nature of the energy supply, solar energy ceases at night, and wind energy is highly sensitive to prevailing winds and local weather. Renewable energy also tends to require large amounts of physical space to produce amounts of energy that are similar to combustion.

It may not be practical or reasonable to expect renewable energy to exist in market isolation; renewable energy must coexist with carbon-based combustion technologies. This leads to something of a paradox; given the technical challenges being faced by renewable energy technologies, how can a company that adopts complete use of renewables hope to compete with a company that does not? The answer to this question, so far, is that renewables cannot yet compete in a free market scenario.[33,34]

However, from a scientific viewpoint, there can be no doubt that reduction in carbon emissions will require the widespread adoption of renewable energy technology.

Is the outlook really as bad as it seems? The analysis so far has shown that energy is only required for biomass gasification. If the calculations in the previous sections were repeated by replacing the combustion with renewables, what improvement in emissions would be achieved? What would it take to replace the combustion with renewable energy?

The two sources of renewable energy investigated in this section are solar and wind. These energy sources could, for example, be used to generate electricity, which in turn could be used to supply the high-temperature heat to the gasifier. There are inefficiencies in converting energy from one form to another, but because we are considering the limits of performance in this analysis, at this stage we used high-level efficiency estimates.

Returning to Table 2.2, the energy required for the partial oxidation of biomass is 367 kJ mol^{-1}. If the basis of 1 mol s^{-1} of biomass gasification is used again, 367 kW of renewable energy supply is needed to power this gasification. How much solar energy is required to meet this requirement?

Maps that show the varying levels of global solar irradiation are available from various sources; one such source is Solargis,[35] a company that collects solar data to provide assistance to researchers and potential solar energy investors. These data show that global irradiation can vary between 2 and 6.4 kW h kWp^{-1} in a day. Before any calculations can be done, those units (kW h kWp^{-1}) need to be understood.

A "kWp" is a kilowatt-peak. It is the rated power production of a standard solar panel operating at full capacity. A standard solar panel is defined as having a maximum capacity of 1 kW m^{-2}. Therefore, a kW h kWp^{-1} is the amount of energy that a standard solar panel produces over an average day.

With that in mind, let us take the average of the variation shown in the data, 4.2 kW h kWp^{-1}. A single standard solar panel will produce 4.2 kW h of energy over a day (24 h). This corresponds to 0.175 kW m^{-2}. In order to supply the 367 kW of energy to gasify 1 mol s^{-1} of the feed ($C_5H_{10}O_5$), some 2100 m^2 of solar panels would be required. To provide some visual context, consider that a soccer pitch has an area of 7140 m^2, and thus this area of solar panels is not unreasonable. (The basis of the calculations in this analysis is for 1 mol s^{-1} of biomass gasified, corresponding to around 583 kg h^{-1} of biomass.)

Figure 2.8 illustrates the minimum amount of solar area needed to provide the energy needed to gasify 1 mol s^{-1} of ($C_5H_{10}O_5$) for different levels of solar radiation. As expected, as the solar radiation intensity increases, the area required decreases. gasification *via* partial oxidation requires significantly more solar area than gasification *via* reforming, and the lower the solar radiation intensity, the bigger the difference in area required. Figure 2.8 scales linearly with the processing rate: if the rate of biomass being sent to gasification is doubled, then the solar area required is also doubled.

The calculations for wind power are similar, but data are not as easily available. Wind is far more dependent on location than solar. Available data[36] indicate that the best locations for wind turbines are in elevated or coastal areas. Regardless of their location, wind turbines have a rated

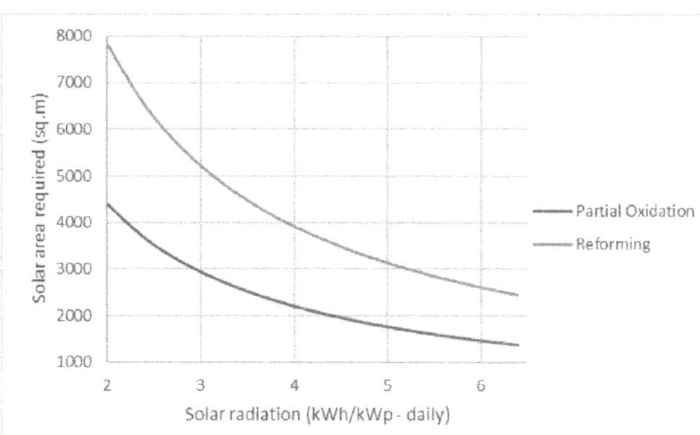

Figure 2.8 Solar area required to provide the energy required to gasify 1 mol s^{-1} of feed, ($C_5H_{10}O_5$).

capacity and an average efficiency of around 33%. Thus in order to supply the energy for gasification of 1 mol s^{-1} of biomass ($C_5H_{10}O_5$) through the partial oxidation route, a wind turbine with a rated power of some 1200 kW is needed; while in order to supply the energy to gasify the same feed rate of biomass *via* the reforming route, a turbine of 2000 kW is needed. Estimates from the U.S. National Renewable Energy Laboratory state the direct land usage for wind turbines to be approximately 4 MW km^{-2}. This corresponds to 0.3 km^2 of area required to gasify 1 mol s^{-1} of feed *via* the partial oxidation system and 0.5 km^2 for the reforming systems (including the 33% efficiency of wind power); this is some 100 times more area than for the solar panels.

In addition to the amount of land required for renewable energy installations, there is also the cost. The U.S. Energy Information Agency[37] has data that summarise the various costs for power facilities; a summary of these data is presented in Table 2.9.

Overnight cost is the cost of the installed technology without interest or depreciation. These data also include the capital costs for building these facilities in the United States. This is a simplification but serves as a sufficient basis for comparison. Table 2.9 already shows the market-based challenges with renewable energy; the cheapest renewable energy is solar energy with photovoltaic cells. While solar photovoltaic (PV) is cheaper than a standard internal combustion engine, it is still more expensive than highly efficient combined cycle power systems.

The biomass system is a biomass-fed power production system; this is the kind of system that would need to be built to supply gasification using biomass combustion – as discussed in the previous section. Power from biomass is one of the more expensive technologies but is still cheaper than solar thermal or offshore wind turbines. Use of this technology, however, produces carbon dioxide emissions, which means that a process using this energy would need to include these emissions when evaluating the processes carbon emissions.

If renewable energy (solar or wind) is used to completely replace the combustion needed to supply the energy for gasification, the emissions from the biomass-to-liquid system are reduced and are summarised in Table 2.10.

Table 2.9 Overnight costs of selected power generation technologies, sourced from the US-EIA.

Technology	Total overnight cost (USD kW^{-1})
Solar thermal	7116
Wind – offshore	5453
Biomass	4078
Wind – inland	1846
Internal combustion	1813
Solar PV	1248
Combined cycle combustion	1082

The largest decrease in carbon dioxide emissions is in the replacement of biomass combustion. On average, the complete replacement of combustion with renewables results in a carbon emission reduction of 29%. Is all of the effort involved with converting to renewables worth a 29% reduction, including the problems with reliability? As long as carbon dioxide has no penalty associated with its emissions, the answer is no, renewables are just too expensive and unreliable ··· at least when it comes to converting biomass into liquid fuels.

Even with all this effort to reduce carbon dioxide emissions, they are still not eliminated. Regardless, this is still an improvement. New biomass can be grown from these carbon dioxide emissions, and carbon neutrality could still be achieved.

2.4 Carbon Neutrality: Biomass Production

2.4.1 Local Carbon Neutrality of the Liquid Fuel Production Process

All the idealised systems analysed thus far have unavoidable carbon dioxide emissions; these are the result of biomass gasification being too lean in hydrogen. The water-gas-shift reaction is used to increase the hydrogen content of the biomass-derived syngas, which produces carbon dioxide as a by-product. This does make it seem like carbon neutrality is beyond reach.

There are two ways to look at carbon neutrality: from a local or process level or from a global or full life-cycle level.

Figure 2.9 illustrates the difference between local and global carbon neutrality. At the local level, only the carbon emissions from the process itself, shown in Table 2.10, need to be converted back to biomass for the system to achieve carbon neutrality. The process itself is then carbon neutral, but this is not the case at the global level. To achieve carbon neutrality on the global level, both the carbon emissions from the liquid production process as well as the emissions produced from burning the liquid fuels themselves (by the consumer) must both be converted back to biomass.

What would it take to achieve such levels of biomass production, and is that even feasible? That is the question this section will seek to answer.

Starting at the local level, consider the best-case carbon dioxide emissions. These are emissions shown in Table 2.10 under the emissions listed for renewables. The average carbon dioxide emissions for the gasification systems are 2.18 mol s^{-1} for every 1 mol s^{-1} of biomass being processed.

Finding data for how much carbon dioxide is absorbed by photosynthetic growth is difficult, but one source provides data[38] for American sycamore trees and water hyacinth.

$$S_{\text{tree}} = 750 \text{ ton}_{CO_2} \text{ per km}^2 \text{ per year} \tag{2.31}$$

$$S_{\text{aqua}} = 6000 \text{ ton}_{CO_2} \text{ per km}^2 \text{ per year} \tag{2.32}$$

where:

- S_{tree} is the amount of carbon dioxide absorbed by American sycamore, and
- S_{aqua} is the amount of carbon dioxide absorbed by water hyacinth.

The water hyacinth is clearly superior for carbon dioxide conversion, but it is important to remember that this is an aquatic plant that requires large amounts of freshwater.

Some simple unit conversions and dimensional analysis show that for every 1 mol s^{-1} of biomass that is converted into liquid fuel (accompanied with full renewable energy facilities), some 4 km^2 of sycamore forest or 0.5 km^2 of open water for water hyacinth is required. Converted into a mass basis, for every 1 kg s^{-1} of biomass conversion, 25 km^2 of sycamore forest or 3.1 km^2 of open water for hyacinth is needed to achieve local carbon neutrality.

These numbers do scale linearly, so it is a simple matter to extend these calculations to cover a range of production rates.

The basis that has been used throughout this chapter (1 mol s^{-1} of biomass) is actually a very small process. To give some perspective to Figure 2.10, there is a point representing the Pearl GTL process. The Pearl GTL facility, based in Qatar, is the largest gas-to-liquid production process in the world. At full capacity, it produces 140 000 barrels of liquid fuels a day. If the material balances and processes, which have been discussed in this chapter, were scaled to match the production rate of the Pearl GTL facility, the area required for growing water hyacinth would need to be 2744 km^2 to achieve local carbon neutrality.

Lake Kivu, in the Democratic Republic of the Congo, has a surface area of 2700 km^2 and would actually be too small for 140 000 barrels per day of methanol fuel. Lake Kariba, in Zimbabwe, has a surface area of 5400 km^2. Half of Lake Kariba's surface would be sufficient for local carbon neutrality.

Table 2.10 Carbon dioxide emission from the energy require replacing combustion with renewable energy.

Fuel	System	Energy source	CO_2 emissions (mol s^{-1})	% Emission reduction
Methanol	Partial oxidation	Biomass combustion	3.2	27.19
		Methanol combustion	2.9	19.66
		Renewable	2.33	—
	Reforming	Biomass combustion	3.54	43.50
		Methanol combustion	3.02	33.77
		Renewable	2	—
FT liquid	Partial oxidation	FT gas combustion	3	18.33
		Renewable	2.45	—
	Reforming	FT gas combustion	3.1	28.71
		Renewable	2.21	—

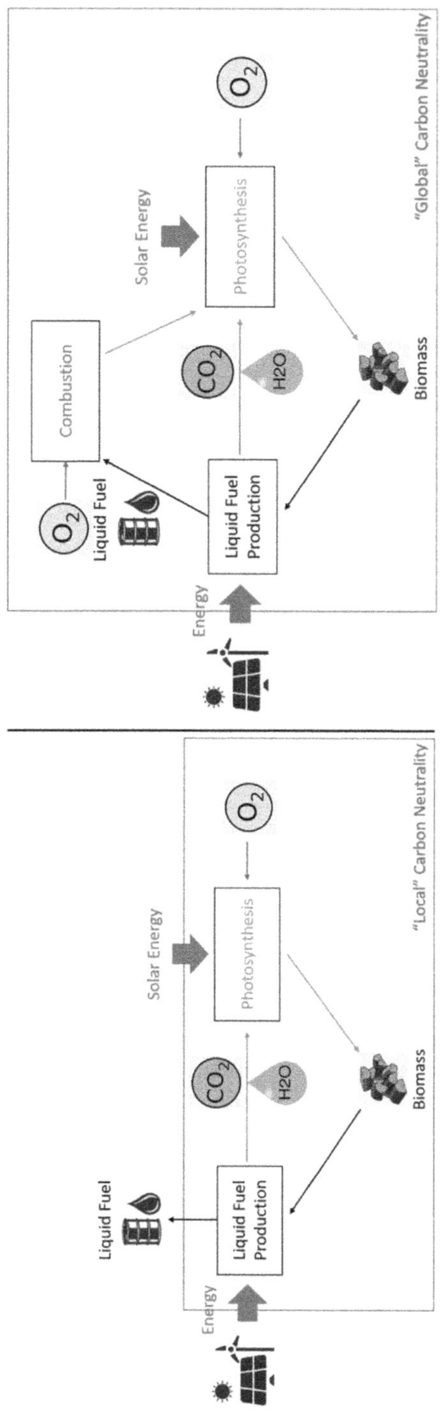

Figure 2.9 Local and global carbon neutrality.

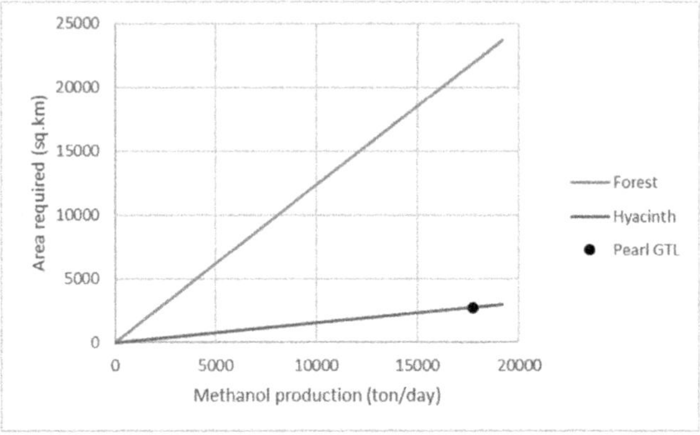

Figure 2.10 Areas of forest or freshwater required to achieve local carbon neutrality for methanol fuel.

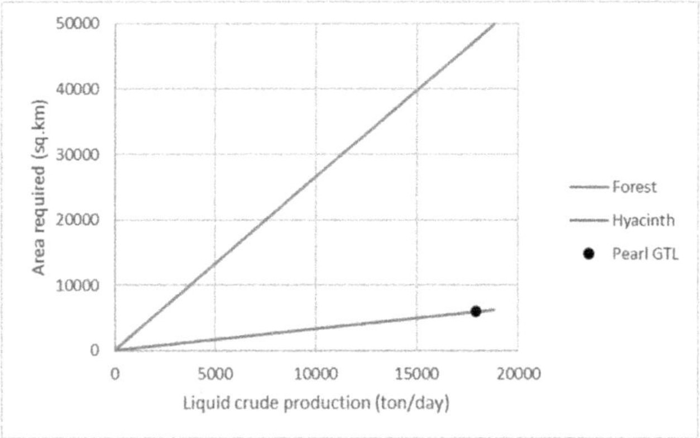

Figure 2.11 Areas of forest or freshwater required to achieve local carbon neutrality for liquid FT fuel.

About a third of the surface of Lake Titicaca, bordering Bolivia and Peru, would also be sufficient.

A similar figure can be drawn showing the area of forest or open water needed for the FT synthesis system.

Figure 2.11 shows the areas required against the production of only the liquid cut from FT synthesis. The gas and wax cuts have not been included in the calculation. This has the effect of increasing the surface area required for carbon neutrality, in comparison to that of methanol.

These surface areas are undoubtedly large, but they are also not completely unachievable. There are already agricultural farms that use more than these amounts of area. The world's largest farm (Mudanjiang City megafarm) has a surface area of around 91 000 km^2, nearly twice the area needed to achieve local carbon neutrality for a liquid fuel process the size of Pearl GTL.

Of course, local carbon neutrality is not a complete accounting of carbon emissions. All of these processes produce a liquid fuel that is burned for energy, and itself releases carbon dioxide during combustion. That carbon dioxide needs to be accounted for to achieve true global carbon neutrality.

2.4.2 Global Carbon Neutrality

Local carbon neutrality could be achieved by using the renewable carbon emissions as shown in Table 2.10, and calculating how much area would be needed to grow the plants needed to absorb that carbon dioxide, as shown in Figures 2.10 and 2.11.

In contrast, in order to achieve global carbon neutrality, the carbon dioxide from both the local (process) emissions as well as that produced by the combustion of the product have to be both taken into account. The methanol production systems produce some 1.84 mol s^{-1} (Figure 2.6) of methanol per 1 mol s^{-1} of biomass being processed, while in FT synthesis, 1 mol s^{-1} of biomass produces some 0.12 mol s^{-1} (Table 2.6) of liquid cut. We need to determine the amount of carbon dioxide produced from the combustion of these products.

For methanol, the material balance that describes the combustion process is

$$1.84CH_3OH + 2.76O_2 \rightarrow 1.84CO_2 + 3.68H_2O \tag{2.33}$$

while for the FT synthesis liquid cut, the material balance for the combustion process is

$$0.12C_{14}H_{30} + 2.58O_2 \rightarrow 1.68CO_2 + 1.8H_2O \tag{2.34}$$

This additional carbon dioxide then needs to be added to the renewable carbon emissions shown in Table 2.10.

Table 2.11 shows that, when the carbon emitted by burning the fuel is taken into account, the total carbon emissions increase by over 70% and nearly double in the case of methanol reforming.

The required surface area needed to recapture the globally emitted carbon dioxide (as shown in Figures 2.10 and 2.11) can be scaled linearly. For example, achieving global carbon neutrality of methanol fuel requires 79% more area if partial oxidation is the selected technology.

Table 2.11 Global carbon emissions for the renewable liquid fuel production systems.

System	Local carbon emissions (mol s^{-1})	Global carbon emissions (mol s^{-1})	% Increase
Methanol – partial oxidation	2.33	4.17	79
Methanol – reforming	2	3.84	92
FT – partial oxidation	2.45	4.22	72
FT – reforming	2.21	3.89	76

If the processes are again scaled to match those of the world's largest liquid fuel facilities (for example, Pearl GTL), the required land now starts to approach the amounts of land being used by the world's largest farms. As massive as these areas are, there is still enough space on earth to accommodate such agriculture.

Unfortunately, the problem does not actually end here. Even though the Pearl GTL process is one of the largest liquid fuel production facilities in the world, it only produces a small fraction of the world's petroleum consumption. Worldwide oil consumption is estimated to be around 90 million barrels per day.[39,40] Extending the previous analysis to account for the carbon dioxide emissions of 90 million barrels of oil per day in order to achieve global carbon neutrality, around 15 million km^2 of forest and/or 1.9 million km^2 of freshwater lake would be required. Existing research[41] puts the total land available for biofuel production at between 0.5 million and 10 million km^2. This is a serious problem; the Earth simply does not have enough arable land available to capture all the carbon dioxide being produced from the combustion of liquid fuels.

Completely converting the world's energy economy over to a globally sustainable biofuel process does not appear to be possible using current technology. However, in order to improve the sustainability of the biofuels-based processes, capturing at least some of the emissions should be targeted.

2.5 Closing Remarks

Biomass as a fuel is often touted as being environmentally sustainable, but is this really the case? In this chapter, a high-level analysis was performed to gain some insight on the reality of such claims. This analysis was highly idealised and conceptual in nature. Aspects such as heat losses, heat exchange and separation have not been considered in this analysis. Such considerations can only be a source of additional inefficiencies. If biomass is not an environmentally sustainable fuel under the most ideal conditions, it will never be sustainable when the conditions are realistic.

One of the biggest challenges for biomass as a fuel is its low energy density. Large amounts of biomass are needed to match the energy content of fuels. If the energy economy was totally switched to biomass, fuel storage would see a dramatic increase in size. It is simply not practical to expect to increase the size of a vehicle's fuel tank by a factor of 10. This necessitates first converting the biomass in liquids to increase the energy density.

This chapter considered two technologies for converting biomass into liquid: partial oxidation and reforming. Furthermore, two liquid fuels were considered: methanol and FT synthetic crude. All of these processes have carbon dioxide emissions that are inherent in the processes themselves. These emissions are unavoidable and are the result of the Law of Conservation of Mass.

Methods for reducing the inherent carbon emissions were investigated, with renewable energy integration being the most effective. Calculations were then performed to quantify how large biomass farming would need to be to compensate for the carbon emissions.

Currently, fossil fuel combustion technologies are cheaper than renewable technologies. This makes it difficult for new and sustainable technologies to be able to compete in an open market. Introducing market controls (such as a carbon tax) would help to bridge this competitive gap. However, introducing such barriers in the marketplace means a *de facto* end of the open market.

In theory, biomass can be produced on a scale that compensates for the emissions produced by fuel combustion. The scale of land needed for renewable energy generation and farming is certainly very large, large enough that there is concern that crops used for energy conversion might begin to supplant crops used for food. But the scale is not completely beyond the capabilities of existing technology and knowledge. Certainly, biomass production is a far more sustainable fuel than the continued use of fossil fuels.

Achieving carbon neutrality for the entire planet is not achievable with biofuels alone. Achieving global carbon neutrality requires 40% more land than estimates say is available for energy crops. These estimates could be improved by using crops engineered specifically for carbon dioxide capture, but worldwide carbon neutrality is currently out of reach.

So what is the answer the question, "Is a renewable-powered biomass energy economy the path towards an environmentally sustainable future"? The answer is yes, at least in theory. In practice, though, not yet. Biomass is definitely preferable to the continued use of fossil fuels, but it is also not going to reverse environmental impact.

Despite these challenges, exploring power and biomass to liquids is still a worthwhile endeavour, where there is some chance for a sustainable future. The continued use of fossil fuels has no chance at all.

Acronyms

ASF	Andersen–Schulz–Flory
FT	Fischer–Tropsch
FTS	Fischer–Tropsch synthesis
PV	Photovoltaic

References

1. J. P. Rodrigue, C. Comtois and B. Slack, *Geogr. Transp. Syst.*, 2016, 1–440.
2. A. Fernihough and K. H. O'Rourke, *Econ. J*, 2021, **131**, 1135–1149.
3. G. Clark and D. Jacks, *Eur. Rev. Econ. Hist.*, 2007, **11**, 39–72.
4. P. Zhai, A. Pirani, S. Connors, C. Péan, S. Berger, N. Caud, Y. Chen, L. Goldfarb, M. Gomis, M. Huang, K. Leitzell, E. Lonnoy, J. Matthews, T. Maycock, T. Waterfield, O. Yelekçi, R. Yu, B. Zhou, N. Bellouin, E. Coppola, R. Jones, G. Krinner, J. Marotzke, V. Naik, G. Plattner, J. Rogelj, M. Rojas, J. Sillmann, T. Storelvmo, P. Thorne, B. Trewin, K. Achuta Rao, B. Adhikary, R. Allan, K. Armour, G. Bala, R. Barimalala,

J. Canadell, C. Cassou, A. Cherchi, W. Collins, W. Collins, S. Corti, F. Cruz, F. Dentener, C. Dereczynski, A. Di Luca, A. Diongue Niang, F. Doblas-Reyes, A. Dosio, H. Douville, F. Engelbrecht, V. Eyring, E. Fischer, P. Forster, B. Fox-Kemper, J. Fuglestvedt, J. Fyfe, N. Gillett, I. Gorodetskaya, J. Gutierrez, R. Hamdi, E. Hawkins, H. Hewitt, P. Hope, A. Islam, C. Jones, D. Kaufman, R. Kopp, Y. Kosaka, J. Kossin, S. Krakovska, J. Lee, J. Li, T. Mauritsen, M. Meinshausen, S. Min, P. Monteiro, T. Ngo-Duc, F. Otto, I. Pinto, K. Raghavan, R. Ranasinghe, A. Ruane, L. Ruiz, J. Sallée, B. Samset, S. Sathyendranath, S. Seneviratne, A. Sörensson, S. Szopa, I. Takayabu, A. Tréguier, B. van den Hurk, R. Vautard, K. von Schuckmann, S. Zaehle, X. Zhang and K. Zickfeld, *The Physical Science Basis. Contribution of Working Group I to the Sixth Assessment Report of the Intergovernmental Panel on Climate Change*, 2021.
5. J. Daniell, M. Köpke and S. D. Simpson, *Energies*, 2012, **5**, 5372–5417.
6. B. Drosg, R. Braun, G. Bochmann and T. Al Saedi, *Biogas Handb. Sci. Prod. Appl.*, 2013, 52–84.
7. W. J. Jewell, R. J. Cummings and B. K. Richards, *Biomass Bioenergy*, 1993, **5**, 261–278.
8. H. G. Moon, Y.-S. Jang, C. Cho, J. Lee, R. Binkley and S. Y. Lee, *FEMS Microbiol. Lett.*, 2016, **363**, 1.
9. B. Schiel-Bengelsdorf, J. Montoya, S. Linder and P. Dürre, *Environmen. Technol.*, 2013, **34**, 1691–1710.
10. O. Konur, *Biodiesel Fuels*, 2021, 153–169.
11. G. Wang, Y. Dai, H. Yang, Q. Xiong, K. Wang, J. Zhou, Y. Li and S. Wang, *Energy Fuels*, 2020, **34**, 15557–15578.
12. A. Habibollahzade, P. Ahmadi and M. A. Rosen, *J. Cleaner Prod.*, 2021, **284**, 124718.
13. A. AlNouss, G. McKay and T. Al-Ansari, *Energy Convers. Manage.*, 2020, **208**, 112612.
14. L. Cao, I. K. M. Yu, X. Xiong, D. C. W. Tsang, S. Zhang, J. H. Clark, C. Hu, Y. H. Ng, J. Shang and Y. S. Ok, *Environ. Res.*, 2020, **186**, 109547.
15. E. V. Jithin, G. K. S. Raghuram, T. V. Keshavamurthy, R. K. Velamati, C. Prathap and R. J. Varghese, *Renewable Sustainable Energy Rev.*, 2021, **146**, 111178.
16. M. Fiore, V. Magi and A. Viggiano, *Appl. Energy*, 2020, **276**, 115415.
17. S. Wang, P. Wang, D. Shi, S. He, L. Zhang, W. Yan, Z. Qin, J. Li, M. Dong, J. Wang, U. Olsbye and W. Fan, *ACS Catal.*, 2020, **10**, 2046–2059.
18. T. Qin, T. Lin, X. Qi, C. Wang, L. Li, Z. Tang, L. Zhong and Y. Sun, *Appl. Catal., B*, 2021, **285**, 119840.
19. H. Wei, Y. Yingting, G. Jingjing, Y. Wenshi and T. Junhong, *Encycl. Sustainable Technol.*, 2017, 601–604.
20. P. K. Das, B. P. Das and P. Dash, *Refining Biomass Residues Sustainable Energy and Bioproducts: Technology, Advances, Life Cycle Assessment and Economics,*2020 , 275–301.
21. K. C. Waugh, *Catal. Lett.*, 2012, **142**(10), 1153–1166.
22. C. D. Rakopoulos and D. C. Kyritsis, *Energy*, 2001, **26**, 705–722.

23. X. Zhen and Y. Wang, *Renewable Sustainable Energy Rev.*, 2015, **52**, 477–493.
24. G. P. van Der laan and A. A. C. M. Beenackers, *Catal. Rev.*, 2011, **41**, 255–318.
25. C. K. Rofer-DePoorter, *Chem. Rev.*, 2002, **81**, 447–474.
26. H. Schulz, *Appl. Catal., A*, 1999, **186**, 3–12.
27. D. Förtsch, K. Pabst and E. Groß-Hardt, *Chem. Eng. Sci.*, 2015, **138**, 333–346.
28. J. S. Albuquerque, F. O. Costa and B. V. S. Barbosa, *Catal. Lett.*, 2019, **149**(3), 831–839.
29. Y. Zhang, Y. Yao, J. Shen, J. Chang, J. Gorimbo, X. Liu and D. Hildebrandt, *Fuel*, 2021, **302**, 121146.
30. T. A. Lanshina, J. A. "Skip" Laitner, V. Y. Potashnikov and V. A. Barinova, *Energy Policy*, 2018, **120**, 600–609.
31. J. Blazquez, R. Fuentes-Bracamontes, C. A. Bollino and N. Nezamuddin, *Renewable Sustainable Energy Rev.*, 2018, **82**, 1–5.
32. Y. Chen and B. Lin, *J. Cleaner Prod.*, 2020, **261**, 121186.
33. W. H. Wang, V. Moreno-Casas and J. H. de Soto, *Energies*, 2021, **14**, 4659.
34. Y. Heath and R. Gifford, *Environ. Behav.*, 2016, **38**, 48–71.
35. Solar resource maps and GIS data for 200 + countries | Solargis, https://solargis.com/maps-and-gis-data/download/world, (accessed 6 October 2021).
36. Global Wind Atlas, https://globalwindatlas.info/, (accessed 6 October 2021).
37. U. Energy Information Administration, Cost and Performance Characteristics of New Generating Technologies, Annual Energy Outlook 2021, 2021.
38. F. J. Dyson, *Energy*, 1977, **2**, 287–291.
39. Global oil consumption in barrels 2020 | Statista, https://www.statista.com/statistics/265239/global-oil-consumption-in-barrels-per-day/, (accessed 13 October 2021).
40. World Oil Statistics – Worldometer, https://www.worldometers.info/oil/, (accessed 13 October 2021).
41. S. Fritz, L. See, M. van der Velde, R. A. Nalepa, C. Perger, C. Schill, I. McCallum, D. Schepaschenko, F. Kraxner, X. Cai, X. Zhang, S. Ortner, R. Hazarika, A. Cipriani, C. Di Bella, A. H. Rabia, A. Garcia, M. Vakolyuk, K. Singha, M. E. Beget, S. Erasmi, F. Albrecht, B. Shaw and M. Obersteiner, *Environ. Sci. Technol.*, 2013, **47**, 1688–1694.

CHAPTER 3

Gasification of Biomass: An Overview

GARIMA CHAUHAN,* NATALIA MONTOYA SANCHEZ, CIBELE MELO HALMENSCHLAGER AND FELIX LINK

Department of Chemical and Materials Engineering, University of Alberta, 9211–116th Street, Edmonton, Alberta T6G 1H9, Canada
*Email: gchauhan@ualberta.ca

3.1 Introduction

Fast-paced demographic and economic developments across the globe contribute significantly to the growing demand for energy and the consumption of natural resources. The size of the global economy doubles approximately every 30 years,[1] which escalates the increase in demand for primary energy input. A report by the U.S. Energy Information Administration projected that global energy consumption will be approximately 949.5 EJ by 2050 – an increase of nearly 50% in energy use across the globe from 2018.[2] Fossil fuels have been one of the primary energy resources to date; however, some of these natural resources are at risk of becoming critically scarce in the coming years, while others have already become less abundant relative to demand. Based on the *R*/*P* ratio (reserves-to-products ratio), it is suggested that oil and natural gas reserves will be available for approximately 55 years (based on the figures for 2012) and that coal will be available for a substantially longer period (>100 years).[3] Although new reserves might be discovered in the future, the fact remains that fossil fuels will not be available for an unlimited time.

Catalysis Series No. 44
Chemicals and Fuels from Biomass *via* Fischer–Tropsch Synthesis: A Route to Sustainability
Edited by Joshua Gorimbo, Xinying Liu, Yali Yao and Diane Hildebrandt

Published by the Royal Society of Chemistry, www.rsc.org

Since it is pragmatically impossible to increase resource consumption efficiency indefinitely, it is equally inconceivable to have infinite economic and industrial growth without an associated increase in the rate of resource depletion. Utilization of alternative renewable energy sources is considered a plausible way to find the balance between energy demand and energy supply. In this regard, renewable resources such as wind, solar, hydropower and biomass are widely considered sustainable energy sources. At the end of 2019, the global renewable energy generation capacity amounted to 2537 GW,[4] while a recent statistical review of world energy suggests that renewables contributed almost 41% to growth in primary energy supply – the largest of any energy source in 2019.[5]

In this chapter, the focus is on biomass as the renewable resource. Biomass has been explored widely for energy generation and as chemical feedstock, due to its abundant availability and sustainable carbon energy. Biomass is organic material that originates from plants or animals and that amounts to approximately 550 GT of biomass carbon across the planet.[6] However, this carbon energy requires processing in order for it to be utilized as bioenergy in the energy sector.

Gasification of biomass is one of the thermochemical processes used to convert biomass into combustible gases (typically a mixture of CO, CO_2, H_2 and H_2O) by substoichiometric addition of gasifying agents at high temperature. A gas rich in hydrogen and carbon monoxide (syngas) is required for the production of fuels and chemicals *via* biomass gasification in combination with Fischer–Tropsch synthesis (FTS). The raw biomass is pretreated *via* size reduction, drying and densification to make it suitable for an efficient gasification. Biomass characteristics influence the gaseous product composition. In addition, the quality of syngas largely depends on the gasifying medium; the selected reactor; the operational parameters such as pressure, temperature and residence time; the effect of the gasifying agent; and the equivalence ratio (ER).

This chapter discusses the process of biomass gasification for the production of syngas that can be used in Fischer–Tropsch synthesis (FTS). It covers the selection of feedstock and its pretreatment, the reaction chemistry of the gasification process, the reactor configuration, the effect of various gasification parameters on the quality of syngas produced and current developments in the field of biomass gasification. The discussion on biomass is divided into two sections, *i.e.* types of biomass (Section 3.2) and pretreatment methods (Section 3.3) used to make biomass a suitable feedstock for efficient gasification operations. Section 3.4 discusses the fundamentals of the gasification process, including reaction chemistry during gasification reactions and the rate controlling factors. Various reactor configurations are discussed in Section 3.5, such as the fluidized bed gasifier and the entrained flow gasifier (EFG). This section also includes a discussion on the impact of operation parameters and type of biomass on the quality of the syngas produced. Major operational challenges are also detailed, to make readers aware of the possible issues that could arise during the biomass gasification process. The last section of this chapter provides an overview of current industrial developments and the outlook for biomass gasification processes.

3.2 Biomass: Feedstock for Gasification

According to European Directive 2009/28/EC, biomass that can be used for renewable energy is defined as: "The biodegradable fraction of products, waste, and residues from biological origin, from agriculture (including vegetable and animal substances), forestry and related industries including fisheries and aquaculture, as well as the biodegradable fraction of industrial and municipal waste".[3] Therefore, any organic material that is produced from animal or plant origin can be regarded as biomass. It can be further categorized as primary and secondary biomass. Primary biomass is formed by autotrophic organisms. These are plants, algae and certain bacteria, which are able to produce their mass through photosynthesis using solar energy. Secondary biomass is produced by the consumption of primary biomass, *e.g.*, in animal production or as organic residue and waste. Of all energy sources, biomass is the third largest energy resource in the world, after coal and petroleum oils.[7] Biomass mainly consists of carbon, oxygen and hydrogen, with traces of nitrogen, sulfur and other compounds such as sodium, potassium and chlorine. This indicates that biomass can be converted to fuel using existing technology to process fossil fuel. However, in order to explore all the potential that biomass offers, issues related to high moisture content, heterogeneous shape and size, and low density must be addressed. The thermochemical conversion behavior of biomass is different from that of coal and petroleum, due to the high water and oxygen content and the lower heating value.[8] Fresh and dry biomass differ in their moisture content.[9] If the feedstock has high moisture content, more energy is needed to gasify the feed, which makes the process less efficient.[10]

3.2.1 Types of Feedstock

Different types of biomass feedstocks can be used for gasification. The two main groups are wood/woody biomass and nonwoody biomass. The main characteristics of these materials are discussed in the following sections.

3.2.1.1 Wood and Woody Biomass

Forests are the biggest contributors on the planet of accessible biomass. Due to competition for available space to cultivate food, the energy sector is focusing on the wood waste biomass that is left behind after forest harvesting.[11]

The woody biomass category includes all nonherbaceous plants with hard stems (70–80%), bark (5–15%) and green leaves (10–15%).[12] Trees, shrubs, cactus and bamboo are the examples of woody biomass.[13] These types of biomass are all-season plants that can survive throughout the year, regardless of weather conditions. Woody biomass is essentially made up of three major polymers: cellulose, hemicellulose and lignin.

Cellulose is the primary structural component of cell walls in woody biomass. Cellulose is represented by the generic chemical formula $(C_6H_{10}O_5)_n$ and is a long-chain polymer with a high degree of polymerization and a high molecular weight. It has a crystalline structure made up of interconnected glucose molecules that gives it high strength and provides the skeletal structure of most terrestrial biomass. It is the dominant component of wood, comprising about 40–44% of the dry weight.[10] Cellulose is insoluble in water, and degradation begins at 240–350 °C, which results in anhydrous cellulose and levoglucosan. The crystalline structure resists thermal depolymerization better than unstructured hemicellulose.[14]

While cellulose has a strong, crystalline structure that is resistant to hydrolysis, hemicellulose has a random, amorphous structure that has little strength. It is a group of carbohydrates with a branched chain structure and a lower degree of polymerization than cellulose. It is represented by the chemical formula $(C_5H_8O_4)_n$ and constitutes about 20–30% of the dry weight of wood biomass.[10] Most hemicellulose contains some simple sugar residue – like D-xylose, D-glucose, D-galactose, L-arabinose, D-glucuronic acid and D-mannose, which tend to yield more gas and less tar during gasification.

Lignin is an integral part of the secondary cell wall of woody biomass. It is a complex, highly branched three-dimensional polymer of 4-propenyl phenol, 4-propenyl-2-methoxy phenol and 4-propenyl-2.5-dimethoxyl phenol. By dry weight, a typical hardwood contains about 18–25% of lignin, while softwood contains 25–35%.[10]

Table 3.1 shows the proximate and elemental composition of 26 species of woody biomass. The data were extracted from the work of Vassilev *et al.*[15] The high heating value (HHV) was calculated based on the elemental analysis (dry basis) using eqn (3.1).[16]

$$\text{HHV} = (0.341 \times \text{C}) + (1.322 \times \text{H}) - (0.12 \times \text{O}) + (0.0686 \times \text{S}) - (0.0153 \times \text{Ash}) \tag{3.1}$$

where: HHV is the higher heating value in MJ kg^{-1}; C is the mass fraction of carbon on a dry basis in wt%; H is the mass fraction of hydrogen on a dry basis in wt%; O is the mass fraction of oxygen on a dry basis in wt%; S is the mass fraction of sulfur on a dry basis in wt%; and Ash is the mass fraction of ash on a dry basis in wt%.

From Table 3.1, it can be generalized that, on a dry basis, the carbon and oxygen content in woody biomass are in the range of 50–55 wt% and 40–44 wt%, respectively. These two compounds, in general, account for over 90 wt% of the elemental composition of the biomass. The average heating value of woody biomass is about 21 MJ kg^{-1}, which indicates that the quality of the syngas produced from different types of woody biomass can be very similar when using the same reaction conditions, provided the initial moisture content is in a similar range.

Table 3.1 Chemical composition of wood and woody biomass.[15]

Wood feedstock	Volatile matter[a] (%)	Fixed carbon (%)	Moisture[a] (%)	Ash[a] (%)	Elemental analysis[b] (%)					HHV[bc] (MJ Kg^{-1})
					C	O	H	N	S	
Alder fir sawdust	36.3	9.1	52.6	2.0	53.2	40.2	6.1	0.5	0.004	21.26
Balsam bark	70.9	18.3	8.4	2.4	54.0	39.5	6.2	0.2	0.100	21.81
Beech bark	67.5	17.0	8.4	7.1	51.4	41.8	6.0	0.7	0.110	20.25
Birch bark	71.9	17.8	8.4	1.9	57.0	35.7	6.7	0.5	0.100	23.93
Christmas tree	46.1	12.9	37.8	3.2	54.5	38.7	5.9	0.5	0.420	21.63
Elm bark	67.0	17.2	8.4	7.4	50.9	42.5	5.8	0.7	0.110	19.72
Eucalyptus bark	68.7	15.1	12.0	4.2	48.7	45.3	5.7	0.3	0.050	18.60
Fir mill residue	30.4	6.5	62.9	0.2	51.4	42.5	6.0	0.1	0.030	20.34
Forest residue	34.5	7.3	56.8	1.4	52.7	41.1	5.4	0.7	0.100	20.05
Hemlock bark	65.9	23.4	8.4	2.3	55.0	38.8	5.9	0.2	0.100	21.84
Land clearing wood	35.4	7.0	49.2	8.4	50.7	42.8	6.0	0.4	0.070	19.79
Maple bark	70.1	17.8	8.4	3.7	52.0	41.3	6.2	0.4	0.110	20.87
Oak sawdust	76.3	11.9	11.5	0.3	50.1	43.9	5.9	0.1	0.010	19.60
Oak wood	73.0	20.0	6.5	0.5	50.6	42.9	6.1	0.3	0.100	20.13
Olive wood	74.3	16.1	6.6	3.0	49.0	44.9	5.4	0.7	0.030	18.33
Pine bark	70.2	23.3	4.7	1.8	53.8	39.9	5.9	0.3	0.070	21.30
Pine chips	66.9	20.0	7.6	5.5	52.8	40.5	6.1	0.5	0.090	21.06
Pine pruning	43.3	7.9	47.4	1.4	51.9	41.3	6.3	0.5	0.010	20.97
Pine sawdust	70.4	14.2	15.3	0.1	51.0	42.9	6.0	0.1	0.010	20.16
Poplar	79.7	11.5	6.8	2.0	51.6	41.7	6.1	0.6	0.020	20.55
Poplar bark	73.6	16.0	8.4	2.9	53.6	39.3	6.7	0.3	0.100	22.36
Sawdust	55.1	9.3	34.9	0.7	49.8	43.7	6.0	0.5	0.020	19.59
Spruce bark	67.3	21.4	8.4	2.9	53.6	40.0	6.2	0.1	0.100	21.62
Spruce wood	75.7	17.1	6.7	0.5	52.3	41.2	6.1	0.3	0.100	20.92
Tamarack bark	63.7	24.1	8.4	3.8	57.0	32.0	10.2	0.7	0.110	28.94
Willow	74.2	14.3	10.1	1.4	49.8	43.4	6.1	0.6	0.060	19.75

[a] As measured.
[b] Dry and ash free (DAF) basis.
[c] Calculated using eqn (3.1).

3.2.1.2 Non-woody Biomass

Due to the lower energy content and high variance in its physicochemical properties, nonwoody biomass is a more challenging feedstock to use in gasification compared to woody biomass. Table 3.2 shows the chemical composition of different nonwoody biomass feedstocks, such as agricultural waste, animal waste, industrial and municipal waste and plastic waste. The HHV of these biomass feedstocks was again calculated using eqn (3.1).

Agricultural waste biomass corresponds to organic material such as corn, straw, plants, farm animal waste and grass.[10] These are different from the woody biomass materials, and are considered herbaceous plants. Unlike non-herbaceous plants, herbaceous plants have vascular bundles with leaves but no persistent wood stems. These plants die each year at the end of the growing season. Wheat and rice are examples of herbaceous plants. These types of plants do not have the thick bark that covers nonherbaceous biomass types such as trees.[17] Nonwoody biomass materials are rich in energy, less expensive than woody biomass materials and available all over the world.

The cell wall of plants is made up of a network of polysaccharides, structural proteins and phenolic components. All agricultural biomass has the same major components, but there is variation in the composition from one species to another.[18]

Animal waste or manure that contains undigested and partially digested dietary nutrients is a resource that benefits plant growth and adds organic matter to soil, which improves its structure. Manure is made of carbohydrates and small amounts of amino acids and fats. The amount of nitrogen and sulfur in these types of waste materials is higher than in woody or herbaceous biomass (Table 3.2), which affects the syngas cleaning process.

Industrial and municipal solid waste and sewage waste also has potential for use as a biomass feedstock in the gasification processes. However, contaminants present in these materials must be considered carefully. For example, while agricultural waste tends to contain a high level of nitrates, the waste produced by the construction industry contains contaminants from protective agents, bonding primers, paint and so on.[19] Sewage waste has a high concentration of nitrogen, phosphorous and heavy metals.[20] Though contaminated waste can be used in gasification, the subsequent gas cleaning process must be able to remove the impurities.

Aquatic biomass such as microalgae and macroalgae are good candidates for the production of fuel because of their higher photosynthetic efficiency, higher biomass production and faster growth compared to lignocellulose biomass. Algae biomass consists of different types of carbohydrates, proteins, lipids and inorganic material, but it does not contain lignin.[21] However, the high cost of harvesting and dewatering this type of biomass becomes challenging, and the process needs to be developed further to ensure it is feasible from a techno-economic point of view.[22]

Unlike woody biomass, nonwoody biomass cannot be generalized in terms of proximate composition, elemental composition or heating value due to

Table 3.2 Chemical composition of nonwoody biomass.[15]

Nonwood feedstock	Volatile matter[a] (%)	Fixed carbon[a] (%)	Moisture content[a] (%)	Ash content[a] (%)	Elemental analysis[b] (%)					HHV[bc] ($MJ\ Kg^{-1}$)
					C	O	H	N	S	
Arundo grass	46.5	9.5	42	2	48.7	44.5	6.1	0.6	0.13	19.22
Bamboo whole	71	15.2	13	0.8	52	42.5	5.1	0.4	0.04	19.32
Bana grass	70.2	15.9	4.5	9.4	50.1	42.9	6	0.9	0.13	19.62
Buffalo gourd grass	73.5	12.3	10	4.2	46.1	44.5	6.5	2.6	0.27	18.61
Kenaf grass	73.5	15.7	7.5	3.3	48.4	44.5	6	1	0.15	18.93
Switchgrass	70.8	12.8	11.9	4.5	49.7	43.4	6.1	0.7	0.11	19.65
Alfalfa straw	71.6	14.3	9.3	4.8	49.9	40.8	6.3	2.8	0.21	20.05
Barley straw	67.4	16.4	11.5	4.7	49.4	43.6	6.2	0.7	0.13	19.65
Corn straw	67.7	17.8	7.4	7.1	48.7	44.1	6.4	0.7	0.08	19.58
Oat straw	73.9	12.5	8.2	5.4	48.8	44.6	6	0.5	0.08	19.08
Rice straw	59.4	14.4	7.6	18.6	50.1	43	5.7	1	0.16	19.04
Wheat straw	67.2	16.3	10.1	6.4	49.4	43.6	6.1	0.7	0.17	19.50
Sugar cane bagasse	76.6	11.1	10.4	1.9	49.8	43.9	6	0.2	0.06	19.59
Chicken litter	43.3	13.1	9.3	34.3	60.5	25.3	6.8	6.2	1.2	25.34
Meat bone meal	61.7	12.4	2.5	23.4	57.3	20.8	8	12.2	1.69	25.90
Currency shredded	79	11.1	4.7	5.2	45.4	46.1	6.3	1.9	0.32	17.99
Greenhouse plastic waste	61	5.5	2.5	31	70.9	16.4	11.2	1.5	0.47	36.38
Refuse-derived fuel	70.3	0.5	4.2	25	53.8	36.8	7.8	1.1	0.47	23.74
Sewage sludge	45	5.3	6.4	43.3	50.9	33.4	7.3	6.1	2.33	21.72
Marine macroalgae	45.1	23.1	10.7	21.1	43.2	45.8	6.2	2.2	2.6	16.77

[a]As measured.
[b]Dry and ash free (DAF) basis.
[c]Calculated using eqn (3.1).

the differences in their origin (Table 3.2). This may cause significant variation in the gaseous product from gasification.

3.3 Biomass Pretreatment

Biomass can be pretreated using various physical or thermochemical methods. Size reduction, drying and densification are the primary pretreatment methods used to prepare the biomass for use in a gasifier. In this chapter, the focus is on physical pretreatment methods. Pretreatment of biomass using thermochemical methods is discussed in section 3.4.

3.3.1 Size Reduction

Mechanical grinding of lignocellulosic substances such as wood typically leads to a finer particle size (1–1000 μm), uniform particle shape and a high surface area compared to the initial biomass.[10] A smaller particle size improves heat transfer in the drying and gasification step, which reduces the amount of energy required for these two processes. Most biomass material passes through two steps of grinding: coarse and fine. The most widely used equipment for grinding dry biomass and wet biomass are the hammer mill[23] and the colloid mill or dissolver,[24] respectively. The energy required to grind biomass depends on the equipment used, the feed flow and the material properties, including the initial size of the biomass feedstock and the moisture content. For instance, when using a hammer mill to grind wheat straw with a moisture content of 4–7% to a sieve dimension of 800 μm, the energy consumption is 185.8 MJ per ton of biomass.[25] In another study, the energy expended to grind biomass to a particular particle size increased from 587 MJ per ton to 718 MJ per ton when the moisture content was increased from 10 wt% to 20 wt%.[26]

3.3.2 Drying

Drying is an energy-intensive process, as each kilogram of moisture requires around 2300 kJ of energy for vaporization.[10] Three elements are involved in drying biomass properly: source of heat, method of removing the water that evaporates, some form of agitation to expose new material for drying.[27]

Raw biomass generally contains 60–70 wt% moisture.[28] This high water content restricts the development of biofuels for direct combustion, pyrolysis or gasification. In gasification, the high moisture content of the biomass increases the energy consumption required for dewatering and promotes the endothermic water-gas-shift reaction. As a result, the consumption of oxygen increases in order to maintain the temperature of the gasifier. The moisture also reduces the heating value of the fuel, due to the increase in non-recoverable heat loss carried by the moist flue gas.[28]

The dryers that are used can generally be divided into two categories, based on how the heat is provided. In direct dryers, the biomass material has direct contact with the material that provides the heat – either hot air or hot steam. In air dryers, the hot air provides some of the heat that causes water to evaporate from the biomass. The air also removes the water vapor formed during the drying process. With superheated steam dryers (SSD), the drying fluid is superheated steam that provides latent heat to the biomass. The process results in a larger amount of steam at a lower temperature at the end of the process. With indirect dryers, the biomass is kept separated from the source of the heat by a heat exchange surface. This method is advantageous because there is the possibility of recovering the latent heat from the water that evaporates from the biomass, since this vapor is not diluted by air.[29]

3.3.3 Densification

Organic residues are a promising biomass feedstock; however, the low density and consequently the high volume make transportation, handling and storage a problem. For instance, when low-density biomass is co-fired with coal, the differences in density cause problems with feeding fuel into the boiler, which reduces the burning efficiency.[7] Densification is one of the preferred methods used to overcome this problem. Biomass densification first became a commercial large-scale process in the second half of the last century. It is used to increase the handling properties of biomass for both energy production and animal feedstock.[30]

Densification can be classified according to the work principles, *i.e.* baling using a baler, pelletization using a pelletizer, briquetting using a roller press or a piston press palletizer.[31,32]

3.3.3.1 Baling

Agricultural biomass, such as corn stover, warm-and-cold season grass and sorghum, are often densified into bales. Large bales can weigh more than 550 kg and can have a rectangular or cylindrical shape. Bale biomass can reach a density of over 200 kg m^{-3}, which helps to lower costs by maximizing the capacity for handling and storing the biomass.[33]

3.3.3.2 Pelletization

There are two important aspects to be considered during pelletization: the ability of particles to form pellets with good mechanical strength; the ability of the process to increase the density of the biomass.[34] Additives are mixed with the biomass to improve pellet durability and physical quality, reduce dust, improve pelletizing efficiency and reduce energy costs. Water, binder and starches are the most common additives used to form pellets.[35]

Wood is the most widely used material used for pellet production. Pellets are produced in a mill, where biomass is forced to flow through the cylindrical press channels and rollers, a high back pressure is built up and heat is generated.[30] With biomass, the most common pellet diameter sizes are 6 and 8 mm.[35]

3.3.3.3 Briquetting

Briquetting is the densification or compaction of residues into a shaped product of higher density than the raw materials.[36] In this process, the raw material is first ground to a smaller size so that it can be compressed properly. Briquettes come in various sizes and shapes but are usually a cylindrical shape, with a diameter between 25 and 100 mm and a length ranging between 10 and 400 mm.[37] The major advantage of the briquetting process is that it does not require extra binder to bind the material together.

3.4 Biomass Gasification

Gasification is the thermochemical conversion of carbonaceous feedstock in the presence of controlled amounts of oxidizing agent, in order to produce gaseous products with usable heating values. Here, the phrase "usable heating value" is used explicitly and in contrast to the combustion process, which releases about 90% of the energy from biomass during the thermal reaction[38] and produces flue gas with virtually no heating value. For syngas applications like FTS, the goal is to obtain a gasification product in which H_2 and CO are the main contributors to the heating value. Noncombustible components like nitrogen and water are also rejected during syngas cleaning after gasification, which leads to an increase in the heating value of the syngas.

3.4.1 Gasification Process

Typical gasification of biomass involves the following sequential steps, as shown in Figure 3.1:

1. Primary decomposition (devolatilization and pyrolysis)
2. Secondary tar-cracking (homogeneous reactions)
3. Char gasification and combustion (heterogeneous reactions)

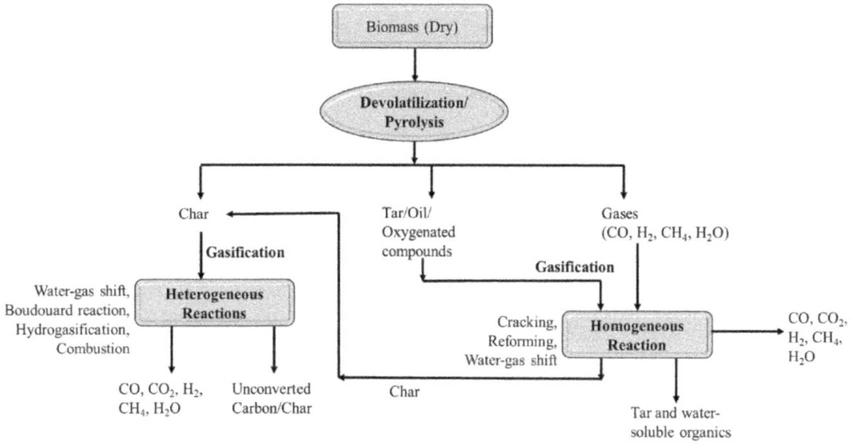

Figure 3.1 General process scheme for biomass gasification.

Biomass is dried up to 10–20% moisture (by weight) before it is fed into the gasifier to minimize energy consumption (Section 3.3). The dried biomass is fed to the gasifier, where the biomass heats up, and the remaining moisture evaporates. As the temperature rises further, pyrolysis of biomass takes place, which causes thermal breakdown of larger hydrocarbon molecules into gas molecules without any major chemical reaction with the gasifying medium. One of the major products of the pyrolysis process is tar, which is a complex mixture of condensable hydrocarbons, including oxygen-containing aromatic rings and complex polyaromatic hydrocarbons.[39] Some other gaseous products and oxygenates also form during thermal decomposition. These pyrolytic products react among themselves as well as with the gasifying medium to form the final gasification product. The unconverted carbon (char) and oxygen-containing hydrocarbons are further gasified at a high temperature through several heterogeneous reactions.

It should be noted that pyrolysis has been mentioned here as one of the important aspects of the gasification process. In practice, it is always necessary to pass through a pyrolytic stage to perform gasification. However, pyrolysis differs from gasification, in that the products of interest are the char and liquid, which retain much of the structure and complexity of the raw material undergoing pyrolysis.[40,41] Gasification of biomass is carried out at a higher temperature than pyrolysis; therefore, in contrast to pyrolysis, the gasification products lose the structural identity of the feed and are solely smaller gaseous molecules.[42]

3.4.2 Gasification Reactions

Some of the major gasification reactions are listed in Table 3.3. There is no sharp boundary to define the sequence of these reactions and gasification steps, as they overlap throughout the process and are influenced by the configuration of the reactor.

Based on the reaction temperature for these reactions, the gasification process can be divided into three different reaction zones: low temperature zone ($200 < T_L < 500$ °C) for devolatilization and primary decomposition; medium temperature zone ($500 < T_M < 900$ °C), where simultaneous pyrolysis and tar/oil gasification can occur; high temperature zone ($T_H > 900$ °C), where unconverted carbon (char) is gasified. Figure 3.2 illustrates the reaction zones and rate controlling step in the different sequential steps of biomass gasification. The reaction chemistry at each reaction zone is discussed in the following subsections.

3.4.2.1 Devolatilization/Pyrolysis (Low Temperature Zone)

Pyrolysis can set in at a low temperature of about 200 °C, in parallel with the biomass heating up, and forms a variety of species such as tar, char and gases, including methane, CO, CO_2, H_2, H_2O *etc.* (see reaction (R1) in Table 3.3). In this temperature range, the cellulose, hemicellulose and lignin

Table 3.3 Major reactions during biomass gasification.

Reaction	Code (name)	ΔH_r^0 at 298 K (kJ mol^{-1})
$C_xH_yO_z \rightarrow aCO_2 + bH_2O + cCH_4 + dCO + eH_2 + \text{char} + \text{tar}$	R1 (pyrolysis)	
Homogenous reactions		
$2CO + 2H_2 \rightarrow CH_4 + CO_2 + CO_2$	R2 (methanation)	−247
$CO + 3H_2 \rightarrow CH_4 + H_2O$	R3 (methanation)	−206
$CO_2 + 4H_2 \rightarrow CH_4 + 2H_2O$	R4 (methanation)	−165
$CO + H_2O \leftrightarrow CO_2 + H_2$	R5 (water-gas shift)	−41.2
$CH_4 + H_2O \leftrightarrow CO + 3H_2$	R6 (steam reforming of methane)	+172.6
$CH_4 + 0.5O_2 \leftrightarrow CO + 2H_2$	R7 (partial oxidation of methane)	−36
$CH_4 + CO_2 \leftrightarrow 2CO + 2H_2$	R8 (dry reforming of methane)	−74.9
$C_nH_m + 2nH_2O \rightarrow \left(2n + \frac{m}{2}\right)H_2 + nCO_2$	R9 (steam-tar reforming)	
$H_2 + 0.5O_2 \rightarrow H_2O$	R10	−242
Heterogeneous reactions		
$C + CO_2 \leftrightarrow 2CO$	R11 (Boudouard reaction)	+172
$C + H_2O \leftrightarrow CO + H_2$	R12 (water-gas reaction)	+131
$C + 2H_2 \leftrightarrow CH_4$	R13 (hydrogasification)	−74.8
$2C + O_2 \leftrightarrow CO$	R14 (partial oxidation)	−111
$C + O_2 \leftrightarrow CO_2$	R15 (in excess of oxygen)	−394

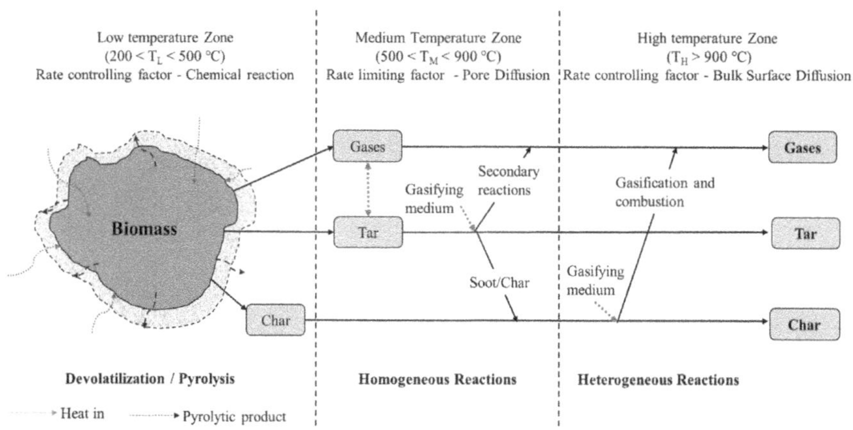

Figure 3.2 Reaction zones and rate controlling step in different sequential steps of biomass gasification.

components of biomass break down into primary tar (which contains oxygenates and condensable hydrocarbons) as a major product:

$$\text{Biomass} \xrightarrow{\Delta(200<T<500)} \text{Tar} + \text{Gases} + \text{Char} \tag{3.2}$$

The rate of pyrolysis is primarily dependent on the rate of heating the biomass particles. Other factors also influence the rate of pyrolysis, such as particle size, reaction temperature, rate of gasification by simultaneous water-gas-shift (WGS) reactions and partial pressure of the gasifying agent.

Devolatilization and pyrolysis processes basically involve physical evaporation of water and thermal decomposition as the dominant processes in this temperature zone. No or little utilization of the gasification agent results in minimal chemical reactions at this gasification stage. These reactions are typically slower than the thermal decomposition reaction and are therefore considered rate-limiting factors.[38]

3.4.2.2 *Homogeneous Reactions (Medium Temperature Zone)*

Products of primary thermal decomposition start reforming into smaller, lighter noncondensable gases and into a series of heavier molecules called secondary tar, at a temperature above 500 °C, by reacting with the gasification medium (reactions (R2)–(R10), Table 3.3). As the temperature increases, secondary tar undergoes further reforming, thermal cracking and steam cracking, which convert tar into an array of smaller and lighter hydrocarbons.

Depending on the rate of heating, tar and other pyrolysis products can react among themselves or with the gasification medium, either in parallel with the pyrolysis reaction or once the pyrolysis reaction is completed. The heating rate can affect the final product composition significantly, as shown in Figure 3.3.

If the heating rate is slow, pyrolysis of biomass starts at about 200 °C. At this temperature, the rate of gasification of volatiles and char is significantly

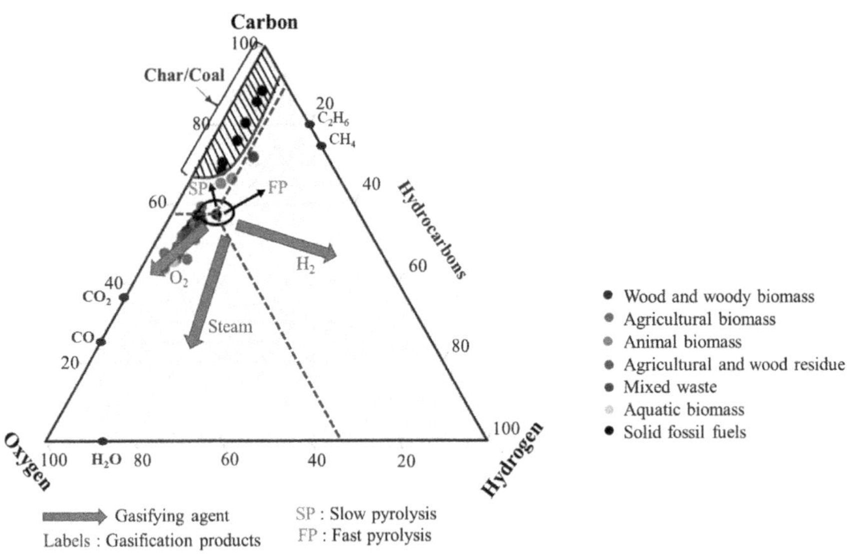

Figure 3.3 C–H–O ternary diagram of various biomass and fossil fuels to illustrate the effect of heating rate on pyrolysis and effect of gasifying medium on gasification.

slower than pyrolysis product formation. This eventually leads to the condensation of volatiles on the external surface of the biomass particles, as shown by the dotted line surrounding the biomass particle in Figure 3.2. In this case, gasification starts only when pyrolysis is completed, and this then results in a large fraction of unreacted condensed volatiles. Consequently, slow pyrolysis moves the product toward carbon through the formation of solid char (Figure 3.3). At high rate of heating (*e.g.* in an entrained flow reactor), pyrolysis and gasification take place in parallel. This limits the possibility of volatile condensation on biomass particles, due to simultaneous gasification of the formed volatile. Fast pyrolysis moves the product toward the C–H axis and away from the oxygen (see Figure 3.3).

Methanation and reforming are the major reactions in this temperature zone. Water formation and the water-gas-shift (WGS) reaction also take place, which may affect the gasifier temperature to some extent. Although the use of a catalyst is not essential for the thermochemical conversion of biomass, it may be used to improve the quality of syngas, depending on product requirements, and to minimize the tar in the final product yield. Reforming (with or without a catalyst) affects the methane concentration in the product gas. Methane reacts with the gasifying agent (steam, O_2 or CO_2) – as listed in reactions (R6), (R7) and (R9) in Table 3.3 – and produces CO and H_2. When the product gas passes over the catalyst particles, the tar or condensable hydrocarbon can be reformed on the catalyst surface in the presence of a gasifying medium (steam or CO_2), to produce additional hydrogen and carbon monoxide (reactions (R8) and (R9), respectively, in Table 3.3). If no catalyst is present, these reactions do occur as well but to a lesser extent. Another important gas phase reaction is the WGS reaction (see reaction (R5) in Table 3.3). The WGS reaction is slightly exothermic, and its equilibrium is shifted to the reactant side with increasing temperature, which leads to a lower yield of H_2. Homogeneous reactions in the medium temperature zone occur much more rapidly than heterogeneous char gasification (see Section 3.4.2.3), during which mass transport limitations play a more important role.

3.4.2.3 Char Gasification

Char gasification is the rate controlling step of the overall gasification process. Its rate is limited due to the external, bulk surface diffusion of the gaseous reactants into solid char. The major reactions are: Boudouard reaction (R11), heterogeneous WGS reaction (R12) and hydrogasification (R13) (see Table 3.3). Although combustion has not been mentioned as a separate sequential process here, it also takes place along with gasification at a high temperature.

In practice, the thermal energy necessary for drying, pyrolysis and endothermic gasification reactions comes from a certain amount of exothermic combustion reactions allowed in the gasifier.[17] Because of simultaneous exothermic and endothermic reactions during gasification, the temperature of the char particle stays nearly the same as the temperature of the gasifier.[43] This prevents the formation of temperature hot spots and material deterioration.

Gasification of biomass char and its corresponding product distribution are largely dependent on the gasifying medium. The C–H–O ternary diagram depicts the product gas composition when different gasifying mediums are used (see Figure 3.3). If oxygen is used as a gasifying medium, the major products are CO and CO_2 with a low yield of H_2 in the product gas. The hydrogasification reaction requires hydrogen in the gasification medium; therefore it can be one of the dominant reactions when steam or hydrogen is being supplied as a gasifying agent. The hydrogen is formed in the homogeneous gas phase reactions (WGS, reforming). Methane is formed as a dominant product that may undergo further reforming in the presence of steam.

Walker *et al.*[44] estimated the relative rates of the char gasification reactions in the presence of oxygen, steam, CO_2 and H_2 at a temperature of 800 °C and a pressure of 0.1 Atm. The reactions of char with oxygen ((R14) and (R15)) were reportedly the fastest among all the gasifying mediums investigated. The availability of relative amounts of fuel, oxidant (air or oxygen) and steam (if used) governs the fraction of carbon or oxygen that enters (R14) or (R15) (see Table 3.3). The relative rate of gasification in the presence of oxygen, steam, CO_2 and H_2 was reported to be as follows:

$$R_{C+O_2} \gg R_{C+H_2O} \gg R_{C+CO_2} \gg R_{C+H_2} \tag{3.3}$$

In another study done by Blasi *et al.*,[45] the rate of the WGS reaction (R12) was found to be 3–5 orders of magnitude slower than that of the char–oxygen reaction. The rate of the Boudouard reaction (R11: $C + CO_2 \rightarrow 2CO$) was 6–7 orders of magnitude lower than (R14)/(R15) and about 2–5 times lower than (R12). The rate of the hydrogasification reaction (R13, char–hydrogen reaction) is the lowest of the other reactions; therefore, it does not play an important role in conventional gasification.

However, other studies[46,47] have reported variation in the reaction rates along with the extent of conversion in different gasifying mediums and with the method used to calculate the rate. Other factors associated with char reactivity also cannot be underestimated, such as heating rate, temperature and pressure of the pyrolysis, feedstock and content/composition of the inorganic matter.[45] Catalytic bed materials (*e.g.* in fluidized bed reactors) and the catalytic effect of ash can also influence the reaction rates. The amount of alkali metals can be considerably higher in waste biomass than in woody biomass, and these metals can form a reactive slag/ash. The reactive nature of the minerals can catalyze the gasification reactions, which interferes with the product gas quality[48,49] during the gasification process.

3.5 Biomass Gasification Technologies

In the last two decades, significant efforts have been focused on the development of advanced gasification technologies for biomass conversion. These gasification technologies rely on existing knowledge for coal gasification processes and seek to accommodate the differences in the biomass feedstock materials and the implications on the overall gasification process.

Currently, a number of options are available, which makes the selection of an appropriate technology – *i.e.* the type of gasifier – a function of the scale (feedstock availability), type of feedstock (quality and form) and desired application (power, heat, synthesis applications and others).[17]

Gasifiers are usually classified based on the gas–solid contacting mode, gasifying agent, temperature, pressure of operation, method of heat supply.[50] Based on the gas–solid contacting mode, gasifiers are divided into three main groups: fixed-bed or moving-bed, fluidized-bed, entrained flow reactor. This is the most widely used gasifier classification system. In terms of the gasifying agent, gasifiers are divided into oxygen-blown, air-blown and steam-blown. The use of air/steam and oxygen/steam mixtures is also common, but most biomass gasification processes use air instead of oxygen. The use of oxygen requires an air separation unit (ASU), which has a significant impact on the process cost;[51] however, the use of oxygen can be justified, depending on the application demand. The temperature of operation, which determines the behavior of the biomass ash (*i.e.* melting), classifies gasifiers as slagging or nonslagging, whereas the pressure of operation divides them as atmospheric or high-pressure types. All types of gasifiers can be designed to operate at either pressure level. The method of heat supply divides gasifiers into autothermal or directly heated and allothermal or indirectly heated types. Autothermal gasifiers use an oxidant that partially oxidizes the biomass to provide the necessary heat for gasification.[50] With indirect gasifiers, pyrolysis and gasification of biomass take place in one reactor, and partial combustion in another. Thus indirect gasifiers use steam, and no oxidant is fed to the gasifier section.[52]

Each type of gasifier is best suited to a particular scale and application (see Figure 3.4). For instance, downdraft fixed-bed gasifiers are suitable for power production at a small scale (10 kWth to 1 MWth), whereas updraft fixed-bed gasifiers are popular for generating heat at the small and medium scales (1–10 MWth). Fluidized-bed gasifiers fall in the middle range capacity (5–100 MWth) and are popular for combined production of heat and power (CHP), as well as for synthesis applications (*i.e.* syngas is transformed into fuels and/or chemicals). EFGs are found at the upper capacity limit (>50 MWth) and

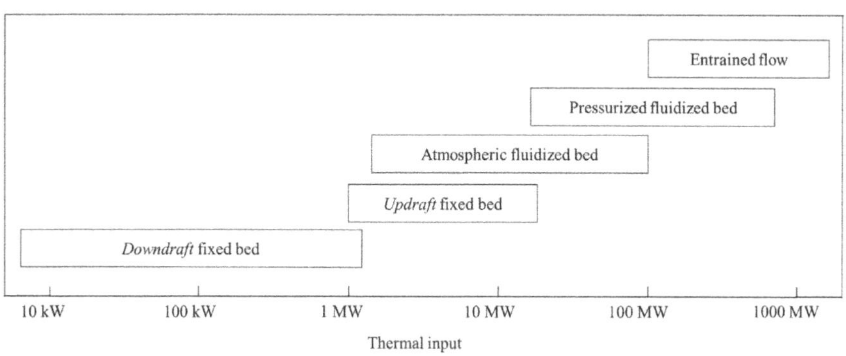

Figure 3.4 Capacity for biomass gasifiers.

are intended for integrated combined cycle plants (IGCC) and synthesis applications.[17,53]

As indicated, this chapter focuses on biomass gasification for fuel production; therefore, fluidized-bed and entrained flow reactors are of particular interest. The operating principles and main characteristics of these gasifiers are detailed in the following section. The effect of different feedstock characteristics and operating parameters on the operation and performance of biomass gasification is discussed thereafter.

3.5.1 Fluidized-bed Gasifiers

Fluidized-bed gasifiers are widely used for biomass gasification. These reactors can process a wide range of biomass feedstock types and are well-known for their excellent mixing and uniform temperature distribution.[17] There are three main types of fluidized-bed reactors: bubbling fluidized-bed (BFB), circulating fluidized-bed (CFB) and dual fluidized-bed (DFB).

In a BFB gasifier, the gasifying agent (*e.g.* oxygen) is introduced at the bottom, through a distribution grid, at a velocity between 1 and 3 m s^{-1} (see Figure 3.5A).[54] Bubbles form and grow larger as they move upward. Biomass is fed in above the distribution grid, and the high velocity of the gas forces it

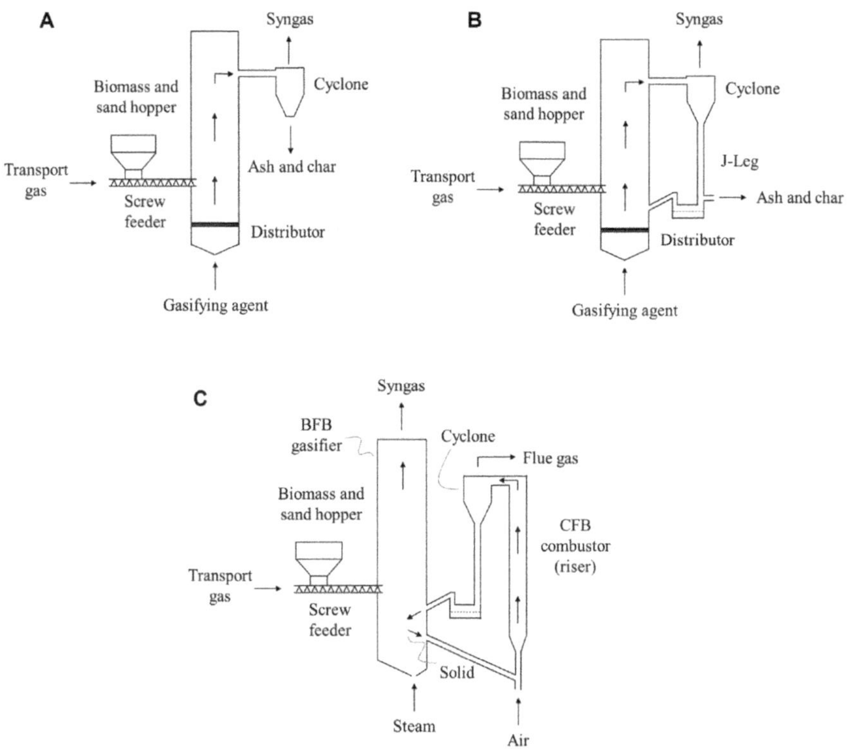

Figure 3.5 Schematic description of (A) BFB gasifier, (B) CFG gasifier, (C) DFB gasifier.

upward through a bed made of a granular, sand-like material. The gas keeps a defined volume of the bed in a semisuspended or fluidized state, and it passes in the form of interstitial gas or bubbles.[17,52,55] Silica particles are typically used as bed material, and these act as a heat carrier and mixer.[17] Biomass particles are quickly heated to the bed temperature by direct contact with the hot bed material, which also disperses the heat released from exothermic reactions. The temperature of the bed is typically between 800 and 1000 °C, and it is controlled by the ratio of oxidizing agent to biomass. Biomass is rapidly pyrolyzed and forms gaseous products, char and tar. The high temperature bed material, which might also contain a tar reforming catalyst (*e.g.* dolomite), further converts the latter to obtain a gas with a low tar content ($<$1–3 $g\,N^{-1}\,m^{-3}$).[55] On top of the bed is a freeboard section, which is typically used for disengagement of particles that are ejected from the bed by the gas bubbles. In some cases, noncondensable gaseous products produced from biomass pyrolysis escape the bed and rise into the cooler freeboard (bubbles allow them to bypass the solids without participating in the gasification reactions) to form tar and char.[17,52]

CFB gasifiers are particularly suitable for feedstocks with a high content of volatiles (*e.g.* biomass) due to the long gas residence time they provide.[17] In CFB gasifiers, solids circulate in a loop, which is characterized by intense mixing. The absence of bubbling in CFB gasifiers avoids the gas bypassing issues seen with BFB gasifiers. CFB and BFB gasifiers have a different hydrodynamic behavior. The gas velocity in a CFB gasifier is higher (5–10 $m\,s^{-1}$)[54] than in a BFB gasifier, and the bed material is carried up in the gasifier shaft by the gas. Some of this material travels radially to the walls and slides down as a wall layer, due to gravity. The remaining solid–gas suspension is carried out by the gas to an external cyclone, which is connected to a recycle line at the bottom of the reactor (see Figure 3.5B). This arrangement separates most of the solid particles and returns them to the base of the gasifier.[52]

The high recycling rate of solids, as well as the high fluidization velocity achieved in a CFB gasifier, leads to a hydrodynamic condition known as fast-fluidized bed and assists in maintaining a uniform temperature in the entire bed, typically in the range of 800–1000 °C, with the temperature being limited by the melting properties of the biomass ash.[50] The temperature of the gas product is similar to the bed temperature, which yields a gas product with a low tar content. Commercial CFB gasifiers have been installed in different countries, the largest of which is in Vaasa, Finland. It is part of a co-firing operation and has a capacity of 140 MWth.

In cases, where dilution of the product gas with nitrogen is an issue, as is the case with syngas production for fuel applications, and the use of oxygen is not practical or economical, DBF gasifiers are an alternative.[56] DFB gasifiers (Figure 3.5C) consist of two interconnected reactors, namely an endothermal gasifier and an exothermal combustor. These gasifiers are operated in allothermal mode, and different reactor configurations are possible. For instance, DFB gasifiers could comprise two BFB or CFB reactors, or a CFB gasifier and a BFB combustor, or a BFB gasifier and a CFB combustor. Figure 3.5C provides a schematic description of the last configuration. Inert bed material circulates between the two reactors and carries heat from the combustor to the gasifier, in order to enable the gasification reactions. Preheated steam (typically at 700–900 °C) is fed into the BFB gasifier and used as the gasifying agent. The tar and residue char generated in this reactor go to the CFB combustor, where they are combusted with air, which generates the temperature required to reheat the bed material. The flue gas contains nitrogen and is discharged after solid particles are recovered in the cyclone. In this way, syngas with a negligible amount of nitrogen is produced. A detailed analysis of different DFB technologies, their characteristics, issues and operational data (H_2 and tar content in the product gas) can be found in various studies.[56,57]

In general, fluidized-bed reactors operate in a temperature range of 800–1000 °C, which is advantageous for fuels forming highly corrosive ash, as is the case for biomass.[58] At these temperatures, ash from biomass typically does not melt and can be removed as dry ash (nonslagging gasification). Depending on the nature of the feed, silica particles might react with the inorganic components in biomass to form low-boiling-point silicates. These silicates melt at higher temperatures and can bind biomass particles together to produce agglomerates,[50] which can potentially affect the fluidized behavior of the bed. It is important to note that fluidized-bed gasifiers are capable of producing syngas with a low tar content ($\sim$10 g N^{-1} m^{-3}), which is particularly important for synthesis applications.[17] Table 3.4 indicates the typical characteristics of fluidized-bed gasifiers.

Table 3.4 Characteristics of fluidized-bed and entrained flow gasifiers. Data from ref. 17, 39 and 58.

Parameter	Fluidized bed	Entrained flow
Feedstock requirements:		
Feedstock particle size	<10 mm	<0.1 mm
Moisture content	—	Low
Ash content	No limit	20–40%[a]
Gasification characteristics:		
Reaction zone temperature	800–1000 °C	1990 °C
Gas exit temperature	800–1000 °C	>1200 °C
Oxygen requirement	Moderate	High
Steam requirement[b]	Moderate	Low
Product properties:		
Pyrolysis products in gas	Possibly	No
Nature of ash produced	Dry	Slag
Carbon conversion	<97%	>99%
Cold gas efficiency	89	80
Application	Medium scale	Large scale

[a]Depends on the type of feed (slurry-feed 20% and dry feed 40%).
[b]Depends on the feed properties and the H_2/CO requirement of the syngas.

3.5.2 Entrained Flow Gasifiers

EFGs are the most successful and most widely used type of reactor for large-scale gasification of coal. Technically, they are also a suitable option for biomass gasification, as long as the feedstock meets certain criteria in terms of particle size, moisture and mineral content (see Table 3.4).

In EFGs, the fuel is injected and partially oxidized in one or more burners, *i.e.* combustion takes place at the entry point, and the gasification of solids and decomposition of tars and other hydrocarbons occurs further downstream, within a few seconds (0.5 to 4.0 seconds).[17,58] For this reason, typical fuels consist of a dry pulverized solid, an atomized liquid fuel or fuel slurry. This can be challenging in terms of biomass feedstocks, as grinding fibrous biomass into such fine particles is not a trivial task. Torrefaction is usually required for this type of gasification to reduce the bulk density and moisture content of the biomass.[59] Figure 3.6 provides a schematic description of a top-fed EFG.

EFGs tend to operate in slagging mode; therefore, they use oxygen rather than air to keep the operating temperature well above the melting point of the biomass ash. Typical temperatures for entrained flow gasification are in the range of 1200–1600 °C. As a result, complete destruction of tars, low methane production and high carbon conversion (close to 100%) are usually attained.[58] The product gas can also be used to produce superheated steam. Biomass with a high content of CaO and a low content of alkali metals (Na, K) results in ash with a high melting point. On the other hand, the melting point of biomass ash with a high alkali content is much lower.[17]

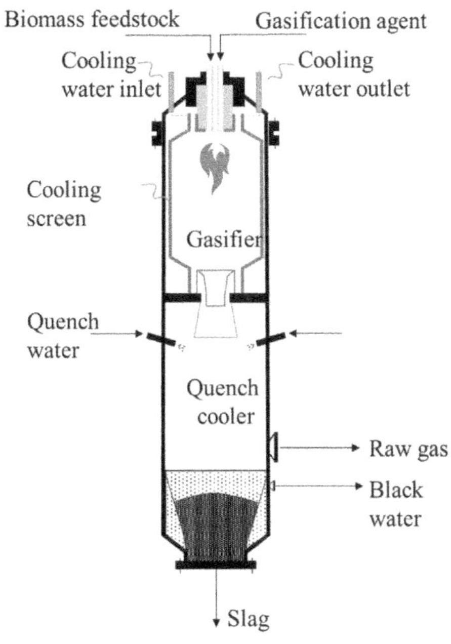

Figure 3.6 Schematic description of a top-fed EFG.

This means that the operating temperature and the oxygen required to reach the required temperature relate to the inorganic components present in the biomass. However, the oxygen requirement for EFGs is higher than for other reactors. It is also worth mentioning that molten biomass ash can be highly corrosive, which affects the life of the gasifier refractory lining.

Besides the high temperature, EFGs are also operated at a high pressure (1–3 MPa for solids and higher pressures, *ca.* 20–70 MPa, for high-energy liquids).[52] These operating conditions, the complexity of the fuel preparation process and the corrosive nature of the molten biomass ash require a more sophisticated reactor design and the use of more sophisticated construction materials.[51] Furthermore, the need to use oxygen is not typically feasible for small facilities due to the high cost. Entrained flow gasification of biomass has not reached commercial scale yet.

Table 3.4 compares the main characteristics of fluidized-bed and EFGs. Similarly, Table 3.5 compares the typical syngas composition from biomass gasification in these types of reactors.

It is worth mentioning that with Fischer–Tropsch (FT) applications, once gasification and gas cleaning have taken place, the H_2/CO ratio of the syngas is typically adjusted using a WGS reactor. This ratio is an important parameter for the FTS step.

3.5.3 Impact of Biomass Feedstock Characteristics and Gasification Operating Parameters on Biomass Gasification

The quality and yield of the syngas product obtained from biomass gasification depend on the gasifier type, the feedstock characteristics and the operating conditions. Feedstock characteristics such as elemental composition, moisture and mineral content are mainly defined by the biomass type and its processing and can lead to differences in the syngas composition and gasification efficiency. Similarly, operating parameters such as temperature, pressure and residence time affect the gasification process and

Table 3.5 Typical syngas compositions from different types of gasifiers. Data from ref. 60.

	Fluidized-bed	Entrained flow
Feedstock	Biomass	Biomass
Oxidizing agent	Steam	Oxygen
Syngas composition (vol%, dry basis)	—	—
Carbon monoxide	25–30	45–55
Carbon dioxide	20–25	10–15
Hydrogen	35–40	23–28
Methane	9–11	0–1
Nitrogen	0–5	0–1
Syngas LHV (MJ N m^{-3})	12–14	10–12

its efficiency. An overview of the main parameters and their influence on biomass gasification is provided in the following section.

3.5.3.1 Feedstock Characteristics

3.5.3.1.1 Biomass Type. When discussing the effect of biomass type on biomass gasification, it is practical to distinguish between lignocellulosic biomass and other biomass types. "Lignocellulosic biomass" refers to woody biomass and other material derived from plants. Examples of other biomass types include sewage sludge, fats and biogenic mixed waste streams. On a dry basis, lignocellulosic woody biomass feedstocks show a similar elemental composition (see Table 3.1). This suggests that if they are processed in a similar manner and have moisture content and particle size in a similar range, then gasification of this type of biomass would yield syngas with a similar composition and heating value. Table 3.6 shows that gasification of hardwood chips, softwood pellets and a mixture of straw/wood pellets

Table 3.6 Feedstock, gasification and product gas parameters for the gasification of different biomass feeds in an air-/steam-fired DFB pilot plant. Data from ref. 61.

	Hardwood chips	Softwood pellets	40% straw 60% wood pellets	Sewage sludge pellets
	Feedstock			
C_{dry} (wt%)	50.2	48.8	49.9	29.7
H_{dry} (wt%)	6.0	5.9	5.7	3.7
O_{dry} (wt%)	43.4	44.1	42.6	20.2
N_{dry} (wt%)	0.1	0.2	0.3	3.9
Ash $(minerals)_{dry}$ (wt%)	0.3	1.0	1.6	41.5
Water (wt%)	6.1	5.7	10.5	11
LHV ($MJ\,kg^{-1}$)	17.5	17	16.2	10.8
	Gasification			
Gasification temperature (°C)	802	807	800	810
Fuel to gasifier fuel to combustor	4.2	3.4	3.0	2.1
Steam to fuel ratio ($kg_{H_2O}\,kg^{-1}{}_{fuel,dry}$)	0.8	1.0	1.1	0.9
	Product gas			
$H_{2,dry}$ (vol%)	35.3	34	38.5	39.2
CO_{dry} (vol%)	29.1	22.4	19.6	16.3
$CO_{2,dry}$ (vol%)	17.3	25.2	23.1	25.4
$CH_{4,dry}$ (vol%)	11.1	11.6	9.9	7.8
Other hydrocarbons, H_2S and $NH_{3\ dry}$ (vol%)	7.2	6.8	8.9	11.3
H_2O (vol%)	40–55			
LHV ($MJ\,m^{-3}$)	14.0	13.4	12.4	11.8
Cold gas efficiency (%)	63	67	73	79
Overall gas efficiency estimated for 100 MW plant (%)	70–75			

(lignocellulosic materials), in a DFB gasifier, using similar operating conditions, yielded syngas with a low heating value of ∼17 MJ kg^{-1} in all cases.

Similarly, the overall gasification efficiency was estimated to be close to 70–75%. Another study done on the gasification of softwood, bagasse, olive pomace, bark and rice husks in a DFB gasifier confirmed that all these feedstocks show comparable gasification behavior.[62] This is also valid for other gasifier types, such as the oxygen-blown CFB gasifier[63] and EFG,[64] when the feedstock water content is similar. The amount of oxygen required for complete conversion in an EFG depends on the biomass feedstock, which can affect the efficiency and economics of the gasification process.[65]

The composition of other biomass feedstocks like biogenic municipal waste or sewage sludge is significantly different from each other and also different from that of lignocellulosic biomass (see Table 3.2). Therefore, general trends are difficult to obtain. Still, it has been proposed that the overall product gas yield and the lower heating value of the product gas are correlated to the mass fraction of volatile matter in the feedstock in a DFB. Higher mass fractions of volatiles are beneficial for both parameters.[62] Table 3.6 suggests that similar overall process efficiencies can be obtained for lignocellulosic and sewage sludge feedstocks. However, high ash content (inerts) in biogenic waste feedstock can lead to practical issues that need to be addressed.

3.5.3.1.2 Moisture Content. The moisture content of biomass feedstocks largely determines the extent of drying required before gasification. Since biomass typically contains a high moisture content (see Section 3.2), partial drying of biomass before gasification is crucial to ensure stable operation of the gasifier.

In order to obtain the required temperature for gasification and to supply the required energy for the endothermic gasification reactions, part of the biomass is oxidized in the gasifier or in the combustor in a DFB gasifier. The water present in the biomass needs to be heated in the gasifier along with the other material. This increases the demand for oxidant and leads to an increased consumption of biomass (*i.e.* fuel), both of which are detrimental to the overall thermal efficiency of the process. Hence, low moisture content is favorable for gasification.[66]

Theoretical analysis of the overall process of drying and gasification has shown that drying biomass feedstock to <20 wt% moisture content strongly increases the overall efficiency of the process (see Figure 3.7). This has been confirmed experimentally for different gasifier types.[67] Simulations suggest that during biomass gasification in a DFB gasifier, it becomes difficult to maintain the desired gasifier temperature with an increase in moisture content of more than 15 wt%.[68] It is important to recognize that some residual moisture in the biomass feedstock can be beneficial to avoid tar formation and increase the H_2 content of the syngas product.[67]

3.5.3.1.3 Mineral Content. The mineral content of biomass can vary strongly, depending on the origin of the feedstock, and the minerals affect

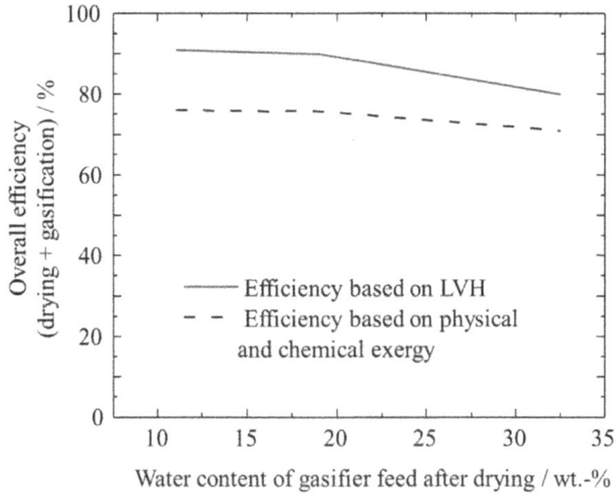

Figure 3.7 Drying of biomass at 100 °C before gasification. Effect of the extent of drying (water content of the feedstock to the gasifier) on overall gasification efficiency. Data from ref. 69.

the design and operation of a gasifier. Although they do not contribute directly to syngas formation, they need to be removed from the gasifier, either as dry ash (nonslagging gasifier) or as molten slag (slagging gasifier). Therefore, it is important to take the mineral content of biomass into consideration when designing the gasification process. This is well established for coal gasification;[38] however, there are some specific challenges with biomass gasification.

Generally, lignocellulosic feedstocks contain lower mass fractions of minerals than coal.[70] However, the higher mass fraction of alkali metals in lignocellulosic feedstocks compared to coal leads to a lower melting point of the ash formed in the gasifier. This has practical implications since with fluidized-bed reactors, softening of the ash needs to be avoided or the fluidization medium needs to be selected carefully, so as to avoid agglomeration of the fluidization medium.[71] The high alkali metal content also leads to a more reactive slag in EFG systems, which can be problematic for the refractory material.[72] The reactive nature of the minerals also has consequences for the gasification reactions, some of which can be catalyzed by the minerals.[71,73,74] The mineral content in waste biomass can be considerably higher (*e.g.* sewage sludge, as per Table 3.3), which lowers the heating value of the fuel and therefore increases fuel consumption.

3.5.3.1.4 Particle Size. Both fluidized-bed and entrained flow gasification require a specific particle size range to ensure proper operation of the gasifier (see Table 3.4). Therefore, particle size reduction is typically carried out before gasification (see Section 3.1). However, even within the particle size boundaries given for the particular type of gasifier, particle size can affect the efficiency of the gasifier. The reason is that the reaction

Table 3.7 Impact of particle size on the gas yield, gas LHV and carbon conversion rate during steam and air gasification of pine sawdust in a BFB at 800 °C. Data from ref. 76.

Biomass particle size	(mm)	0.6–0.9	0.45–0.6	0.3–0.45	0.2–0.3
Gas yield	($m^3 kg^{-1}$)	1.53	1.93	2.37	2.57
Gas LHV	($MJ m^{-3}$)	6.98	7.94	8.71	8.74
Carbon conversion	(%)	78	84	91	95

rate of the pyrolysis and gasification reactions are limited by mass and heat transfer effects rather than by chemical kinetics, even for the lower temperature fluidized-bed reactors (see Section 3.4).[75] Smaller particles offer a higher surface-to-volume ratio and consequently a higher mass and heat transfer rate, which is typically beneficial for the gasification process. This can be seen in Table 3.7 for a fluidized-bed reactor. With a smaller particle size, a higher portion of the biomass feedstock is converted to gaseous products, which leads to a higher gas yield and higher heating value of the product gas. The formation of tar is also mitigated by a smaller feedstock particle size.[75]

Simulation work has indicated that a broader biomass particle size distribution could increase the gasification efficiency in fluidized-bed gasifiers slightly. However, this is not supported experimentally.[77] For entrained flow gasification, a very small particle size (<0.1 mm) is required. Mass and heat transfer resistance are responsible for limiting the reaction rates. A smaller particle size leads to a more pronounced release of volatiles in the pyrolysis zone of the gasifier. The char that is formed is therefore more reactive, and hence the heterogeneous gasification reactions occur more rapidly, which leads to an increase in carbon conversion with a decrease in particle size.[78] The reduction in particle size is beneficial for gasification but has a cost factor that needs to be taken into account in the overall process design.

3.5.3.2 Influence of Gasification Operating Parameters

The usual ranges for the operating conditions of the different gasifier types are presented in Table 3.4. Within these ranges, certain parameters can be manipulated to influence the gasification operation and syngas composition, *e.g.* temperature, ER, steam addition and residence time. Some of these parameters – such as temperature and ER – are interdependent, and the effect of the biomass composition previously discussed is superimposed, which makes the selection of operating conditions a complex decision. Nevertheless, the general effects of the operating parameters are discussed in the following subsections.

3.5.3.2.1 Temperature. Temperature affects the kinetics and thermodynamics of a gasification process. A higher temperature leads to a higher mass transfer and reaction rate. The equilibrium constants of the homogeneous gasification reactions, like the WGS reaction and the reforming reactions are also determined by temperature (see Section 3.4). Thus the temperature of the gasification process has a strong effect on the final syngas composition. Since a change in temperature is typically achieved by adding more oxidizing agent, the temperature and the equivalence ratio (ER) are interdependent.

With the fluidized-bed process, the upper limit for the temperature is set according to the softening point of the ash to prevent agglomeration. Within this temperature limit, it is generally reported that an increase in temperature could lead to improved cold gas efficiency and a higher heating value for the syngas.[79] The char and tar conversion rate increases at a higher temperature.[80–82] The equilibrium of the WGS reaction is determined by the temperature, which influences the ratio of H_2/CO in the syngas.[83] The temperature in EFGs needs to be kept above the slagging temperature of the ash for slag removal. The upper temperature limit is determined by the refractory material. From an energy efficiency perspective, a lower temperature is desired to minimize the requirements for an oxidizing agent. However, CH_4 increasingly converts to CO and H_2 *via* steam reforming with an increase in temperature, and this contributes to an increase in the syngas yield.[64]

Since most of the gasification processes used to produce syngas for FTS are run autothermally and use oxygen as an oxidizing agent, a higher temperature is obtained by adding more oxygen to the feed. Hence, the ER is increased. The effect of the ER is discussed in detail in the following section.

3.5.3.2.2 Equivalence Ratio. Figure 3.8 shows the flow rate of oxidant relative to the flow rate of oxidant required for complete oxidation of the fuel to CO_2. It is one of the most flexible aspects of using a gasifier, and the choice

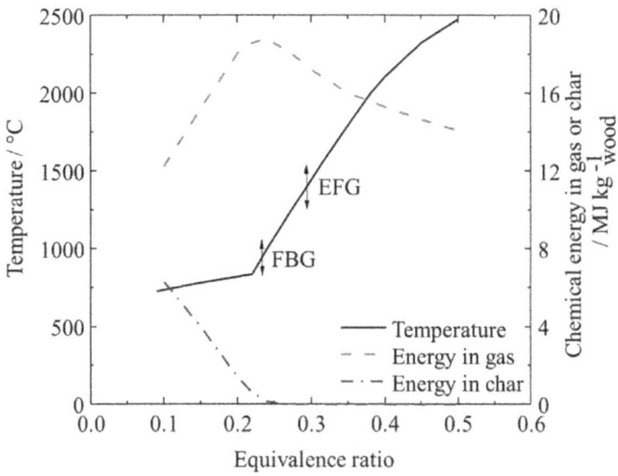

Figure 3.8 Adiabatic temperature and energy content of the gas and char product at thermodynamic equilibrium during gasification of dry wood (LHV = 20.9 MJ kg^{-1}) with oxygen as a function of the ER. The typical operation temperature range of fluidized-bed gasifiers (FBG) and EFGs are indicated with arrows. Data from ref. 84.

of ER affects the temperature in the gasifier and the syngas composition, as well as other factors. An increase in ER leads to more pronounced combustion of the fuel (biomass) and consequently to an increase in temperature. At a low ER, some of the chemical energy of the fuel is retained in the solid phase as char. With an increase in ER, the char is converted, and the amount of chemical energy in the gas phase product goes through a maximum. At an even higher ER, the amount of chemical energy in the gas phase decreases, since part of the energy is converted to sensible heat.

In Figure 3.8, the maximum chemical energy in the gas phase product is found at an ER of 0.225. In real gasifiers, heat loss and biomass composition (water, ash, *etc.*) shift this maximum. Industrial gasifiers operate in an ER range of 0.2–0.4.[17] With fluidized-bed gasification, there is a temperature window in which the biomass ash softens, and this needs to be avoided. In the case of entrained flow gasification of biomass, it has been proposed that adding an ash fluxing agent to the feedstock could enable the gasifier to operate at a slightly lower ER, which can increase the cold gas efficiency rate.[85]

3.5.3.2.3 Steam Addition. Steam or water is added to the gasifier in coal gasification to increase the hydrogen yield in the syngas. Increasing levels of water shift the equilibrium of the WGS reaction (see Table 3.3), and this generates more hydrogen.[64] Adding water leads to an overall lower temperature in the gasifier due to the increased thermal inertia in the reactor, and a higher ER is then required to reach the same temperature than was required without water being added to the gasifier.

With fluidized-bed reactors that operate at a relatively low temperature, adding water has the advantage of decreasing the formation of tars and increasing the carbon conversion.[63,81] However, gasification with pure steam leads to an increase in tar formation compared to gasification in air or oxygen-steam mixtures.[86]

3.5.3.2.4 Pressure. The pressure of a gasification process is typically determined by the downstream applications. The FTS process operates at approximately 2 MPa; therefore gasification would be operated at 2.5–3.0 MPa, to consider the drop in pressure afterward. A benefit of pressurized gasification is the lower volume required for a similar residence time. Gasification reactors have a high capital cost, and saving reactor volume can be economically beneficial. However, the system for feeding solid biomass to the gasifier also becomes more complex with an increase in system pressure. Lock hoppers and piston feeding systems have been used for feeding biomass to gasifiers.[67]

An increase in pressure shifts the methane reforming and heterogeneous gasification reaction equilibrium toward a lower CO and H_2 generation rate in fluidized-bed gasifiers;[87,88] however, this thermodynamic effect is less pronounced in EFGs.[85] Tar formation is favored by an increased pressure;[89] therefore, other tar mitigation strategies (such as a high temperature) need to be employed, or downstream tar removal measures need to be designed.

3.5.3.2.5 Residence Time. Generally, the residence time needs to be sufficient to ensure almost complete conversion of the feedstock, but this also depends on the temperature and the properties of the biomass feedstock. Longer residence time in the gasifier leads to a closer approach of thermodynamic equilibrium conditions. This typically means a higher syngas yield and a lower tar yield in fluidized-bed reactors.[68] However, in practice, too long a residence time can lead to operational problems and heat loss and consequently to a decrease in the efficiency of the gasifier.[90] Similarly, increasing residence time leads to higher conversion and increased efficiency in EFGs.[78,91] However, with the same thermal load of the gasifier, increasing residence time requires a higher reactor volume, which increases the cost of the reactor.

3.5.4 Operational Considerations

Gasification is carried out at a high temperature and typically at an elevated pressure, both of which are challenging in terms of designing and operating the gasifier. Engineered solutions for challenges like pressurized feed introduction, temperature measurement, heat management, slag removal and syngas cooling have been developed and have proven to be effective in practice. These technologies have also been reviewed in the literature.[38] However, gasification of biomass has to address several other specific challenges. Two of these are discussed in the following sections, *i.e.* the stability of the refractory material and the removal of tars.

3.5.4.1 Material Stability

The high temperature in a gasifier is challenging in terms of the insulation material. Particularly with EFGs, the insulation material is exposed to molten slag at temperatures of >1000 °C. Two types of insulation material are commonly used, *i.e.* either a refractory lining or a cooling screen that is cooled by water.[38] The surface of the cooling screen needs to be protected by ceramic material that can tolerate a high temperature. A layer of molten or frozen slag covers the ceramic material and acts as an insulating layer that protects the ceramic material. The challenges with biomass gasifiers are that the slag is more reactive than the slag in coal gasification and that the total amount of slag formed is lower than in coal gasification. This poses a challenge to the material's stability.[72,74,92] While no data on commercial biomass EFGs are available, operating an EFG with biomass feedstock over a long period of time has been successfully demonstrated, without evidencing any issues with the insulation material.[39,93]

3.5.4.2 Tar Removal

Tar formation during biomass gasification is a major issue, since condensed tars can lead to fouling of the downstream equipment. The most effective ways to remove tar are the primary tar mitigation strategies, *i.e.* methods that aim to reduce tars in the gasifier; these include a suitable choice of operating parameters, as well as proper selection of bed materials that catalyze tar-cracking in fluidized-bed reactions.[39] Since tar formation cannot be avoided completely using a primary strategy, a secondary tar removal step is also necessary. These strategies are also often used to remove other solid particles in the product gas. Secondary tar removal strategies include catalytically coated filters and water or oil scrubbers.[94,95] These are part of the syngas cleaning and conditioning process that is required downstream of the gasifier.

3.6 Current and Future Trends for Biomass Gasification

Gasification is a flexible and mature technology that has gained increasing attention for biomass conversion in recent years. Extensive discussions on its technical feasibility and environmental benefits have already been produced and are readily available. However, the extent of biomass gasification has largely been limited to small-scale CHP applications as well as some co-firing applications that are driven by regional and local regulations (or considerations) on climate and economic development.[96,97] Currently, at a global level, thousands of small- to medium-scale gasifiers in the range of 0.02–20 MWth are being used for power and CHP applications. There are also tens of stationary and CFB gasifiers in the range of 10–140 MWth, which use biomass residue and waste to produce syngas for use as fuel in cement kilns and for co-firing with coal in power plants.[96]

However, biomass gasification for advanced applications, such as the production of transportation fuels, has been hampered by competition with fossil fuels, policy uncertainty and insufficient incentives to stimulate strong partnerships among developers, industry and other stakeholders.[96,97] Technology development goes through laboratory, bench, pilot and demonstration phases before reaching the point of commercial implementation. Gasification projects for the production of transportation fuels from biomass feedstock require large-scale facilities; therefore, the last steps in the technology development process entail significant technical and economic risks, which are difficult to overcome with the current level of support and market systems. While there is a good foundation due to the available coal gasification technologies, modifying the existing technology and understanding the different challenges due to the different characteristics of the feedstock materials are slow in coming and not always easy.

Although a number of large-scale biomass gasification facilities have been closed for economic reasons, a number of projects are still currently operational or under construction (see Table 3.8).

Even though there is a trend toward the electrification of road transport, using syngas from biomass gasification for the production of fuel products – and particularly aviation (jet) fuel – seems to be promising. Jet fuel produced *via* biomass gasification in combination with FTS could potentially ensure a greater reduction in greenhouse gas than competing pathways.[99]

Table 3.8 Gasification projects for the production of transportation fuels from biomass feedstock. Data from ref. 98.

Project name	Product	Feedstock	Gasification technology	Location	Status	Start-up	Scale
LLC thermal reformer synthesis	FT liquids	Forest residues, wood	DFB	U.S.	Operational	2007	Demo
Bioliq	Dimethyl ether[a]	Slurry from lignocellulosic crops, straw[d]	EFG	DE	Operational	2012	Pilot
Red Rock biofuels	FT liquids[bc]	Forestry residues	Pyrolysis + steam reforming	U.S.	Under construction	2021	Commercial
Fulcrum bioenergy	FT liquids	Waste	Indirect fluidized bed	U.S.	Under construction	2021	Commercial
BioTfuel demo	FT liquids[c]	Torrefied biomass[d]	EFG (PRENFLO)	FR	Under commissioning	—	Demo

[a]Further processed to gasoline.
[b]Diesel.
[c]Jet fuel.
[d]Densified biomass-derived feedstock.

This indicates the potential for this technology to reach economic viability and contribute to greenhouse gas reduction in the future.

It is worth mentioning that, in recent years, there has been a change in the preferred type of feedstock for biomass gasification. Some facilities are no longer focused on the use of clean woody biomass due to the annual increases in price; thus the use of waste materials is gaining significant attention, *e.g.* sewage sludge, plastics and agricultural residue.[100] Densification before gasification *via* a pretreatment step, such as pyrolysis, hydrothermal liquefaction or torrefaction, can offer some benefits, including easier transport and homogenization of the waste feedstock. This is discussed in more detail in Chapter 4. Similarly, the combination of thermal gasification with other renewable energies, *e.g.* wind power and photovoltaic (PV), brings new possibilities for electric grid balancing and/or energy storage.[97]

Acronyms

ASU	Air separation unit
BFB	Bubbling fluidized-bed
CFB	Circulating fluidized-bed
DFB	Dual fluidized-bed
EFG	Entrained flow gasifier
ER	Equivalence ratio
FT	Fischer–Tropsch
FTS	Fischer–Tropsch synthesis
HHV	High heating value
SSD	Superheated steam dryers
WGS	Water-gas shift

References

1. A. B. M. Abdul Malek, M. Hasanuzzaman, N. A. Rahim and Y. A. Al–Turki, *Energy Environ.*, 2021, **32**, 295.
2. U.S. Energy Information Administration, September 2019, https://www.eia.gov/outlooks/ieo/pdf/ieo2019.pdf (Accessed April 2021).
3. W. de Jong and J. Ruud van Ommen, *Biomass as a sustainable energy source for the future: Fundamentals of conversion processes*, John Wiley & Sons, Hoboken, New Jersey, 2015.
4. International Renewable Energy Agency (IRENA), March 2020, https://www.irena.org/-/media/Files/IRENA/Agency/Publication/2020/Mar/IRENA_RE_Capacity_Highlights_2020.pdf (Accessed April 2021).
5. BP Statistical Review of World Energy 2020, https://www.bp.com/content/dam/bp/business-sites/en/global/corporate/pdfs/energy-economics/statistical-review/bp-stats-review-2020-full-report.pdf (Accessed April 2021).
6. Y. M. Bar-On, R. Phillips and R. Milo, *Proc. Natl. Acad. Sci. U. S. A.*, 2018, **115**, 6506.

7. J. S. Tumuluru, C. T. Wright and J. R. Hess, *Biofuels, Bioprod. Biorefin.*, 2011, **5**, 683.
8. H. Yang and H. Chen, in *Woodhead Publishing Series in Energy, Gasification for Synthetic Fuel Production*, ed. R. Luque and J. G. Speight, Elsevier, Amsterdam, The Netherlands, 2015, pp. 241–275.
9. M. Lippe, I. Lewandowski and R. Unseld, in *The Origin of Biomass, Bioeconomy for Beginners*, ed. J. Pietzsch, Springer, Berlin, Germany, 2020, pp. 11–66.
10. E. R. Widjaya, G. Chen, L. Bowtell and C. Hills, *Renewable Sustainable Energy Rev.*, 2018, **89**, 184.
11. C. A. Gonçalves, I. Malico and M. O. Sousa, in *Renewable Resources and Biorefineries*, ed. E. Jacob-Lopez and L. Queiroz Zepka, InTechOpen, 2019, pp. 23–46.
12. R. Nosek, M. Holubcik and J. Jandacka, *BioResources*, 2016, **11**, 44.
13. Woody Plants, https://www.thespruce.com/woody-plants-meaning-examples-2131128 (Accessed April 2021).
14. J. S. Tumuluru, S. Sokhansanj, J. R. Hess, C. T. Wright and R. D. Boardman, *Ind. Biotechnol.*, 2011, 7, 384.
15. S. V. Vassilev, D. Baxter, L. K. Andersen and C. G. Vassilev, *Fuel*, 2010, **89**, 913.
16. T. A. Milne, A. H. Brennan and B. H. Glenn, *Sourcebook of Methods of Analysis for Biomass and Biomass Conversion Processes*, United Kingdom, N.P. 1990, ISBN 1-85166-527-7.
17. P. Basu, *Biomass Gasification, Pyrolysis and Torrefaction*, Elsevier, San Diego, USA, 3rd edn, 2018.
18. O. A. Kusi, P. Duangporn, S. Premjet and J. Pertanika, *Tropic. Agri. Sci.*, 2018, **41**, 19.
19. J. Moskalik, J. Skvaril, O. Stelcl, M. Balas and M. Lisy, *Polytechnica*, 2012, **52**, 77.
20. S. Adhikari, H. Nam and J. P. Chakraborty, in *Waste Biorefinery*, ed. T. Bhaskar, A. Pandey, S. V. Mohan, D. J. Lee and S. K. Khanal, Elsevier, The Netherlands, 2018, pp. 239–268.
21. P. J. de Wild, in *Industrial Biorefineries and White Biotechnology*, ed. A. Pandey, R. Hofer and C. Larroche, Elsevier, The Netherlands, 2015, pp. 341–368.
22. P. Halder and A. K. Azad, in *Advanced Biofuels – Applications, Technologies and Environmental Sustainability*, ed. A. K. Azad and M. Rasul, Woodhead Publishing, Elsevier, 2019, pp. 167–179.
23. S. J. Tumuluru and D. J. Heikkila, *Bioengineering*, 2019, **6**, 12.
24. M. N. Aftab, in *Biomass for Energy – Recent Trends and Future Challenges*, ed. A. E. Abomohra, Intech Open Limited, United Kingdom, 2019, 15–22.
25. G. Moiceanu, G. Paraschiv, G. Voicu, M. Dinca, O. Negoita, M. Chitoiu and P. Tudor, *Sustainability*, 2019, **11**, 2477.
26. K. V. Probst, R. P. Kingsly Ambrose, R. L. Pinto, R. Bali, P. Krishnakumar and K. E. Ileleji, *Trans. Am. Soc. Agric. Biol. Eng.*, 2013, **56**, 1025.

27. Report on Biomass rying Technology 1998, https://www.nrel.gov/docs/fy99osti/25885.pdf (Accessed April 2021).
28. H. T. Luk, T. Y. G. Lam, A. O. Oyedun, T. Gebreegziabher and C. W. Hui, *Energy*, 2013, **63**, 205.
29. Report on Biomass Drying and Dewatering for Clean Heat & Power 2008, http://northwestchptap.org/NwChpDocs/BiomassDryingAndDewateringForCleanHeatAndPower.pdf(AccessedApril2021).
30. W. Stelte, A. R. Sanadi, L. Shang, J. K. Holm, J. Ahrenfeldt and U. B. Henriksen, *BioResources*, 2012, **7**, 4451.
31. I. N. Manickam and D. Subramanian, *Cogener. Distrib. Genera. J.*, 2006, **21**, 33.
32. J. S. Tumuluru, *ASABE Annual International Meeting*, Pittsburgh, USA, 2010.
33. M. J. Darr and A. Shah, *Biofuels*, 2012, **3**, 321.
34. L. G. Tabil and S. Sokhansanj, *Powder Handl. Process.*, 1996, **8**, 117.
35. A. Garcia-Maraver and J. Perez-Jimenez, Biomass Pelletization, *Wit Pr/Computational Mechanics*, United Kingdom, 2015.
36. A. Kaur, M. Roy and K. Kundu, *Int. J. Recent Sci. Res.*, 2017, **8**, 20561.
37. S. Y. Kpalo, M. F. Zainuddin, L. A. Manaf and A. M. Roslan, *Sustainability*, 2020, **12**, 4609.
38. C. Higman and M. van der Burgt, *Gasification, Gulf Professional Publishing*, Elsevier, 2008.
39. L. Devi, K. J. Ptasinski and F. J. J. G. Janssen, *Biomass Bioenergy*, 2003, **24**, 125.
40. M. Balat, *Energy Sources, Part A*, 2008, **30**, 636.
41. A. Demirbaş, *Energy Sources*, 2005, **27**, 339.
42. S. Şensöz and M. Can, *Energy Sources*, 2002, **24**, 347.
43. A. Gómez-Barea, B. Leckner, D. Santana and P. Ollero, *Chem. Eng. J.*, 2008, **141**, 151.
44. P. L. Walker, F. Rusinko and L. G. Austin, *Adv. Catal.*, 1959, **11**, 133.
45. C. Di Blasi, *Prog. Energy Combust. Sci.*, 2009, **35**, 121.
46. W. F. de Groot and G. N. Richards, *Carbon*, 1989, **27**, 247.
47. Y. Yang, PhD Thesis, Gasification Kinetics of Western Canadian Coals. University of British Columbia, 1991.
48. K. Fürsatz, J. Fuchs, F. Benedikt, M. Kuba and H. Hofbauer, *Energy*, 2021, **219**, 119650.
49. K. Umeki, A. Moilanen, A. Gomez-Barea and J. Konttinen, *Chem. Eng. J.*, 2012, **207–208**, 616.
50. T. Kan and V. Strezov, in *Biomass Processing Technologies*, ed. V. Strezov and T. J. Evans, CRC Press, Boca Raton, FL, 2015, ch. 4, pp. 81–121.
51. R. Luque and J. Speight, *Gasification and Liquid Fuel Production: Fundamentals, Processes, and Applications*, Woodhead Publishing, Cambridge, 2015.
52. Swedish Gasification Centre, http://www.sfc-sweden.se/ (Accessed February 2021).

53. IEA Bioenergy Annual Report 2019, https://www.ieabioenergy.com/wp-content/uploads/2020/05/IEA-Bioenergy-Annual-Report-2019.pdf (Accessed February 2021).
54. A. Molino, S. Chianese and D. Musmarra, *J. Energy Chem.*, 2016, **25**, 10.
55. Y. Richardson, M. Drobek, A. Julbe, J. Blin and F. Pinta, in *Recent Advances in Thermochemical Conversion of Biomass*, ed. A. Pandey, T. Bhaskar, M. Stocker and R. K. Sukumaran, Elsevier, Amsterdam, 2015, ch. 8, pp. 213–250.
56. K. Göransson, U. Söderlind, J. He and W. Zhang, *Renewable Sustainable Energy Rev.*, 2011, **15**, 482.
57. J. Corella, J. M. Toledo and G. Molina, *Ind. Eng. Chem. Res.*, 2007, **46**, 6831.
58. J. G. Speight, *Handbook of Gasification Technology: Science, Processes, and Processes*, Wiley-Scrivener, Hoboken, NJ, 2020.
59. N. Escalona, R. Garcia and P. Reyes, in *Advanced Biofuels and Bioproducts*, ed. J. W. Lee, Springer, New York, NY, 2013, ch. 12, vol. 1, pp. 185–209.
60. International Energy Agency, April 2016, https://www.ieabioenergy.com/wp-content/uploads/2017/12/Status-report-Task-33-2016.pdf (AccessedApril2021).
61. J. C. Schmid, U. Wolfesberger, S. Koppatz, C. Pfeifer and H. Hofbauer, *Environ. Prog. Sustainable Energy*, 2012, **31**, 205.
62. F. Benedikt, J. C. Schmid, J. Fuchs, A. M. Mauerhofer, S. Müller and H. Hofbauer, *Energy*, 2018, **164**, 329.
63. M. Siedlecki and W. de Jong, *Biomass Bioenergy*, 2011, **35**, S40.
64. K. Qin, W. Lin, P. A. Jensen and A. D. Jensen, *Fuel*, 2012, **93**, 589–600.
65. M. Kremling, L. Briesemeister, M. Gaderer, S. Fendt and H. Spliethoff, *Energy Fuels*, 2017, **31**, 3949.
66. H. Ghassemi and R. Shahsavan-Markadeh, *Energy Convers. Manage.*, 2014, **79**, 18.
67. I. L. Motta, N. T. Miranda, R. Maciel Filho and M. R. Wolf Maciel, *Renewable Sustainable Energy Rev.*, 2018, **94**, 998.
68. J. He, K. Göransson, U. Söderlind and W. Zhang, *Biomass Convers. Biorefin.*, 2012, **2**, 1.
69. K. J. Ptasinski, M. J. Prins and A. Pierik, *Energy*, 2007, **32**, 568.
70. D. Mallick, P. Mahanta and V. S. Moholkar, *Fuel*, 2017, **204**, 106.
71. M. Zevenhoven-Onderwater, R. Backman, B. Skrifvars and M. Hupa, *Fuel*, 2001, **80**, 1489.
72. M. Carlborg, F. Weiland, C. Ma, R. Backman, I. Landälv and H. Wiinikka, *J. Eur. Ceram. Soc.*, 2018, **38**, 777.
73. J. J. Hernández, G. Aranda-Almansa and C. Serrano, *Energy Fuels*, 2010, **24**, 2479.
74. P. Carlsson, C. Ma, R. Molinder, F. Weiland, H. Wiinikka, M. Öhman and O. Öhrman, *Energy Fuels*, 2014, **28**, 6941.

75. K. R. Gaston, M. W. Jarvis, P. Pepiot, K. M. Smith, W. J. Frederick and M. R. Nimlos, *Energy Fuels*, 2011, **25**, 3747.
76. P. M. Lv, Z. H. Xiong, J. Chang, J. C. Z. Wu, Y. Chen and J. X. Zhu, *Bioresour. Technol.*, 2004, **95**, 95.
77. T. Zhou, S. Yang, Y. Wei, J. Hu and H. Wang, *Renewable Energy*, 2020, **148**, 534.
78. J. J. Hernández, G. Aranda-Almansa and A. Bula, *Fuel Process. Technol.*, 2010, **91**, 681.
79. X. Meng, W. De Jong, N. Fu and A. H. M. Verkooijen, *Biomass Bioenergy*, 2011, **35**, 2910.
80. H. Chen, B. Li, H. Yang, G. Yang and S. Zhang, *Energy Fuels*, 2008, **22**, 3493.
81. B. Hejazi, J. R. Grace, X. Bi and A. Mahecha-Botero, *Energy Fuels*, 2017, **31**, 12141.
82. F. Kirnbauer, V. Wilk and H. Hofbauer, *Fuel*, 2013, **108**, 534.
83. A. Aghaalikhani, J. C. Schmid, D. Borello, J. Fuchs, F. Benedikt, H. Hofbauer, F. Rispoli, U. B. Henriksen, Z. Sárossy and L. Cedola, *Renewable Energy*, 2019, **143**, 703.
84. R. Desrosiers, in *A Survey of Biomass Gasification*, ed. T. B. Reed, Solar Energy Research Institute, Colorado, 1979, vol. 2, pp. 133–176.
85. F. Weiland, H. Wiinikka, H. Hedman, J. Wennebro, E. Pettersson and R. Gebart, *Fuel*, 2015, **153**, 510.
86. J. Gil, J. Corella, M. P. Aznar and M. A. Caballero, *Biomass Bioenergy*, 1999, **17**, 389.
87. C. Berrueco, D. Montané, B. Matas Güell and G. del Alamo, *Energy*, 2014, **66**, 849.
88. F. Feng, G. Song, L. Shen and J. Xiao, in *Simulation of Bio-syngas Production from Biomass Gasification via Pressurized Interconnected Fluidized Beds*, ed. T. Zhang, P. Ouyang, S. Kaplan and B. Skarnes, Proceedings of the 2012 International Conference on Applied Biotechnology (ICAB 2012), Springer, Berlin, 2013, pp. 1145–1155.
89. M. Mayerhofer, P. Mitsakis, X. Meng, W. de Jong, H. Spliethoff and M. Gaderer, *Fuel*, 2012, **99**, 204.
90. X. Fan, L. Yang and J. Jiang, *Renewable Energy*, 2020, **158**, 32.
91. J. Zhou, Q. Chen, H. Zhao, X. Cao, Q. Mei, Z. Luo and K. Cen, *Biotechnol. Adv.*, 2009, **27**, 606.
92. H. Ji, Y. Zhou, Z. He, X. Wu and B. Dai, *Energy Fuels*, 2021, **35**, 3867.
93. S. M. Cavagnol, K. Covella and M. Müller-Hagedorn, *Fuel*, 2018, **228**, 369.
94. J. Han and H. Kim, *Renewable Sustainable Energy Rev.*, 2008, **12**, 397.
95. M. Nacken, L. Ma, S. Heidenreich and G. V. Baron, *Appl. Catal., B*, 2009, **88**, 292.
96. IEA Bioenergy – Emerging Gasification Technologies for Waste & Biomass, 2021, https://www.ieabioenergy.com/wp-content/uploads/2021/02/Emerging-Gasification-Technologies_final.pdf (Accessed April 2021).

97. EIA Bioenergy – The past, present and future of gasification, https://www.ieabioenergy.com/wp-content/uploads/2020/01/The_past_present_and_future_for_biomass_gasification_30_Jan_2020_webinar.pdf (Accessed April 2021).
98. EIA Bioenergy - Annex 3. Gasification facilities for fuel synthesis – operational, under construction, under commissioning, http://task33.ieabioenergy.com/app/webroot/files/file/publications/T33%20Projects/FS-op.pdf (Accessed April 2021).
99. S. De Jong, K. Antonissen, R. Hoefnagels, L. Lonza, M. Wang, A. Faaij and M. Junginger, *Biotechnol. Biofuels*, 2017, **10**, 1.
100. EIA Bioenergy – Status report on thermal gasification of biomass and waste 2019, http://task33.ieabioenergy.com/download.php?file=files/file/publications/T33%20Projects/Status%20report%20final.pdf (Accessed April 2021).

CHAPTER 4

Gasification of Bio-oil and Torrefied Biomass: An Overview

NATALIA MONTOYA SANCHEZ,* FELIX LINK, CIBELE MELO HALMENSCHLAGER AND GARIMA CHAUHAN

Department of Chemical and Materials Engineering, University of Alberta, 9211–116th Street, Edmonton, Alberta T6G 1H9, Canada
*Email: nmontoya@ualberta.ca

4.1 Introduction

Production of transportation fuels (*i.e.* biofuel) and chemicals from biomass and waste feedstocks has been an area of intense research and development during the last decade. The primary drivers are the continuous growth in global energy demand (and the corresponding search for security of energy supply), the political drive to shift from fossil resources to renewable feedstocks and increasingly stringent emissions standards. Therefore, significant efforts have been focused on the production of high-quality synthetic fuels from synthetic gas (syngas) *via* biomass gasification. The two major approaches involve (1) fluidized-bed gasification coupled with catalytic reforming, with both processes working at temperatures close to 900 °C, and (2) entrained flow gasification, which works at temperatures close to 1300 °C.[1] In the case of fluidized-bed gasification, technology has been developed and demonstrated, using biomass as feedstock, in small- and medium-scale combined heat and power (CHP) applications.[2] Typically, the product gas contains not only CO and H_2 (syngas) but also significant amounts of CO_2, H_2O and other hydrocarbons, such as CH_4 or tars.[3] Therefore, process development

Catalysis Series No. 44
Chemicals and Fuels from Biomass *via* Fischer–Tropsch Synthesis: A Route to Sustainability
Edited by Joshua Gorimbo, Xinying Liu, Yali Yao and Diane Hildebrandt

Published by the Royal Society of Chemistry, www.rsc.org

has focused on the catalytic reforming step. In the case of entrained flow gasification, technology has been developed at commercial scale and used for coal gasification in coal-to-liquid (CTL) processes. Thus challenges have centered on accommodating the different characteristics of biomass feedstocks through feed pretreatment or feed preparation steps. The production of bio-oil and torrefied biomass have been explored as biomass pretreatment options for entrained flow gasification.[3]

Independent of the gasification technology, the back end of the process in fuel synthesis applications (namely, Fischer–Tropsch synthesis) and the downstream operations benefit greatly from economy of scale. In other words, synthesis and upgrading steps need to be carried out at the largest feasible scale to ensure the economic viability of biofuel production.[4] It has been calculated that only large biomass-to-liquids (BTL) plants with a biofuel production capacity of at least 1 000 000 tons per year (*i.e.* at least 6 000 000 tons per year of dry lignocellulosic biomass as input) would profit from meaningful economies of scale.[5] As biomass – particularly residual biomass from forestry or agriculture – is not typically concentrated in such large quantities in a single location, the front end of BTL processes needs to accommodate an adequate supply of biomass feedstock. Transporting significant quantities of biomass, which is a low density and bulky material of low energy density (typically < 20 MJ kg^{-1})[6,7] to a central refinery increases the cost and carbon footprint of the overall process. Therefore, from a logistical point of view, there is benefit in the conversion of biomass to other more energy-dense feedstock forms, *i.e.* bio-oil or torrefied biomass. Spoke-and-hub concepts that rely on feeding a central, large-scale refinery with such energy dense feedstocks, *e.g.* bio-oil, have been proposed.[8,9] Figure 4.1 illustrates this type of concept.

This chapter centers on the front end of BTL processes and explores strategies for biomass energy densification. Therefore, the focus is on entrained-flow gasification of bio-oil and torrefied biomass. The processes and technologies available for the production of these energy dense feedstocks are discussed. Particular consideration is given to the gasification of bio-oil and torrefied biomass, including feeding systems, reaction chemistry, and ash behavior in entrained flow gasifiers (EFG). The status of this gasification technology is highlighted, as well as the opportunities and challenges for its implementation in BTL processes.

4.2 Biomass Pretreatment for Syngas Production

Thermochemical processes, such as pyrolysis, hydrothermal liquefaction (HTL) and torrefaction, transform biomass feedstocks to a homogeneous and more energy-dense product that can be used as the feed for syngas production (*i.e.* gasification). Pyrolysis and HTL processes convert biomass to bio-oil, while torrefaction converts it to a solid similar to coal. The production pathways of these materials are discussed in the following sections.

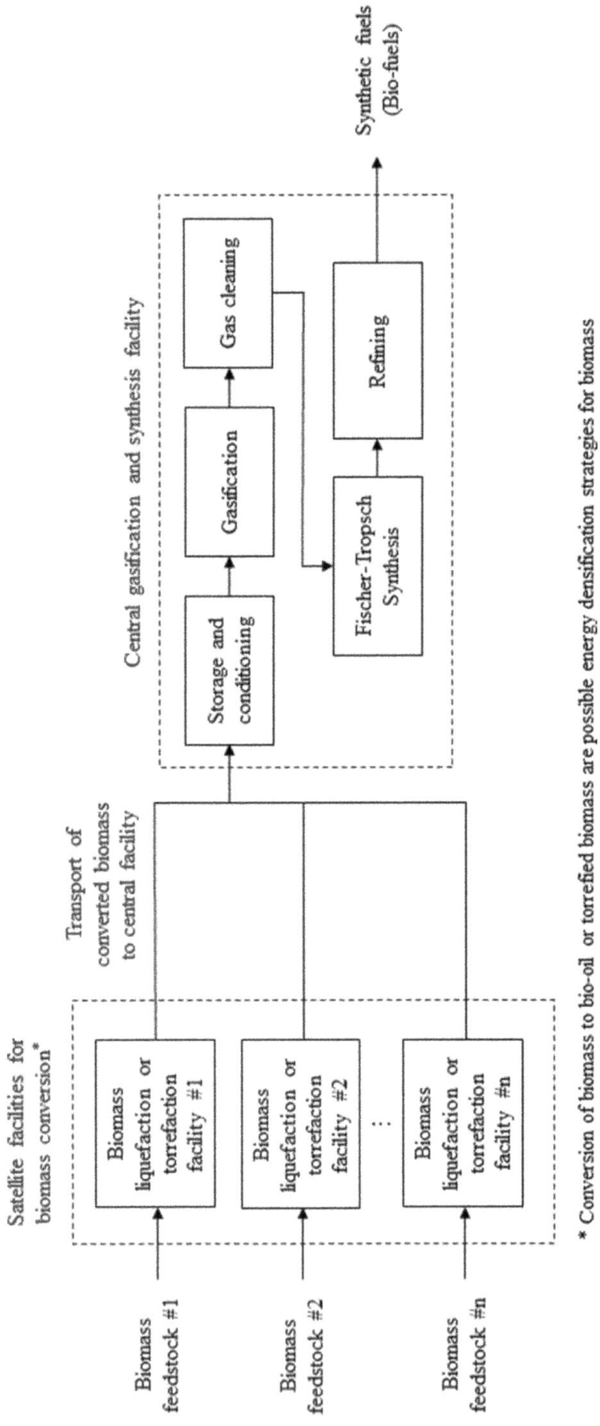

Figure 4.1 Simplified block flow diagram of a biomass-to-liquids process.

4.2.1 Production of Bio-oil

4.2.1.1 Pyrolysis

Pyrolysis refers to the thermal decomposition of biomass, in the absence of oxygen or in an oxygen-deficient environment, to produce a black tarry liquid known as pyrolysis oil, bio-oil or pyrolytic oil. Gases and char are obtained as by-products. Pyrolysis oils consist of a mixture of complex hydrocarbons including alkanes, aromatic hydrocarbons, phenol derivatives, ketones, esters, ethers, sugars, amines and alcohols.[10] Their elemental composition (on a dry basis) lies in the range of $w_C = 54–58$ wt%, $w_H = 5.5–7.0$ wt% and $w_O = 35–40$ wt%. In addition, the water content is about 15–30 wt%, and the ash content is typically up to 0.2 wt%.[11] The ash content depends on the type of feedstock used and the pretreatment performed before feeding the biomass to the pyrolysis reactor; therefore, different values are reported in literature. Pyrolysis oils are acidic in nature, and their pH value is typically in the range of 2.5–3.2. Table 4.1 presents some properties of pyrolysis oils.

The type of biomass feedstock used for pyrolysis plays an important role in the quality and yield of pyrolysis oil. Commonly used biomass feedstocks include forestry residues,[16,17] land biomass residues such as rapeseed and sunflowers,[18] sugarcane bagasse,[19] coconut shell,[20] rice husk,[21] microalgae,[22] urban waste[17] and sludge.[23,24] Biomass feedstocks vary in terms of their physicochemical characteristics, *i.e.* the composition of the major components (cellulose, hemicellulose and lignin), the elemental composition (C, H, O, N and S), the ash content, the water content, and the heating value. Detailed information on the typical ranges for these properties for different biomass feedstocks can be found in Chapter 3.

During pyrolysis, biomass is rapidly heated to ~600 °C, which leads to decomposition of the major components. The decomposition temperature for each component is different and lies in the range of 150–350 °C for hemicellulose, 275–350 °C for cellulose and 250–500 °C for lignin.[25] Decomposition of hemicellulose leads to noncondensable gases, whereas the cellulose fraction mainly contributes to the formation of condensable gases. Lignin, on the other hand, is difficult to decompose, and the lignin decomposition temperature covers a broad range. Lignin decomposition is the main contributor to char formation that primarily consists of carbon (~85 wt%), along with small

Table 4.1 Physicochemical properties of pyrolysis oils.

Feedstock	Elemental composition[a] (wt%) C	O	H	N	S	Water content (wt%)	Ash content (wt%)	HHV[a] (MJ kg^{-1})	Ref.
Sewage sludge	65.2	17.4	8.6	8.4	0.4	10.9	0.00	36.4	12
Trommel fines	54.7	30.9	6.6	6.4	1.8	5.8	—	17.4	13
Oak and pine	58.6	34.1	6.6	0.7	—	8.0	—	25.2	14
Miscanthus	53.8	39.6	6.2	0.4	—	5.2	—	22.7	14
Hard wood	55.5	37.7	6.7	0.1	0.0	20.5	0.13	17.5	15

[a]Dry basis.

fractions of oxygen, hydrogen and some inorganic ash, depending on the type of feedstock used in the process. The heating value of the char is in the range of 25–32 MJ kg^{-1} (dry basis), which is substantially higher than that of the parent biomass or its corresponding pyrolysis oil.[3] Gases produced during pyrolysis are a mixture of low-molecular-weight noncondensable gases, such as carbon dioxide, carbon monoxide, methane, ethane and ethylene. The lower heating value of these gases is in the range of 11–20 MJ kg^{-1}.

The condensable gases are the desired product in pyrolysis and form the pyrolysis oil upon condensation. The higher heating value (HHV) of pyrolysis oil on a dry basis is generally in the range of 22–26 MJ kg^{-1}. The condensable gases can undergo secondary cracking reactions to produce liquid, a small fraction of char and noncondensable gases (see Figure 4.2).[3,26] These secondary reactions lead to a decrease in the yield of pyrolysis oil due to the formation of char along with the condensed oil.

Drying and size reduction are common pretreatments used before the pyrolysis of biomass. The water content in biomass directly affects the water content of pyrolysis oils, which in turn affects their energy density and stability. Pyrolysis oils with a low water content exhibit a higher energy density, while pyrolysis oils with a water content of >30–40 wt% can experience phase separation.[27] Some authors recommend reducing the water content of the biomass feedstock to below 10 wt% before introducing it into the pyrolysis reactor.[13,28] However, the particle size of the biomass controls the rate of primary pyrolysis. Finer biomass particles offer less resistance to the volatilization of condensable gases, which suppresses secondary cracking of these components.[3] Pyrolysis of larger feedstock particles has led to lower liquid yields due to the effect of particle size on the secondary reactions of the volatiles.[29]

The temperature used for pyrolysis has a direct impact on the yield of pyrolysis oil, char and gas. A low reaction temperature results in pronounced char production, whereas with a reaction high temperature, the possibility of secondary cracking reactions results in an increased fraction of noncondensable gases.[30,31] Furthermore, the heating rate affects the product distribution of the pyrolysis process. Rapid heating to a moderate

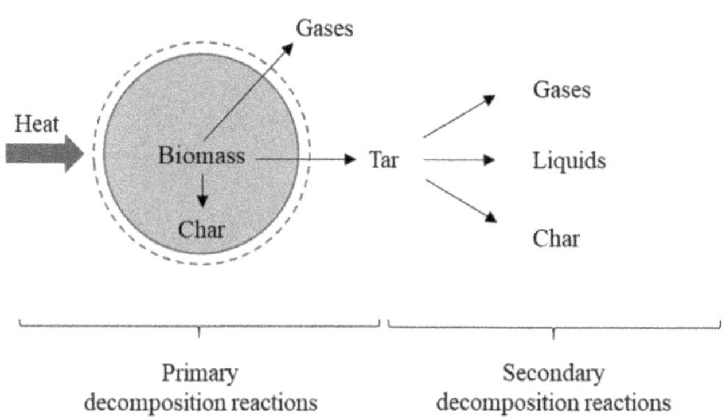

Figure 4.2 Schematic representation of biomass pyrolysis. Based on ref. 3.

temperature (400–600 °C) yields higher fractions of condensable volatiles and hence more pyrolysis oil product, while slower heating to that temperature favors char production. During slow heating, volatiles are not removed as rapidly from the reactor as happens during rapid heating process.[3] This increases the possibility of secondary reactions between gases and char and leads to secondary char formation. The use of a high heating rate, a moderate final temperature (450–600 °C) and a short gas residence time are recommended to increase the liquid product yield.[3]

4.2.1.1.1 Pyrolysis Processes. Pyrolysis processes can be categorized into conventional (slow) pyrolysis, fast pyrolysis, flash pyrolysis and ultrarapid pyrolysis, depending on the heating rate, temperature and residence time used during reaction (see Table 4.2).

Although the temperature range can be varied from 300 to 650 °C, slow pyrolysis of biomass is typically performed at about 400 °C. The vapor residence time varies from 10 to 100 min, and the heating rate is in the range 0.1–1.0 °C per s. In this way, components in the vapor phase continue to react with one another, as the solid char and liquid are being formed. Typical product yields during slow pyrolysis are 30 wt% bio-oil, 35 wt% char, and 35 wt% noncondensable gases.[32,33] Slow pyrolysis can be referred to as carbonization because char is the major product of the reaction. Pyrolysis oil from slow pyrolysis tends to undergo phase separation during storage due to inefficient water removal from the product.[34] In addition, the high amount of char affects the heating value of the final product. These characteristics make slow pyrolysis oils less attractive for gasification.

Compared to conventional pyrolysis, the heating rate used during fast pyrolysis is much higher (typically 10–200 °C per s), which aids volatilization. Rapid heating at temperatures between 400 and 500 °C favors the formation of the intermediate pyrolysis liquid product. Furthermore, rapid cooling and condensation of the pyrolysis vapors/aerosols is carried out to prevent secondary cracking and tar formation. Vapor residence times are short (typically <2 s). Depending on the biomass feedstock, fast pyrolysis processes produce on average 50–60 wt% of pyrolysis oil, 15–20 wt% of solid char and 20–30 wt% of

Table 4.2 Characteristics of thermal and hydrothermal decomposition processes.

Process	Residence time	Heating rate	Temperature (°C)	Products
Fast pyrolysis	<2 s	High	~500	Bio-oil
Flash pyrolysis	<1 s	Very high	<650	Bio-oil, chemicals
Ultrapaid pyrolysis	<0.5 s	Very high	~1000	Chemicals, gas
Hydropyrolysis	<10 s	High	<500	Bio-oil
Torrefaction	10–60 min	Very small	280	Torrefied biomass
Carbonization	Days	Very low	>400	Charcoal
Supercritical HTL	1–60 min	High	>374	Bio-oil
Subcritical HTL	10–90 min	High	280–370	Bio-oil

noncondensable gases.[32] Pyrolysis oils from fast pyrolysis are suitable for gasification.[35,36]

Flash pyrolysis and ultrarapid pyrolysis processes are technologies that are currently under development, and these differ from fast pyrolysis in terms of their temperature and heating rate. The residence time in flash pyrolysis is about 1 s at a temperature of 500–650 °C, and the heating rate is greater than 1000 °C per s. The product yield distribution of biomass flash pyrolysis is about 75 wt% pyrolysis oil, 12 wt% char and 13 wt% gas.[32,33] Ultrarapid pyrolysis involves extremely fast heating of biomass to a temperature range of 900–1300 °C with a residence time of less than 0.5 s.[33]

In some of the studies, pyrolysis was performed in the presence of a medium such as hydrogen (hydropyrolysis) or steam (steam pyrolysis). These processes are advantageous in producing bio-oil with a lower oxygen content over the conventional pyrolysis process. Furthermore, an increase in the volatile yield is also expected in these processes, due to sufficient hydrogenation of free-radical fragments and minimizing the possibility of repolymerization.[3,37]

4.2.1.1.2 Reactor Technologies for Pyrolysis of Biomass. Among the various types of reactors, fluidized beds have been widely used for fast pyrolysis of biomass due to their excellent heat and mass transfer characteristics, simplicity of operation and relative ease of scale-up. Rotating cone reactors are advantageous in terms of maintaining a high throughput and rapid heat transfer without the use of an inert carrier gas. Biomass is mechanically mixed with a solid heat transfer medium in a rotating cone. The auger reactor uses twin screws for mechanical mixing of biomass with a solid heat transfer medium. Even though this reactor configuration is currently used at pilot scale only, it has potential for scaling up because it is a robust system that is capable of continuous processing and requires low amounts of processing gas. Bubbling- and fluidized-bed pyrolysis reactors, as well as rotating cone reactors, have been commercialized in plants with a capacity above 100 tons of feedstock per day (see Table 4.3). Other types of reactors, including entrained flow reactors, ablative reactors and microwave pyrolysis reactors are in the early phase of technology development. Detailed information on pyrolysis reactor technologies is available in various reviews.[28,38]

Although pyrolysis is capable of producing a more energy-dense, liquid product from biomass that can be used in gasification, pyrolysis oils have a significant water content, which reduces their energy density. Furthermore, the complete evaporation and condensation of water that take place during pyrolysis are energetically unfavorable. Therefore, hydrothermal liquefaction (HTL) processes that also produce bio-oil and partly overcome these issues have been developed and are discussed in the next section.

4.2.1.2 Hydrothermal Liquefaction

HTL is an alternative to pyrolysis, which allows the conversion of wet feedstocks[44] such as biomass to a bio-oil (also known as HTL oil or bio-crude),

Table 4.3 Reactor technologies for fast pyrolysis. Based on ref. 39.

Reactor type	Development status	Yield (%)	Complexity	Feed size specification	Inert gas requirement	Specific reactor size	Scale-up	Pyrolysis oil producer (plant feedstock capacity)
Bubbling fluidized bed	Commercial	75	Medium	High	High	Medium	Easy	Dynamotive Corp.[a] (200 tons per day)[40]
Circulating fluidized bed	Commercial	70	High	High	High	Medium	Easy	Ensyn Corp.[b] (216 tons per day)[41]
Rotating cone	Commercial	70	High	High	Low	Low	Easy	BTG BV (120 tons per day)[42]
Entrained flow	Lab scale	60	Medium	High	High	Medium	Easy	
Auger	Pilot	60	Medium	Medium	Low	Low	Medium	KIT (12 tons per day)[43]

[a]Out of operation.
[b]Commissioning.

with properties that are similar to those of crude oil. The process is carried out at a moderate temperature (280–420 °C) and high pressure (7–30 MPa) in hydrothermal media. The temperature of the HTL process is sufficient to initiate pyrolytic mechanisms in the biomass feedstock, while the pressure is sufficient to maintain a liquid water processing phase. The residence time of the biomass feedstock at these conditions is between 10 and 60 min, and alkaline catalysts are often added. Catalysts assist certain base-catalyzed condensation reactions that lead to HTL oil formation and compete with undesired acid-catalyzed polymerization reactions to minimize the formation of solid by-product.[45] The major products of the HTL process are HTL oil, gas and water with dissolved organics. The produced HTL oil has a high heating value in the range of 30–36 MJ kg^{-1} and is a complex mixture of oxygenated compounds that extend over a wide range of molecular weights, including acids, alcohols, cyclic ketones, phenols and methoxy-phenols (guaiacols from softwood lignin), as well as more condensed structures, like naphthols and benzofurans.[45] Table 4.4 summarizes the physicochemical properties of various HTL oils.

As mentioned, HTL is suitable for liquefaction of biomass feedstock with a significant water content. This applies to feedstock such as agricultural residues, waste from food processing and municipal and agricultural sludge. Algae biomass has also been at the center of a significant amount of research work in this area, due to its high water content.[49,50] Biomass feedstocks used for HTL typically have heating values in the range of 10–20 MJ kg^{-1} and an oxygen content of 30–50 wt%.[51] HTL does not require drying of the biomass, which gives it flexibility for the processing of wet or dry feedstock materials. However, depending on the feedstock, a reduction of the particle size may be required. Woody biomass generally requires grinding prior to processing, whereas algal biomass, manure and sludge are a suitable size for direct processing.

During HTL, water is added to the biomass feedstock to obtain concentrations of 5–35 wt% of dry biomass in water.[45] The slurry is pressurized to 7–30 MPa and heated to temperatures between 280 and 420 °C. Liquefaction of biomass occurs in three main steps: (1) depolymerization of biomass into monomers; (2) decomposition of monomers through cleavage, dehydration, decarboxylation and deamination, which results in the formation of unstable and reactive light fragments of small molecules; and (3) condensation, cyclization and polymerization reactions, which result in the recombination of the light fragments into new compounds.[52] Depolymerization of biomass involves a sequential reduction of the chain length of biogenic macromolecules, *e.g.* hydrolysis of cellulose and hemicellulose to form mono- and oligosaccharides. In the decomposition step, dehydrogenation and decarboxylation reactions lead to a significant decrease in the oxygen content in the hydrolyzed biomass fragments by forming H_2O and CO_2, while nitrogen is removed *via* deamination. The products of this step are furans, phenols, aldehydes and other reactive fragments. These reactive species can recombine to form heavier products or polymerize to form char.[53]

Table 4.4 Physicochemical properties of HTL oils.

Feedstock	Process	Elemental composition[a] (wt%)					Water content (wt%)	Ash content (wt%)	HHV[a] ($MJ\ kg^{-1}$)	Ref.
		C	O	H	N	S				
Alkaline pretreated wood chip	Supercritical	77.0	13.0	8.9	1.1	—	—	5.30	36.0	46
Woody biomass	Supercritical	80.5	10.7	8.4	0.4	0.1	—	—	37.3	47
Pine wood	Subcritical	83.0	10.0	6.7	0.2	0.0	17	0.07	35.9	48
Chlorella sp.	Subcritical	79.0	3.7	10.6	5.5	0.6	12	0.47	39.6	48
Sewage sludge	Subcritical	77.0	8.1	10.1	4.3	0.6	13	0.33	37.8	48

[a]Dry basis.

The oxygen content in HTL oils is reported to be lower (10–20 wt%) than in pyrolysis oils, which provides better stability (Section 4.2.1.1).[48,54,55] The lower oxygen content leads to the relatively nonpolar nature of HTL oils. Furthermore, the effective phase separation between the HTL oil and water phase at high-pressure conditions results in oils with a low water content (typically, 10–16 wt%[48,55]). Nevertheless, a fraction of lighter oxygenates, and therefore a small fraction of the heating value of the feedstock is lost to the aqueous phase.

Feedstock composition, reaction temperature, pressure and heating rate are some of the most important parameters that affect the HTL process, as well as the properties and yield of HTL oils. The chemical composition of lignocellulosic biomass affects the product distribution. High cellulose and hemicellulose contents lead to a high yield of HTL oil; they also lead to increased production of aqueous phase products. Lignin, on the other hand, seems to be detrimental to the yield of HTL oil. Oxygen and water contents are usually high in HTL oils obtained from liquefaction of loosely structured biomass species. This lowers the HTL oil quality and HHV significantly.[56]

Temperature can be a critical parameter in HTL processing, as water properties change rapidly at near supercritical conditions.[57–59] At temperatures <280 °C, incomplete decomposition of individual biomass components suppresses the yield of HTL oil. An increase in temperature increases the yield of HTL oil.[59,60] However, after reaching the maximum oil yield, a further increase in temperature inhibits biomass liquefaction due to the occurrence of secondary decomposition reactions at high temperatures. This leads to the formation of gases, particularly above the critical point of water.[59] Furthermore, free radicals present in high concentration under supercritical conditions tend to recombine and result in char formation.

Pressure also plays an important role during HTL. High pressure ensures a single-phase reaction, which helps to avoid the large enthalpy inputs required for water evaporation. The ionic product of water is dependent on pressure and temperature close to the critical point. It indirectly impacts the rate of ionic hydrolysis reactions *vs.* radical reactions that lead to gas formation during HTL.[59]

Heating rate has a less pronounced effect on product distribution in hydrothermal liquefaction than in pyrolysis due to better dissolution and stabilization of fragmented species under hydrothermal conditions.[61] Nevertheless, a higher heating rate promotes bulk fragmentation of biomass and inhibits char formation. A slower heating rate favors char formation due to secondary reactions.[62]

4.2.1.2.1 HTL Processes. HTL processes are divided into subcritical and supercritical water processes. The conditions are determined relative to the critical point conditions of water at 374 °C and 22.1 MPa.[63] HTL processes operating above the critical point of water, at a relatively high pressure (30–35 MPa) and temperature (390–420 °C), are considered supercritical HTL processes, whereas reaction conditions below the critical point of water lead to subcritical HTL processes.[59] In the region above or near the critical point, properties like the dielectric constant, the ionic product and the density of water change strongly with the operating conditions. The dielectric constant affects the solubility of nonpolar biomass hydrolysis fragments, and the ionic product determines the rate of ionic *vs.* radical reactions.[59] Consequently, the properties of water determine the HTL oil yield.[64]

4.2.1.2.2 Reactor Technologies for HTL of Biomass. Extensive laboratory research has been done on HTL and scale-up to pilot and demonstration scale has been conducted by various companies and research institutes.[59] Batch autoclave reactors are commonly used in lab-scale HTL research; however, commercial operation of batch reactors is uneconomical due to loading and unloading times. For larger-scale operations, externally heated tubular plug-flow type reactors are typically used because continuous HTL operation is preferred. These are relatively easy to scale up, but the need for high-pressure equipment increases the investment cost. Commercial applications of HTL are not established yet, and HTL processes that use tubular reactors have only been scaled up to a feedstock capacity of 100 $L\,h^{-1}$.[65]

During continuous HTL processes that are close to supercritical conditions, a transition from liquefaction to a gasification regime can occur, which leads to a change in density and causes instability in the process. In order to mitigate this instability, the use of higher pressures (30–35 MPa) than what is commonly used (20–25 MPa) has been suggested.[66] An additional benefit of operating at a higher pressure is that the ionic product is high enough to reduce radical formation. Radical formation can lead to unwanted polymerization and coke formation.[67]

Another challenge with the scale-up of HTL technology is the wastewater generated in the process, which contains lighter oxygenated (C_1–C_4) organic molecules. Recirculation of this wastewater benefits the overall process, since it decreases the amount of freshwater that needs to be added, and the lighter organic fraction can partly be converted to HTL oil during recycling.[67] However, excess wastewater still needs to be treated to remove organics. Other issues with scale-up include storage of the slurry feed and material corrosion.[53] Despite these challenges, Steeper Energy/Silva and Genifuel are developing demonstration plants with a feedstock capacity of 4000 $kg\,day^{-1}$ of wood[66] and 2000 $kg\,day^{-1}$ of sewage sludge,[64] respectively. Mura Technologies plans to process 1 million tons of plastic waste per year *via* HTL by 2025,[68] which indicates the trend toward commercialization of this technology.

4.2.2 Bio-oil Stability

Once biomass is converted to bio-oil, through fast pyrolysis or HTL processes, the stability of the bio-oil becomes important for storage, transport and handling purposes. Viscosity, water content and acidity are the main properties used to evaluate bio-oil stability.

Compared to conventional oils, the high amount of oxygen-containing compounds in bio-oils make them chemically and thermally less stable. During storage, such instability is perceived as changes in their physico-chemical properties due to aging. This phenomenon refers to the polymerization, esterification and condensation reactions between the volatile and oxygenated compounds present in bio-oils.[69–71] Aging can lead to a slow and gradual increase in bio-oil viscosity, as well as the appearance of additional phases. The formation of tars, sludge, waxes and water is possible.[72] In the case of pyrolysis bio-oils, it has been demonstrated that the presence of char fines has a significant effect on the rate at which viscosity increases. Alkali metals present in the char particles promote reactions linked to viscosity increase.[73] Adding solvents (*e.g.* methanol or ethanol) and filtration of bio-oils have been explored as strategies to control the increase in viscosity.[74] Hot vapor filtration of pyrolysis bio-oils can significantly reduce the ash and alkali content (<0.01 wt% and <10 ppm, respectively) at the expense of a liquid yield loss of up to 20 wt%.[74] HTL bio-oils tend to be more stable than pyrolysis bio-oils, due to the lower content of reactive oxygenated species.[75]

The heating of bio-oils also promotes reactions that result in an increase in viscosity. Depending on the severity of the heating process (temperature and time), bio-oil changes can vary from an increase in viscosity and phase separation due to water formation to the formation of heavy compounds such as tars or even coke.[75,76] The viscosity of bio-oils is an important property in pump and pipeline design. It also plays a central role in the quality of bio-oil atomization to an EFG.

The amount of water in conventional fuels is regulated due to phase separation, emulsion formation and poor combustion problems. Bio-oils are not much different in this respect and require regulation of the water content because of these issues. For instance, it has been reported that when the water content is higher than 30 wt%, pyrolysis oils can experience phase separation.[27] As mentioned, pyrolysis oils tend to have a much higher water content than HTL oils. In both cases, additional water can be produced because of aging *via* condensation reactions, which may cause phase separation. An undetected bio-oil phase separation can cause significant problems during gasification. For instance, proper atomization might not be achieved. Injector nozzles are designed to operate with feeds that have consistent properties. Furthermore, feeding a bio-oil in which phase separation is observed can lead to an inhomogeneous syngas product and temperature fluctuations.[77] It is worth mentioning that adding water reduces the viscosity of bio-oil, which is beneficial; however, it also decreases the heating value, which means that additional bio-oil feed is required to obtain the same temperature during gasification in the gasifier.[74]

The acidity of bio-oils is mainly derived from volatile acids, *e.g.* formic and acetic acids. Other compound groups that also contribute to an increase in bio-oil acidity include phenolics, fatty acids and resins, and hydroxyl acids.[78] The acidity of bio-oils is typically determined and reported as pH or total acid number (TAN). Typical pH values for bio-oils are about 2–3 (Section 4.2.1.1).

From a chemical point of view, the presence of acids in bio-oil is detrimental to their stability due to their role in polymerization reactions that lead to an increase in viscosity during storage.[79] From a material point of view, the presence of acids and water in bio-oils leads to corrosiveness, especially at high temperatures. This increases the cost of construction materials for bio-oil storage and handling because corrosion-resistant materials are required.

4.2.3 Production of Torrefied Biomass

Torrefaction is a slow pyrolysis process that transforms solid biomass into a brittle and easy-to-pulverize material, which can be treated similarly to coal. Liquid and gaseous products are also obtained as part of the process (Figure 4.3).[80] Torrefaction takes place in an inert or oxygen-limited environment, at temperatures between 225 and 300 °C.[3,81,82] Removing oxygen and hydrogen from the biomass feedstock leads to a final solid product (char) with higher carbon-to-oxygen (C/O) and carbon-to-hydrogen (C/H) ratios, a lower moisture content and a higher heating value. Table 4.5 presents some of the physicochemical properties of raw biomass and torrefied

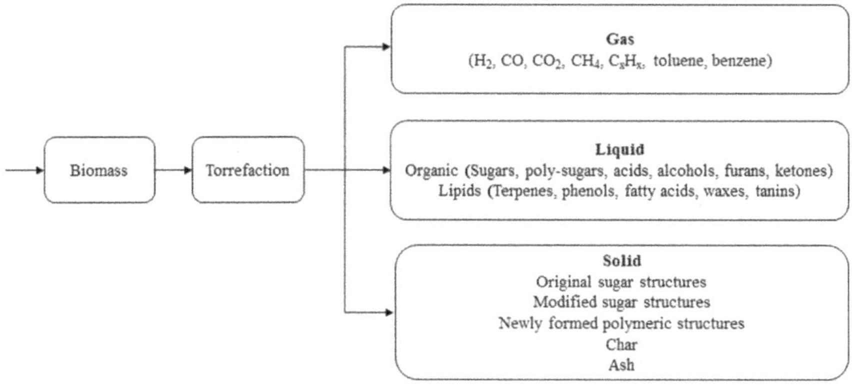

Figure 4.3 Products from biomass torrefaction. Based on ref. 3.

Table 4.5 Properties of raw wood and torrefied wood. Based on ref. 84.

Property	Raw wood	Torrefied wood
Moisture content (wt%)	35	3
Lower heating value (MJ kg^{-1})	10.5	19.9
Higher heating value (MJ kg^{-1})	17.7	20.4
Bulk density (kg m^{-3})	550	230
Bulk energy density (GJ m^{-3})	5.8	5.6
Dust/ash formation	Moderate	High
Hygroscopic nature	Water uptake	Hydrophobic
Biological degradation during storage	Possible	Not possible
Seasonal influence	High	Poor

Table 4.6 Solid yield and heating values of different biomass feedstocks before and after torrefaction at different temperatures and residence times. Based on ref. 86.

Biomass type	Process	Temperature (°C)	Residence time (min)	Solid yield (%)	HHV (MJ kg^{-1})
Wood briquettes	Raw	—	—	—	20.02
	Torrefied	220	60	90	20.99
	Torrefied	250	60	65	22.06
	Torrefied	270	60	54	22.98
Wheat straw	Raw	—	—	—	18.9
	Torrefied	250	30	82.6	19.82
	Torrefied	270	30	71.5	20.7
	Torrefied	290	30	55.1	22.6
Rice husk	Raw	—	—	—	14.32
	Torrefied	250	60	77.5	15.89
	Torrefied	270	60	74.25	16.07
	Torrefied	300	60	58.25	17.59
Coffee residue	Raw	—	—	—	20.37
	Torrefied	240	30	89.41	21.78
	Torrefied	240	60	84.01	22.87
	Torrefied	270	30	74.13	25.34
	Torrefied	270	60	64.12	28.06

biomass. A torrefied solid typically contains up to 90% of the initial energy content of the biomass and only 70% of its initial weight.[83]

Torrefaction is suitable for processing a variety of biomass feedstocks, including commercial timber, agricultural waste and energy crops that have similar physical and chemical properties. Table 4.6 presents the yield of solid product and the heating values for different biomass feedstocks before and after torrefaction. Seasonal influences on the properties of biomass feedstocks are reduced by torrefaction,[85] which makes the process useful in pretreating biomass.

The torrefaction process can be divided into distinct stages: heating, drying, torrefaction and cooling.[3] Initial heating results in moisture evaporation, and further heating leads to decomposition (Table 4.7). In wood torrefaction, hemicellulose degrades to a high extent in the heating stage and forms acetic acid and water as the main condensable liquids.[87] Slow decomposition of hemicellulose is observed at temperatures <200 °C, and this becomes significant when the temperature exceeds 200 °C.[80] Furthermore, when a temperature of about 250 °C[84] is reached, extensive devolatilization of hemicellulose is observed.[87] Lignin and cellulose require a higher temperature (250 °C and above) to react *via* limited devolatilization and carbonization.[3,88] The loss of tenacity and the fibrous structure of biomass during torrefaction has been related to cellulose decomposition.[3] Depolymerization and devolatilization/softening of lignin are also observed during torrefaction.[89] The softening of lignin contributes to densification of the biomass. Degradation of OH-groups reduces the ability of biomass to form hydrogen bonds with water, which decreases the hydrophilicity.[84]

Table 4.7 Processes taking place at different temperatures during wood torrefaction. Based on ref. 3.

Temperature (°C)	Process	Regime	Main product
20–150	Moisture evaporation	Nonreactive process.	Bone dry wood
110–200	Removal of moisture traces, start of decomposition	Reactive drying. Biomass is deformed by breaking H and C bonds and depolymerization.	Preheated dry wood
200–270	Release of volatiles	Torrefaction of hemicellulose. Breakdown of most inter- and intramolecular hydrogen, C–C, and C–O bonds. Formation of condensable liquids and noncondensable gases.[a]	Mildly torrefied biomass
270–300	Exothermic decomposition releasing condensable and non-condensable vapors	Extensive decomposition of hemicellulose into volatiles and solid products. Limited devolatilization and carbonization of cellulose and lignin. Complete destruction of biomass cells, making the product brittle and nonfibrous.	Severely torrefied biomass
—	Cooling process	—	Solid brittle char

[a]Species such as acetic acid, methanol, CO, CO_2 are produced.

During torrefaction, exothermic condensation reactions compete with endothermic decomposition reactions.[90] For instance, pyrolysis of hemicellulose and lignin, as well as charring reactions, are exothermic in nature, while pyrolysis of cellulose is endothermic.[86,91,92] The different reaction rates are influenced by a range of parameters, with consequences for the torrefaction process and thereby the final product characteristics. These parameters include temperature, heating rate, residence time, operating pressure and type of reactor. The increase in reaction temperature results in an increase in the C/O ratio and consequently in a higher energy density in the solid product. However, torrefaction at temperatures above 300 °C are not recommended due to (1) extensive devolatilization and carbonization, which is undesirable due to yield loss; (2) a high loss of lignin, which would make it difficult to form pellets from torrefied products for transport purposes; and (3) fast thermal cracking of cellulose, which leads to tar formation.[3]

The heating rate is also crucial during torrefaction. A slow heating rate allows for maximizing the solid yield. Typically, the heating rate during torrefaction is less than 50 °C per min.[80] A higher heating rate would increase the liquid yield at the expense of the solid product, similar to pyrolysis for bio-oil production.[3] The third important parameter affecting torrefaction is residence time. A longer residence time results in a lower yield of solid product but contributes to an increase in the C/O and C/H ratios, giving a product with a higher energy density.[3,85] Extended residence times will result in energy loss, due to a significant release of volatiles, similar to a carbonization process.

The control of parameters – such as heating rate, temperature and residence time – during torrefaction is complex, due to the variability in the properties of biomass feedstocks. A poorly controlled process can lead to a low-quality product and decreased energy efficiency. Different reactors and technologies for biomass torrefaction have been developed in order to achieve good control and productivity. Conceptually, these technologies can be classified as dry torrefaction, oxidative torrefaction and wet torrefaction. An overview of the torrefaction technologies and reactor types involved is provided in the following sections.

4.2.3.1 Dry and Oxidative Torrefaction

Dry torrefaction is the most common torrefaction process. The process is carried out in a nonoxidative environment, using CO_2 or N_2 as the inert gas.[87] Carbonization and devolatilization of the cellulose causes the biomass to become more brittle. Degradation of the hemicellulose causes the nature of the biomass to change from hydrophilic to hydrophobic and the heating value to increase. The lignin portion of the biomass is not significantly affected by torrefaction.[93]

Heat integration during torrefaction is important for an economical process design. Energy requirements for drying and torrefaction depend on the process conditions. It has been suggested that autothermal operation of dry torrefaction is possible.[94] Commercially, the gaseous torrefaction by-product is combusted in an external reactor to supply the heat for biomass drying.[95] Alternatively, oxygen can be supplied directly to the torrefaction reactor.[96] This concept is called oxidative torrefaction. Oxidative torrefaction is characterized by shorter reaction times and a lower solid yield compared to dry torrefaction.[81] The energy yield of oxidative torrefaction is lower compared to dry torrefaction; however, the difference is low enough to consider oxidative torrefaction to be potentially feasible.[97]

4.2.3.2 Wet Torrefaction

Torrefaction can also be carried out in wet mediums such as water or diluted acid. Wet torrefaction – also known as hydrothermal torrefaction or carbonization – is normally operated at temperatures between 180 and 260 °C.[98] Increased pressure of up to 4.6 MPa is required to keep the water in the liquid phase, and residence time is typically between 5 and 240 min.[99] Wet torrefaction is suitable for lower-cost biomass feedstocks such as forestry residue, agricultural waste and municipal waste, which have a higher water content.[81,98] The reaction mechanism in wet torrefaction is different from that in dry torrefaction. It consists of hydrolysis processes (due to the presence of compressed water) that result in the cleavage of ether bonds between saccharides.[93] Considering that wet torrefaction happens in a wet environment, no predrying is necessary. However, the product separation process in wet torrefaction is more complex compared to dry torrefaction because the char is immersed in the water phase.[81] Wet torrefaction requires a higher capital investment than dry torrefaction due to the need for pressure vessels that can deal with the pressurized water, and the overall process is less energy efficient.

4.2.3.3 Torrefaction Technologies

A typical torrefaction plant includes several units such as biomass handler, preparation, dryer, torrefier and product cooler. The most common torrefaction technologies include fluidized-bed reactors, rotating drums and screw reactors. These concepts are used in commercial-scale torrefaction plants with a capacity of up to 80 000 tons per year.[100]

Moving, fixed and fluidized-bed reactors are convective reactors for biomass torrefaction. In fixed- and moving-bed reactors, the heat carrier is a hot gas percolating through or flowing past biomass particles, heating them primarily by convection.[3,101] In fluidized-bed reactors, biomass is fluidized in the heat carrier medium, which leads to an excellent gas–solid mixing and uniform temperature profile.

Rotary drum reactors circulate biomass in the presence of inert gas.[102] The rotary drum can either be heated directly or indirectly, *i.e.* using hot gas or *via* the reactor tube. Rotary drum reactors have been used for other applications in industry and are therefore a commercially proven technology. However, scaling up and controlling the temperature of the process are difficult.[102] The screw or stationary shaft reactor uses spirals to transport the biomass through the reactor tube,[102] and heat is provided indirectly *via* the reactor wall or hollow screws.[75] This type of reactor is relatively inexpensive but can exhibit poor heat exchange, uneven biomass torrefaction and high maintenance costs.[102] Other possible torrefaction technologies include microwave reactors,[3,103] entrained flow reactors[3] and multiple hearth furnaces.[100] Details on torrefaction reactors are provided in the literature.[3,78,100,104]

4.2.4 Stability of Torrefied Biomass

As torrefaction is a pretreatment process used to obtain a solid feed suitable for gasification, the storage and transport properties of the product are important considerations. In contrast to raw biomass, storage issues like off-gassing and self-heating are insignificant in torrefied biomass because most of the raw biomass components that are chemically and microbiologically active are removed during the torrefaction process.[105] The hydrophobicity of torrefied biomass leads to a significantly lower water uptake compared to raw biomass. This mitigates further biological degradation, similar to coal.[105]

Torrefaction occurs in an oxygen-deficient environment, and the product is potentially reactive when in contact with oxygen, which can result in self-ignition.[106] Rapid quenching of the torrefied product is used in order to mitigate these risks. However, even during transport and storage of torrefied biomass, there is a risk of explosion,[107] since the brittle torrefied material can form fine char dust due to abrasion. Torrefied biomass has a higher explosion reactivity than certain types of coal.[108] Therefore, precautions and safety systems may be needed when handling and transporting torrefied biomass.

4.3 Syngas Production at Large Scale

Since producing biofuels *via* Fischer–Tropsch synthesis makes use of large quantities of syngas, pressurized entrained flow gasification has gained significant attention for the production of syngas from biomass feedstocks. This technology falls in the high-capacity range (>50 MW_{th}), and can be used to meet scale requirements.[3,77]

EFG operate at high temperatures, typically above 1200 °C. At such conditions, both thermodynamic equilibrium and practically complete conversion are rapidly achieved. Vapor tars are efficiently gasified to give a syngas that is virtually tar free. Furthermore, soot formation *via* the Boudouard or methane decomposition reactions (eqn (4.2) and (4.4), respectively) is not thermodynamically favored.[109] The formation of methane is also limited (<0.5 mol%).[110] The syngas composition depends on homogeneous equilibrium reactions (Table 4.8), mainly the water-gas-shift reaction (eqn (4.7)), and can be estimated from thermodynamic equilibrium calculations with acceptable accuracy.[111] When compared to other gasification technologies such as fluidized-bed gasification, the high operating temperatures of

Table 4.8 Main gasification reactions. Based on ref. 3.

Reaction types	Eqn
Heterogeneous reactions:	
Pyrolysis	
$C_xH_yO_z \rightarrow aCO_2 + bH_2O + cCH_4 + dCO + eH_2 + \text{char} + \text{tar}$	(4.1)
Gasification	
$C + CO_2 \leftrightarrow 2CO$ (Boudouard)	(4.2)
$C + H_2O \leftrightarrow CO + H_2$ (water-gas)	(4.3)
$C + 2H_2 \leftrightarrow CH_4$ (hydrogasification)	(4.4)
Combustion	
$C + O_2 \rightarrow CO_2$	(4.5)
$C + 1/2O_2 \rightarrow CO$	(4.6)
Homogeneous reactions:	
$CO + H_2O \leftrightarrow CO_2 + H_2$ (water-gas shift)	(4.7)
$2CO + 2H_2 \rightarrow CH_4 + CO_2$ (methanation)	(4.8)
$CH_4 + H_2O \leftrightarrow CO + 3H_2$ (steam reforming of methane)	(4.9)
$CH_4 + 0.5O_2 \rightarrow CO + 2H_2$ (partial oxidation of methane)	(4.10)
$CH_4 + CO_2 \leftrightarrow 2CO + 2H_2$ (dry reforming of methane)	(4.11)
$C_nH_m + 2nH_2O \rightarrow (2n + m/2)H_2 + nCO_2$ (steam reforming of tar)	(4.12)

entrained flow gasification increase the oxygen requirement and decrease the cold gas efficiency. EFG operates at pressures that vary between 1 and 3 MPa for solids and up to 70 MPa for liquids.[110,112] For synthesis applications, a pressure of up to 10 MPa is common. High-pressure operation accelerates the gasification rate of solid particles with (approximately) the square root of the operating pressure.[77] It also allows for a higher specific throughput, reduces the syngas volume to be treated in downstream operations and avoids expensive intermediate syngas compression steps.

Pressurized entrained flow gasification is a flexible technology that can accommodate different feedstocks. In the case of biomass, obtaining a feed with a suitable particle size (typically < 100 μm), which would allow using modified versions of existing and commercially proven coal-entrained flow gasification technology, is an energy-intensive process.[1,3] Therefore, biomass pretreatment options (Section 4.2) to yield a more energy-dense product (bio-oil or torrefied biomass) suitable for feeding into an EFG have been explored. The use of these materials can reduce the transportation cost per energy unit ($ per MJ) to supply biomass from different satellite locations to a central gasification facility. Compared to coal or biomass, the different characteristics of bio-oil or torrefied biomass have some implications in terms of feeding to the gasifier, the chemistry of gasification and the ash behavior, all of which are discussed in the following section.

4.3.1 Feeding of Pretreated Biomass in Entrained Flow Gasifiers

4.3.1.1 Feeding of Bio-oils

Efficient operation of an EFG involves complete conversion of the feed to a nearly tar-free syngas, while operating at minimum oxygen consumption to maximize the cold gas efficiency.[113] Bio-oil atomization plays a critical role, as the droplet size distribution influences feed conversion and syngas quality. A rather large fraction of fine droplets (about 100 μm in diameter) is required to achieve stable ignition at different fuel loads.[114] Nevertheless, a relatively small fraction of larger droplets (about 1000 μm in diameter) is tolerable in maintaining a high degree of conversion. The larger surface-area-to-volume ratio of smaller droplets facilitates their vaporization. This improves fuel/air mixture properties and ignition characteristics, which results in increased flame stability.[115,116] The high viscosity and surface tension of pyrolysis oils, compared to fuel oils, result in the formation of larger droplets.[117,118]

Droplet-size distribution depends on the feed properties (viscosity, density, surface tension), operating conditions (temperature and pressure) and atomizer nozzle design.[119] In general, the droplet size of a spray increases with the increase in viscosity, density and surface tension. Preheating of bio-oil to reduce the value of these properties and to facilitate atomization during EFG is recommended. However, the low thermal stability of bio-oils limits this preheating step. Configurations have been evaluated in which bio-oils (or bio-oil slurries) are initially preheated to temperatures not exceeding 80–100 °C for mixing and homogenization purposes and then are heated to a temperature above 100 °C just before injection into the gasifier.[1] In this way, problems associated with preheating of bio-oil (*e.g.* plugging of the feed line and spray burner nozzles[117]) are mitigated. The recommended viscosity for bio-oil atomization in spray combustion applications is in the range of 4–25 cSt.[117]

Atomization nozzles such as twin fluid spray burners, which introduce both the liquid fuel and the oxidizing agent into the gasifier, have been evaluated for gasification of bio-oil in EFG.[117,119] In such nozzles, the bio-oil is atomized by a jet of gasifying agent (oxygen and/or steam) that can have a velocity of 100 m s^{-1} or higher. As gasification reactions take place at a low stoichiometry, the amount of oxygen (and steam) available for atomization is limited, and atomizers operate at a low gas-to-liquid ratio.[113] This can be challenging (particularly at high pressure) because the volume and velocity of the gasifying agent decrease with the increase in pressure. Despite the significant commercial experience with fluidization of heavy oils for entrained flow gasification, the need for systematic studies to evaluate the influence of operational parameters (such as high pressure) on twin-fluid atomization of bio-oil has been highlighted.[33] Successful atomization of bio-oil slurries (pyrolysis oil combined with the char fraction) on a demonstration scale has been achieved as part of the technology development work on the Bioliq concept, led by the Karlsruhe Institute of Technology (KIT).[120]

4.3.1.2 Feeding of Torrefied Biomass

Feeding solids to a pressurized EFG is a challenging task, and solutions have been developed for coal as feedstock but not for less brittle, fibrous materials such as biomass.[121] Dedicated systems for raw biomass feeding are not easily available. However, the characteristics of torrefied biomass similar to those of coal enables the use of feeding systems similar to the ones used for coal. This is often one of the major reasons why torrefaction is considered, despite the loss of energy during the process. Feeding systems for coal and torrefied biomass consist of lock hoppers that pressurize the feedstock using compressed inert gas. Torrefied biomass has a lower bulk energy density than coal. Therefore, more pressurized inert gas is required to feed the same energy equivalent of torrefied biomass to a gasifier, compared to coal. It has been estimated that the amount of compressed inert gas and consequently the electricity requirement for compression are about three times higher for torrefied biomass than for coal.[122] Hence pressurization using lock hoppers is possible, but it comes with an energy penalty when using torrefied biomass rather than coal.[122]

After pressurization, the feed is stored in a fluidized bed operated in a smooth regime by using inert gas as the fluidization medium. From there, the solids are pneumatically fed to the EFG.[80] Smooth fluidization is the most crucial aspect of the feeder, as it enables a continuous, stable flow of fuel particles to the stationary flame present in the EFG.[80] To obtain such a fluidization regime, small particle sizes are required.[123] For coal-based EFG, particle sizes are in the range of 50–100 μm, and feeding similar particles of torrefied biomass is realistic.

4.3.2 Chemistry of Gasification

In principle, gasification of bio-oil or torrefied biomass involves the same steps observed in gasification of solid biomass: (1) devolatilization and pyrolysis, (2) secondary tar-cracking, and (3) char gasification and combustion. A complete description of these steps can be found in Section 4.3. However, some differences in the way in which gasification takes place are pointed out in the following sections.

4.3.2.1 Gasification of Bio-oil

Bio-oil is introduced into the gasifier as liquid droplets rather than as fine solid particles. This has implications for the mass transfer, particularly during the first step of the gasification process, *i.e.* devolatilization and pyrolysis.

As discussed, bio-oil essentially consists of a complex mixture of partly oxygenated molecules with a broad boiling point distribution and water. The majority of the organic molecules exhibit a significantly higher boiling point than water, and the heaviest organic molecules decompose or form coke rather than volatilizing.[124] After introduction of a bio-oil droplet into the gasification chamber, the droplet heats up rapidly, leading to volatilization of water and light boiling compounds on the droplet surface (see Figure 4.4).

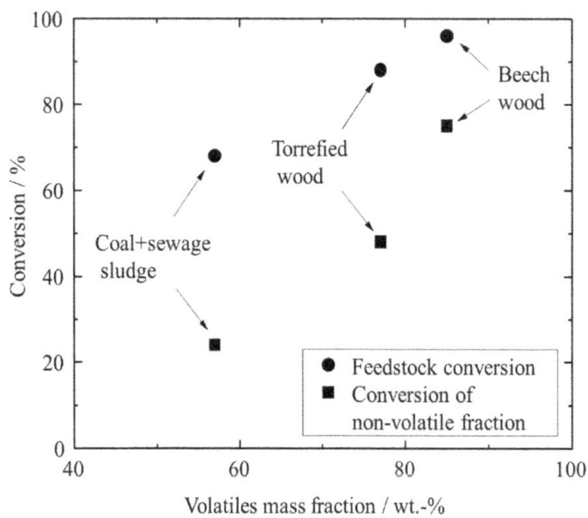

Figure 4.4 Schematic representation of bio-oil droplet evaporation and microexplosion in an EFG. Based on ref. 111.

The bulk volume of the drop heats up as well. Due to the viscous nature of the bio-oil, the liquid mass transfer within the bulk volume is limited. Water and light boiling compounds in the bulk volume are not readily transported to the surface, where they can volatilize. As a result, vapor pockets are formed inside the droplet that lead to the droplet swelling.[125] This eventually causes the droplet surface to rupture, which is often referred to as micro-explosion.[111] The resulting smaller droplets exhibit a higher surface area, which is beneficial in terms of mass and heat transfer in the gasifier.[126]

Pyrolysis of the organic feed molecules, which leads to the formation of tar, char and gases (eqn (4.1)) occurs at the same time as volatilization. With a further increase in the temperature (above 500 °C), the tar cracks and undergoes reforming reactions (homogeneous reactions). This produces short-chain hydrocarbons, as well as CO, CO_2 and H_2. The reactions are the same as with biomass gasification and are shown in Table 4.8.

The char that is formed during pyrolysis requires the highest temperature in order for it to be gasified because this is a heterogeneous reaction that exhibits stronger mass transfer limitations. The reactions that take place during char gasification correspond to the Boudouard reaction (eqn (4.2)), the heterogeneous water-gas reaction (eqn (4.3)), hydrogasification (eqn (4.4)) and partial and total oxidation (eqn (4.5) and (4.6)). The main products are CO, CO_2, H_2 and CH_4. The gaseous products then undergo homogeneous reforming and partial oxidation reactions to form mainly CO, CO_2, H_2 and H_2O. The reaction chemistry for the secondary tar-cracking and char gasification steps is similar to that in biomass gasification (or fossil fuel gasification).

4.3.2.2 *Gasification of Torrefied Biomass*

Char gasification is the rate-controlling step in gasification processes.[3] In the case of torrefied biomass – which is essentially char that consists of carbon and a small fraction of hydrocarbons – char gasification reactions dominate the reaction chemistry during gasification.

The rate of char gasification depends on the gasifying medium, as well as the char reactivity. The highest rate for char gasification is observed in reactions with oxygen (eqn (4.5)), followed by reactions with steam (eqn (4.6)) and carbon dioxide (eqn (4.2) and (4.3)).[3] On the other hand, char properties depend strongly on pyrolysis conditions, particularly on the heating rate.[127] This has implications for the overall reaction rate of gasification. Feedstocks with a higher volatiles content (such as biomass) show an overall higher gasification reaction rate compared to feedstocks with a lower volatiles content (such as torrefied biomass or coal). During biomass gasification, rapid pyrolysis of biomass results in the fast release of volatiles. This, in turn, deforms the biomass structure to create chars with a large macropore surface area and high relativity.[127] However, the slow heating rate used during torrefaction of biomass results in the release of a fraction of the volatile species, which leads to a less reactive char and consequently an overall lower gasification rate of torrefied biomass compared to pulverized biomass.

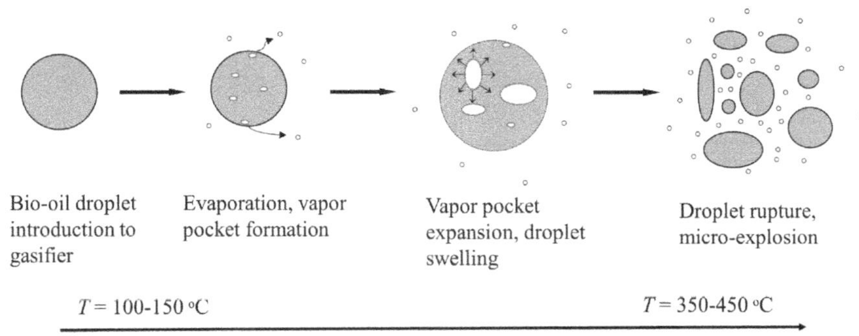

Figure 4.5 Conversion of different feedstocks in a pilot-scale EFG ($T = 1400$ °C) after 0.1 s of residence time as a function of the feedstock volatiles content. Conversion is given on a dry- and ash-free basis.[1]

Similarly, coal has a lower content of volatiles than biomass or torrefied biomass; therefore, it exhibits a lower reaction rate during gasification (see Figure 4.5). The nature of the biomass feedstock also influences the reactivity of the torrefied biomass. A higher hemicellulose content in the initial biomass leads to a torrefied product with a higher reactivity. For instance, the hemicellulose content in the initial biomass leads to the following order of reactivity for torrefied biomass: larch < willow/beech < straw.[80]

In gasification, char reactivity determines the conversion that can be obtained at a given residence time. This has implications for the gasifier design because the process needs to be designed to obtain close-to-complete conversion.

4.3.3 Ash Behavior in Entrained Flow Gasification

The management of ash or slag is one of the determining factors for the operation of EFGs. In principle, an EFG can be operated in either slagging or nonslagging mode. Slagging occurs when the ash in the feedstock melts. The molten ash is referred to as slag. In a nonslagging operation, the ash needs to remain solid. If a nonslagging gasifier is operated above the ash deformation temperature, the ash softens, becomes sticky and attaches to the wall of the gasifier. This causes operational problems because it can lead to clogging. Raw biomass and bio-oil can differ significantly in their ash content, which affects the gasification.

The ash deformation temperature has been determined to be around 1175 °C for torrefied biomass and between 1180 and 1280 °C for various biomass samples.[128,129] Hence, operation significantly below 1200 °C is required for nonslagging gasification of bio-oil.

While nonslagging operation has been studied for bio-oil gasification,[130] a number of studies and most commercial entrained flow gasification systems operate in slagging mode.[36,131] This is because slagging gasifiers are more flexible in terms of fuel, and even at a relatively low temperature, partial softening of ash particles cannot be avoided.[1,132] Nevertheless, the strong decrease in the ash content of pyrolysis oils compared to raw biomass has been suggested as an opportunity to use the less costly nonslagging gasifier technology for pyrolysis oil gasification.[133] Contrarily, the feasibility of using the nonslagging operation for gasification of torrefied biomass has been questioned. Although melting ash from torrefied wood is difficult at typical temperatures used with an EFG, some melting cannot be avoided. It is nearly impossible to keep ash particles separated from the gasifier walls, and once a sticky layer is created, more particles can easily be trapped.[122]

Since operation of an EFG in slagging mode requires temperatures above the ash fusion temperature, the addition of an ash fluxing agent (such as limestone) is required when the ash fusion temperature is above 1400 °C.[134] The wall of the gasifier is covered with a layer of molten slag that slowly moves down the wall toward the quench zone, and the layer is constantly recovered. This layer acts as an additional thermal barrier that minimizes heat loss and protects the gasifier wall material. Slagging gasifiers require an ash content of around 6 wt% for sufficient constant wall coverage.[1] This is already an issue with raw biomass, since it typically contains lower mass fractions of ash forming compounds. Bio-oils contain even less ash forming minerals, since they are concentrated in the char during pyrolysis or in the water phase during HTL (Section 4.2.1.2). This needs to be overcome by either gasification of the pyrolysis char along with the pyrolysis oil[36] or by external addition of slag-forming minerals. Ash forming minerals can be added in the form of coal slag, by recirculation of slag from bio-oil gasification or by adding ash fluxing agents such as CaO or FeO.[1,134] This can be advantageous because it facilitates control of the slag viscosity.

4.3.4 Technology Development for Gasification of Bio-oils and Torrefied Biomass in an EFG

Entrained flow gasification of torrefied biomass has been demonstrated at the laboratory scale[135–138] and has been compared to raw biomass gasification.[138,139] A higher syngas yield and a higher gasification temperature were obtained when using torrefied biomass than when using raw biomass.[138,139] Gasification of torrefied biomass has also been investigated in a pilot-scale EFG (270 kW_{th}).[140] The product gas from torrefied wood gasification showed a higher content of H_2 and CO compared to that of raw wood at comparable H_2/CO ratios (see Table 4.9). The difference in the gas composition was related to the lower oxygen content in torrefied biomass compared to raw biomass.[140] Furthermore, the overall plant efficiency reported for syngas production from different torrefied wood feedstocks ($\eta = 0.62$–0.79) was higher than that for raw wood ($\eta = 0.49$), mainly due to the reduced milling energy requirements. The cost and energy requirements for torrefaction were not considered in this study.

In terms of bio-oils, entrained flow gasification at the laboratory scale has mainly been carried out using pyrolysis oils as the feedstock.[141–143] When compared to raw biomass gasification, higher cold gas efficiency (CGE) and

Table 4.9 Dry and ash-free syngas composition after oxygen-blown entrained flow gasification of raw wood and torrefied wood. Based on ref. 140.

Parameter[a]	Raw wood[b]	Torrefied wood[c]
CO (mol%)	50.0	54.6
H_2 (mol%)	26.2	27.2
CO_2 (mol%)	23.3	18.6
CH_4 (mol%)	1.8	1.60
H_2/CO	0.52	0.50
Carbon conversion	0.89	1.10

[a]Gasification parameters: $P = 200$ kPa, λ (equivalence ratio) = 0.44, $T = 1200$–1350 °C.
[b]Dry and ash-free composition: C: 50.9 wt%, H: 6.2 wt%, O: 41.8 wt%, N: 0.2 wt%; moisture: 3.0 wt%.
[c]Dry and ash-free composition: C: 54.4 wt%, H: 6.0 wt%, O: 38.7 wt%, N: 0.2 wt%; moisture: 2.4 wt%.

Table 4.10 Dry and ash-free syngas composition after oxygen-blown gasification of pyrolysis oil in a pilot-scale EFG. Based on ref. 133.

Parameter[a]	Wood oil[b]	Straw oil[c]
CO (mol%)	45.6	46.4
H_2 (mol%)	30.1	30.3
CO_2 (mol%)	22.5	23.1
C_xH_y (mol%)	≈ 2	≈ 2
Carbon conversion	0.96	0.89

[a]Gasification parameters: $P = 400$ kPa, λ (equivalence ratio) = 0.44, $T = 1200$–1350 °C.
[b]Dry and ash-free composition: C: 57.4 wt%, H: 6.6 wt%, O: 35.0 wt%, N: >1 wt%; moisture: 21.1 wt%.
[c]Dry and ash-free composition: C: 61.4 wt%, H: 6.6 wt%, O: 24.0 wt%, N: 8.0 wt%; moisture: 24.0 wt%.

lower tar content were observed during pyrolysis oil gasification.[141] Syngas production *via* pyrolysis oil gasification is expected to be more economical than pyrolysis oil steam reforming.[141] This is due to the endothermic nature of reforming and reforming catalyst deactivation *via* coke formation. Pilot-scale entrained flow gasification (400 kW_{th}) has also been carried out using pyrolysis oil, and a 2-day continuous operation has been demonstrated.[133] The syngas composition (see Table 4.10) is close to temperature-corrected equilibrium predictions.[133] The energy efficiency of the gasification of wood-derived pyrolysis oil was 68%, while that of straw-derived pyrolysis oil was 60%.[133] Optimized atomization was identified as a key parameter to increase gasification efficiency.

Further scale-up of the gasification of torrefied biomass and pyrolysis oil was carried out in two landmark projects. In the BioTfuel project, the overall process of torrefaction, entrained flow gasification and Fischer–Tropsch synthesis for the production of aviation fuel, synthetic biodiesel and bionaphtha from lignocellulosic biomass was validated recently. The consortium leading the project announced that their demonstration plant torrefied and gasified various woody biomass feeds in a 5 MW_{th} gasifier for several weeks and produced syngas that later was converted into Fischer–Tropsch products.[144] The Bioliq project concept is based on entrained-flow gasification of pyrolysis oil-char slurries with subsequent liquefaction to dimethyl ether, which is then converted to gasoline. In the 5 MW_{th} demonstration-scale gasifier, continuous gasification with N_2 free syngas compositions in the range of 26–36 mol% H_2, 27–39 mol% CO, 14–28 mol% CO_2 and <0.1 mol% CH_4 were demonstrated.[120] At this scale, practical issues relevant to commercial operation are being addressed, *e.g.* slag coverage of the gasifier membrane wall.[145] In addition, 2100 L of renewable-based gasoline was synthesized from 2017 to 2019, proving the overall process chain for biomass to fuel.[146] The ongoing efforts to demonstrate entrained-flow gasification of bio-oils and torrefied biomass at a larger scale indicate that commercialization of this technology to generate syngas for Fischer–Tropsch synthesis is technically feasible and realistically achievable in the future.

4.4 Perspectives and Conclusions on BTL Processes

The political drive toward the production of fuels and chemicals with a lower carbon footprint requires commercial implementation of biomass-to-liquid processes. Several projects that are using different concepts for the production of sustainable fuel and chemicals are now at a stage at which they are close to establishing first-of-a-kind commercial-scale plants.[147] The economic viability of any process route is crucial for its success. Gasification of biomass feedstocks with subsequent liquefaction *via* Fischer–Tropsch synthesis is one of several routes that are being discussed in that context. The front end of the process, *i.e.* biomass pretreatment and gasification, is a key component that determines the overall process economics.

If used directly, biomass pretreatment would entail drying, grinding and pressurization for feeding purposes. Grinding wood to particles of 200 μm in size may require using about 10% of the primary energy contained in the feedstock,[122] which is a significant amount of energy. Although lock hopper systems similar to the ones used in coal gasification are suitable for pulverized biomass, pneumatic feeding is not possible. Dedicated feeding systems for pulverized biomass into commercial-scale EFG require development. It has been estimated that feeding biomass to an EFG could use up to 5% of the primary feedstock energy.[122] Gasification of larger biomass particles (*e.g.* up to 1 mm) has been considered, as this would require less energy for grinding and could benefit from pneumatic feeding.[122] However, this has not been confirmed with a large-scale EFG.

Alternatively, an energy densification step *via* torrefaction, pyrolysis or hydrothermal liquefaction before EFG could be used for biomass pretreatment. This additional pretreatment step combined with gasification offers several benefits compared to raw biomass gasification: (1) homogenization of the feedstock, (2) reduced effort for feeding and (3) reduced transport

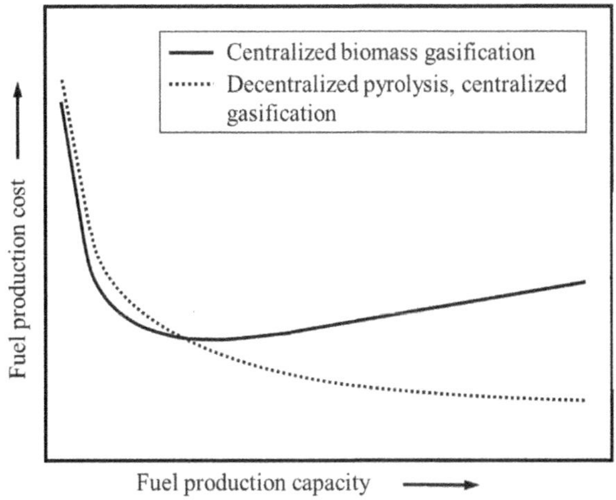

Figure 4.6 Qualitative relationship between fuel production cost and fuel production capacity from biomass gasification and Fischer–Tropsch synthesis for a centralized biomass gasification facility and a spoke-and-hub concept with decentralized pyrolysis and centralized pyrolysis oil gasification.[150] Note: underlying assumptions for the data are found in ref. 150.

cost. In the case of torrefaction, about an 80% grinding energy reduction is feasible[148] due to the more brittle nature of torrefied biomass compared to raw biomass. Furthermore, feeding using a lock hopper system is possible. In the case of pyrolysis or HTL, the bio-oil product requires pumping, which is a more efficient system than lock hoppers.

Apart from the technical feasibility, the reduced transport cost associated with the use of an energy densification step for biomass needs to be considered in relation to the scale of the application. Due to the high capital cost associated with indirect liquefaction processes, economies of scale are used to make them profitable. Most commercial FT plants, which use coal-derived syngas as feedstock, have a capacity of >200 million gallons per year.[149] While the fuel production cost of a BTL plant decreases with increasing scale, the opposite is true for the feedstock transportation cost. The low energy density and high moisture content of raw biomass lead to significant transport costs over a long distance. For a 550-million-gallon biomass-based FT fuel production plant, it has been estimated that the cost of biomass transportation is about 25% of the total FT fuel production cost.[150] The transport cost is expected to increase further with increasing plant size; however, energy densification strategies can mitigate this cost.

Figure 4.6 qualitatively relates the fuel production cost *via* entrained flow gasification followed by Fischer–Tropsch synthesis and the plant production capacity for (1) centralized biomass gasification and (2) small-scale decentralized pyrolysis and centralized gasification of the pyrolysis oil (*i.e.* spoke-and-hub concept). By increasing the production capacity, the fuel cost of a centralized biomass gasification plant goes through a minimum and then increases at higher capacity levels, due to the higher transport cost. At a low capacity, the cost of fuel produced *via* the spoke-and-hub concept is higher than that of fuels produced in a centralized biomass gasification facility, due to the lower efficiency of the process. However, with increasing fuel production capacity, the spoke-and-hub concept offers lower fuel production costs than centralized biomass gasification. The analysis done by Wright *et al.* indicated that decentralized pyrolysis and centralized gasification is more cost efficient only at a capacity of 550 million gallons of fuel per year.[150] It can be expected that the cost will differ greatly, based on the boundary conditions – *e.g.* biomass availability, supply chain and transport costs – and the underlying efficiency of the pyrolysis and gasification processes.

The selection of biomass pretreatment technologies (pyrolysis, HTL and torrefaction) needs to take into consideration the biomass source, scale and type of application. Key challenges for the implementation of such technologies are process control of the densification process and the lack of efficiency data from large-scale facilities from which to assess economic viability reliably. Landmark projects that include energy densification as a pretreatment step before gasification – such as the Bioliq and the BioTfuel projects – are key to demonstrating the technology on a larger scale and mitigating the risk associated with commercial application of the biomass-to-liquid process.

References

1. A. van der Drift and H. Boerrigter, *Synthesis gas from biomass for fuels and chemicals*, ECH-C-06-001, IEA Bioenergy Task 33, 2006.
2. IEA Bioenergy Annual Report, 2019, https://www.ieabioenergy.com/wp-content/uploads/2020/05/IEA-Bioenergy-Annual-Report-2019.pdf (accessed June 2021).
3. P. Basu, *Biomass Gasification, Pyrolysis and Torrefaction: Practical Design and Theory*, Elsevier Academic Press, London, 3rd edn, 2018.
4. N. Dahmen, E. Henrich, A. Kruse and K. Raffelt, in *Biomass to Biofuels: Strategies for Global Industries*, ed. A. A. Vertés, N. Qureshi, H. P. Blaschek and H. Yukawa, John Wiley & Sons, United Kingdom, 1st edn, 2010, ch. 5, pp. 91–122.
5. E. Henrich, N. Dahmen and E. Dinjus, *Proceedings of the 15th European Biomass Conference*, Berlin, Germany, 2007.
6. S. Kumar and R. K. Sani, *Biorefining of Biomass to Biofuels: Opportunities and Perception*, Springer, Switzerland, 2018.
7. R. Luque and J. Speight, *Gasification for Synthetic Fuel Production: Fundamentals, Processes and Applications*, Woodhead Publishing, Cambridge, 2015.
8. P. Lan, L. Lan, T. Xie and A. Liao, *Energy Sources, Part A*, 2014, **36**, 242.
9. A. de Klerk, R. Sehdev and R. R. Lehoux, *A Process for Producing Synthetic Jet Fuel*, *USA Pat.*, WO2020154810, 2020.
10. S. Zhang, X. Yang, H. Zhang, C. Chu, K. Zheng, M. Ju and L. Liu, *Molecules*, 2019, **24**, 2250.

11. M. Crocker and E. Santillan-Jimenez, *Chemical Catalysts for Biomass Upgrading*, John Wiley & Sons, Wiley VCH, Germany, 2019.
12. R. O. Arazo, D. A. D. Genuino, M. de Luna and S. Capareda, *Sustainable Environ. Res.*, 2017, **27**, 7.
13. J. Eke, J. Onwudili and A. Bridgewater, *Waste Biomass Valorization*, 2020, **11**, 3711.
14. J. Martin and A. Boateng, *Fuel*, 2014, **133**, 34.
15. Y. Solantausta, N. Nylund, M. Westerholm, T. Koljonen and A. Oasmaa, *Bioresour. Technol.*, 1993, **46**, 177.
16. J. Carrasco, S. Gunukula, A. Boateng, C. Mullen, W. DeSisto and M. Wheeler, *Fuel*, 2017, **193**, 477.
17. R. Ware, S. Rowland, R. Rodgers and A. Marshall, *Energy Fuels*, 2021, **31**, 8210.
18. A. Casoni, M. Bidegain, M. Cubitto, N. Curvetto and M. Volpe, *Bioresource*, 2015, **177**, 406.
19. S. Rabiu, M. Auta and A. Koo, *Egypt. J. Pet.*, 2018, **27**, 590.
20. X. Wei, X. Xue, L. Wu, H. Yu, J. Liang and Y. Sun, *Energy*, 2020, **212**, 188692.
21. M. AbuBakar and J. Titiloye, *J. Anal. Appl. Pyrolysis*, 2013, **103**, 362.
22. N. Zainan, S. Srivatsa, F. Li and S. Bhattacharya, *Fuel*, 2018, **223**, 12.
23. M. Haghighat, N. Majidian, A. Hallajisani and M. Samipourgiri, *Sustainability Energy Technol. Assess.*, 2020, **42**, 100870.
24. J. V. Milato, R. J. Franca and M. R. C. Marques, *Environ. Technol.*, 2021, **42**, 1013.
25. U.S. Department of Agriculture Forest Service, Forest Products Laboratory, Madison, Wis. 1970. Thermal Degradation of Wood. FPL 130, https://www.fpl.fs.fed.us/documnts/fplrp/fplrp130.pdf (accessed July 2021).
26. K. Rajendran, R. Lin, D. M. Wall and J. D. Murphy, in *Sustainable Resource Recovery and Zero Waste Approaches*, ed. M. J. Taherzadeh, K. Bolton, J. Wong and A. Pandey, Elsevier, Missouri, 1st edn, 2019, 5, pp. 65–78.
27. A. Oasmaa and S. Czernik, *Energy Fuels*, 1999, **13**, 914.
28. M. I. Jahirul, M. G. Rasul, A. A. Chowdhary and N. Ashwath, *Energies*, 2012, **5**, 4952.
29. J. L. Klinger, T. L. Westover, R. M. Emerson, C. L. Williams, S. Hernandez, G. D. Monson and J. C. Ryan, *Appl. Energy*, 2018, **228**, 535.
30. F. Abnisa, W. M. A. Wan Daud, W. N. W. Husin and J. N. Sahu, *Biomass Bioenergy*, 2011, **35**, 1863.
31. P. T. Williams and N. Nugranad, *Energy*, 2000, **25**, 493.
32. A. Kumar and R. Anand, in *Advances in Eco-Fuels for a Sustainable Environment*, ed. K. Azad, Woodhead Publishing Series, United Kingdom, 1st edn, 2019, pp. 119–155.
33. Y. Zhang, P. Cui, S. Chen, N. Liu, K. Zhou, L. Ding, M. Fan, Y. Min, Y. Cheng, Y. Wang, B. Liu, B. Li and R. Ruan, in *Sustainable Resource*

Recovery and Zero Waste Approaches, ed. M. J. Taherzadeh, K. Bolton, J. Wong and A. Pandey, Elsevier, Missouri, 1st edn, 2019, 14, pp. 193–206.
34. A. Pawar, N. Pawar and B. Salvi, *J. Mater. Cycles Waste Manage.*, 2020, **22**, 1712.
35. Karlsruhe Institute of Technology, Germany, The Bioliq process, https://www.bioliq.de/english/55.php (accessed July 2021).
36. N. Dahmen, E. Dinjus, T. Kolb, U. Arnold, H. Leibold and R. Stahl, *Environ. Prog. Sustainable Energy*, 2012, **31**, 176.
37. T. Bhaskar, B. Bhavya, R. Singh, D. Naik, A. Kumar and H. Goyal, in *Biofuels*, ed. A. Pandey, C. Larroche, S. Ricke, C. Dussap and E. Gnansounou, Academic Press, The Netherlands, 2011, 3, pp. 51–77.
38. J. Garcia-Nunez, M. Pelaez-Samaniego, M. Garcia-Perez, I. Fonts, J. Abrego, R. Westerhof and M. Garcia-Perez, *Energy Fuels*, 2017, **31**, 5751.
39. K. Wang and R. Brown, in *Fast Pyrolysis of Biomass: Advances in Science and Technology*, ed. K. Wang and R. Brown, Royal Society of Chemistry, United Kingdom, 1st edn, 2017, 1, pp. 1–11.
40. R. Venderbosch and W. Prins, *Biofuels*, 2010, **4**, 178.
41. http://www.ensyn.com/quebec.html (accessed July 2021).
42. https://bio4products.eu/wp-content/uploads/2019/11/2.-Gerhard-Muggen_Market-status-of-pyrolysis-plants.pdf (accessed July 2021).
43. C. Pfitzer, N. Dahmen, N. Troger, F. Weirich, J. Sauer, A. Gunther and M. Muller-Hagedorn, *Energy Fuels*, 2016, **30**, 8047.
44. A. R. K. Gollakota, N. Kishore and S. Gu, *Renewable Sustainable Energy Rev.*, 2018, **81**, 1378.
45. D. Elliot, P. Biller, A. Ross, A. Schmidt and S. Jones, *Bioresour. Technol.*, 2015, **178**, 147.
46. I. Sintamarean, I. Grigoras, C. Jensen, S. Toor, T. Perdersenm and L. Rosendahl, *Biomass Convers. Biorefin.*, 2017, 7(4), 425.
47. S. Bjelic, J. Yu, B. Iversen, M. Glausius and P. Biller, *Energy Fuels*, 2018, **32**, 3579.
48. J. Jarvis, J. Billing, R. Hallen, A. Schmidt and T. Schaub, *Energy Fuels*, 2017, **31**, 2896.
49. O. Djandja, Z. Wang, L. Chen, L. Qin, F. Wang, Y. Xu and P. Duan, *Energy Fuels*, 2020, **34**(10), 11723.
50. N. Latif, M. Ong and S. Nomanbhay, *Eng. Life Sci.*, 2019, **19**, 246.
51. S. N. Sahu, N. K. Sahoo, S. N. Naik and D. M. Mahapatra, in *Bioreactors*, L. Singh, A. Yousuf and D. M. Mahapatra, Elsevier, United Kingdom, 1st edn, 2020, 12, pp. 195–213.
52. H. Huang and X. Yuan, *Prog. Energy Combust. Sci.*, 2015, **49**, 59.
53. S. Toor, L. Rosendahl and A. Rudolf, *Energy*, 2011, **26**, 2328.
54. A. Demirbas, *Energy Sources*, 2005, **27**, 1235.
55. P. Bulsink, F. de, M. Mercader, L. Sandstrom, B. van de Beld, F. Preto, A. Zacher, A. Oasmaa, N. Dahmen, A. Funke and B. Bronson, *Fuels*, 2020, **34**(9), 11123.

56. J. Akhatar and N. Amin, *Renewable Sustainable Energy*, 2011, **32**, 459.
57. J. Akhatar and N. A. S. Amin, *Renewable Sustainable Energy Rev.*, 2011, **15**, 1615.
58. Z. Zhu, L. Rosendahl, S. S. Toor, D. Yu and G. Chen, *Appl. Energy*, 2015, **137**, 183.
59. C. U. Jensen, J. K. Rodriguez Guerrero, S. Karatzos, G. Olofsson and S. B. Iversen, *Biomass Convers. Biorefin.*, 2017, 7, 495.
60. B. Hao, D. Xu, G. Jiang, T. A. Sabri, Z. Jing and Y. Guo, *Green Chem.*, 2021, **23**, 1562.
61. M. Jindal and M. Jha, *Rev. Chem. Eng.*, 2016, **32**, 459.
62. S. Brand, F. Hardi, J. Kim and D. Suh, *Energy*, 2014, **68**, 420.
63. B. E. Poling, J. P. Prausnitz and J. P. O'Connell, *The Properties of Gases and Liquids*, McGraw-Hill, New York, 5th edn, 2001.
64. K. Anastasakis, P. Bille, R. B. Madsen, M. Glasius and L. Johnnsen, *Energies*, 2018, **11**(10), 2695.
65. A. Peterson, F. Vogel, R. Lachance, M. Froling, M. Antal Jr and J. Tester, *Energy Environ. Sci.*, 2008, **1**, 32.
66. N. Montesantos and M. Maschietti, *Energies*, 2020, **13**, 1600.
67. D. Castello, T. Pedersen and L. Rosendahl, *Energies*, 2018, **11**, 3165.
68. Licella, https://www.licella.com.au/news/dow-invests-in-mura-technology-using-licellas-cat-htr-technology-at-its-core-to-accelerate-roll-out-of-advanced-recycling-solution-for-plastics/ (accessed July 2021).
69. D. Chen, J. Zhou, Q. Zhang and X. Zhu, *Renewable Sustainable Energy Rev.*, 2014, **40**, 69.
70. T.-S. Kim, J.-Y. Kim, K.-H. Kim, S. Lee, D. Choi, I.-G. Choi and J. Choi, *J. Anal. Appl. Pyrolysis*, 2012, **95**, 118.
71. J. Diebold, *A Review of the Chemical and Physical Mechanisms of the Storage Stability of Fast Pyrolysis Bio-Oils*, Report NREL/SR-570-27613, NREL, Golden, 2000.
72. M. Ringer, V. Putsche and J. Scahill, *Large-scale Pyrolysis Production: A Technology Assessment and Economic Analysis*, NREL/TP-510-37779, Golden, 2006.
73. F. A. Agblevor, S. Bessler and R. J. Evans, in *Proceedings of Biomass Pyrolysis Oil Properties and Combustion Meeting*, Milne, T.A., 77–89, 1994.
74. A. Bridgewater, *Biomass Bioenergy*, 2012, **25**, 68.
75. U. Jena and K. Das, *Energy Fuels*, 2011, **25**, 5472.
76. J. Lehto, A. Oasmaa, Y. Solantausta, M. Kyto and D. Chiaramonti, *Fuel oil quality and combustion of fast pyrolysis bio-oils*, 2013, Available at: https://www.vttresearch.com/sites/default/files/pdf/technology/2013/T87.pdf (accessed July 2021).
77. E. Henrick, N. Dahmen and A. Niebel, *Study on energy carrier use for entrained flow gasification*, 2015, Available at: https://bioboost.eu/uploads/files/bioboost_d5.8_study_on_energy_carrier_use_for_entrained_flow_gasification_ver (accessed November 2021).
78. A. Oasmaa, D. Elliot and J. Korhonen, *Energy Fuels*, 2010, **24**, 6548.

79. K. Svoboda, M. Pohořelý, M. Hartman and J. Martinec, *Fuel Process. Technol.*, 2009, **90**, 629.
80. P. Bergman, A. R. Boersma, R. Zwart and J. Kiel, *Torrefaction for Biomass Co-firing in Existing Coal-Fired Power Stations*, ECN-C-05-013, Netherlands, 2005.
81. W.-H. Chen, Y.-Y. Lin, Y.-S. Chu, A. T. Ubando, P. Show, H. Ong, J.-S. Chang, S.-H. Ho, A. Culaba, A. Petrissans and M. Petrissans, *Prog. Energy Combust. Sci.*, 2021, **82**, 100887.
82. B. Acharya, I. Sule and A. Dutta, *Biomass Convers. Biorefin.*, 2012, **2**, 349.
83. C. ZhiWen, W. Minfeng, R. Yongzhi, J. Enchen, J. Yang and L. Weizhen, Biomass torrefaction: A promising pretreatment technology for biomass utilization, *IOP Conf. Ser.: Earth Environ. Sci.*, 2018, **113**, 012201–012207.
84. P. Bergman, J. Kiel and H. Veringa, *Combined Torrefaction and Pelletization: The Top Process*, ECN-C-05-073, Netherlands, 2005.
85. Q. Chen, J. S. Zhou and B. J. Liu, *Chi. Sci. Bull.*, 2011, **56**, 1449.
86. W.-H. Chen, J. Peng and X. Bi, *Renewable Sustainable Energy Rev.*, 2015, **44**, 847.
87. M. J. C. van Der Stelt, H. Gerhauser, J. H. A. Kiel and K. J. Ptasinski, *Biomass Bioenergy*, 2011, **35**, 3748.
88. S. Hill, W. Grigsby and P. Hall, *Biomass Bioenergy*, 2013, **56**, 92.
89. T. Mamvura and G. Danha, *Heliyon*, 2020, **6**, e03531.
90. J. Kiel, F. Verhoeff, H. Gerhauser and B. Meuleman, BO2 Technology for biomass upgrading into solid fuel - pilot scale testing and market implementation, in *European Biomass Conference & Exhibition*, Netherlands, 2008.
91. H. Yang, R. Yan, H. Chen, D. Lee and C. Zheng, *Fuel*, 2007, **86**, 1781.
92. C. Gomez, E. Velo, F. Barontini and V. Cozzani, *Ind. Eng. Chem. Res.*, 2009, **48**, 10222.
93. B. Acharya, A. Dutta and J. Minaret, *Sustainability Energy Technol. Assess.*, 2015, **12**, 26.
94. H. Yun, Z. Wang, R. Wang, X. Bi and W.-H. Chen, *Energy Res.*, 2021, **0**, 97.
95. K. Lampe, G. Grund and R. Erpelding, Biocoal preparation - biomass a suitable fuel for industrial processes, in *Proceedings of the 6th European Metallurgical Conference*, Dusseldorf, 2011.
96. C. Wang, J. Peng, H. Li, X. Bi, R. Legros, C. Lim and S. Sokhansanj, *Bioresour. Technol.*, 2013, **127**, 318.
97. A. Alvarez, I. Gutierrez, C. Pizarro, A. G. Lavin and J. L. Bueno, *Energy Sustain*, 2017, **224**, 247.
98. Q.-V. Bach and K.-Q. Tran, *Energy Procedia*, 2015, **75**, 150.
99. C. He, C. Tang, C. Li, J. Yuan, K.-Q. Tran, Q.-V. Bach, R. Qiu and Y. Yang, *Renewable Sustainable Energy Rev.*, 2018, **91**, 259.
100. M. Cremers, J. Koppejan, J. Middelkamp, J. Witkamp, S. Sokhansanj, S. Melin and S. Madrali, *Status overview of torrefaction technologies:*

A review of the commercialisation status of biomass torrefaction, IEA Bioenergy: Task 32, 2015.
101. P. Basu, S. Rao and A. Dhungana, *Can. J. Chem. Eng.*, 2013, **91**, 466.
102. S. Negi, G. Jaswal, K. Dass, K. Mazumder, S. Elumalai and J. Roy, *Rev. Environ. Sci. Bio/Technol.*, 2020, **19**, 463.
103. V. Gude, P. Patil, E. Martinez-Guerra, S. Deng and N. Nirmalakhandan, *Sustainable Chem. Processes*, 2013, **1**(5), 1.
104. P. Stępień, J. Pulka and A. Białowiec, in *Pyrolysis*, ed., M. Samer, IntechOpen, Rijeka, 1st edn, 2017, p. 37.
105. S. J. Tumuluru, S. Sokhansanj, J. R. Hess, C. T. Wright and R. D. Boardman, *Ind. Biotechnol.*, 2011, **7**, 384.
106. B. Evangelista, P. Arlabosse, A. Govin, S. Salvador and O. Bonnefoy, *Fuel*, 2018, **214**, 590.
107. A. Boskovic, P. Basu and P. Amyotte, *Can. J. Chem. Eng.*, 2015, **93**, 658.
108. C. Huescar Medina, H. N. Phylaktou, G. E. Andrews and B. M. Gibbs, *Biomass Bioenergy*, 2015, **79**, 116.
109. K. van Heek and H. Muhlen, in *Fundamental Issues in Control of Carbon Gasification Reactivity*, ed., J. Lahaye and P. Ehrburger, Springer, Dordrecht, 1991, pp. 1–34.
110. J. Speight, *Handbook of Gasification Technology: Science, Processes, and Applications*, John Wiley & Sons, USA, 2020.
111. Y. Chhiti and S. Salvador, *Gasification for Practical Applications*, ed., Y. Yun, InTech, 2012.
112. Swedish Gasification Centre, http://www.sfc-sweden.se/ (accessed February 2021).
113. T. Kolb, M. Aigner, R. Kneer, M. Muller, R. Weber and N. Djordjevic, *J. Energy Inst.*, 2016, **89**, 485.
114. M. Marklund, R. Tegman and R. Gebart, *IFRF Combustion J.*, 2008, 200801.
115. A. Lefebvre and V. McDonell, *Atomization and Sprays*, CRC Press, Boca Raton, 2nd edn, 2017.
116. K. Kuo and R. Acharya, *Fundamentals of Turbulent and Multiphase Combustion*, John Wiley & Sons, USA, 2012.
117. J. Lehto, A. Oasmaa, Y. Solantausta, M. Kyto and D. Chiaramonti, *Appl. Energy*, 2014, **116**, 178.
118. M. Broumand, M. Shahzeb Khan, S. Yun, Z. Hong and M. Thompson, *Fuel*, 2020, **276**, 118035.
119. M. Gullberg and M. Marklund, *At. Sprays*, 2012, **22**, 897.
120. M. Eberhard, U. Santo, B. Michelfelder, A. Gunther, P. Weigand, J. Matthes, P. Waibel, V. Hagenmeyer and T. Kolb, *Chem. Bio. Eng. Rev*, 2020, 7, 106.
121. K. Svoboda, M. Pohorely, M. Hartman and J. Martinec, *Fuel Process. Technol.*, 2009, **90**, 629.
122. A. van der Drift, H. Boerrigter, B. Coda, M. K. Cieplik and K. Hemmes, *Entrained Flow Gasification of Biomass – Ash Behaviour, Feeding Issues, and System Analyses*, ECN-C-04-039, Netherlands, 2004.

123. G. Rovero, M. Curti and G. Cavaglia, in *Advances in Chemical Engineering*, ed. Z. Nawaz, Intech, Rijeka, 1st edn, 2012, p. 405.
124. P. Abdelnur, B. Vaz, J. Rocha, M. de Almeida, M. Teixeira and R. Pereira, *Energy Fuels*, 2013, **27**, 6646.
125. M. Wornat, B. Porter and N. Yang, *Energy Fuels*, 1994, **8**(5), 1131.
126. J.-L. Zheng, M.-Q. Zhu, K. Kang and R.-C. Sun, *Sustainable Energy Fuels*, 2019, **3**, 1600.
127. C. Di Blasi, *Prog. Energy Combust. Sci.*, 2009, **35**(2), 121.
128. A. Tremel, J. Stemann, M. Herrmann, B. Erlach and H. Spliethoff, *Fuel*, 2012, **102**, 396.
129. D. Bushnell, C. Haluzok and A. Dadkhah-Nikoo, *Biomass Fuel Characterization: Testing and Evaluating the Combustion Characteristics of Selected Biomass Fuels, Final Report*, Portland, Oregon, 1990.
130. M. Patah, PhD thesis, University of Canterbury, 2016.
131. C. Higman and M. van de Burgt, *Gasification*, Elsevier, United Kingdom, 2nd edn, 2008.
132. P. Carlsson, C. Ma, R. Molinder, F. Weiland, H. Wiinikka, M. Ohman and O. Ohrman, *Energy Fuels*, 2014, **28**, 6941.
133. E. Leijenhorst, D. Assink, L. van de Beld, F. Weiland, H. Wiinikka, P. Carlsson and O. Ohrman, *Biomass Bioenergy*, 2015, **79**, 166.
134. A.-G. Collot, *Int. J. Coal Geol.*, 2006, **65**, 191.
135. K. Dhanavath, K. Shah, M. Islam, A. Ronte, R. Parthasarathy, S. Bhargava and S. Bankupalli, *J. Environ. Chem. Eng.*, 2018, **6**, 1242.
136. L. Liao, J. Zheng, Y. Zhang, C. Li and C. Yuan, *Fuel*, 2021, **289**, 119919.
137. X. Ku, J. Lin and F. Yuan, *Energy Fuels*, 2016, **30**, 4053.
138. C. Couhert, S. Salvador and J.-M. Commandre, *Fuel*, 2009, **88**, 2286.
139. W.-H. Chen, C.-J. Chen, C.-I. Hung, C.-H. Shen and H.-W. Hsu, *Appl. Energy*, 2013, **112**, 421.
140. F. Weiland, M. Nordwaeger, I. Olofsson, H. Wiinikka and A. Nordin, *Fuel Process. Technol.*, 2014, **125**, 51.
141. J.-L. Zheng, M.-Q. Zhu, J.-L. Wen and R.-C. Sun, *Bioresour. Technol.*, 2016, **211**, 164.
142. J. Marda, J. Dibenedetto, S. McKibben, R. Evans, S. Czernik, R. French and A. Dean, *Int. J. Hydrogen Energy*, 2009, **34**, 8519.
143. A. Bach-Oller, K. Kirtania, E. Furusjo and K. Umeki, *Fuel*, 2017, **202**, 46.
144. IFPEN, IFP Energies Nouvelles, 2021, Available at: https://www.ifpenergiesnouvelles.com/article/biotfuelr-successful-advanced-biofuels-production-woody-biomass-demonstration-units (accessed July 2021).
145. S. Cavagnol, K. Covella and M. Muller-Hagedorn, *Fuel*, 2018, **228**, 369.
146. N. Dahmen, B. Zimmerlin and M. Eberhard, Bioliq – Pilot plant for the preparation of synthetic fuel – operating experience, in *European Biomass Conference and Exhibition Proceedings*, Lisbon, 2019.

147. I. Hannula and E. Kurkela, Liquid transportation fuel via large-scale fluidised-bed gasification of lignocellulosic biomass, VTT Technology 91, http://www.vtt.fi/publications/index.jsp (accessed July 2021).
148. P. Bergman and J. Kiel, Torrefaction for biomass upgrading, in *Proceedings 14th European Biomass Conference*, Paris, 2005.
149. S. Gunukula, A. Daigneault, A. Boateng, C. Mullen, W. DeSisto and M. Wheeler, *Fuel*, 2019, **249**, 326.
150. M. Wright, R. Brown and A. Boateng, *Biofuels, Bioprod. Biorefin.*, 2008, **2**, 229.

CHAPTER 5

Syngas Conditioning (Catalyst, Process: Sulfur and Tar Cl, F)

PARVIN KIANI, MARYAM MESHKSAR, MOHAMMAD-AMIN MAKAREM AND MOHAMMAD-REZA RAHIMPOUR*

Department of Chemical Engineering, Shiraz University, Shiraz, Iran
*Email: rahimpor23@gmail.com

5.1 Introduction

Currently a lot of research and development is happening with synthesis gas (syngas), and it is receiving too much attention. Syngas is currently burned in boilers and power generation sectors and is also used when synthesizing different chemicals, including methanol (MeOH), ethanol, methane, *etc.* In order to ensure a long life of heat and power generators, hydrogen sulfide (H_2S) should be eliminated. The safe quantity should be fewer than 200 parts per million (ppm v/v),[1] which means syngas needs to be improved. This involves significant operating costs and energy consumption. The first step is to determine what the gas is made of, so that the right technologies can be used to remove the pollutants.[2]

If biomass-derived gas (either syngas or bio-syngas) is to be used for biofuel production, it must first be thoroughly cleaned to eliminate impurities and potential catalyst-toxic materials, and it must have the required qualitative composition before being sold for use in various applications. Syngas can be made in different ways (*i.e.* by anaerobic digestion, gasification, pyrolysis, reforming, *etc.*).[3] The composition of syngas produced

Catalysis Series No. 44
Chemicals and Fuels from Biomass *via* Fischer–Tropsch Synthesis: A Route to Sustainability
Edited by Joshua Gorimbo, Xinying Liu, Yali Yao and Diane Hildebrandt

Published by the Royal Society of Chemistry, www.rsc.org

from the gasification process, as per different sources, is summarized in Table 5.1. The produced syngas is usually saturated by water vapor and includes varying levels of H_2S in addition to various amounts of other pollutants. The pollution source depends on the type of biomass as well as the process conditions through which syngas product is generated.[4,5]

Contaminants have qualities that make the gas useless and potentially toxic in certain processes, if they are not removed. For instance, H_2S is a poisonous gas that has a distinct, unpleasant odor reminiscent of rotten eggs.[6] When combined with water vapor in the syngas, it forms acidic compounds that can cause corrosion of devices. To avoid this and the negative influences of other impurities, the syngas must be dewatered, and all contaminants should be eliminated before it is transported to a conversion unit or sold.[10,11]

H_2S, H_2O, CO_2 and halogenated chemicals are the most common (but not always the only) impurities that may need to be removed in gas cleaning equipment. Desulfurization is used to protect pipelines and other equipment from corrosion and to reduce harmful hydrogen sulfide levels in the workplace and in the usage units.[12,13] Furthermore, if carbonyl sulfide (COS) and other sulfur-containing substances – like thiol compounds (RSH, also commercially known as mercaptans) and H_2S – are not eliminated, burning of which creates sulfur trioxide and sulfur dioxide, which are much more toxic than H_2S.[14] Simultaneously, the existence of H_2S in the flow gas reduces the stack gas dew point (the temperature at which gas steam is chilled for being saturated with H_2O in the vapor phase). H_2SO_3 and H_2CO_3 are another two corrosive impurities of syngas that are produced by the reaction of water and SO_2 or water and CO_2, respectively, as shown in eqn (5.1) and (5.2):

$$SO_2 + H_2O \rightarrow H_2SO_3 \tag{5.1}$$

$$CO_2 + H_2O \rightarrow H_2CO_3 \tag{5.2}$$

Removing water from syngas is therefore critical not only because of the risk of H_2O reacting with pollutants in the gas stream but also due to the risk of H_2O vapor condensation in the line and due to the toxic acidic solution that forms as a result of dissolving H_2S in water. In addition, when syngas is kept

Table 5.1 Composition of produced syngas, as per various sources.[6–9]

Content	Vol% 6	Mole fraction 7	Vol% 8	Mole fraction 9
H_2	40.000	0.293	80.667	0.109
CH_4	10.000	0.000	4.527	0.045
CO	25.000	0.287	1.176	0.222
CO_2	21.000	0.118	1.571	0.115
N_2	1.500	0.030	—	0.509
C_2H_4	2.500	0.272	1.884	—
C_2H_6	—	—	1.867	—
C_3H_6	—	—	7.433	—

at a high pressure, a low dew point is ensured to prevent freezing and condensation.[15] Furthermore, if noncombustible CO_2 is not eliminated, the energy content of the syngas is reduced, and the presence of this pollutant may have an impact on the environment.[16]

The selectivity with which one acidic element in gas is removed from the process in preference to another is referred to as "process selectivity". Some methods remove both H_2S and CO_2, while others are designed to remove just H_2S. Recognizing the process selectivity for, say, H_2S removal *vs.* CO_2 removal to ensure minimal element concentrations in the product necessitates considering the CO_2-to-H_2S ratio in the gas stream.[17,18]

Besides these syngas impurities, the produced syngas contains some contaminants, such as particulate matter (PM), alkali metals (*i.e.* Na, K), condensable hydrocarbon derivatives (*i.e.* tar products), nitrogen-containing derivatives (*i.e.* NH_3, HCN), sulfur-containing derivatives (*i.e.* H_2S, COS, CS_2) and halogen-containing derivatives (*i.e.* HCl, HBr, HF) that should be eliminated before the syngas is used.[19] The problems caused by each of these impurities, as well as their maximum allowable amounts for the Fischer–Tropsch (FT) process, are indicated in Table 5.2.[20]

Due to the obvious complicated nature of the feedstock and the existence of one or more of the corrosive or poisonous contaminants in the table, the feedstock is typically not directly applied as fuel in the gas processing business,[26] as it is critical to prevent any gas from entering the numerous reactors from the contaminants. After being created, the basic raw syngas is usually processed into chemical or physical alterations (refining). In this chapter, the different methods used for syngas processing are discussed in detail.

5.2 Syngas Impurity Types and Purification Methods

A schematic of different syngas processing units is provided in Figure 5.1. As can be seen, after removing the corrosive components, solid particles are separated from the syngas produced, after which nitrogen-containing components, alkali components, and methane, tar and trace contaminants are eliminated. In the following subsections, different produced syngas impurities are described in detail, in addition to the elimination techniques used.

5.2.1 Acid Gas Removal Processes

COS, H_2S, and CO_2 are the most common acidic gases produced during the gasification or reforming process. Current methods for eliminating these gases from raw syngas usually require counter-current absorption in an absorber column using a regenerating solvent. There are three different types of solvent processes:[25,27]

- Chemical solvents like alkanolamine-base solvents.
- Physical solvents, such as MeOH, polyethylene glycol dimethyl ethers, *etc.*

Table 5.2 Syngas impurities, their maximum allowable amounts for the FT process and the problems caused.[20–25]

Impurity calssification	Impurities	Ref. 24 and 25	Ref. 23	Ref. 22	Ref. 20	Problems[21]
Sulfur-containing compounds	H_2S COS CS_2	$H_2S + COS + CS_2 < 0.06$ mol ppm	$H_2S + COS < 0.01$ ppm[a]	$H_2S + COS < 1$ mol ppm	$H_2S + COS + CS_2 < 1$ ppm	Could be a harmful pollutant and cause metals to corrode in acid. Sulfur compounds irritate the catalyst.
Nitrogen-containing compounds	HCN NH_3 NO_x	< 0.01 mol ppm < 10 mol ppm < 0.1 mol ppm	$HCN + NH_3 < 0.02$ ppm[a] —	$HCN + NH_3 < 1$ mol ppm —	$HCN + NH_3 < 1$ ppm —	During combustion, NO_x components will be produced, which leads to pollution issues. In addition, the catalysts are vulnerable to nitrogen molecules.
Alkali-metals-containing compounds	Na K	—	$Na + K < 0.01$ ppm[a]	$Na + K < 0.01$ mol ppm	$Na + K < 10$ ppbv	Can result in high-temperature corrosion of metal and bed defluidization. The vapor phase is where alkali metals occur.
Halide-containing compounds	HF HCl HBr	$HF + HCl + HBr < 0.01$ mol ppm	$HCl < 0.01$ ppm[a]	$HF + HCl + HBr < 0.01$ mol ppm	$HF + HCl + HBr < 10$ ppbv	—

[a]Mass or mol ppm not specified.

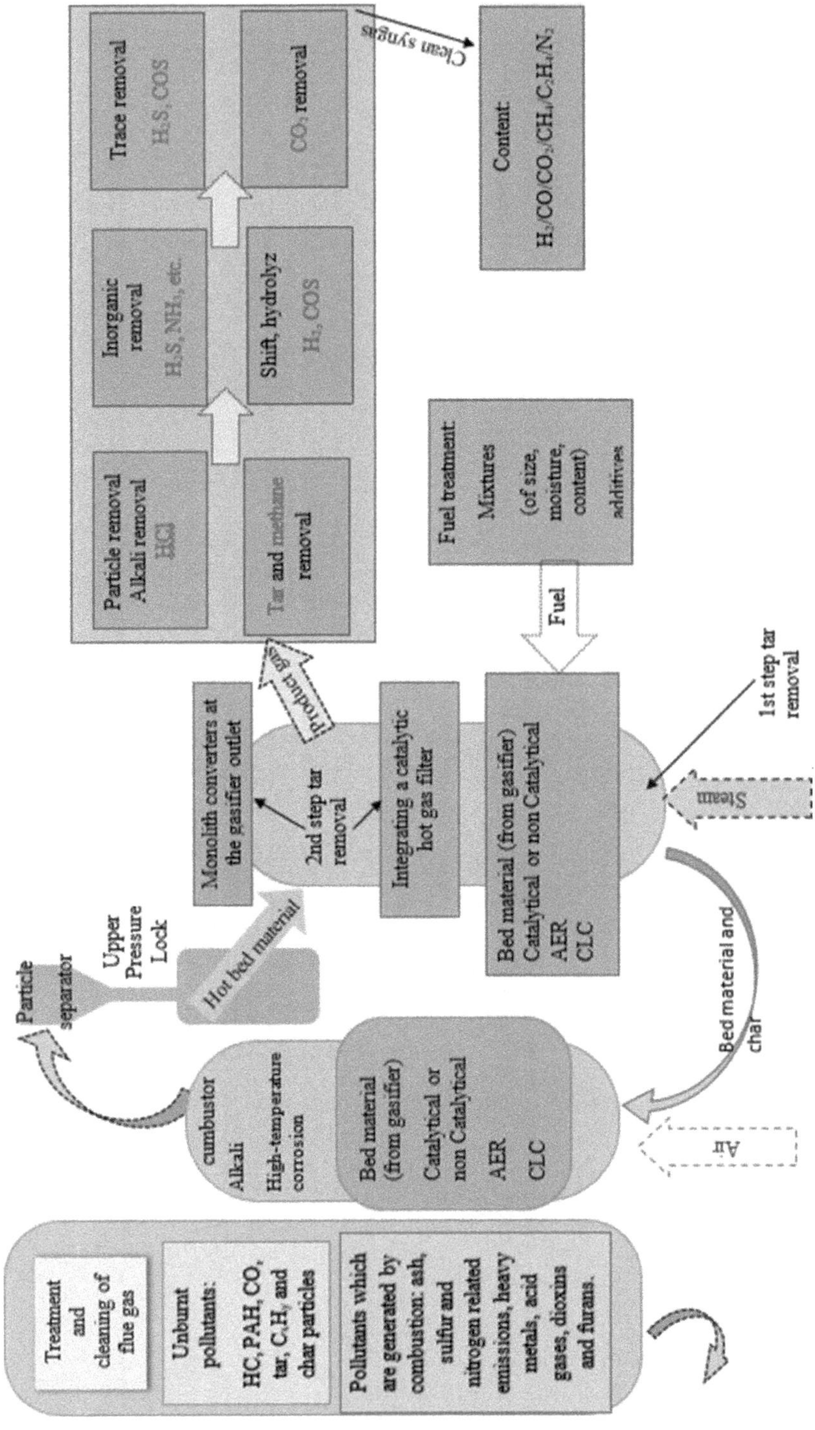

Figure 5.1 Schematic of different syngas processing units.

- Hybrid solvents (physical and chemical solvents), including alkanolamine (*e.g.* DIPA) and sulfolane.

Figure 5.2 shows each of these types of solvents and the process used for acid gas removal (AGR) from syngas.

Other AGR technologies include adsorbents, membranes and cryogenic fractionation (see Figure 5.2). The primary goal of these techniques is to decrease the acid gas content to as low a level as is feasible while ensuring the process remains cost-effective. To ensure complete sulfur removal and the low sulfur percentage criterion necessary to use as an FT synthesis catalyst, a final purification step is demanded, regardless of the AGR technology used.[18]

5.2.1.1 Physical Absorption Processes for CO_2 and H_2S Removal

As previously mentioned, absorption processes using appropriate solvents are the major industrial technologies for AGR. Given the definition of the driving force for the absorption process as the difference between the partial pressure of acid gas and the applied liquid phase, high pressure is needed to increase AGR. Therefore, it is economical to use physical solvents gas flow

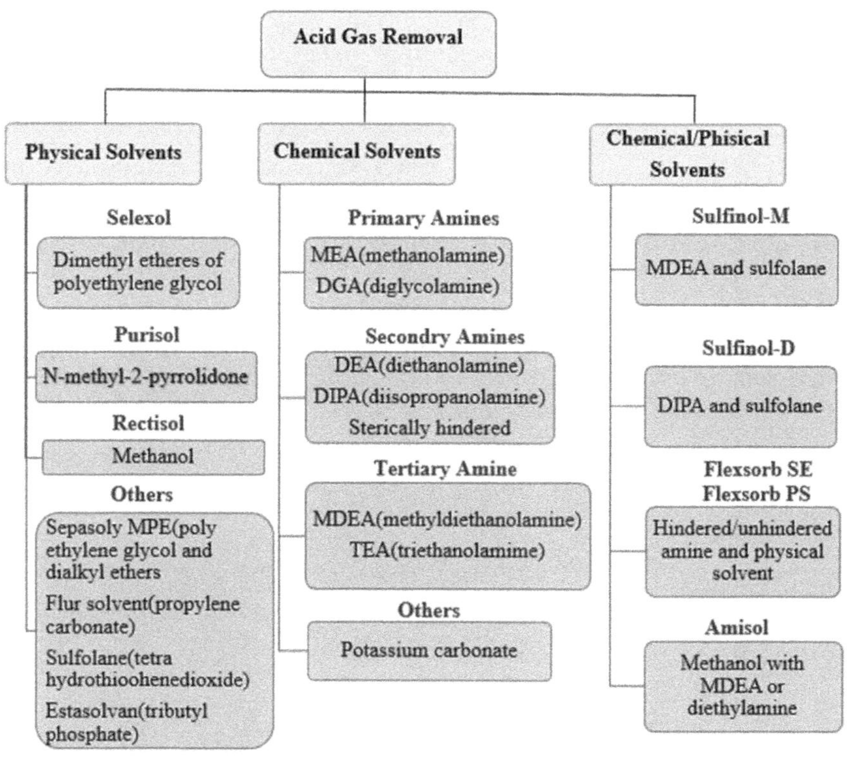

Figure 5.2 AGR solvents and the processes used.

streams with high pressure or high acidic component concentrations.[21] Besides the operating pressure, heavy hydrocarbon concentration in the feedstream is another factor to be considered when choosing a solvent. As in high heavy hydrocarbon concentration, co-absorption increases between hydrocarbons, so physical solvents are not a good choice. Due to the low quantity of heavy hydrocarbons in the synthesis gas stream produced, physical absorption can be used to purify syngas from acid components.[28]

The four most commonly used physical solvents are dimethyl ether of polyethylene glycol (DEPG), MeOH, *N*-methyl-2-pyrrolidone (NMP), and propylene carbonate (PC). Many reported studies indicate how and where they can be used.[29] Table 5.3 presents the physical properties and the solubility of different gases in these physical solvents.[29,30] The main benefits of physical solvents can be summarized as their higher COS and H_2S absorption selectivity than CO_2, low required heat for regeneration, low circulation rates and good stability.[31] Despite the mentioned advantages of physical solvents, they have some drawbacks, such as high capital cost in addition to low operational temperatures and no complete hydrogen sulfide removal achievement.[32]

DEPG is used as a physical absorber for removing CO_2, H_2S and mercaptans from the syngas stream. Coastal Chemical Company (Coastal AGR), UOP (Selexol) and Dow (Selexol) all have a license to manufacture DEPG.

Table 5.3 Physical properties and solubility of different gases in different physical solvents.[29,30]

Solvent	DEPG	PC	NMP	MeOH
Process name	Selexol or Coastal	Fluor solvent	Purisol	Rectisol
Molecular weight	280	102	99	32
Viscosity at 298.15 K (Pa s)	0.0058	0.003	0.00165	0.0006
Specific gravity at 298.15 K (kg m^{-3})	1030	1195	1027	785
Freezing point (K)	245.15	225.15	249.15	181.15
Vapor pressure at 298.15 K (Pa)	0.10	11.33	53.33	16665.25
Thermal conductivity (W (m K)$^{-1}$)	0.19	0.21	0.16	0.21
Boiling point at 101324.72 Pa (K)	548.15	513.15	475.15	338.15
Specific heat at 298.15 K (J (kg K)$^{-1}$)	0.49	0.34	0.40	0.57
Maximum operating temperature (K)	448.15	338.15	—	—
CO_2 solubility at 101 325 Pa and 298.15 K (ppm)	3.63	3.57	3.53	13.46[a]
CO solubility at 101 325 Pa and 298.15 K (ppm)	0.10	0.07	0.07	0.27[a]
H_2 solubility at 101 325 Pa and 298.15 K (ppm)	0.05	0.02	0.03	0.07[a]
COS solubility at 101 325 Pa and 298.15 K (ppm)	8.46	9.73	15.10	52.75[a]
CH_4 solubility at 101 325 Pa and 298.15 K (ppm)	0.24	0.26	0.13	0.67[a]
H_2S solubility at 101 325 Pa and 298.15 K (ppm)	32.40	36.40	11.20	147.00[a]

[a]Data at −25 °C and 2 atm.

Clariant GmbH produces similar solvents that are named Genosorb. Clariant solvents are a family of dialkyl ethers of polyethylene glycol that fall under the Genosorb name.[28] DEPG is used for removing H_2S selectively to yield an H_2S rich syngas stream to the Claus unit and to the bulk CO_2 elimination unit. Two absorptions and regeneration columns are required for removing H_2S selectively with CO_2 removal; H_2S is removed at the first absorber using a lean liquid solvent with steam, and in the second column CO_2 is eliminated. After absorption, the solvent should be regenerated using flash tanks or air/N_2 for removing CO_2. Besides using DEPG to absorb H_2S and CO_2, it can also be used for gas dehydration or removing HCN molecules. The higher viscosity of DEPG than in other solvents causes a reduction in the mass transfer rates, as well as tray efficiency, which increases the tray or packing requirements, typically at low temperatures, as it is necessary to decrease the absorption temperature because of the enhanced acid gas solubility. The low vapor pressure of DEPG means there is no requirement for water for solvent recovery. Therefore, the operating temperature of the absorption process with DEPG is in the range of −18 to 175 °C.[29]

The Rectisol and Ifpexol processes are licensed by Lurgi AG and Prosernat, respectively. These are two different MeOH processes for eliminating acid gas from syngas. The Rectisol process is the earlier and more complex commercial process that works at an extremely low temperature. It is used for the purification of syngas derived from coal or for heavy oil gasification rather than CH_4 treatment applications. The Ifpexol process is applicable for CH_4 applications with two stages: absorption of water and condensable hydrocarbons for removing acid gas.[33] Even though other solvents are available, the Rectisol process offers flexibility in terms of possible flow schemes. For recovering used MeOH, washing effluent streams with water is often applied. The high vapor pressure of MeOH at normal conditions is the reason for deep refrigeration in order to prevent solvent loss, which increases operating cost. Since the operating temperature of the absorption process when using MeOH is in the range of −40 to −62 °C, stainless steel equipment should be used in the refinery plant.[28] Despite the increase in the capital cost of MeOH processes due to the low-temperature operation, gas solubility can be enhanced significantly and solvent losses reduced. The advantages of the MeOH processes are higher selectivity of MeOH for absorbing H_2S over CO_2 and the capability to remove COS. Another notable point is the higher solubility of COS and H_2S into MeOH than DEPG, which reduces the required flow rate for solvent circulation.

NMP is a solvent produced by means of the Purisol process (licensed by Lurgi AG) and the flow schemes are the same as those for DEPG. The process temperature ranges from ambient temperature to −15 °C. The higher vapor pressure of NMP than PC or DEPG means there is a possibility of using water for washing the treated gas or absorbed acid gases from the used solvent. Therefore, if NMP is washed with water, it cannot be applied for gas dehydration. At subambient temperatures with the Purisol process, it is not essential to recover NMP with water.[34] The selectivity of NMP for H_2S

removal over CO_2 is higher than other previously mentioned physical solvents, which are suitable for purifying high-pressure streams and rich CO_2 syngas for integrated gasification combined cycle (IGCC) turbines.[35] The COS solubility in NMP is not as much as hydrogen sulfide, but hydrolysis of COS occurs in NMP solvent.[29]

PC is a solvent produced *via* the Fluor Solvent process (license: Fluor Daniel, Inc.) It is used in processes in which a low concentration of hydrogen sulfide or none is present in the syngas stream, and therefore just CO_2 must be removed. The low solubility of hydrogen and light hydrocarbons in syngas and natural gas streams, respectively, causes the lower need for gas compression as well as loss of hydrocarbons in the carbon dioxide vent gas. The gas volume required for recompression, product loss and operating costs can be decreased by using intermediate pressure in the absorber of the Fluor Solvent process.[36] Another effective way to reduce hydrocarbon absorption is to chill the syngas feedstream to -18 °C so as to condense the hydrocarbons. Furthermore, the holding capacity of the acid gases in the PC solvent increases, which reduces the plant cost and solvent circulation rate. The instability of the PC solvent at a high temperature causes lower selectivity of this solvent (compared with NMP and DEPG) for eliminating H_2S from a CO_2-containing syngas stream, as a high temperature is needed for stripping H_2S from the used solvent. Therefore, it is recommended that the PC solvent be applied in streams with trace concentrations of H_2S (<20 ppm). The low vapor pressure of PC means there is no water requirement for solvent recovery. Compared with DEPG, MeOH and NMP, PC is not completely solvable in water, and it reacts irreversibly and more slowly with H_2O and CO_2 at 90 °C. This makes it inappropriate for controlling water using atmospheric distillation. For this purpose, glycol is injected at the absorber upstream, in order to control hydrate formation in the chilled syngas stream.[29]

5.2.1.2 *Chemical Absorption Processes for CO_2 and H_2S Removal*

Primary, secondary and tertiary amines used as chemical solvents for AGR from syngas act *via* the formation of weak chemical bonds with acid gases dissolved using eqn (5.3) and (5.4). Further information about the kinetics of AGR using amine solvents are available in Vaidya and Kenig's review paper.[37] The properties and the molecular structure of used amines are summarized in Table 5.4.[38,39] Whereas primary amines consist of one N atom attached directly to two H atoms, secondary and tertiary amines have one attached H atom and no H atom, respectively. Despite the rapid reaction of primary and secondary alkanolamines with carbon dioxide for carbamate formation, the associated heat of absorption is high, which increases the cost of solvent regeneration. Furthermore, the maximum carbon dioxide loading capacity for these amines is 0.5 mol CO_2 per mol amine. As tertiary amines do not have H atoms attached to the N atom, the carbonation reaction does not take

Table 5.4 Properties and molecular structure of the used amines.[38,39]

Property	AEEA	MDEA	MEA	PZ	AMP	DETA
Full name	Aminoethyl-ethanolamine	Methyldi-ethanolamine	Mono-ethanolamine	Piperazine	2-Amino-2-methyl-1-propanol	Diethanolamine
Chemical formula	$C_4H_{12}N_2O$	$C_5H_{13}NO_2$	C_2H_7NO	$C_4H_{10}N_2$	$C_4H_{11}NO$	$C_4H_{13}N_3$
Density (kg m^{-3})	1029	1038	1012	1100	934	955
Molecular weight (g mol^{-1})	104.15	119.16	61.08	86.14	89.14	103.17
Boiling point (K)	513	243	443	420	438	207
Vapor pressure at 293 K (kPa)	0.00015	0.0013	0.0085	0.1066	0.1347	0.02
Vapor pressure at 393 K (kPa)	0.96	n/a	15.90	41.66	n/a	n/a
Activation energy (kJ mol^{-1})	n/a	44.30	46.70	35	n/a	n/a
Absorption capacity (mol CO_2 mol^{-1} amine)	1.00	1.00	0.50	1.00	1.00	1.00
Pseudo first order rate constant at 298 K (m^3 kmol s^{-1})	12 100	3.50	7000	53 700	681	49 740
Type of amine	Primary and secondary	Tertiary	Primary	Secondary	Primary	Secondary
Molecule structure	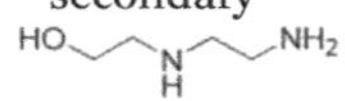	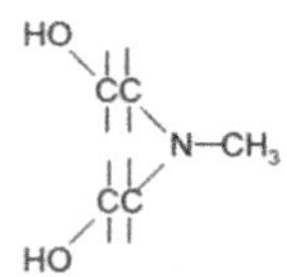	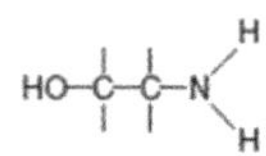	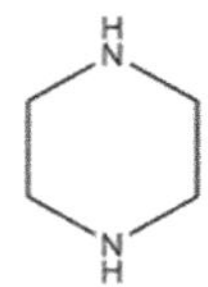	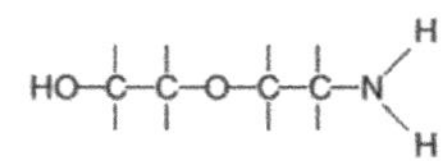	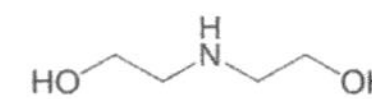

place in these amines, which results in low reactivity with carbon dioxide molecules. Instead, these alkanolamines form bicarbonates as a result of facilitation in carbon dioxide hydrolysis reaction, which releases lower heat than the carbamate formation reaction. Thus the cost of tertiary amine regeneration is lower than for primary and secondary amines. Furthermore, the maximum carbon dioxide loading capacity for tertiary amines is 1 mol CO_2 per mol amine, which is twice that of the other mentioned alkanolamines, and it can be increased by adding small amounts of secondary or primary amines.[40]

$$R_1R_2R_3N + H_2S \leftrightarrow R_1R_2R_3NH^+HS^- \tag{5.3}$$

$$CO_2 + H_2O + R_1R_2R_3N \leftrightarrow R_1R_2R_3NH^+HCO_3^- \tag{5.4}$$

2-Amino-2-methyl-1-propanol (AMP) and 2-piperidineethanol (PE) are two types of sterically hindered amines (a novel class of amines that have a lower regeneration cost than conventional primary and secondary amines) in which the amino group has been attached, respectively to the primary and secondary amines. Low stable carbamates are formed *via* the reaction of these amines with acidic components, as they have large attached groups with N atoms, which causes the same carbon dioxide loading capacity as that of tertiary amines.[37] The literature reports that the amine-based material piperazine (PZ) has also been used alone or when mixed with other amines to remove acid gas components.[41,42]

Despite the benefits of the amine solvents – including inexpensive and low co-absorption of the main components into the syngas stream – they suffer from poor COS absorption, high degradation, high equipment corrosion and a high heat requirement for regeneration because of the high heat reaction values.[21] A basic diagram of the amine-based unit is provided in Figure 5.3. After absorption of acid gases with the amine solution, the liquid phase goes through a regeneration process, in order for it to be refreshed and then reused.[21]

Potassium carbonates are another type of chemical solvent for AGR, with the process known as the Benfield process. The exothermic reaction between potassium carbonate and carbon dioxide is suitable for low CO_2 concentrations in which bicarbonate components are formed. Based on the solution pH values, different reactions will take place.[43] At a value higher than 8 pH, the reaction of water and carbon dioxide can be eliminated; therefore, HCO_3^- will be produced as a reaction of carbon dioxide with OH^-(eqn (5.5)), and it is then consumed for the production of CO_3^{2-}, as per eqn (5.6).[44]

$$CO_2 + OH^- \leftrightarrow HCO_3^- \text{ (fast)} \tag{5.5}$$

$$HCO_3^- + OH^- \leftrightarrow CO_3^{2-} + H_2O \text{ (instantaneous)} \tag{5.6}$$

However, at a pH value lower than 8, dissolved CO_2 is hydrated to produce carbonic acid (eqn (5.7)) followed by CO_3^{2-} production as a result of the HCO_3^- and OH^- reaction (eqn (5.8)).[43]

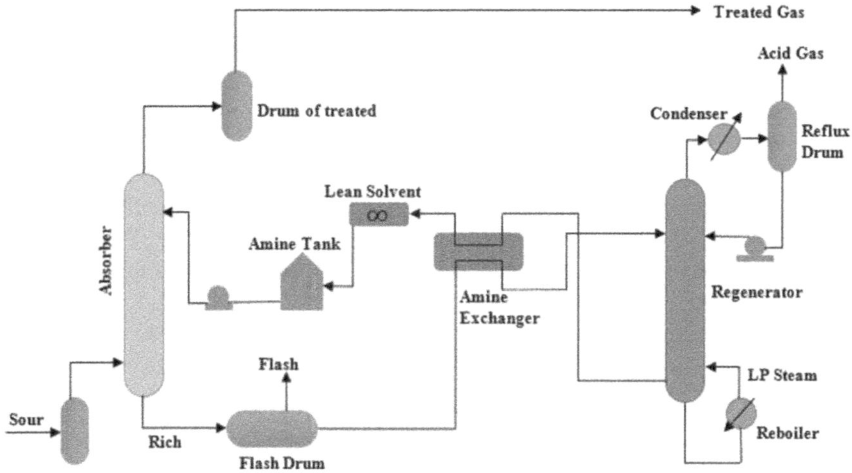

Figure 5.3 Flow diagram of the amine-based unit.

$$CO_2 + 2H_2O \leftrightarrow HCO_3^- + H_3O^+ \text{ (fast)} \tag{5.7}$$

$$HCO_3^- + H_2O \leftrightarrow CO_3^{2} + H_3O^+ \text{(instantaneous)} \tag{5.8}$$

The advantages of syngas purification using the Benfield process are high CO_2 solubility, easy regeneration, low cost and toxicity, and low degradation tendency. However, there are some limitations, including a slow reaction rate as a result of the low mass transfer rate, which can be reduced by adding an appropriate promoter; the severe equipment corrosion problem, which is less severe than the corrosion effect of amines; and crystal accumulation and fouling of the solvent in the reboilers.[44]

5.2.1.3 *Adsorption Processes for CO_2 and H_2S Removal*

The adsorption process refers to physical or chemical interaction between gas/liquid molecules and solid-phase surfaces called "adsorbents". A good adsorbent has a high capacity for adsorbing maximum amounts of molecules and is easily regenerated. Therefore, porous granular solids that have a high surface area per adsorbent unit mass are typically used as adsorbents. Potassium iodide impregnated on an activated carbon surface has the ability to react with the H_2S and O_2 molecules for sulfur and H_2O at a 50–70 °C reaction temperature and 100–120 psi pressure. After the adsorption process, activated carbon-based adsorbents should be regenerated by using hot dried fuel gas passed through the adsorbent bed. Another type of adsorbents are molecular sieves that can efficiently remove acid gases from the syngas stream if only low traces of H_2S and CS_2 exist in the syngas, which is then regenerated using steam or hot air.

As metal compounds or metal oxides are capable of reacting with H_2S to create the equivalent metal sulfides, they are widely employed in selective

H_2S removal methods. Manganese oxide, copper oxide, zinc oxide, iron oxide, nickel oxide and lead oxide are the most prevalent metal oxides used in this application. Although they can be regenerated, they lose their activity after each regeneration cycle. According to Westmoreland and Harrison,[45] corresponding sulfidation processes are thermodynamically quite favorable.

Iron oxide is the most applicable and the oldest adsorbent used for syngas sweetening using the process known as the iron sponge method. This is usually applied with gases with a low to medium sulfur content (300 ppm v/v). However, the iron sponge process is not used to remove significant amounts of CO_2. The produced syngas passes through a wood chips bed that is impregnated with iron active sites or pellets of iron oxide for adsorbing sulfur *via* eqn (5.9). This adsorption reaction should be performed at temperatures less than 43 °C, in a slightly alkaline media with a pH value of 8–10, obtained by adding diluted caustic soda.[46]

$$Fe_2O_3 + 3H_2S \rightarrow Fe_2S_3 + 3H_2O \tag{5.9}$$

For regeneration of the produced ferric sulfide as a reaction of H_2S and Fe_2O_3, oxidation by air is needed. Due to the exothermic nature of the regeneration reaction (eqn (5.10)), air should be added slowly to the adsorption media for purposes of dissipating the heat released from the reaction. As not all sulfur compounds detach from the bed after the addition of an air stream, they convert to the deposited coke on the Fe_2O_3 particles after several air cycles, which then decreases the reactivity of the bed. Generally, the bed contents should be removed completely and new adsorbents loaded after approximately 10 cycles.[47]

$$2Fe_2S_3 + 3O_2 \rightarrow 2Fe_2O_3 + 6S \tag{5.10}$$

$$2S + 2O_2 \rightarrow 2SO_2 \tag{5.11}$$

ZnO is another metal oxide that is widely used for the syngas sweetening process. Kinetics and sulfidation mechanisms have been documented for a variety of metal oxides, such as ZnO. Studies on the sulfidation of ZnO crystals demonstrate that the ZnS phase grows outward, with Zn diffusion occurring by the resultant sulfide layer, according to the Kirkendall mechanism.[48] Because it has fewer side reactions than other oxides and is less expensive, zinc oxide is one of the most extensively utilized H_2S trapping catalysts.[49] ZnO has a high performance for removing H_2S, and the reaction equation is shown in eqn (5.12).

$$ZnO + H_2S \leftrightarrow ZnS + H_2O \tag{5.12}$$

Because of the exothermic nature of acid gas adsorption on the catalyst surface, it is better to keep the reaction media temperature at its lowest value. However, in order to boost the rate of a reaction, the process usually occurs at a high temperature (typically range from 100 to 350 °C, with a maximum of 500 °C). The maximum theoretical sulfur concentration on purified ZnO is about 39.5 mass%, which corresponds to complete

ZnO-to-ZnS conversion. On average, the capacity reached is about 25% of the whole mass.[50] The rate of sulfur removal is further influenced by the sorbent's textural qualities, specific surface area and porosity. Furthermore, the presence of CO, H_2O and CO_2 in the atmosphere could affect H_2S absorption on ZnO.[51]

- At equilibrium condition, the concentration of H_2O presence in a gas line may have an impact on the trap's desulfurization efficiency. However, at temperatures below 300 °C, this is unlikely to have a significant influence on S removal.
- The effect of competitive adsorption among CO_2 and H_2S on S trapping has been documented.[15,52] The reaction between CO_x and H_2S may generate COS, the adsorptive interaction of which on ZnO surface is weak leading to low adsorption capacity. Based on the reactant, COS formation is preferred at high or low temperatures. The incidence of those reactions is nevertheless dependent on the amount of H_2O or H_2 in the synthesis gas.

$$CO + H_2S \leftrightarrow COS + H_2 \tag{5.13}$$

$$CO_2 + H_2S \leftrightarrow COS + H_2O \tag{5.14}$$

The existence of H_2 or CO as a consequence of the working parameters and various side reactions has been documented.[53] From 500 °C upward, ZnO is said to be reduced to volatile Zn (g). It then leads to the creation of high-temperature reduction-resistant ZnO-based trapped media (like zinc titanate compounds). Cu/ZnO-based sorbents could be employed for MeOH synthesis and water-gas-shift reactions because they are similar to the catalysts used for such reactions. Because of the possibility of nickel carbonyl production, nickel-based sorbents are not recommended for synthesis gas purification usages.[54,55]

The pressure swing adsorption (PSA) technology that benefits from fixed-bed adsorption columns uses a constant temperature and changeable pressure. CO_2 is removed from the syngas with this method by sorption on a surface at high pressure. Before the column is loaded, the adsorbing substance – commonly zeolites or activated carbon – is renewed by a progressive drop in pressure.[6] Using this method, an upgrading facility typically includes four, six or nine containers that operate simultaneously. The feed gas is shifted to the next vessel, where the adsorbing substance has been renewed after the adsorbing material in one vessel reaches saturation.[6] The pressure is gradually reduced throughout regeneration. The regenerated gas in the next pressure reduction stage is either routed to the next column or discharged to the atmosphere if it is almost entirely devoid of methane. The adsorption of H_2S on the adsorbent occurs irreversibly if it is available in the raw gas. Also, water in the raw gas has the potential to degrade the material's structure. As a result, hydrogen sulfide and water must be eliminated before the PSA column can be used.[56] In the PSA column, CO_2 and syngas can be separated. However, additional chemicals like oxygen, nitrogen and carbon monoxide can also be separated in this process. Silica gel, zeolite and activated carbon are some of the packing materials that can be used to remove CO_2.

The PSA process is made up of many different fixed-bed columns that run in alternate stages of adsorption, regeneration and pressure buildup in order to ensure continual efficiency. The regeneration process is sometimes worked under vacuum, which is referred to as vacuum-swing adsorption (VSA). This process necessitates the use of a vacuum pump near the output.[57] Furthermore, the increasing temperature during the regeneration stage aids the desorption of chemicals like hydrogen sulfide. Temperature swing adsorption (TSA) is a method that involves raising the temperature while maintaining a steady pressure. In gas sweetening, the TSA technique is used. The PSA, VSA and TSA procedures all require dry syngas; thus the water content should be lowered before these processes are initiated.[58]

Kinetic separation or equilibrium could be used in the PSA process. In equilibria and kinetic separation, the more firmly adsorbed and quicker diffusing chemicals are kept on the packing material. PSA with desorption thermally at 25 °C in the pressures between 45 and 100 psi has been utilized to remove H_2S and CO_2 from natural zeolites.[53] However, differences in CO_2 concentration can have an impact on the efficiency of the PSA process. For example, a 5% increase in CO_2 flow can result in a 5% rise in packed bed temperature at its top; however, by using a thermocouple, the controlling strategy can be improved.[59]

Nitrogen and methane could also be separated using the PSA approach by utilizing an aperture-sized adsorbent fairly near the molecular dimension of the bigger species (CH_4), which allows N_2 to seep through to the adsorbent. As a result, a filtered gas flow is produced that meets pipeline criteria. Then the adsorbent is renewed, resulting in a N_2 flow that is extremely clean. The PSA technique is a versatile nitrogen rejection mechanism that can be used at both low and high flow rates.[60]

Adsorption systems have several restrictions, but the most important are minimization of PM and the condensation of liquids on their surface, which then masks their surface and therefore reduces their efficiency.

5.2.1.3.1 Claus Process. The basic principles of the Claus plant as a process for the disposal of H_2S in low purification are to burn most of the H_2S as flare gas or fuel gas for producing highly toxic SO_2 (sulfur dioxide). Therefore, H_2S is removed from the refinery *via* an olamine process, in which an acid gas is formed and then treated to produce water and elemental sulfur. Actually, about one-third of the H_2S is burnt to SO_2 and then reacted with the remaining H_2S in the fixed-bed packing of cobalt molybdenum, activating alumina catalysts to form elemental sulfur.

$$2H_2S + 3O_2 \rightarrow 2SO_2 + 2H_2O \tag{5.15}$$

$$2H_2S + SO_2 \rightarrow 3S + 2H_2O \tag{5.16}$$

Split-flow and once-through are two flow configurations for achieving the optimal flow ratio of H_2S/SO_2 into the reactors. With the split-flow configuration, one-third of the H_2S is combusted entirely at first, and the produced gases are then reacted with noncombusted H_2S. In the once-through configuration, one-third of the H_2S is partially combusted by enough O_2 into the combustion chamber.

Depending on the required H_2S conversion level, two or three reactors are required. Additional stages yield less (not increased) H_2S conversion than the previous one. Generally, 96–97% v/v H_2S/elemental sulfur conversion is achievable in the Claus plant. If air quality regulations cannot be achieved, a tail gas Claus process treater is used for removing the entire remaining H_2S that is essentially in the tail gas feedstream, from the Claus plant, by absorbing H_2S and then converting it into elemental sulfur.

5.2.1.4 COS Removal

This section reviewed COS removing techniques that are based on adsorption and catalytic reactions from produced syngas. A review of several other existing COS elimination procedures is also available in the literature.[61]

COS has a lower reactivity than CO_2 because of its low acidity and polarity. As a result, the number of COS adsorption processes is limited, and they are a priori unfavorable. However, there are a few research studies on activated carbons coated with copper or iron salts, as well as metal hydroxides. If the feedstock to be processed does not require water or CO_2, commercial options are available.[62]

COS can also conduct catalytically favorable hydrolysis and hydrogenation processes. The following is a formula for the COS hydrolysis reaction:

$$COS + H_2O \leftrightarrow CO_2 + H_2S \tag{5.17}$$

The hydrogenation process of COS is endothermic and thermodynamically not favored.

$$COS + H_2 \leftrightarrow CO + H_2S \tag{5.18}$$

Thus, at low temperatures, COS hydrolysis is thermodynamically favored over the hydrogenation process (or sour gas shift reaction), which has a smaller equilibrium constant.[63] The method most often used to enhance reaction kinetics is to prefer COS hydrolysis in the existence of a catalyst. In terms of this objective, metal oxides like Al_2O_3, TiO_2, ZnO and ZrO_2 are the most often reported catalysts in the literature. Catalyst activity, in particular, appears to be linked to the basicity of the catalyst surface. However, despite the fact that ZnO is well-known for its fundamental surface characteristics, there are some contradictory findings concerning this material, such as those reported by Aboulayt *et al.*,[64] who claim that ZnO has no activity in the COS hydrolysis process. COS may also be removed using TiO_2 and alumina catalysts, which are similar to those used in the Claus procedure. Gamma alumina compounds are more effective than TiO_2 materials in terms of activation energy and hence should prefer COS hydrolysis at lower temperatures ($T < 200$ °C). Water is required for the process but might adsorb on the surface of the catalyst and decrease the catalytic activity, according to certain experimental findings.[65] Catalysts made of alumina, which has relatively strong hydrophilic characteristics, appear to be more impacted than materials made of TiO_2. As a result, TiO_2-based catalysts appear to be more productive compared to alumina catalysts at working circumstances.[66,67] This was observed, for example, in a comparison study using commercial alumina and titania catalysts.[68,69] The surface morphology of any catalyst is a critical metric, with a larger catalyst surface favoring the rate of COS conversion.[63]

5.2.1.5 *Removal of Trace Contaminants*

Due to the extreme sensitivity of the FT catalyst, high syngas purity levels are needed, which is practically impossible to achieve with traditional solvent-based procedures. Even after removing acid gases *via* the methods indicated, as well as using water scrubbing equipment, trace pollutants still exist in the syngas.

Tail gas cleaning refers to the processes used for removing remaining sulfur compounds in the gas stream after the sulfur recovery process. Therefore, developing and implementing reliable, cost-effective methods to achieve environmental requirements and sales specifications are necessary. The Claus process is one efficient process for sulfur recovery: it has 90–96% H_2S per elemental sulfur w/w efficiency in a two-stage plant and 95–98% in a three-stage plant. Due to the 98.5–99.9% sulfur removal efficiency required by most environmental agencies, the tail gas exiting from the Claus plant should be further purified. The gas produced by the Claus process usually has small quantities of CS_2, H_2S, SO_2, COS and sulfur vapor, in addition to H_2, CO and CO_2, which can be removed by first converting it to H_2S and then absorbing it using an efficient solvent. After that, the cleaned gas is recycled or vented into the vessel for further purification.

Gas treatment procedures based on chemical adsorption, chemical reaction or physical adsorption onto a solid carrier bed are employed to ensure the deep removal of pollutants. However, there isn't a reaction between the fluid molecules and the solid adsorbent in the physisorption or physical adsorption. The chemical reaction happens when the interactions between the solid and the gases are greater than the interactions inside the gas itself.[70] Physisorption is an exothermic method that begins with the development of a monolayer and progresses to the formation of multilayers in some situations. Weak intermolecular interactions keep the adsorbed molecule on the adsorbent surface. *In situ* bed regeneration is possible with some of these techniques, such as molecular sieve dryers or PSA. They are not selective procedures, since there is a gap between the impurities that are to be eliminated and the surface of the solid. These operations are also run at low temperatures (<50 °C).[71]

Shell Claus off-gas treatment (SCOT) is a process for tail gas purification based on hydrotreating the H_2S feedstream, followed by absorbing it using an amine solution. While diisopropanolamine was used as a solvent in the early SCOT plants, methyl diethanolamine-based Sulfinol (Sulfinol-M) was applied to elevate H_2S separation and let selective CO_2 elimination into the absorber. Later still, a super-SCOT process was introduced to achieve an even lower H_2S content in the gas stream, with the Sulfinol-M absorber being regenerated in a two-step process. While partly stripped amine solvent is injected into the absorption chamber middle, the stripped solvent enters the absorption chamber from above and is then cooled below the absorber for use in the SCOT conventional technology. In this method, tail gas that contains H_2S and SO_2 is reduced into the hydrotreating reactor and forms H_2O and H_2S in contact with cobalt/molybdenum supported on alumina catalysts. After being cooled in the water contactor, the gas goes to the H_2S selective amine-based absorber that is segregated from other amine-based refinery systems. Besides high selectivity toward H_2S, an appropriate amine should not be affected by high amounts of CO_2 in the gas. Nevertheless, the SCOT unit is usually integrated with the AGR removal unit at its upstream point.

Chemical adsorption – also known as chemisorption – occurs when a fluid molecule combines chemically with a solid. In most cases, this reaction is delayed and irreversible. It frequently occurs at high temperatures and is followed by heat release. Chemical changes cause the adsorbed molecules to attach powerfully and irreversibly to the adsorbent. Chemisorption separation methods are often quite selective.[72]

Other purification methods rely on a chemical reaction among gaseous compounds to be eliminated and the applied solid as a trapping medium. With chemisorption and solid-state diffusion mechanisms, these reactions include bulk solid transformations. In addition to being slow and irreversible, these reactions are also highly selective. The next section details some trace pollutant removal procedures that are relevant to syngas purification use.[73,74]

5.2.2 Removal of Nitrogen-containing Compounds

5.2.2.1 HCN Removal

A number of activated carbons appear to be accessible industrially for use in the hydrogen cyanide (HCN) absorbing process. These materials were initially investigated for use as filters to remove pollutants in gas masks, but they don't appear to have any industrial use in gas treatment.[67,75] Numerous studies have impregnated transition metals like Mo, Cu, Zn, Ag, Cr and Co on activated carbon support to absorb HCN components from the syngas stream. In addition to activated carbon, HCN can be adsorbed on mesoporous SiO_2 and NaX zeolites impregnated with metallic ions.[76,77]

HCN can be subjected to hydrogenation and hydrolysis processes *via* the following equations:

$$HCN + H_2O \leftrightarrow CO + NH_3 \tag{5.19}$$

$$HCN + 3H_2 \leftrightarrow CH_4 + NH_3 \tag{5.20}$$

HCN hydrolysis catalysts are similar to COS hydrolysis catalysts, allowing for simultaneous COS and HCN removal.[78] The following oxides have been found to have increased activity in the hydrolysis of HCN:

$$SiO_2 \ll Al_2O_3 < ZrO_2 < TiO_2$$

5.2.2.2 *NH_3 Removal*

High N_2 content biomass produces a gas containing HCN and ammonia (NH_3), resulting in the production of nitrogen oxides (NO_x) after combustion. The percentage of ammonia in the product gas of biomass gasification (to create synthesis gas) is determined not only by the structure of the raw material utilized but also by the gasifier design specifications and operation circumstances.[59]

The wet scrubbing technique, which is also widely employed in biomass gasification operations can eliminate ammonia, and the ammonia can also be deleted after the gas is dried. Therefore, there is frequently no need for a separate removal process.[79]

In comparison to the wet scrubbing method, a hot-gas cleaning innovation – benefiting catalysts – is much less energy intensive, since it avoids the requirement for cooling the produced gas and then reheating it for syngas use. By reducing ammonia to nitrogen, water and hydrogen, catalytic reactions can effectively eliminate ammonia. Some catalysts have poor ammonia potential, while nickel-based catalysts have better activity.[80]

5.2.3 Tar Removal

Tar is a liquid substance with high viscosity that is usually produced in heat treatment processes and that condenses in the apparatus zones with low temperatures (pyrolyzer or gasifier), which disrupts the gas stream and causes system failure. Tar is the least desired by-product of high-temperature processes, as high residual tar levels preclude it from being used.[81] Tar is a condensable hydrocarbon derivative that is composed of aromatic chemicals with a higher boiling point than benzene. The number of rings and the boiling point distribution and physical properties are the main variables used to classify tar constituents.[82] In fact, the composition and amount of tar produced are mainly dependent on the fuel reactions used in the gas phase, as well as the operational conditions.

Three categories are used to divide the different types of tar, based on their formation operational temperature range (see Table 5.5). The first category of tars (primary tars) consists of mixed oxygenates produced as a product of

Table 5.5 Classification of different tars, their properties and formation temperatures.[46]

Classification	Generation temperature (°C)	Constituents
Primary	400–600	Phenolic ethers Mixed oxygenates
Secondary	600–800	Heterocyclic ethers Alkyl phenols
Tertiary	800–1000	Phenolic ethers Polynuclear aromatic hydrocarbons

the pyrolysis process. Due to the higher operating temperature of the gasification process, decomposition of primary tars to tertiary and secondary tars and to light gases occurs.[83] Actually, tertiary and secondary tar compounds are formed when primarily oxygenated biomass decomposition products react further, which leads to the formation of components that did not exist in the original material.[84] While secondary tar is composed of one- or two-ring alkylated aromatic compounds (*i.e.* heterocyclic compounds), aromatic hydrocarbon derivatives (*i.e.* naphthalene, benzene, polycyclic aromatic hydrocarbon derivatives (PAHs)) make up tertiary tar. Tertiary tars are the most stable tar category. These are created from cellulose and lignin compounds of the used fuel. Nevertheless, fuels rich in lignin cause the formation of heavier tertiary aromatics to occur more quickly than in other fuels.

Two different technologies exist for reducing tar: (1) tar reducing *in situ*, called primary tar removal, in which the tar formed in the reactor is omitted or at least reduced; (2) tar reduction *via* post-gasification, called secondary tar reduction, in which tar is eliminated of the product gas stream.[58] Cracking and physical removal (catalytic or thermal) are the two post-gasification processes generally used.[82] These are discussed in more detail later in the chapter. Combining the postgasification and *in situ* tar reduction methods improves tar removal over using one technology only. Biological methods using active enzymes for tar removal are promising methods for use in the future. However, despite their high conversion rates, they need further research for industrial applications due to tar complexity, micro-organism activity and high capacity for treating used enzymes.[85]

Catalytic cracking is an effective procedure for converting high-molecular-weight chemicals (including crude oil, extra or medium heavy oil, as well as tar sand bitumen) into liquids and also light gases, which is used at numerous refinery facilities.[86] Recovering the tar energy content has the advantages of allowing for use of the catalytic cracking method in which the molecular weight of the tar components is reduced to gaseous products like hydrogen and hydrocarbon derivative products at a high temperature (up to 1200 °C) or by using chemical reactions at a lower temperature (up to 800 °C). If the feed gas consists of nitrogen, oxygen or sulfur, H_2S, H_2O, and NH_3 compounds are formed in the catalytic cracking process.[12] These should then be removed from the product gas, so that the hydrocracking product is free of nitrogen and sulfur impurities and mostly consists of naphthenic and paraffinic hydrocarbons.[12]

Numerous tar conversion techniques have been used: nonmetallic catalysts such as zeolite, calcite ($CaCO_3$) and dolomite ($CaCO_3{\cdot}MgCO_3$): metallic catalysts like Ni-, Ni/Co-, Ni/Mo-, Pt-, Mo-, and Ru-based catalysts.[87] Catalytic reformation could also be used to remove tar, with the added benefit of removing the carbon in the tar that is accessible and can be converted into fuel. Using catalytic bed materials in the gasification reactor is another alternative for reducing the tar content in synthesis gas. Olivine sand has been proven to lower the amount of tar in synthesis gas produced by steam gasification.[88]

Char is also used in the tar removal catalytic cracking process, due to its advantages, including a porous structure, low cost, high reliability, easy recovery after deactivation and natural production in a gasifier.[89] Nunes *et al.*[90] simulated a char bed layer that was adopted in the gasifier, with the results showing 98% tar removal efficiency. It was also reported that nearly all produced tar can be transformed into fuel gas at 950 °C when using charcoal.[91] Different metallic active sites were loaded on char to enhance its efficiency in removing tar, and 92.3% tar removal conversion was achieved at 800 °C using low-cost Ni-Fe active sites on a rice husk char catalyst.[92] Ni/char catalysts with 5%,[93] 6%[94] and 15%[95] loadings have also been used for biomass tar removal in various studies, showing 96.6%, 83% and 97% conversion, respectively, at 800 °C. A summary of the different char-supported catalysts used and their advantages and disadvantages in terms of tar removal are provided in Table 5.6.[52,96–98]

5.2.4 Particulate Removal

The product gas that contains PM could be a major concern for some end users, as it has a deleterious impact on the catalysts used for purifying product gases. Different methods are used to remove particulates from the gas stream.[99] Selecting the most suitable innovation for gasification of biomass relies on the required effective particle separation in order to attain optimum particulate loading in the purified syngas. Thermal integration and pressure drop in pollutant removal operations are additional important design elements that need to be considered. Particulate removal procedures are also influenced by the tars formed throughout the biomass gasification process.[100] To prevent tar condensation and to avoid particle matter from becoming agglomerated and sticky, most particulate removal systems should operate at temperatures higher than the tar dew point.

The tar is processed as mist or dust particles and then separated from the gas stream using barrier filters, cyclones, wet electrostatic precipitators (ESPs) and wet scrubbers *via* the physical tar removal method or the particulate removal method. The device application is determined by the load percentages of tar and PM, the distribution of the tar particle sizes and downstream consumers' tolerance of the particulate.[101]

Table 5.6 Advantages/disadvantages of different catalysts applied on char support for tar removal.[52,96–98]

Type	Advantage	Disadvantage	Ref.
Olivine	✓ Plentiful and inexpensive ✓ Abrasion performance good	✓ Catalytic activity is low.	52
Clay	✓ Plentiful and inexpensive ✓ No reprocessing after inactivation	✓ Catalytic activity is lower than olivine. ✓ High temperatures are not recommended.	96
Limestone	✓ Plentiful and inexpensive ✓ Excellent catalytic properties ✓ A safe bed for costly catalysts	✓ Critical pulverization abrasion	97
Iron ore	✓ Plentiful and inexpensive	✓ Catalytic activity is lower than olivine. ✓ Inactivation is easy in a reducing atmosphere.	98
Alkaline earth metal and alkaline metal	✓ Produced in gasification ✓ No reprocessing after inactivation	✓ High-temperature sintering ✓ Poorer catalytic ability than olivine	96
Char without active sites	✓ Plentiful and inexpensive ✓ Generate from gasifier continuously without regeneration ✓ Activity better than olivine ✓ Neutral alkalescent	✓ Gasification and tar catalytic activity must be balanced ✓ Catalytic property is unstable	96
Activated aluminum oxide	✓ Activity better than olivine	✓ Quick activation after soot	98
Transition metal	✓ Higher operation temperature ✓ H_2 and CO production are effectively promoted	✓ Expensive ✓ Simple inactivation after poisoning ✓ Regeneration is difficult	98

The most popular type of inertial collector is the cyclone collector, which works by contacting particles into the syngas stream by a liquid to remove coarser portions of particulates. As a result of the integration of tar particles into the liquid bath, they become significantly larger; hence it is easier to collect them from the liquid stream. The particle-laden stream of syngas tangentially enters from a cylindrical segment and travels downward *via* a conic section in this operation. Particles immigrate due to the centrifugal force that is created by exposing the carrier gas to a vortex-like spin, which moves the particles to the wall and removes them across a seal devised at the upper section of the inverted cone. The top center opening of the cyclone discharges a vortex with reverse direction that flows upward *via* the cyclone.[102,103] Cyclones can be constructed to remove particles of a certain size distribution efficiently, including a smaller size limit, while multiple cyclones of various designs can be utilized in sequence to achieve near-micron size particle removal with great efficiency.[104,105]

As little tar droplets may not be effectively removed by cyclones, block filters are used. Block filters are porous materials that trap a specified quantity of tar when the produced gas flows by them. To aid in the simultaneous removal of tar and particulate matter, catalyst particles could be inserted as a permanent bed into the filter. Nondisposable filtration bags are commonly used with filtration systems. Particulate matter (PM) is gathered on the bag side as a dust cake by passing dust-containing gas *via* a filter media (usually Teflon, polypropylene, cotton or fiberglass).[106] Fabric filters have a collection efficiency of up to 99.9% and several additional benefits, but they do suffer from a few drawbacks.

Wet scrubbers are machines that separate particles from an airflow using a countercurrent spraying liquid. Plate, packed bed, orifice, venturi scrubbers and spray towers are examples of operating systems that can be used individually or in different combinations. In these systems, an appropriate solvent is used for dissolving tar materials and therefore removing it from the syngas stream. Wet scrubbers often achieve great collection efficiency under conditions in which a pressure drop is not permitted due to the high cost. However, wet scrubbers have certain limitations, including solvent recovery and boost cost of solvent preparation.

A type of wet scrubber is a foam scrubber wherein particle-laden gas is routed *via* a foam generator that encloses the gas stream and the particles in microscopic foam bubbles.[107] Other methods include electrostatic scrubbers or high-energy required venturi scrubbers in which particulates or H_2O droplets are varied; also, fluxes force/condensation scrubbers, which interface a warm humid flow gas with steam or a cooled liquid injected into the saturated gas. The water vapor carries the particles to the cold surface of the water in the latter scrubber (diffusiophoresis), and the H_2O vapor condensation on the solid particles leads to their sizes growing, which aids in collecting fine particulates.[108]

ESPs work by imbuing particles in the incoming airflow with an electrical charge, which is subsequently collected on an opposite charge plate all over a high-voltage zone. The most challenging particles to gather are those with high resistance. To reduce resistance, conditioning chemicals like sulfur trioxide (SO_3) have been utilized. The configuration of the electrodes, the spacing of the collecting plates, air channeling reduction, and collection-electrode tapping techniques are all important considerations that are used to dislodge particles.[10] High-voltage pulse energy used to improve electron-beam ionizing, particle charging and wide plate spacing is among the techniques being investigated. However, under optimal circumstances, electrical precipitators can achieve an efficiency rate of more than 99%, but their performance in new or uncontrolled situations is hard to predict. Wet electrostatic precipitator devices offer high collection effectiveness (mostly in the order of 90%) with a low pressure drop across the whole particle size spectrum, down to 0.5 mm. Wet scrubbers also may attain excellent collection yield (about 90%).[109,110]

5.2.5 HCl Removal

Although in AGR and water wash cleaning units, some HCl is separated from the syngas flow, the gas purification is likely to be insufficient for protecting the FT catalysts if no extra chlorine preserver bed is installed following the AGR stage. Single-bed chloride preserver beds are the most common configuration used for HCl removal. These are nonregenerative devices with a service life between six months and many years. The size of these traps is determined by the capacity of the adsorbent chloride, the process stream and the intended life cycle. In gas phase settings, the adsorbent chloride capability generally ranges from 10 to 30 mass percent.[111] There are different adsorption processes for catching HCl on adsorbents, as illustrated by the reactions detailed in Table 5.7 (physisorption and chemisorption).

Molecular sieves and activated alumina were once widely employed for physisorption-based chloride removal. Adsorbents based only on adsorption physically, generally, do not behave well in chloride trap applications due to their limited chloride loading capability (5–10 mass percent). As a result, promoters such as alkaline earth-based metal oxides are added to boost their performance.[113] Calcium oxide and sodium oxide are the main promoters used for this purpose. They can be applied in a variety of concentrations but generally not more than 14 percent, so as to maintain the mechanical properties of the finished product.

Adsorption of HCl on promoted alumina can take two forms: chemisorption and physisorption.[111] Furthermore, superpromoted alumina (with the greatest promoter concentration) has been developed to function primarily by chemisorption. These materials have a higher chloride removal capability (up to 24 mass percent) than boosted alumina, which is mostly determined by the quantity of promoter used.[64,114]

The amount of Cl picked up may be influenced by the operation temperature. Although higher temperatures are preferred for chemisorption *vs.* physisorption in materials that use both chemical adsorption and physical processes (promoted alumina), lower temperatures are sufficient for physisorption (but not chemisorption). To avoid competition for adsorption, a rise in temperature may be necessary.[115] Because ZnO bulk oxide is also recognized as an adsorbent of sulfur (selective adsorption), the development of the sulfide site is preferred thermodynamically on the creation of the chloride phase over 150 °C. For Cl deletion on ZnO materials, temperatures below 150 °C are recommended.[116]

Table 5.7 Physisorption and chemisorption reactions for HCl adsorption on the adsorbents.[112]

Physisorption	$M_xO_y + HCl \rightarrow M_xO_{y-1}\text{-O-HCl}$ (M = Al, Si)
Chemisorption	$2HCl + M_2O \rightarrow 2MCl + H_2O$ (M = Na)
	$2HCl + MO \rightarrow MCl_2 + H_2O$ (M = Ca, Mg, Zn)

5.2.6 Hg Removal

The most productive innovation for Hg expulsion relies on catching the media utilized to protect the bed when mercury responds to make steady, hazardless compounds (chemisorption of solid bulk phase change). In synthesis gas, the mercury removal process must work in reducing circumstances (*i.e.* H_2). Many of the adsorbents have an active phase composed of sulfur-based components, and when the gas passes through the guard bed, mercury and sulfur interact to create solid-state sulfide mercuric (HgS), which is known as metacinnabar or cinnabar. Other sorbents employ active phases, such as silver, that can combine with Hg to produce an amalgam.[117] The used adsorbent is delivered to a dangerous chemicals disposal facility once the active phase has been depleted. There are two types of trapping media. Either the active phase (bulk component) is combined *via* a binder, or the activated site is impregnated into the porous structure of the solid phase (such as zeolites, activated carbon, Al_2O_3).[118]

Depositing active phases into the porous support is the most widely used process for mercury removal, with their being reacted chemically with Hg to create amalgams or cinnabar. The combination of elemental sulfur with Hg occurs in a redox process for oxidizing Hg and reducing sulfur through eqn (5.21) on the bed of the active site. This active phase is often used for purifying dry gas streams because sulfur becomes soluble when it is contacted by the liquid hydrocarbon, and this reduces the lifetime of the guard bed.[119] Sulfur atoms are frequently generated on activated carbons. Mercury can decrease the metal sulfide or form amalgams with metal, as illustrated in eqn (5.22) and (5.23), respectively. With amalgamation, some metals (like Ag) can specifically capture mercury (eqn (5.24) and (5.25)). Silver is frequently swapped for zeolites, which have the dual function of dehydrating and removing mercury. When mercury is heated during the regeneration cycle, it is also emitted through mercury vapor, which is subsequently treated using a traditional mercury removal guarding bed.[120,121]

$$\mathrm{Hg + S \rightarrow HgS} \tag{5.21}$$

$$\mathrm{Hg + 2CuS \rightarrow Cu_2S + HgS} \tag{5.22}$$

$$\mathrm{Hg + 2Ag_2S \rightarrow Ag_2Hg + Ag_2HgS_2} \tag{5.23}$$

$$\mathrm{Hg + 2Ag^{+} \rightarrow 2Ag + Hg^{2+}} \tag{5.24}$$

$$\mathrm{Hg^{2+} + Hg \rightarrow Hg_2{}^{2+}} \tag{5.25}$$

Some side reactions must be avoided, since they could produce chemical kinds that are not reactive with mercury (*e.g.* Cu_2S-reduced metal sulfide and/or H_2S and COS). Based on different kinetics associated with each component creation, side reactions could result in lower trapping performance and risk.[122] Capillary condensation, in particular to side reactions, may be a significant factor in the proficiency of supported active phases. Indeed, materials with mesopores or micropores are more likely to generate liquid-filled holes in the porous structure, which increases mass transition resistance to mercury diffusion inside the support.[108] As a result, the gas frequently skips the guard bed, and Hg is absorbed in the trapping media, which leads to untreated synthesis gas being produced. Because of the enormous microporous capacity, activated carbons, as well as zeolites, are individually susceptible to capillary condensation; however, alumina supports are largely mesoporous and can avoid this technical issue. Based on the solid texture quality, bulk phase entrapment media could also be sensitive to capillary condensation.[123]

In order to improve the reaction kinetics of both mercury and the active phases, trapping media are commonly performed at temperatures over 25 °C. Raising the temperature may favor the possibility of adverse reactions. Furthermore, the trapping medium is often not worked above 150 °C because of the gas heating cost. Various active phases can have stability issues, such as elemental sulfur melting at temperatures exceeding 115 °C. As a result, the optimal working temperature is determined by the technological solution selected, particularly the structure of the potential solid support and the active phase.[124]

5.2.7 Removal of Metal Carbonyls

Various activated carbons and zeolites have been used for capturing metal carbonyls, suggesting that they could be employed in industrial settings. The structure of zeolites is determined by the size of the $Fe(CO)_5$ and $Ni(CO)_4$ molecules, which have a diameter of 6.7 Å and 7.0 Å, respectively.[125,126] Faujasites are the most researched type of zeolites. These have pores in the range of 7–9 Å, depending on the cations used in the synthesis section. In contrast to faujasite, Linde L and TMA-omega (MAZ), other zeolites (including mordenite and silicalite) are not appropriate for adsorption purposes due to their small pore size.[125]

Various research studies have been done on gas purification using adsorbents, one of which compared five adsorbents that may be employed in the TSA process.[127] The results indicated that activated carbon with 1100 $m^2 g^{-1}$ and HY zeolite with 900 $m^2 g^{-1}$ had the highest CO_2 adsorption capacity because of their high surface areas. When the adsorption temperature was reduced from 38 to 22 °C, sorption of metal carbonyl was also favored. The development of pyrophoric metal particles following sorbent regeneration poses a difficulty for sorbent use to an industrial level (used sorbent unload and probable exothermic side reactions during the sorbent function). The ability of several zeolites for adsorbing metal carbonyls in a syngas stream with other impurities – like H_2O, H_2S and CO_2, which may influence metal carbonyl sorption – have been compared in numerous publications[128] This research demonstrated that sorption capability is mostly determined by material microporosity (volume and pore size), with the ideal pore size being 7–8 Å. The utilization of faujasite with a strong ratio of Si/Al (dealuminated zeolites) appears to be well suited to the removal of metal carbonyls in a synthesis gas, including water and, finally, H_2S. The metal carbonyl adsorption capacity is similarly reduced when CO_2 is present. It is also possible to employ activated carbon, but there are no details in the literature about the effect of water on metal carbonyl sorption[112]

5.3 Conclusion

"Syngas cleanup" is a general term used to describe the process of eliminating undesired pollutants from the gas produced by gasification. It often entails a multistep, integrated strategy that is dependent on the final use of the output gas. In addition to water and condensable components, N_2 and other nitrogen-containing compounds (HCN, NH_3), CO_2, and sulfur-containing gases (H_2S, COS), HCl, Hg, tar compounds as organic hydrocarbons and particulates are among the gas-phase contaminants that need to be removed from syngas, with the concentration mainly depending on the feedstock type. After acid gas components like CO_2, H_2S and COS (which cause a corrosive problem in pipelines and equipment), tar compounds are the main bottleneck in processes that use syngas feedstock, especially at low to medium operational temperatures. In addition to absorption processes using amine-based and carbonate solutions for AGR, adsorbents like ZnO, Fe_2O_3 and CuO are also used for low acid gas concentrations. For tar removal as a type of particulate in syngas, physical methods that use cyclones, filters, wet ESPs and wet scrubbers that use absorbents are widely used.

In addition, thermal and catalytic cracking methods are also applicable for removing tar and effectively converting it to fuel. With these processes, a catalyst bed of char, zeolites, mesoporous silica, activated carbon or alumina with or without metallic active sites is placed into the high-temperature reactor.

Abbreviations

AGR	Acid gas removal
AMP	2-Amino-2-methyl-1-propanol
COS	Carbonyl sulfide
DEPG	Dimethyl ether of polyethylene glycol
ESPs	Electrostatic precipitators
FCC	Fluid catalytic cracking
FT	Fischer–Tropsch
IGCC	Integrated gasification combined cycle
MeOH	Methanol
NMP	*N*-Methyl-2-pyrrolidone
PAHs	Polycyclic aromatic hydrocarbon derivatives
PC	Propylene carbonate
PE	2-Piperidineethanol
PM	Particulate matter
PZ	Piperazine

PSA	Pressure-swing adsorption
SCOT	Shell Claus off-gas treating
Sulfinol-M	Methyl diethanolamine-based Sulfinol
TSA	Temperature-swing adsorption
VSA	Vacuum-swing adsorption

References

1. M. Ajhar, M. Travesset, S. Yüce and T. Melin, Siloxane removal from landfill and digester gas–a technology overview, *Bioresour. Technol.*, 2010, **101**(9), 2913–2923.
2. A. I. Adnan, M. Y. Ong, S. Nomanbhay, K. W. Chew and P. L. Show, Technologies for biogas upgrading to biomethane: A review, *Bioengineering*, 2019, **6**(4), 92.
3. S. Anis and Z. Zainal, Tar reduction in biomass producer gas via mechanical, catalytic and thermal methods: A review, *Renewable Sustainable Energy Rev.*, 2011, **15**(5), 2355–2377.
4. P. Cherosky and Y. Li, Hydrogen sulfide removal from biogas by bio-based iron sponge, *Biosyst. Eng.*, 2013, **114**(1), 55–59.
5. N. S. El-Gendy and J. G. Speight, *Handbook of Refinery Desulfurization*, CRC Press, 2015.
6. K. Göransson, U. Söderlind, J. He and W. Zhang, Review of syngas production via biomass DFBGs, *Renewable Sustainable Energy Rev.*, 2011, **15**(1), 482–492.
7. V. Liso, Y. Zhao, W. Yang and M. P. Nielsen, Modelling of a solid oxide fuel cell CHP system coupled with a hot water storage tank for a single household, *Energies*, 2015, **8**(3), 2211–2229.
8. H. Heiskanen, I. Virkajärvi and L. Viikari, The effect of syngas composition on the growth and product formation of Butyribacterium methylotrophicum, *Enzyme Microb. Technol.*, 2007, **41**(3), 362–367.
9. M. Hrabovsky, M. Hlina, V. Kopecky, A. Maslani, O. Zivny and P. Krenek, *et al.*, Steam plasma treatment of organic substances for hydrogen and syngas production, *Plasma Chem. Plasma Process.*, 2017, **37**(3), 739–762.
10. J. G. Speight, *The Chemistry and Technology of Petroleum*, CRC Press, 1999.
11. W. Zhang, Automotive fuels from biomass via gasification, *Fuel Process. Technol.*, 2010, **91**(8), 866–876.
12. J. Huang, K. G. Schmidt and Z. Bian, Removal and conversion of tar in syngas from woody biomass gasification for power utilization using catalytic hydrocracking, *Energies*, 2011, **4**(8), 1163–1177.
13. B. Ozturk and F. Demirciyeva, Comparison of biogas upgrading performances of different mixed matrix membranes, *Chem. Eng. J.*, 2013, **222**, 209–217.

14. M.-B. Hägg and L. Deng, Membranes in gas separation, in *Handbook of Membrane Separations: Chemical, Pharmaceutical, Food, and Biotechnological Applications*, 2015, pp. 143–180.
15. D. Nikolic, E. S. Kikkinides and M. C. Georgiadis, Optimization of multibed pressure swing adsorption processes, *Ind. Eng. Chem. Res.*, 2009, **48**(11), 5388–5398.
16. S. Rapagnà, K. Gallucci, M. Di Marcello, M. Matt, M. Nacken and S. Heidenreich, *et al.*, Gas cleaning, gas conditioning and tar abatement by means of a catalytic filter candle in a biomass fluidized-bed gasifier, *Bioresour. Technol.*, 2010, **101**(18), 7123–7130.
17. C. Ratnasamy and J. Wagner, Catalysis Reviews, *Sci. Eng.*, 2009, **51**, 325.
18. V. Repellin, A. Govin, M. Rolland and R. Guyonnet, Energy requirement for fine grinding of torrefied wood, *Biomass Bioenergy*, 2010, **34**(7), 923–930.
19. W. Rongwong, S. Boributh, S. Assabumrungrat, N. Laosiripojana and R. Jiraratananon, Simultaneous absorption of CO2 and H2S from biogas by capillary membrane contactor, *J. Membr. Sci.*, 2012, **392**, 38–47.
20. H. Boerrigter, H. Calis, D. Slort, H. Bodenstaff, A. Kaandorp and H. Den Uil, *et al.*, Gas cleaning for integrated Biomass Gasification (BG) and Fischer-Tropsch (FT) systems, *Energy Research Centre of the Netherlands (ECN)*, Petten, The Netherlands, 2004, report C – 04-056.
21. D. Chiche, C. Diverchy, A.-C. Lucquin, F. Porcheron and F. Defoort, Synthesis gas purification, *Oil Gas Sci. Technol.*, 2013, **68**(4), 707–723.
22. H. Boerrigter, H. Den Uil and H.-P. Calis, Green diesel from biomass via Fischer-Tropsch synthesis: new insights in gas cleaning and process design, *Pyrolysis and Gasification of Biomass and Waste*, 2003, 1.
23. M. J. Tijmensen, A. P. Faaij, C. N. Hamelinck and M. R. van Hardeveld, Exploration of the possibilities for production of Fischer Tropsch liquids and power via biomass gasification, *Biomass Bioenergy*, 2002, **23**(2), 129–152.
24. O. Mathieu, E. L. Petersen, A. Heufer, N. Donohoe, W. Metcalfe and H. J. Curran, *et al.*, Numerical study on the effect of real syngas compositions on ignition delay times and laminar flame speeds at gas turbine conditions, *J. Eng. Gas Turbines Power*, 2014, **136**(1), 011502–011511.
25. B. S. Turk, T. Merkel, A. Lopez-Ortiz, R. P. Gupta, J. W. Portzer, G. N. Krishnan, B. D. Freeman and G. K. Fleming, *Novel Technologies For Gaseous Contaminants Control, Final*, National Energy Technology Lab., 2001.
26. T. Lieuwen, V. Yang and R. Yetter, *Synthesis Gas Combustion: Fundamentals and Applications*, CRC Press, 2009.
27. S. Yuan, Z.-J. Zhou, J. Li, X.-L. Chen and F.-C. Wang, HCN and NH3 released from biomass and soybean cake under rapid pyrolysis, *Energy fuels*, 2010, **24**(11), 6166–6171.
28. F. Biasca, N. Korens, B. Schulman and D. Simbeck *Process Screening Study of Alternative Gas Treating and Sulfur Removal Systems for IGCC*

(Integrated Gasification Combined Cycle) Power Plant Applications, SFA Pacific, Inc., Mountain View, CA (USA), 1987.
29. B. Burr and L. Lyddon, A comparison of physical solvents for acid gas removal, in *Gas Processors' Association Convention*, Grapevine, TX, 2008.
30. G. Ranke and V. H. Mohr, The rectisol wash: new developments in acid gas removal from synthesis gas, in *Acid and Sour Gas Treating Processes*, 1985, pp. 80–111.
31. K. Liu, C. Song and V. Subramani, *Hydrogen and Syngas Production and Purification Technologies*, John Wiley & Sons, 2010.
32. Ø. Borg, N. Hammer, B. C. Enger, R. Myrstad, O. A. Lindvåg and S. Eri, *et al.*, Effect of biomass-derived synthesis gas impurity elements on cobalt Fischer–Tropsch catalyst performance including in situ sulphur and nitrogen addition, *J. Catal.*, 2011, **279**(1), 163–173.
33. J. Larue, A. Minkkinen and S. Patel, IFPEXOL for environmentally sound gas processing, *Proceedings of the Seventy-first Annual Convention*, 1992.
34. G. Hochgesand, Rectisol and purisol, *Ind. Eng. Chem.*, 1970, **62**(7), 37–43.
35. D. Bhattacharyya, R. Turton and S. E. Zitney, *Acid gas removal from syngas in IGCC plants. Integrated Gasification Combined Cycle (IGCC) Technologies*, Elsevier, 2017, pp. 385–418.
36. E. O. Agiaye and M. Othman, CO2 capture and usage: harnessing the CO2 content in natural gas for environmental and economic gains, *Oil Gas Facil.*, 2016, **5**(03), 1–18.
37. P. D. Vaidya and E. Y. Kenig, CO2-alkanolamine reaction kinetics: a review of recent studies, *Chem. Eng. Technol.*, 2007, **30**(11), 1467–1474.
38. C.-H. Yu, C.-H. Huang and C.-S. Tan, A review of CO2 capture by absorption and adsorption, *Aerosol Air Qual. Res.*, 2012, **12**(5), 745–769.
39. Ö. Yildirim, A. A. Kiss, N. Hüser, K. Leßmann and E. Y. Kenig, Reactive absorption in chemical process industry: A review on current activities, *Chem. Eng. J.*, 2012, **213**, 371–391.
40. G. Sartori and D. W. Savage, Sterically hindered amines for carbon dioxide removal from gases, *Ind. Eng. Chem. Fundam.*, 1983, **22**(2), 239–249.
41. A. Abotaleb, M. H. El-Naas and A. Amhamed, Enhancing gas loading and reducing energy consumption in acid gas removal systems: A simulation study based on real NGL plant data, *J. Nat. Gas Sci. Eng.*, 2018, **55**, 565–574.
42. A. A. Abd, S. Z. Naji and A. Barifcani, Comprehensive evaluation and sensitivity analysis of regeneration energy for acid gas removal plant using single and activated-methyl diethanolamine solvents, *Chin. J. Chem. Eng.*, 2020, **28**(6), 1684–1693.
43. G. Astaria, D. W. Savage and A. Bisio, *Gas Treating with Chemical Solvents*, 1983.
44. T. N. G. Borhani, A. Azarpour, V. Akbari, S. R. W. Alwi and Z. A. Manan, CO2 capture with potassium carbonate solutions: A state-of-the-art review, *Int. J. Greenhouse Gas Control*, 2015, **41**, 142–162.

45. P. R. Westmoreland and D. P. Harrison, Evaluation of candidate solids for high-temperature desulfurization of low-Btu gases, *Environ. Sci. Technol.*, 1976, **10**(7), 659–661.
46. J. G. Speight, *Synthesis Gas: Production and Properties*, John Wiley & Sons, 2020.
47. A. M. Montebello, M. Fernández, F. Almenglo, M. Ramírez, D. Cantero and M. Baeza, *et al.*, Simultaneous methylmercaptan and hydrogen sulfide removal in the desulfurization of biogas in aerobic and anoxic biotrickling filters, *Chem. Eng. J.*, 2012, **200**, 237–246.
48. J. P. Ciferno and J. J. Marano, *Benchmarking Biomass Gasification Technologies for Fuels, Chemicals and Hydrogen Production*, US Department of Energy National Energy Technology Laboratory, 2002.
49. E. A. Efthimiadis and S. V. Sotirchos, Reactivity evolution during sulfidation of porous zinc oxide, *Chem. Eng. Sci.*, 1993, **48**(5), 829–843.
50. L. Neveux, D. Chiche, D. Bazer-Bachi, L. Favergeon and M. Pijolat, New insight on the ZnO sulfidation reaction: Evidences for an outward growth process of the ZnS phase, *Chem. Eng. J.*, 2012, **181**, 508–515.
51. I. Bezverkhyy, J. Skrzypski, O. Safonova and J.-P. Bellat, Sulfidation mechanism of pure and Cu-doped ZnO nanoparticles at moderate temperature: TEM and in situ XRD studies, *J. Phys. Chem. C*, 2012, **116**(27), 14423–14430.
52. L. Devi, K. J. Ptasinski and F. J. Janssen, A review of the primary measures for tar elimination in biomass gasification processes, *Biomass Bioenergy*, 2003, **24**(2), 125–140.
53. M. Fernández, M. Ramírez, R. M. Pérez, J. M. Gómez and D. Cantero, Hydrogen sulphide removal from biogas by an anoxic biotrickling filter packed with Pall rings, *Chem. Eng. J.*, 2013, **225**, 456–463.
54. G. Israelson, Results of testing various natural gas desulfurization adsorbents, *J. Mater. Eng. Perform.*, 2004, **13**(3), 282–286.
55. M. Liang, H. Xu and K. Xie, Bench-scale testing of zinc ferrite sorbent for hot gas clean-up, *J. Nat. Gas Chem.*, 2007, **16**(2), 204–209.
56. J. Corella, J.-M. Toledo and G. Molina, Biomass gasification with pure steam in fluidised bed: 12 variables that affect the effectiveness of the biomass gasifier, *Int. J. Oil, Gas Coal Technol.*, 2008, **1**(1–2), 194–207.
57. R. Zwart *Gas Cleaning Downstream Biomass Gasification*, SenterNoven, Energy Research Center of the Netherlands, 2009, pp. 1–65.
58. H. Leibold, A. Hornung and H. Seifert, HTHP syngas cleaning concept of two stage biomass gasification for FT synthesis, *Powder Technol.*, 2008, **180**(1–2), 265–270.
59. G. M. Al Kaisy, M. I. A. Mutalib, M. A. Bustam, J.-M. Leveque and N. Muhammad, Liquid-Liquid extraction of aromatics and sulfur compounds from base oil using ionic liquids, *J. Environ. Chem. Eng.*, 2016, **4**(4), 4786–4793.
60. S. Newman *Acid and sour gas treating processes*, 1985.
61. S. Watson, R. Kimmitt and R. B. Rhinesmith, Study compares COS-removal processes, *Oil Gas J.*, 2003, **101**(36), 66–73.

62. D. E. Sparks, T. Morgan, P. M. Patterson, S. A. Tackett, E. Morris and M. Crocker, New sulfur adsorbents derived from layered double hydroxides: I: Synthesis and COS adsorption, *Appl. Catal., B*, 2008, **82**(3–4), 190–198.
63. P. D. Svoronos and T. J. Bruno, Carbonyl sulfide: a review of its chemistry and properties, *Ind. Eng. Chem. Res.*, 2002, **41**(22), 5321–5336.
64. A. Aboulayt, F. Mauge, P. Hoggan and J. Lavalley, Combined FTIR, reactivity and quantum chemistry investigation of COS hydrolysis at metal oxide surfaces used to compare hydroxyl group basicity, *Catal. Lett.*, 1996, **39**(3), 213–218.
65. R. Rauch, H. Hofbauer, K. Bosch, I. Siefert, C. Aichernig and K. Voigtlaender, *et al.*, Steam gasification of biomass at CHP plant Guessing-Status of the demonstration plant, *Citeseer*, 2004, 1–4.
66. D. Haffad, U. Kameswari, M. Bettahar, A. Chambellan and J. Lavalley, Reduction of benzaldehyde on metal oxides, *J. Catal.*, 1997, **172**(1), 85–92.
67. T. M. Oliver, K. Jugoslav, P. Aleksandar and D. Nikola, Synthetic activated carbons for the removal of hydrogen cyanide from air, *Chem. Eng. Process.: Process Intensif.*, 2005, **44**(11), 1181–1187.
68. S. Tong, I. Dalla Lana and K. Chuang, Kinetic modelling of the hydrolysis of carbonyl sulfide catalyzed by either titania or alumina, *Can. J. Chem. Eng.*, 1993, **71**(3), 392–400.
69. S. Tong, I. G. Dalla Lana and K. T. Chuang, Effect of catalyst shape on the hydrolysis of COS and CS2 in a simulated Claus converter, *Ind. Eng. Chem. Res.*, 1997, **36**(10), 4087–4093.
70. S. V. Vassilev, D. Baxter, L. K. Andersen and C. G. Vassileva, An overview of the chemical composition of biomass, *Fuel*, 2010, **89**(5), 913–933.
71. R. E. Melchers and R. Emslie, Investigations for structural safety assessment of corroded cast iron bridge piers, *Aust. J. Struct. Eng.*, 2016, **17**(1), 55–66.
72. K. Froment, F. Defoort, C. Bertrand, J. Seiler, J. Berjonneau and J. Poirier, Thermodynamic equilibrium calculations of the volatilization and condensation of inorganics during wood gasification, *Fuel*, 2013, **107**, 269–281.
73. M. Nrel, *Survey and Down-Selection of Acid Gas Removal Systems for the Thermochemical Conversion of Biomass to Ethanol with a Detailed Analysis of an MDEA System*, Efficiency & Renewable Energy, operated by the Alliance for Sustainable Energy, LLC, San Francisco, California, 2011.
74. D. Burns and W. Echt, *Chemical solvent-based processes for acid gas removal in gasification applications*, 1997.
75. R. Nickolov and D. Mehandjiev, Comparative study on removal efficiency of impregnated carbons for hydrogen cyanide vapors in air depending on their phase composition and porous textures, *J. Colloid Interface Sci.*, 2004, **273**(1), 87–94.
76. R. Kotdawala, N. Kazantzis and R. W. Thompson, Molecular simulation studies of adsorption of hydrogen cyanide and methyl ethyl ketone on zeolite NaX and activated carbon, *J. Hazard. Mater.*, 2008, **159**(1), 169–176.

77. M. Naderi, J. L. Pickett, M. J. Chinn and D. R. Brown, Modified mesoporous silicates for the adsorption and decomposition of toxic gases, *J. Mater. Chem.*, 2002, **12**(4), 1086–1089.
78. H. M. Huisman, *The Hydrolysis of Carbonyl Sulfide, Carbon Disulfide and Hydrogen Cyanide on Titania Catalysts*, Uitgever niet Vastgesteld, 1994.
79. R. Curry, *Fundamentals of Natural Gas Conditioning*, 1981.
80. P. Gislon, S. Galli and G. Monteleone, Siloxanes removal from biogas by high surface area adsorbents, *Waste Manage.*, 2013, **33**(12), 2687–2693.
81. J. S. Dennis and S. A. Scott, In situ gasification of a lignite coal and CO_2 separation using chemical looping with a Cu-based oxygen carrier, *Fuel*, 2010, **89**(7), 1623–1640.
82. S. V. Van Paasen, M. Cieplik and N. Phokawat, *Gasification of Non-woody Biomass. Economic and Technical Perspectives of Chlorine and Sulphur Removal from Product Gas (Non-confidential Version),* 2006.
83. M. N. P Santos, C. A. Grande and A. E. Rodrigues, Dynamic study of the pressure swing adsorption process for biogas upgrading and its responses to feed disturbances, *Ind. Eng. Chem. Res.*, 2013, **52**(15), 5445–5454.
84. J. G. Speight, *Handbook of petroleum refining*, CRC Press, 2016.
85. X. Zeng, Y. Ueki, R. Yoshiie, I. Naruse, F. Wang and Z. Han, *et al.*, Recent progress in tar removal by char and the applications: A comprehensive analysis, *Carbon Resour. Convers.*, 2020, **3**, 1–18.
86. Q. Jing, H. Lou, J. Fei, Z. Hou and X. Zheng, Syngas production from reforming of methane with CO_2 and O_2 over Ni/SrO–SiO_2 catalysts in a fluidized bed reactor, *Int. J. Hydrogen Energy*, 2004, **29**(12), 1245–1251.
87. P. Graham, *Building Ecology: First Principles for a Sustainable Built Environment*, John Wiley & Sons, 2009.
88. Z. Abu El-Rub, E. A. Bramer and G. Brem, Review of catalysts for tar elimination in biomass gasification processes, *Ind. Eng. Chem. Res.*, 2004, **43**(22), 6911–6919.
89. Y.-G. Wang, J.-L. Sun, H.-Y. Zhang, Z.-D. Chen, X.-C. Lin and S. Zhang, *et al.*, In situ catalyzing gas conversion using char as a catalyst/support during brown coal gasification, *Energy Fuels*, 2015, **29**(3), 1590–1596.
90. S. Monteiro Nunes, N. Paterson, D. Dugwell and R. Kandiyoti, Tar formation and destruction in a simulated downdraft, fixed-bed gasifier: Reactor design and initial results, *Energy Fuels*, 2007, **21**(5), 3028–3035.
91. S. K. Chembukulam, A. S. Dandge, N. L. K. Rao, K. Seshagiri and R. Vaidyeswaran, Smokeless fuel from carbonized sawdust, *Ind. Eng. Chem. Prod. Res. Dev.*, 1981, **20**(4), 714–719.
92: Y. Shen, P. Zhao, Q. Shao, D. Ma, F. Takahashi and K. Yoshikawa, In-situ catalytic conversion of tar using rice husk char-supported nickel-iron catalysts for biomass pyrolysis/gasification, *Appl. Catal., B*, 2014, **152**, 140–151.
93. Y. Shen and K. Yoshikawa, Tar conversion and vapor upgrading via in situ catalysis using silica-based nickel nanoparticles embedded in

rice husk char for biomass pyrolysis/gasification, *Ind. Eng. Chem. Res.*, 2014, **53**(27), 10929–10942.
94. M. Hu, M. Laghari, B. Cui, B. Xiao, B. Zhang and D. Guo, Catalytic cracking of biomass tar over char supported nickel catalyst, *Energy*, 2018, **145**, 228–237.
95. D. Wang, W. Yuan and W. Ji, Char and char-supported nickel catalysts for secondary syngas cleanup and conditioning, *Appl. Energy*, 2011, **88**(5), 1656–1663.
96. A. S. Al-Rahbi and P. T. Williams, Hydrogen-rich syngas production and tar removal from biomass gasification using sacrificial tyre pyrolysis char, *Appl. Energy*, 2017, **190**, 501–509.
97. C.-Z. Li, Importance of volatile–char interactions during the pyrolysis and gasification of low-rank fuels–a review, *Fuel*, 2013, **112**, 609–623.
98. N. B. Klinghoffer, M. J. Castaldi and A. Nzihou, Catalyst properties and catalytic performance of char from biomass gasification, *Ind. Eng. Chem. Res.*, 2012, **51**(40), 13113–13122.
99. R. N. Maddox, *Gas Conditioning and Processing, Gas and Liquid Sweetening*, 1982, vol. 4.
100. W. S. Epling, G. B. Hoflund and D. M. Minahan, Reaction and surface characterization study of higher alcohol synthesis catalysts: VII. Cs-and Pd-promoted 1: 1 Zn/Cr spinel, *J. Catal.*, 1998, **175**(2), 175–184.
101. L. Dong, G. Xu, T. Suda and T. Murakami, Potential approaches to improve gasification of high water content biomass rich in cellulose in dual fluidized bed, *Fuel Process. Technol.*, 2010, **91**(8), 882–888.
102. L. Wei, S. Xu, J. Liu, C. Lu, S. Liu and C. Liu, A novel process of biomass gasification for hydrogen-rich gas with solid heat carrier: preliminary experimental results, *Energy Fuels*, 2006, **20**(5), 2266–2273.
103. L. P. Rabou, R. W. Zwart, B. J. Vreugdenhil and L. Bos, Tar in biomass producer gas, the Energy research Centre of the Netherlands (ECN) experience: an enduring challenge, *Energy Fuels*, 2009, **23**(12), 6189–6198.
104. J. Iranmahboob and D. O. Hill, Alcohol synthesis from syngas over K2CO3/CoS/MoS2 on activated carbon, *Catal. Lett.*, 2002, **78**(1), 49–55.
105. S. D. Phillips, J. K. Tarud, M. J. Biddy and A. Dutta, Gasoline from woody biomass via thermochemical gasification, methanol synthesis, and methanol-to-gasoline technologies: a technoeconomic analysis, *Ind. Eng. Chem. Res.*, 2011, **50**(20), 11734–11745.
106. M. Aramendıa, V. Borau, C. Jiménez, J. Marinas, J. Ruiz and F. Urbano, *et al.*, Catalyst development for the gasification of biomass in the dual-bed gasifier 169, *Appl. Catal., A*, 2003, **255**, 365–367.
107. R. Zhang, R. C. Brown, A. Suby and K. Cummer, Catalytic destruction of tar in biomass derived producer gas, *Energy Convers. Manage.*, 2004, **45**(7–8), 995–1014.
108. S.-J. Doong, R. A. Jadhav and F. Lau, *Membrane reactor for H_2S, CO_2 and H_2 separation*, *Google Pat.*, 2011, 7, 938, 893.

109. S. Mokhatab, W. A. Poe and J. Y. Mak, *Handbook of Natural Gas Transmission and Processing: Principles and Practices*, Gulf Professional Publishing, 2018.
110. C. S. Hsu and P. R. Robinson, *Springer Handbook of Petroleum Technology*, Springer, 2017.
111. M. Erfan, Chloride removal in refineries, *Pet. Technol. Q.*, 2011, **16**(2), 43–51.
112. R. Quinn, T. A. Dahl, B. W. Diamond and B. A. Toseland, Removal of arsine from synthesis gas using a copper on carbon adsorbent, *Ind. Eng. Chem. Res.*, 2006, **45**(18), 6272–6278.
113. P. V. Broadhurst, Removal of chloride compounds, *Pet. technol. Q.*, 2003, **8**(3), 127–135.
114. M. Zhao, N. H. Florin and A. T. Harris, Mesoporous supported cobalt catalysts for enhanced hydrogen production during cellulose decomposition, *Appl. Catal., B*, 2010, **97**(1–2), 142–150.
115. C. D. Blasi, C. Branca and A. Galgano, Products and global weight loss rates of wood decomposition catalyzed by zinc chloride, *Energy Fuels*, 2008, **22**(1), 663–670.
116. H. Yang, S. Kudo, K. Norinaga and J.-I. Hayashi, Steam–Oxygen Gasification of Potassium-Loaded Lignite: Proof of Concept of Type IV Gasification, *Energy Fuels*, 2016, **30**(3), 1616–1627.
117. E. Nfon, I. T. Cousins, O. Järvinen, A. B. Mukherjee, M. Verta and D. Broman, Trophodynamics of mercury and other trace elements in a pelagic food chain from the Baltic Sea, *Sci. Total Environ*, 2009, **407**(24), 6267–6274.
118. A. Chalkidis, D. Jampaiah, P. G. Hartley, Y. M. Sabri and S. K. Bhargava, Mercury in natural gas streams: a review of materials and processes for abatement and remediation, *J. Hazard. Mater.*, 2020, **382**, 121036.
119. C. Cameron, P. Courty, J.-P. Boitiaux, P. Varin and G. Leger, *Method of eliminating mercury or arsenic from a fluid in the presence of a mercury and/or arsenic recovery mass. Google Pat.*, 1993, 5, 245, 106.
120. H. Nishino, T. Aibe and K. Noguchi, *Process for removal of mercury vapor and adsorbent therefor*, *Google Pat.*, 1985, **4**, 500, 327.
121. T. M. Matviya, R. S. Gebhard and M. Greenbank, *Mercury adsorbent carbon molecular sieves and process for removing mercury vapor from gas streams*, *Google Pat.*, 1987, 4, 708, 853.
122. A. Loppinet-Serani, C. Aymonier and F. Cansell, Current and foreseeable applications of supercritical water for energy and the environment, *ChemSusChem*, 2008, **1**(6), 486–503.
123. J. N. Kuhn, Z. Zhao, L. G. Felix, R. B. Slimane, C. W. Choi and U. S. Ozkan, Olivine catalysts for methane-and tar-steam reforming, *Appl. Catal., B*, 2008, **81**(1–2), 14–26.
124. T. Van Gerven and A. Stankiewicz, Structure, energy, synergy, time The fundamentals of process intensification, *Ind. Eng. Chem. Res.*, 2009, **48**(5), 2465–2474.

125. T. Bein and P. A. Jacobs, Interaction between zeolites and cluster compounds. Part 1.—Adsorption of iron pentacarbonyl on zeolites, *J. Chem. Soc., Faraday Trans. 1*, 1983, **79**(8), 1819–1831.
126. N. K. Indu, H. Hobert, I. Weber and J. Dátka, Adsorption of nickeltetracarbonyl on zeolites studied by infrared spectroscopy, *Zeolites*, 1995, **15**(8), 714–718.
127. T. C. Golden, T. H. Hsiung and K. E. Snyder, Removal of trace iron and nickel carbonyls by adsorption, *Ind. Eng. Chem. Res.*, 1991, **30**(3), 502–507.
128. J. A. Moulijn, A. Stankiewicz, J. Grievink and A. Górak, Process intensification and process systems engineering: a friendly symbiosis, *Comput. Chem. Eng.*, 2008, **32**(1–2), 3–11.

CHAPTER 6

Fischer–Tropsch Reactions and Product Distribution Suits for Downstream

WENPING MA,*[a] WILSON D. SHAFER[b] AND MICHELA MARTINELLI[a]

[a] Center for Applied Energy Research, University of Kentucky, 2540 Research Park Drive, Lexington KY 40511, USA; [b] Asbury University, One Macklem Drive, Wilmore KY 40390, USA
*Email: wenping.ma@uky.edu

6.1 Introduction

The Fischer–Tropsch Synthesis (FTS) reaction is a surface catalytic polymerization process that utilizes syngas derived from natural resources such as coal, natural gas and biomass to form a broad spectrum of hydrocarbons. The FTS reaction is described in eqn (6.1). It generally takes place on transition metal catalysts and some noble metal-based catalysts, *i.e.* iron, cobalt, nickel and ruthenium.[1] In addition to the FTS reaction, the water-gas-shift (WGS) reaction shown in eqn (6.2) is accompanied by these catalysts and produces additional H_2 to participate in the FTS reaction. This is particularly important for the process of syngas with a low H_2/CO ratio (0.5–1.4) derived from coal or biomass. Apparently, the FTS reaction is extremely complicated due to the hundreds of reactions taking place simultaneously and because

Catalysis Series No. 44
Chemicals and Fuels from Biomass *via* Fischer–Tropsch Synthesis: A Route to Sustainability
Edited by Joshua Gorimbo, Xinying Liu, Yali Yao and Diane Hildebrandt

Published by the Royal Society of Chemistry, www.rsc.org

the FTS products, which are constrained by eqn (6.1), are significantly influenced by eqn (6.2).

$$nCO + \left(n + \frac{m}{2}\right)H_2 \rightarrow C_nH_m + nH_2O \tag{6.1}$$

$$CO + H_2O \rightleftarrows CO_2 + H_2 \tag{6.2}$$

The FTS reaction over iron and cobalt catalysts has received extensive attention from academia and industry in the past century due to the high activity and low cost of the catalysts.[1–13] To date, FTS technology has been commercialized and classified as a three-temperature model, *i.e.* low-temperature FT (LTFT) at 220 to 260 °C; medium-temperature FT (MTFT) at 265–290 °C; and high-temperature FT (HTFT) at 300–350 °C.[5–10] FTS has already played a critical role in providing increased energy needs in some nations, *e.g.*: Sasol coal to liquid (CTL) plants in South Africa (160 000 BPD); Sasol-Oryx gas to liquid (GTL) in Qatar (34 000 BPD); Shell GTL plants in Bintulu Malaysia (16 000 BPD) and Qatar (34 000 BPD); Escravos GTL in Nigeria (34 000 BPD); and CTL plants in Erduoshi (3425 BPD) and Nixia in China (85 600 BPD).[6–13]

To better understand the Fischer–Tropsch (FT) process, it is important to establish a mathematical relationship between the conversion and product distribution, in addition to developing kinetic models. In the earlier studies done by Ledakowicz *et al.*[14] and Bukur *et al.*,[15] equations for various reactant conversion processes, the exit H_2/CO ratio (E) and the average H/C ratio (β) were correlated by using the H_2/CO feed ratio (F) and the H_2/CO usage ratio (U), which assume a WGS reaction that approaches equilibrium. This provides a way to calculate the water and CO_2 selectivity when the WGS approaches equilibrium status. However, in most scenarios of the FTS reaction on iron or cobalt catalysts, WGS is far from equilibrium (CO_2 selectivity 5–40%); thus using the models of Ledakowicz *et al.* and Bukur *et al.* could result in a large error in the estimation of U and the mean H/C ratio (β) due to the narrow conditions applied for the models. Therefore, building generic models that can predicate these parameters is needed.

The ideal FTS product distribution was described by the Anderson–Schulz–Flory distribution (ASF).[1,2] It can be mathematically described as per eqn (6.3) and (6.4):

$$M_n = (1 - a)a^{n-1} \tag{6.3}$$

$$a = r_p/(r_p + r_t) \tag{6.4}$$

where M_n represents the mole fraction of the hydrocarbon with a carbon number of n; α is the chain growth probability; r_p and r_t are the chain growth and chain termination rates, respectively.

Furthermore, in many simulation studies done on the FTS process, the hydrocarbons were lumped together as one hydrocarbon pool with an average carbon number of $\bar{n}$. When the composition of oil and wax are absent – which is a common situation in most studies – models are required to estimate the $\bar{n}$ and α values based on the basic data collected during the FTS reaction, such as gas analysis results and product mass. However, such $\bar{n}$ and α models are rarely reported on as yet.

The FTS hydrocarbon distribution often deviates from the ideal ASF distribution.[16–23] The deviation often occurs in the lower hydrocarbons at C_1, C_2 or C_7–C_{10}, which has been explained as being caused by two different chain growth mechanisms;[16] dual active sites;[17–20] the secondary reaction of 1-olefin re-adsorption;[20–23] wax accumulation;[24] chain-length-dependent olefin desorption.[25,26] These reported distributions mainly reflect on the chain growth of α-olefins and *n*-paraffins, which were defined as the main products of FTS.[22–24] Apart from the main products obtained on iron and cobalt catalysts, an abundance of branching hydrocarbons are produced in the FTS, which could reflect more reaction pathways other than those for the dominant products. These minor products are essential in adjusting the physical property of the FT oil, *i.e.* cetane number and cold flow properties for fueling vehicles. Recently, several studies have investigated the formation of branching products on iron and cobalt catalysts.[27–32] The independence of iso-paraffin content from carbon number on iron catalysts was reported.[30,31]

The objectives of this work were (1) to build models for the average C/H ratio (β) and average carbon number ($\bar{n}$) using the basic gas analysis results and product mass (without the need for detailed compositions of oil and wax) and to establish the relationship between chain growth probability (α) and C_{5+} selectivity; (2) to study the overall hydrocarbon distribution of iron and cobalt catalysts to further understand the reaction mechanism during FTS; (3) to study the formation of individual 2–5 methyl branches over iron and cobalt catalysts and further to explore the effects of the promoters K (on Fe), Zr, and Pt (on Co) and the poisons HBr (on Fe) and COS (on Co) and methyl formation to understand its formation mechanism.

6.2 Experimental

6.2.1 Catalyst Preparation

Potassium promoted iron catalysts, 100Fe/5.1Si/2Cu/*x*K (part per atom) – where $x = 1$ or 3 – were used to investigate the CO conversion effect. A base Fe catalyst (100Fe/5.1Si) was first prepared using a continuous precipitation method. The 100Fe/5.1Si/2Cu/*x*K catalyst was prepared by sequential impregnation of an aqueous solution of the respective metal nitrates (KNO_3, 99.9% purity) and solutions containing Cu ($Cu(NO_3)_2 \cdot 2.5H_2O$, 99.9% purity). Between each impregnation step, the catalyst was dried under vacuum in a rotary evaporator at 80 °C, and the temperature was then slowly increased to 95 °C. After the impregnation/drying step, the catalyst was calcined under air flow at 350 °C for 4 h and then sieved down to a size range of 45–90 μm. Details on catalyst preparation can be found elsewhere.[33,34] The BET surface area of the Fe–K catalysts was ~110 $m^2\,g^{-1}$.

The 0.5%Pt (or Re, Zr)–25%Co/Al_2O_3 catalysts used in this study were prepared using a slurry phase impregnation method. The catalyst support was Sasol Catalox-150 γ-Al_2O_3. Cobalt nitrate hexahydrate ($Co(NO_3)_2 \cdot 6H_2O$) (Alfa Aesar) was employed to make the Co precursor solution. The two-step method for cobalt nitrate impregnation was applied. Between each step, the catalyst was dried under vacuum in a rotary evaporator. After the last step of cobalt addition, a platinum or ruthenium precursor solution (tetra-amine platinum(II) nitrate or rhenium oxide (Re_2O_7) and/or zirconyl nitrate) was added to the Co/Al_2O_3 catalyst by incipient wetness impregnation. The catalysts were then calcined under air flow at 350 °C for 4 h. A more detailed description of these preparation procedures can be found elsewhere.[34]

6.2.2 Fischer–Tropsch Synthesis Reaction

The 100Fe/5.1Si/xK/2.0Cu ($x = 1, 3$) catalysts (5–10 g) were activated *in situ* by 30 NL h^{-1} CO in 1 L CSTR, at 270 °C atmospheric pressure, for 24 h. The catalyst performance was evaluated at a temperature of 230–270 °C, a pressure of 1.3 MPa, an H_2/CO ratio of 0.67–0.77 and a CO conversion of 10–90%, with and without the poison HBr. Additionally, the 0.5–5%M–Co/Al_2O_3 (M = Zr, Pt or Re) catalysts used were evaluated in a 1 L CSTR. The Co catalysts (~15 g) were activated *ex situ* at 350 °C using a 60 NL h^{-1} H_2/He gas mixture with a mole ratio of 1/3 for 15 h. The reduced Co catalysts were injected into a CSTR using inert N_2 gas. The catalysts were further reduced *in situ* in 40 NL h^{-1} H_2 at 230 °C for 24 h. The FTS reaction over the Co catalysts was performed under the following conditions: 220–230 °C, 1.5–2.2 MPa, H_2/CO = 2.0, 10–90% CO conversion with and without H_2S or COS poison. In all tests, total mass closure and elemental mass closure during the mass balance period were 100 ± 3.

The selectivity to CH_4 and C_{5+} used in this paper was the carbon atom basis. 1-C_4 or 2-C_4 olefin selectivity is defined as 100×(rate of 1-C_4 or 2-C_4 alkenes)/(rate of all C_4 alkenes). The fraction of individual branching is defined as 100×rate of methyl paraffin/rate of all paraffins per carbon number.

6.2.3 GC–MS Sample Preparation and Analysis

Iron and cobalt catalysts produce a broad spectrum of hydrocarbons, which causes difficulty with the separation of branching hydrocarbons. A hydrogenation process for each oil sample allows for more selective quantification of each singly branched isomer.[27,28] For comparison purposes, the oil samples collected at similar CO conversions from different iron and cobalt FT runs were chosen. About 6 g of FTS products (oil and wax) and 0.5 g of 5%Pd/carbon catalyst (ion exchange method) were placed in a glass hydrogenation vessel. The reaction mixture was hydrogenated under a pressure of 0.16 MPa at room temperature, along with stirring until normal alkenes were not detected in the reaction mixture. The starting yellow oil came out clear. The reaction mixture was then filtered to remove catalyst particles, before GC/FID & MS analysis.

The low pressure used in this specific chamber allows only the olefins to be hydrogenated. Thus the normal-to-iso-products ratio was the same in the original sample and in the sample after hydrogenation.[27,28,31,32]

A 6890 Agilent (Santa Clara, CA) GC-FID (flame ionization detector) was used to analyze the FT liquids. Two microliters of the sample were injected into an Agilent JW Scientific DB-5 GC Column (Agilent, Santa Clara), held at 35 °C for 60 min, with a linear increase of 4 °C per min to 325 °C. The specific branched paraffins were validated by means of an Agilent 5975N MSD (mass selective detector) connected directly to the Agilent 6890 GC, using the same GC conditions and column as with the 6890 Agilent GC-FID. Thus the retention times and patterns of the specific compounds in the two instruments theoretically remained the same.

6.3 FTS Stoichiometry and Model Development

In this section, various conversions (X_i, i = CO, H_2 and CO + H_2) and the H_2/CO exit ratio (E), which is based on the stoichiometry of the FTS and WGS reactions provided in eqn (6.1) and (6.2), are defined. Second, a model of the mean *m*/*n* ratio (β) and a model of $\bar{n}$ are built to estimate the parameters' values using basic gas analysis results and product mass. Finally, a relationship between C_{5+} selectivity and chain growth probability (α) is established.

6.3.1 FTS Stoichiometry

CO conversion (X_{CO}) is commonly used in kinetic modeling and process simulation. It is a percentage of converted CO in total CO in feed and is defined in eqn (6.5).

$$X_{CO} = \frac{(N_{co}^0 - N_{CO})}{N_{co}^0} \tag{6.5}$$

Assuming that the H_2/CO molar ratio in the feed is $F\left(N_{H_2}^0/N_{CO}^0\right)$ and that the usage ratio U is defined as consumed moles of H_2 to consumed moles of CO, the relationship between H_2 conversion (X_{H_2}) or syngas conversion (X_{CO+H_2}) and U, F and CO conversion are obtained using eqn (6.6) and (6.7):

$$X_{H_2} = \frac{U}{F} X_{CO} \tag{6.6}$$

$$X_{CO+H_2} = \frac{(1+U)}{(1+F)} X_{CO} \tag{6.7}$$

Alternatively, syngas conversion (X_{CO+H_2}) can also be correlated with H_2 conversion (X_{H_2}) using eqn (6.8):

$$X_{CO+H_2} = \frac{F(1+U)}{U(1+F)} X_{H_2} \tag{6.8}$$

If H_2 and syngas conversions are not available, the conversions could easily be obtained from eqn (6.6)–(6.8). In two extreme scenarios, when CO or H_2 conversion is limited (Ledakowicz,[14] Bukur[15]), the maximum syngas conversion, based on eqn (6.7) and (6.8), are, respectively:

$$(X_{CO+H_2})_{max} = \frac{(1+U)}{(1+F)} \quad \text{or} \quad (X_{CO+H_2})_{max} = \frac{F(1+U)}{U(1+F)}$$

The molar ratio of H_2/CO at the reactor exit (E) is an important factor in influencing FTS product selectivity. The E value can be calculated using eqn (6.9), which is a function of H_2 and CO conversion.

$$E = \frac{F(1 - X_{H_2})}{(1 - X_{CO})} \tag{6.9}$$

Bukur *et al.*[15] simulated E and U values over precipitated iron and fused iron catalysts at WGS equilibrium conditions and at various temperatures between 220 and 265 °C. The study showed a large gap between the experimental values and the predicted values at high H_2/CO ratios, *i.e.* greater than 1.0. The difference may be due to the WGS reaction on iron catalysts at actual FTS conditions being away from the equilibrium, *i.e.* $1\text{–}40 < K_{eq} = 65\text{–}142$. Therefore, the utilization of experimental data to calculate U and E is necessary.

6.3.2 Models Used for the Mean H/C Ratio (β) and Mean Carbon Number ($\bar{n}$)

In reactor modeling or kinetic simulation studies, one usually lumps various hydrocarbons together as one component to simplify the calculation process, *i.e.* $(CH_\beta)_{\bar{n}}$, and where $\beta = \bar{m}/\bar{n}$ is the mean H/C ratio for the hydrocarbon, and $\bar{n}$ *and* $\bar{m}$ are average C and H atoms, respectively.

Therefore, obtaining the β and $\bar{n}$ values is of vital importance. If the compositions of oil and wax are available, it is easy to obtain these parameter values; however, they cannot be obtained directly due to the lack of analytical data regarding oil and wax products in most public studies. Thus using the empirical values has been proposed, for example β: 2.1–2.4 and the average carbon number $\bar{n}$: 6–10. Here, we develop models of β and $\bar{n}$ to predict the values at various process conditions.

Assuming N^0_{CO} moles of CO and $N^0_{H_2}$ moles of H_2 participate in the FTS reaction $\left(F = N^0_{H_2}/N^0_{CO}\right)$, it forms N_{CO_2} moles of CO_2, N_{H_2O} moles of H_2O and $N_{C\bar{n}H\bar{m}}$ moles (or m_{HC} g) of hydrocarbons, along with unreacted N_{CO} moles of CO and N_{H_2} moles of H_2. The C and O balance can then be obtained in eqn (6.10) and (6.11), respectively, in terms of stoichiometry described in eqn (6.1) and (6.2).

$$\text{C balance: } N_{\text{CO}}^{\text{in}} - N_{\text{CO}} = N_{\text{CO}_2} + \bar{n}N_{\text{C}\bar{n}\text{H}\bar{m}} \tag{6.10}$$

$$\text{O balance: } N_{\text{CO}}^{\text{in}} - N_{\text{CO}} = 2N_{\text{CO}_2} + N_{\text{H}_2\text{O}} \tag{6.11}$$

When substituting the O consumption in eqn (6.11) into the C consumption in eqn (6.10), it yields the equation of β described in eqn (6.12):

$$\beta = \frac{\bar{m}}{\bar{n}} = \frac{m_{\text{HC}}}{r_{\text{FT}}} - 12 \tag{6.12}$$

where $m_{\text{HC}} = m_{\text{gasHC}} + m_{\text{oil}} + m_{\text{wax}}$; $r_{\text{FT}} = N_{\text{CO}_2} + N_{\text{H}_2\text{O}}$.

Eqn (6.12) allows obtaining the mean H/C ratio (β) of the hydrocarbons formed by FTS in terms of only hydrocarbon mass and gas results, which are basic data in the FTS experiment. This provides a way to calculate the value of β value without requiring the detailed composition of heavier hydrocarbons.

In order to build a model for the mean carbon number ($\bar{n}$) of the hydrocarbons formed, we consider that eqn (6.1) is a volume contraction reaction that arises by the polymerization reaction to form high-molecular-weight hydrocarbons. Assuming that 1 mol of CO and F mol of H_2 participate in the reaction, by correlation of reaction volume contraction $\delta(\Delta V/V)$ with the stoichiometry described in eqn (6.1) and (6.2), a model of $\bar{n}$ is established in eqn (6.13):

$$\bar{n} = \frac{1}{1 + \frac{\beta}{2} - \frac{\delta(1+F)}{(r_{\text{CO}} - r_{\text{CO}_2})}} \tag{6.13}$$

where β is the mean H/C ratio of the FT hydrocarbons that can be obtained from eqn (6.12); F is the H_2/CO feed ratio; δ is the reaction volume contraction; r_{CO} and r_{CO_2} are the CO consumption and CO_2 formation rates, respectively.

6.3.3 Model for C_{5+} Selectivity and Chain Growth Probability (α)

When FTS product distribution follows the ASF law, as described in eqn (6.3) and (6.4), especially at higher hydrocarbon numbers above C_{10}, determining the chain growth probability (α) is very important to understand the total hydrocarbon product distribution. However, similar to the C/H ratio (β) and mean carbon number ($\bar{n}$) discussed previously, the chain growth probability (α) value cannot be directly obtained in most cases, due to the oil and wax compositions being missing. Next is a model we developed to establish the relationship between chain growth probability (α) and C_{5+} selectivity ($S_{C_{5+}}$), which can be obtained from the gaseous compositions. This allows the chain growth probability to be predicted using the C_{5+} selectivity value. CH_4

selectivity is not used for the prediction of α due to the deviations at C_1 and C_2 (as addressed in the previous section).

$S_{C_{5+}}$ is the fraction of the hydrocarbons with carbon numbers equal to and greater than C_5, which can be described by eqn (6.14).

$$S_{C_{5+}} = \frac{\sum_{5}^{n} (i \times N_{C_i})}{\sum_{1}^{n} (i \times N_{C_i})} \tag{6.14}$$

where i is the carbon number of hydrocarbon C_i ($i = 1, 2, 3...n...$); N_{C_i} is moles of hydrocarbon C_i.

By considering the stoichiometric constraints shown in eqn (6.1) and (6.2) and substituting eqn (6.3) into eqn (6.14), the relationship between C_{5+} selectivity and chain growth probability α is established, as expressed in eqn (6.15).

$$S_{C_{5+}} = 5\alpha^4 - 4\alpha^5 \tag{6.15}$$

Experimental $S_{C_{5+}}$ selectivity can be obtained from the basic gas results using eqn (6.16), which lumps together the rates of r_{C_i}, ($i = 1, 2, 3, 4$), CO (r_{CO}) and CO_2 (r_{CO_2}). Thus, the chain growth probability α can be worked out conveniently by solving the high order eqn (6.15).

$$S_{C_{5+}} = 1 - \sum_{1}^{4} S_{C_i} = 1 - \sum_{1}^{4} (i \times r_{C_i})/(r_{CO} - r_{CO_2}) \tag{6.16}$$

In summary, the models developed for the mean H/C ratio (β), mean carbon number $\bar{n}$ and the relationship between C_{5+} selectivity ($S_{C_{5+}}$) and chain growth probability (α) were explained in this section. These models allow predicting these important parameters using only basic experimental data such as gas analysis results and oil and wax mass.

6.3.4 Case Study to Examine the Models for β and $\bar{n}$

To examine the models of β and $\bar{n}$ shown in eqn (6.12) and (6.13), experimental data on several cobalt and iron catalysts were used. The experimental and calculated β and $\bar{n}$ values are summarized in Table 6.1. The calculated β and $\bar{n}$ were obtained using only gas data and the overall oil and wax mass. The experimental data for β and $\bar{n}$ were obtained from the full product analysis, including gas, oil and wax. The results shown in Table 6.1 indicate that the predicted and experimental β and $\bar{n}$ values are consistent with each other, which suggests that the models developed can be applied generically in a wide range of conditions.

Table 6.1 Experimental and predicated β and $\bar{n}$ values on iron and cobalt catalysts.[a]

		Calculated value		Experimental value	
Catalyst	CO conversion (%)	β	Ave n	β	Ave n
Co 1	25.97	2.02	4.31	2.34	4.67
	24.55	2.09	5.33	2.31	5.16
Co 2	39.92	2.81	5.52	2.40	5.45
	51.94	2.99	5.49	2.35	5.93
Fe 1	76.64	2.18	7.93	2.15	8.22
	35.51	2.24	6.12	2.22	5.74
Fe 2	55.94	2.13	4.32	2.32	4.10
	13.23	2.17	4.71	2.29	4.40

[a]Reaction conditions: 220–270 °C, 1.3–2.2 MPa, $H_2/CO = 0.67–2.1$.

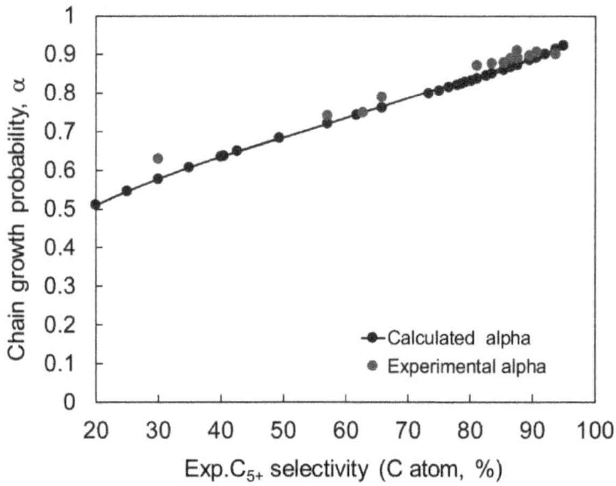

Figure 6.1 Experimental and predicted chain growth probability from eqn (6.15).

6.3.5 Case Study to Verify the Models for $S_{C_{5+}}$ and α

The predicated α values from eqn (6.15) are plotted in Figure 6.1. The experimental α values, *i.e.* 0.6–0.9, were obtained by analyzing the full range of FTS products – including gas, oil and wax hydrocarbons – from various iron and cobalt catalysts. The experimental α results are also plotted in Figure 6.1. The predicated α values using eqn (6.15) are consistent with the experimental α values. This suggests that eqn (6.15) is adequate to estimate chain growth probability, based on the C_{5+} selectivity derived from the gas results.

Weivearmp[35] developed an empirical model (eqn (6.17)) to calculate the chain growth probability of an iron catalyst operated in a fluidized-bed reactor. Using eqn (6.17), the chain growth probability is about 0.65–0.70, which deviates significantly from the results of the model deduced by the current work, 0.82–0.92, as shown in eqn (6.15). Therefore, the empirical model (6.17) adapts only to certain conditions, while the model (6.15) developed in this study is generic in nature.

$$a = 0.606 + 0.012n \tag{6.17}$$

6.4 Hydrocarbon Distribution and Promoter Effect

6.4.1 GC–MS Chromatography Results

Typical GC chromatographs of cobalt and iron catalysts are shown in Figure 6.2a and b. α-olefins, β-olefins and *n*-paraffins are displayed in the largest retention time (RT) or total ion current (TIC) for each carbon number. Iso- and *n*-paraffins were identified using a mass selective detector (MSD) and quantified with FID. Typical MSD results of *n*-paraffins and iso-paraffins are shown in Figure 6.3a–c. According to the mass spectrum of the representative *n*-C_{15} compound, the *n*-paraffins are characterized by clusters of peaks separated by 14 *m*/*z* with no loss of $[M–CH_3]^+$ being seen (see Figure 6.3a). The most intense peak is at C_5, and the intensity decreases with

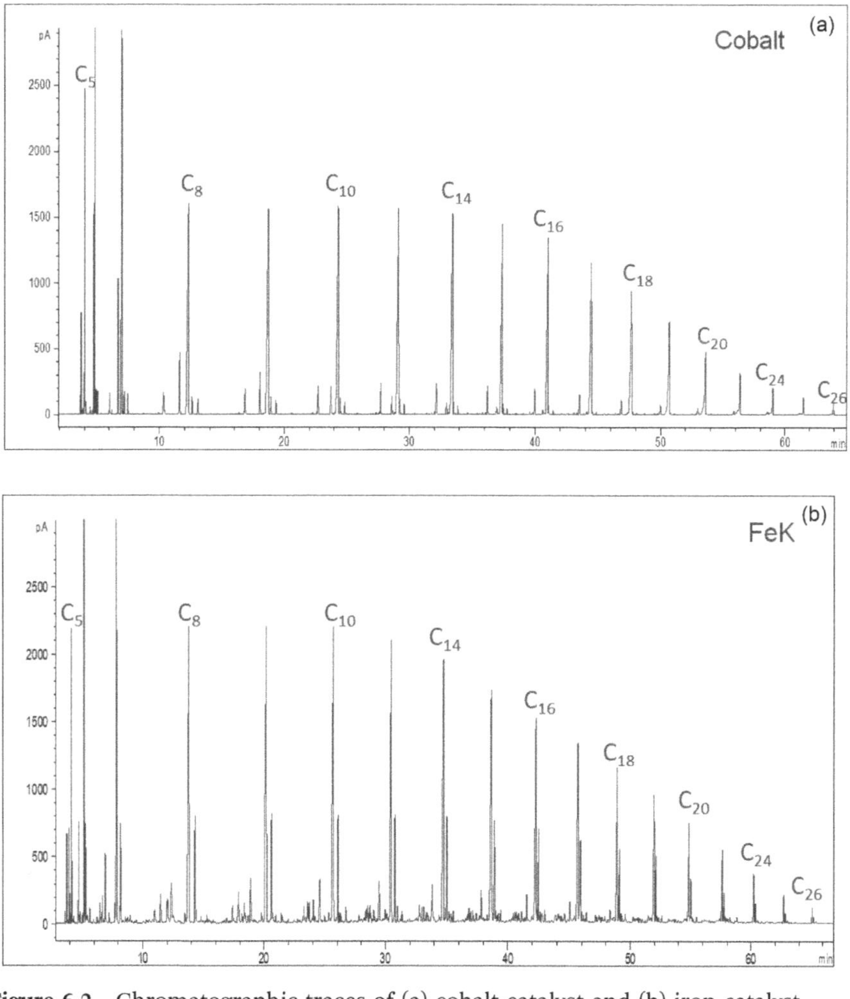

Figure 6.2 Chromatographic traces of (a) cobalt catalyst and (b) iron catalyst.

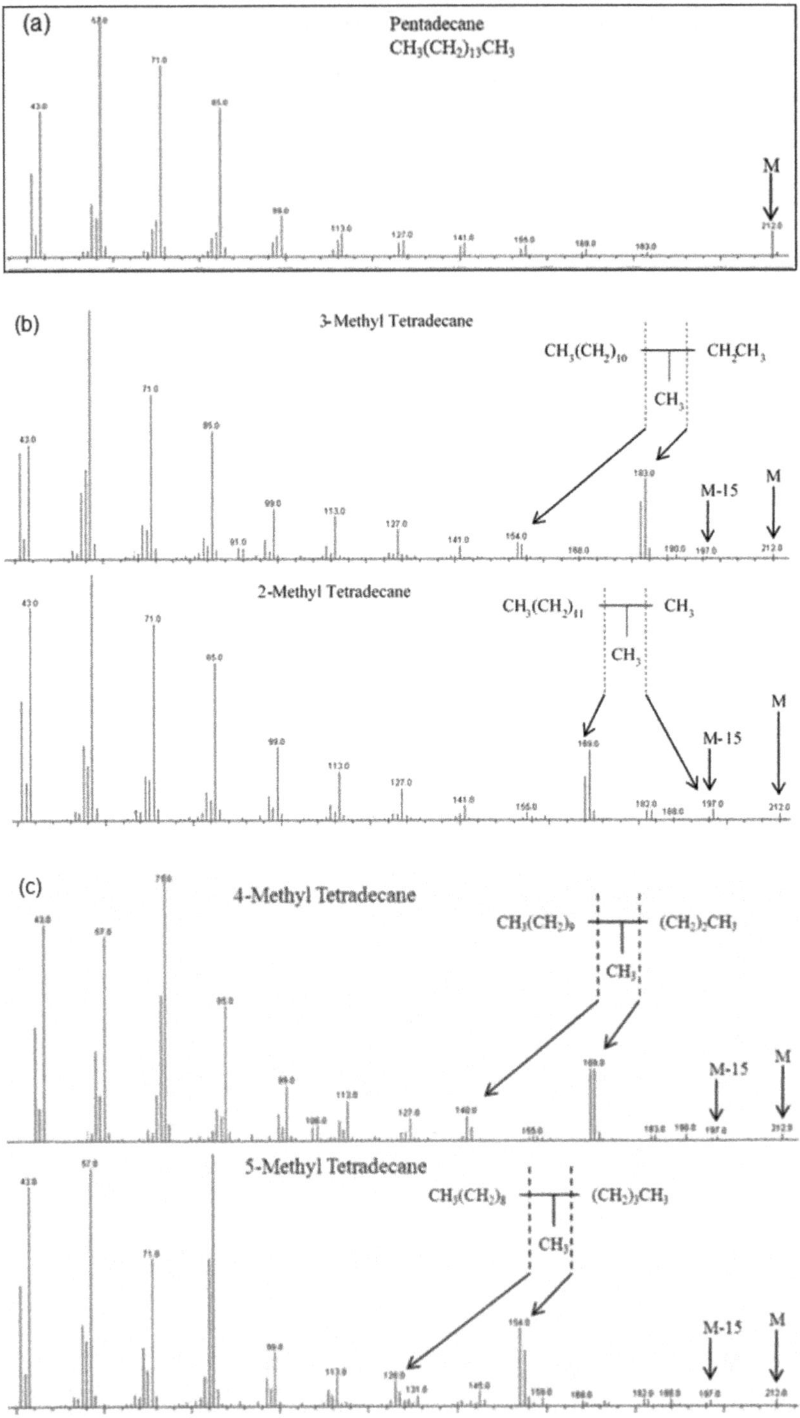

Figure 6.3 MSD fragmentation pattern of (a) C_{15} *n*-alkane, where M is the parent peak; (b) 2-methyl and 3-methyl pentadecane; (c) 4-methyl and 5-methyl pentadecane.

an increase in the carbon number. The fragmentation pattern of methyl-C_{15} (see Figure 6.3b and c) is similar to that of n-C_{15}, except for the loss of the methyl group ($M{-}CH_3$) at the tertiary carbon. This is consistent with the earlier studies done with iso-paraffins and olefins reported by the Davis group.[27,29,30]

Representative identification of iso-paraffins in C_8–C_{15} and C_{16}–C_{21} are shown in Figure 6.4a and b. Except for the 2-methyl and 3-methyl isomers, the

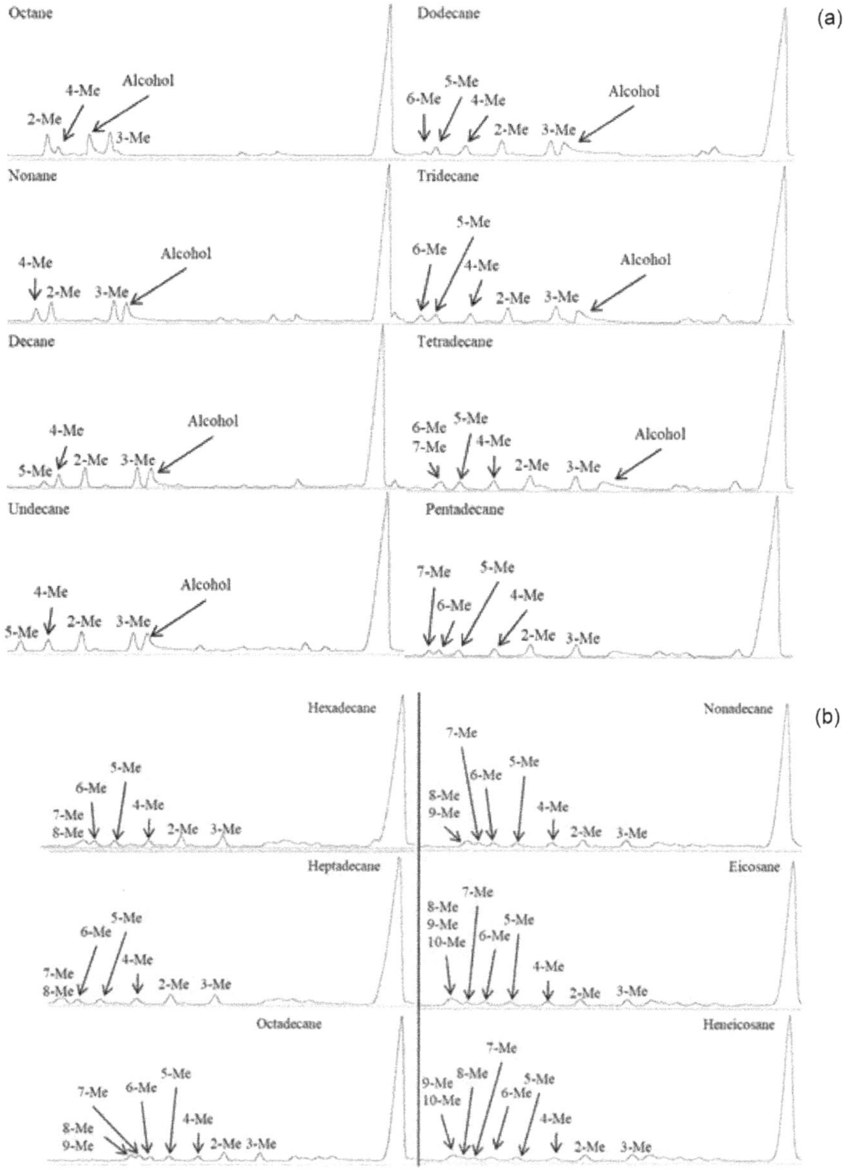

Figure 6.4 Identification of iso-paraffins for (a) C_8–C_{15}, (b) C_{16}–C_{21} hydrocarbons.

branched paraffins were eluted in the order that was based on the location of the methyl group in the hydrocarbon chain. With carbon chains smaller than C_9, 2-methyl was eluted before 4-methyl. The iso-paraffin with the methyl group attached to the center carbon was eluted first, and the process progresses toward the end of the chain until the *n*-paraffin is eluted from the column.[29,30]

Note that the number of branched paraffins rises with an increase in the carbon number, but the peak intensity becomes smaller at higher carbon numbers, *i.e.* C_8 produces *n*-octane and 2–4 methyl iso-octanes, but C_{20} produces *n*-eicosane and 2-10 methyl iso-nonadecanes, and the iso-paraffin peaks become too small to analyze after C_{21}. In this study, only the iso-paraffins results below C_{15} were studied to ensure the accuracy of the data discussed.

6.4.2 Overall Hydrocarbon Distribution

The hydrocarbon distribution over the unsupported 25%Co/Al_2O_3 catalyst, the 0.5%Pt and 0.5%Re promoted 25%Co/Al_2O_3 catalysts, and the Fe/Si/Cu catalysts promoted with 1% K and 3% K were examined. The results obtained in 1 L-CSTR at the industrially relevant FTS conditions for the Co and Fe catalysts are plotted in Figure 6.5a and b, respectively. The experimental hydrocarbon distribution over various cobalt catalysts (220–230 °C, 2.1 MPa, H_2/CO = 2.0 and 40–50% CO conversion) followed a single alpha ASF distribution, except for the slight deviations at C_1 and C_2 (see Figure 6.5a). Contrarily, two alpha ASF distributions (230–270 °C, 1.3–2.1 MPa, H_2/CO = 0.67–2.0 and 40–75% CO conversions) were found on both the 1 K and 3 K iron catalysts. One is displayed at the low carbon range C_1–C_4, and the other is in the high carbon range above C_5. In both cases, outliers at C_5–C_8 can be explained by the volatilization of the components during product collection and sample preparation for GC analysis. There are two abnormal peaks at about C_{28} and C_{35} in the hydrocarbon distribution of the iron catalysts, which came from the residual of the start-up solvent C_{30} oil (see Figure 6.5b).

With both the cobalt and iron catalysts, the hydrocarbon chain growth probability is more sensitive to the addition of a promoter (type and amount), compared to the mild influence of temperature. According to Figure 6.5a and b, the chain growth probability had a relatively larger change with the addition of 0.5%Pt and 0.5%Re promoters on cobalt (0.87–0.9). Similarly, with the iron catalysts, K was found to increase α from 0.82 to 0.89 when increasing the loading from 1 to 3%. However, there were smaller changes (*i.e.* 0.89–0.90) with the cobalt catalysts and the iron catalysts (0.89–0.92) even though the temperature had increased remarkably, *i.e.* by 10 °C and 40 °C, respectively.

The different hydrocarbon distributions observed on the cobalt and iron catalysts may imply dissimilar chain growth mechanisms on these two types of catalysts. Figure 6.5a and b suggest that one chain growth monomer was mainly involved in the chain prolongation on the cobalt catalyst, while more than one chain growth monomer might participate in the chain growth on the Fe–K catalysts, leading to single and double alpha hydrocarbon distributions on the cobalt and iron catalysts, respectively.

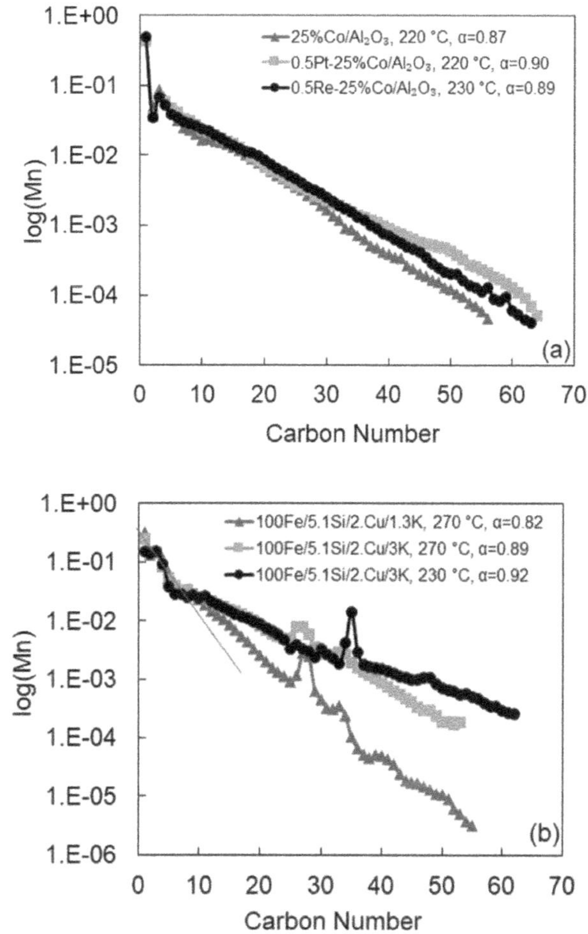

Figure 6.5 ASF distribution of (a) CO catalysts (220–230 °C, 2.1 MPa, $H_2/CO = 2.0$ and $X_{CO} = 40–50\%$), (b) iron catalysts (230–270 °C, 1.3–2.1 MPa, $H_2/CO = 0.67–2.0$ and $X_{CO} = 40–75\%$).

Three popular FTS mechanisms have been proposed and widely accepted to date, including carbide by Fischer *et al.*,[36] enol by Anderson and Emmett[1,37] and CO insertion mechanisms by Pichler and Schuz.[38] These mechanisms vary by chain starter and chain growth monomer. For example, CH_2-s and CO-s were proposed as the chain growth monomers for the carbide and CO insertion mechanism, respectively, while CHOH-s was assumed to be the monomer in the enol mechanism. It is not easy to draw a definitive conclusion regarding the mechanism from the overall distribution shown in Figure 6.5a and b. However, if primarily one chain growth monomer participated in the chain growth on the cobalt catalyst (as previously discussed), the CO insertion plus H-assistant dissociation might be more plausible pathways (CO-s + 2H-s → HCHO-s → CH_3-s + OH-s, R-s + CO-s → RCO-s →

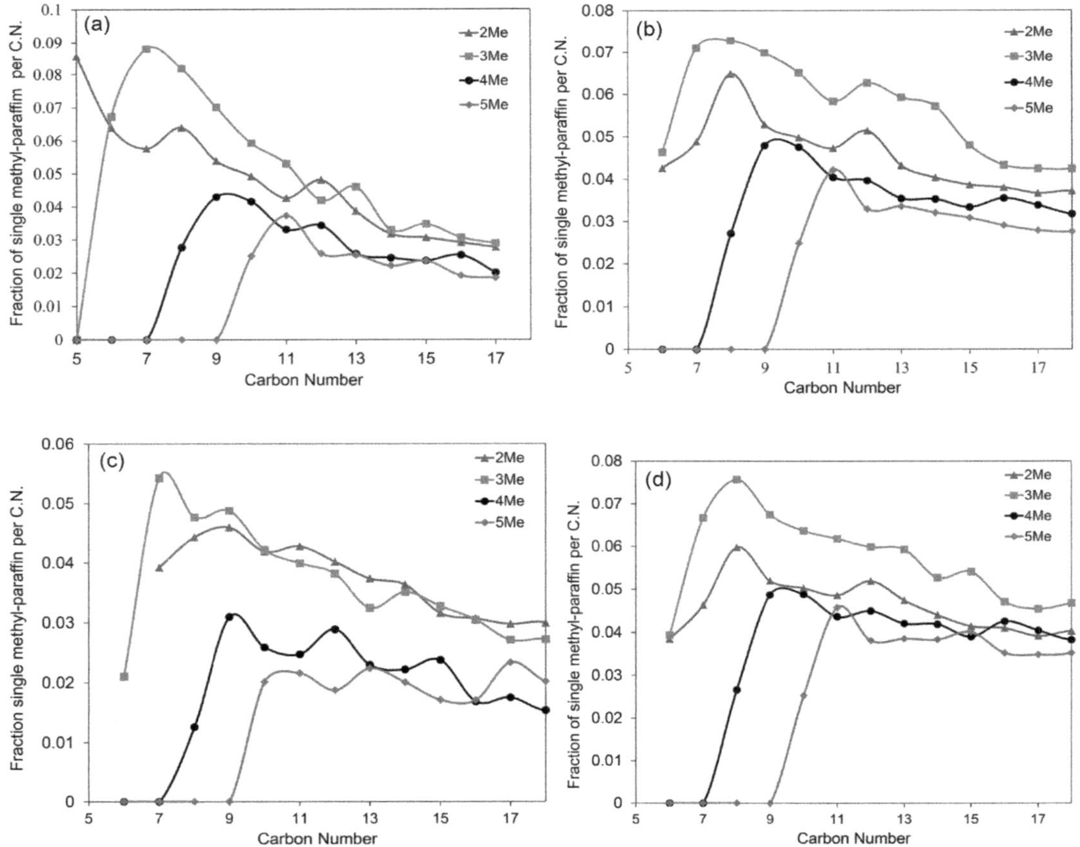

Figure 6.6 Change in the individual 2–5 methyl fraction with carbon number over (a) 3K catalyst-clean condition at 270 °C, (b) 3K catalyst-HBr at 270 °C, (c) 3K catalyst-230 °C, (d) 1K catalyst-clean condition at 270 °C. Legend *x*Me ($x = 2, 3, 4, 5$) refers to the 2–5 methyl branching hydrocarbons.

RCHO-s → RCH_2-s → olefin and paraffins), as this can explain the formation of both hydrocarbons and oxygenates (small amounts), based on the RCHO-s intermediates proposed. The carbide mechanism with CH_2-s or CH-s being the possible chain growth monomers[27,39] is able to interpret the single alpha ASF on the cobalt catalysts, but it fails to account for the oxygenate formation. Recently, a detailed kinetic study on the Re–Co catalyst was conducted, based on the carbide and CO insertion mechanisms plus the chain-length-dependence of α-olefin desorption.[23,24] These studies indicated that both carbide and CO insertion mechanisms adequately explained the kinetic data, but the latter one is slightly superior. The results support the preceding hypothesis for the single α product distribution on cobalt catalysts.

Several studies proposed two types of FT mechanism to explain the double α hydrocarbon distribution on iron catalysts.[19,40–42] Patzlaff *et al.*[41] explained the double α distribution with kinetic variation of P_{CO} and P_{H_2} and ignored the secondary reaction of 1-olefins. The CH_2 insertion was presumed to be responsible for the first chain growth, whereas the second chain growth was assumed to follow a different pathway. Dictor *et al.*[42] and Shi *et al.*[24] reported that the reactor holdup was the cause of the deviation for the second chain growth. Huff *et al.*,[19] Donnelly *et al.*[17] and Bukur *et al.*[18] have proposed two sites (*i.e.* with K-containing and no K-containing sites) accounting for the double α distribution of the iron catalysts. The two insertion mechanisms – CH_2 and CO – might account for the current two-alpha distribution over the iron catalysts, since changing the K loading from 1 to 3% and changing the temperature primarily altered the chain growth at higher carbons (see Figure 6.5b).

6.4.3 Isoparaffins Distribution for Iron Catalysts: Effects of K Loading, Poison HBr and Temperature

The mole fraction of individual branching (2-methyl, 3-methyl, 4-methyl and 5-methyl) for the 1K and 3K iron catalysts was obtained at 230 and 270 °C in the clean and 20 ppm HBr poisoning runs. The results are present in Figures 6.6a–d. In all cases, the fraction of methyl-paraffins at C_5–C_{15} tend to follow the order 3-methyl > 2-methyl > 4-methyl > 5-methyl. This is consistent with the results of Schulz[27,28] and Davis *et al.*[31,32] Moreover, the methyl fractions decrease with an increase in the carbon number, which is consistent with the literature.[27,32] The increase in isomerization at carbon 14 and 15 was due to the overlap of the peaks from the 3-methyl and alcohol compounds. It is difficult to interpret the impact of HBr on the formation of these individual methyls due to the low difference in magnitude. However, when comparing the total fraction of iso-paraffin under the clean and HBr poisoning conditions shown in Figure 6.7, it is clear that adding 20 ppm HBr led to improved iso-paraffin formation, and the total methyl percentage increased by 1.5–5% in C_8–C_{15}. Undoubtedly, the fractions of individual 2–5 methyl decreased by 1–4%, when the reaction temperature was decreased from 270 °C to 230 °C. This result could indicate that the addition of HBr during the

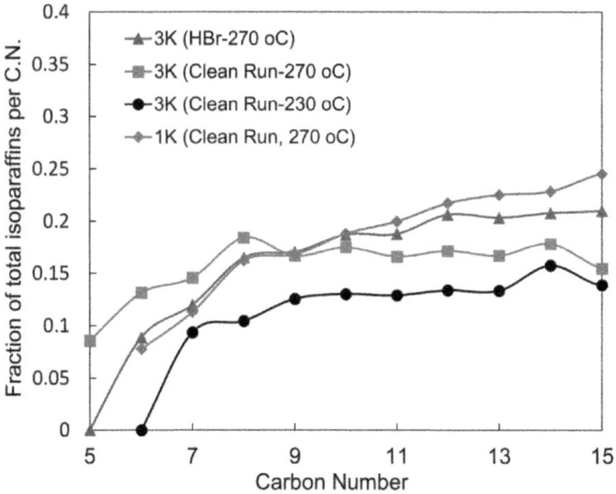

Figure 6.7 Fraction of total methyl as a function of carbon number and temperature over iron catalysts. Other reaction conditions: 1.3 MPa, $H_2/CO = 0.7$, $X_{CO} \sim 70\%$.

Table 6.2 Change in 1-C_4 and 2-C_4 olefin selectivity with K loading on 100Fe/5.1Si/2Cu/xK catalysts.[a]

		Olefin content (%)	
K loading (%)	X_{CO} (%)	1-C_4	2-C_4
1	75	65	17
3	75	80	9

[a]Reaction conditions: 270 °C, 1.3 MPa, $H_2/CO = 0.67$, ~165 h.

reaction increased the surface acid properties of the iron catalysts (for example H^+ ion) and promoted isomerization of the paraffins, whereas the low temperature promoted chain growth and kinetically inhibited isomerization of olefins.

Figure 6.6 shows slightly higher fractions of 2-, 3-, 4-, and 5-methyl on the 1K iron catalyst than on the 3K iron catalyst – 0.11–0.24 *vs.* 0.1–0.15, and Figure 6.7 shows a slight increase in total iso-paraffins with an increase in carbon number in the range of 5–15. These trends are not consistent with the study done by Luo *et al.*,[31] which reported a positive effect of K loading on the iso-paraffin formation and independence of total iso-paraffin content from carbon number on Fe–K catalysts in terms of the data at different times at 230 °C. K has been reported as a structure promoter or an electronic promoter (back-donation to CO) to Fe in order to enhance CO adsorption and dissociation, consequently improving 1-olefin formation.[1,2] Meanwhile, increasing K loading raised the basicity of the iron catalyst surface,[1,2] which should adversely impact branching formation. This result is consistent with the gas phase olefin results shown in Table 6.2, in which a higher fraction of 2-C_4 olefin (17 *vs.* 9%) and low 1-C_4 olefin (65 *vs.* 80%) was obtained on the 1K catalyst relative to that of the 3K iron catalyst at the same conditions. The isoparaffin fraction shown in Figure 6.7 shows a rapid increase with an increase in the carbon number, when the carbon number is smaller than C_9. In the higher carbon number range, the fraction increment is slower. The increase in methyl along with an increase in carbon number could be due to a higher concentration of alkyl intermediates and a longer residence time for heavier hydrocarbons.

6.4.4 Iso-paraffins Distribution for Cobalt Catalysts: Effects of Promoters Zr and Pt, and Poison COS

Figure 6.8a–c shows the plots of the fractions of individual 2-methyl, 3-methyl, 4-methyl and 5-methyl over unpromoted 5% Zr and 2% Pt promoted 25%Co/Al_2O_3 catalysts at the same process conditions: 220 °C, 2.2 MPa, H_2/CO = 2.0 and about 50% CO. The effect of the addition of 1 ppm COS on the product distribution is summarized in Figure 6.8d. In all cases, the fraction of the methyl-paraffins in the C_8–C_{15} range essentially follows the same order as that of the iron catalyst: 3-methyl > 2-methyl > 4-methyl > 5-methyl paraffins. However, the fractions of methyl paraffins on the cobalt catalysts are 5–7 times lower than on the iron catalysts, suggesting low activity for the secondary reactions of the cobalt catalyst. After the addition of the Zr and Pt promoters, the fraction of the methyl paraffins decreased from 0.5–1% to 0.4–0.6% and 0.1–0.6%, respectively. The addition of 1 ppm COS in the feed of the cobalt test runs resulted in a decrease in the fractions of 2-, 3-, 4-, 5-methyl paraffins as well, *i.e.* from 0.5–1% to 0.2–0.8% compared to the "clean" results of the unpromoted Co catalyst. Interestingly, all individual methyl fraction did not change significantly with carbon number in all four cobalt test runs (see Figure 6.8 a–d). This is different from what was observed with the iron catalysts (see Figure 6.6a–d).

The distribution of total iso-paraffin with carbon number is illustrated in Figure 6.9. It is consistent with the results of the individual distribution of 2–5 methyl paraffins with carbon numbers. The addition of Zr led to the largest decrease in the formation of iso-paraffins – from 2.8 to 1.0%, in the C_7–C_{15} range – whereas co-feeding 1 ppm COS or adding 0.5%Pt decreased the methyl to a lesser extent, *i.e.* to 1.5–2% and to 2–2.5%, respectively (see Figure 6.9). This result was accompanied by: an increase in the fraction of 1-olefin with the addition of Zr, a decrease in 1-olefin formation with the addition of Pt, and nearly unchanged 1-olefin formation with the addition of COS in C_5–C_{15} (see Figure 6.9b). According to Figure 6.9, the total fractions of iso-paraffins on the cobalt catalysts were determined to be 1–3%. In contrast, it was 15–22% on the Fe catalysts (see Figure 6.7). The results recorded in this study are in line with the study done by Shi *et al.* (5% on cobalt and 25% on the Fe catalyst).[30]

6.4.5 Discussion

The decrease in 2–5 methyl paraffins with carbon number on the iron catalysts (Figure 6.6a–d) is in line with the studies done by Schulz[28] and

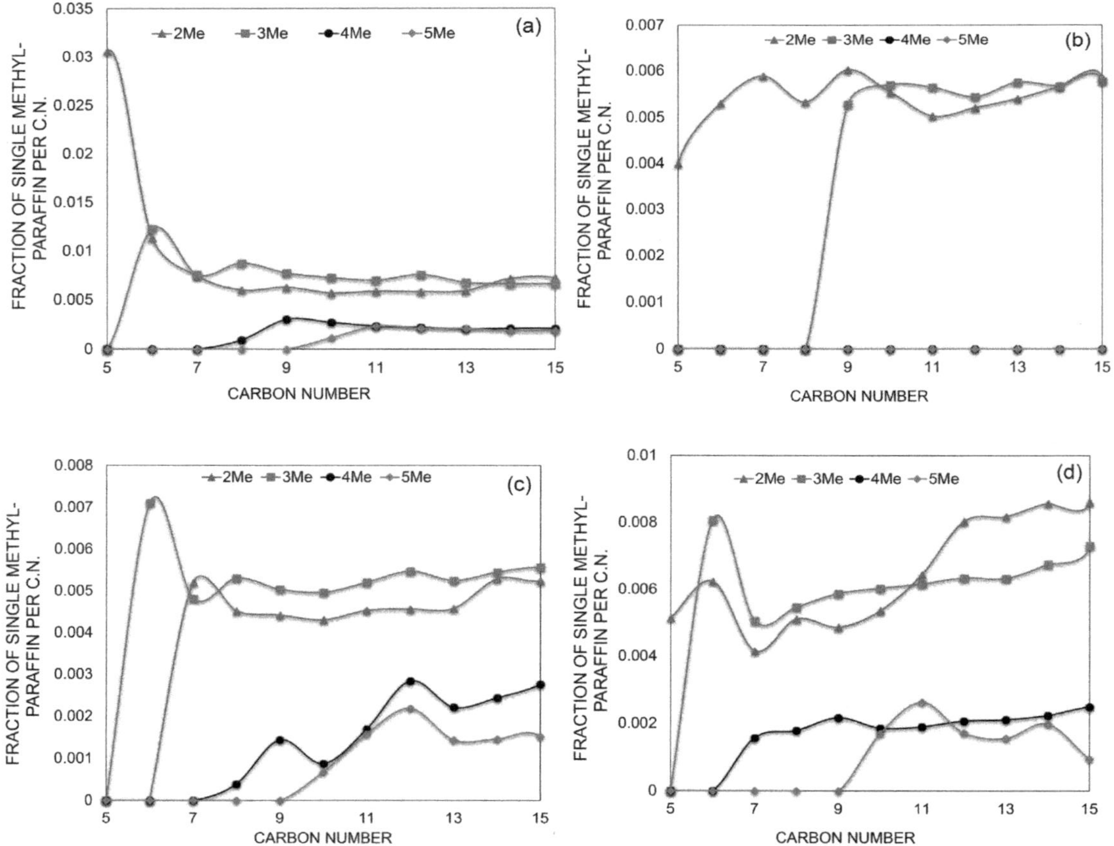

Figure 6.8 Change in individual 2–5 methyl fraction with carbon number over cobalt catalysts: (a) clean run with unprompted 25%Co/Al_2O_3, (b) clean run with 5%Zr–25%Co/Al_2O_3, (c) clean run with 0.5%Pt–25%Co/Al_2O_3, (d) COS poisoning run on 0.5%Pt–25%Co/Al_2O_3.

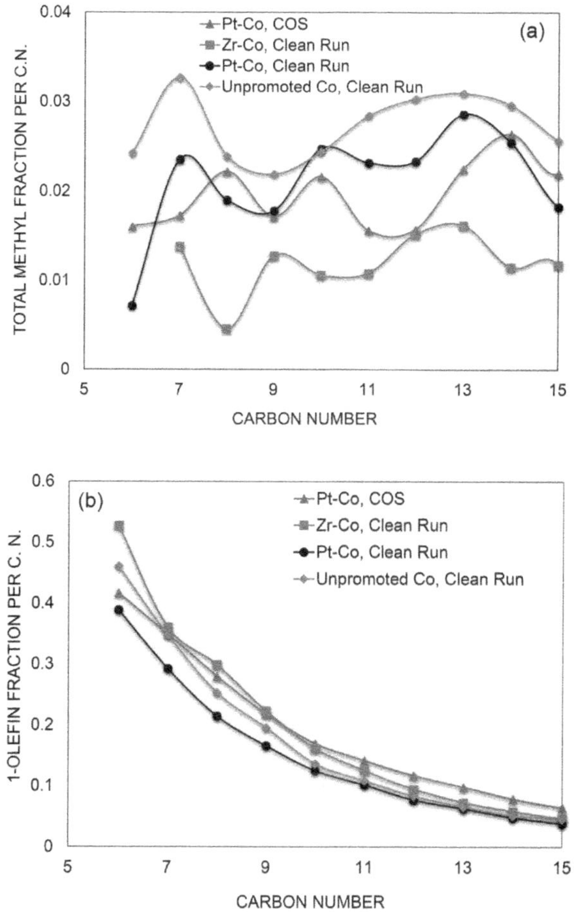

Figure 6.9 Fraction of (a) total methyl and (b) 1-olefin, as a function of carbon number over different cobalt catalysts. Reaction conditions: 220 °C, 2.0 MPa, $H_2/CO = 2.0$, $X_{CO} = 42–48\%$.

Shi *et al.*[29] Schulz[28] reported that C_4–C_9 branching products declined almost linearly with carbon number and peaked at C_4. Shi *et al.*[29] reported a decline in the fraction of the branched products at C_7–C_{14} with carbon number on an iron catalyst – a slight increase in the fraction of the branched products with carbon number on a cobalt catalyst. The latter conclusion is consistent with current results for methyl over various cobalt catalysts, as shown in Figure 6.8a–d. Both Schulz[27,28] and Shi *et al.*[29] proposed an alkylidene mechanism for the formation of methyl-branched hydrocarbons only, *i.e.* once 1-olefin formed, it was readsorbed and converted to 1-alkylidene or 2-alkylidene species, which can prompt chain growth with the monomer of methyl or methylene species and enhance methyl-branched products (see Scheme 6.1). Certainly, 2-olefins could also be formed from the alkylidene species *via* β-elimination. Therefore, the chain-size-dependent branching

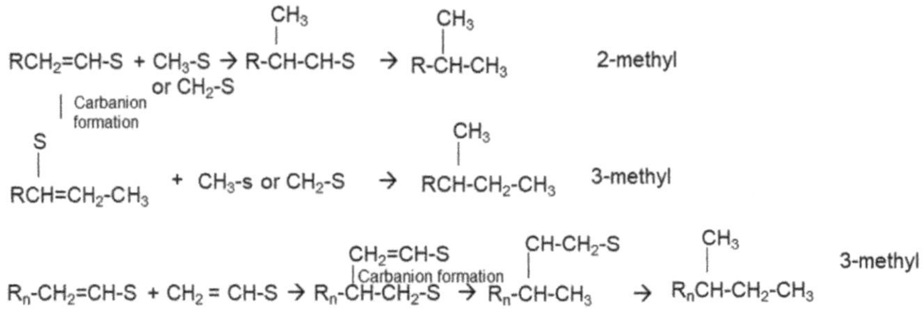

Scheme 6.1 Methyl formation from methylene or ethylidene monomer.

products were more likely due to a kinetic reason, *i.e.* the concentration of intermediates including olefins, methyl, methylene and alkylidene species with the carbon number. This should be different from the reaction pathways for the formation of the main 1-olefin and *n*-paraffin products, which has been proposed through CO insertion or carbide mechanisms ($R–CH_2–M + CH_2$ (CO)).[25,26] An additional explanation of the decrease in the trend of methyl with carbon number could be a greater steric hindrance for the formation of larger molecular-branched products in the catalyst pores.

The fraction of 3-methyl branching was generally greater than that of 2-methyl and other methyls on both the iron and cobalt catalysts (see Figures 6.6 and 6.8). This might suggest that ethylidene ($CH_2=CH–M$), which formed *via* the reabsorption of most of the active ethene species, served as a chain growth monomer and induced the branched hydrocarbon chain growth (Scheme 6.1), which led to higher amounts of 3-methyl on both the iron and cobalt catalysts.

The difference in the individual 2–5 methyl branched distribution with carbon number on the iron (*i.e.* decreasing trend) and cobalt catalysts (relatively stable) could suggest a different chain growth mechanism for the production of minor products on the cobalt and iron catalysts. The high fraction of isoparaffins on the iron catalysts relative to the cobalt catalysts (5–7 times higher) – at 220–230 °C – may reflect a steric hindrance during the branching growth and the lesser effect on the precipitated iron catalysts due to the reactions on the surface of the iron catalysts. The higher temperatures used for the iron catalysts could be another reason for the higher amount of methyl products on the iron catalysts, even though the temperature (230 *vs.* 270 °C) did not have a significant effect on iso-paraffin formation (see Figure 6.7).

The promoter or poison changing 1-olefin and methyl fractions on iron and cobalt catalysts could also be due to interaction with cobalt and/or assisting in the adsorption and dissociation of CO. Potassium has long been known to assist in the adsorption and dissociation of CO when used as a promoter.[1,2] An active iron carbide FTS catalyst yields poor back-donation; CO dissociation is then kinetically unfavorable as the major products are light *n*-paraffins, methane and isomerized hydrocarbons.[43] As the K promoter increases the ability of CO to bind, a selective increase in the 1-olefin formation is promoted, and isomerization is suppressed. The addition of 20 ppm HBr increased the fraction of iso-paraffins (Figure 6.7), which could be associated with HBr modifying the surface acidity of the iron catalysts, consequently the enhanced isomerization.

As an active FTS catalyst, cobalt is itself sufficient to provide the necessary back-donation needed for cleaving the CO bond. The dominant products are 1-olefin, *n*-paraffin and linear alcohol.[22–24,43] Beneficial results have been seen with the addition of a promoter.[43–45] Zr appears to be a weak promoter, in terms of the ability to provide a low level of back-donation to the active metal. Most promoters assist cobalt in the initial reduction. If the Zr promotes H_2 adsorption, the total amount of olefin will most likely be deviated.[45] Moreover, Zr may act more as an anchor for the cobalt, diverting the interaction between it and the support with the different-pore-sized alumina-supported catalysts.[45] This interaction might foster a close proximity between the cobalt and zirconia catalysts. So, even if zirconia is a weak donor and promotes only a small increase in back-donation, its close proximity is indicated by a selective increase in the 1-olefin (Figure 6.9b) and a decrease in the methyl content (Figure 6.9a). A Pt promoter has been known to promote the hydrogenation of 1-olefin and increase paraffins through the H_2 spillover role.[44] Minor iso-paraffins were decreased with the addition of Pt (Figure 6.8c), which could be due to a kinetic effect (lowered 1-olefin pressure).

The influence of sulfur on the performance of the cobalt catalyst appears to be more complicated. In the initial reaction period – for example in the first 10 h in this study – sulfur may serve as a promoter and increase the electron back-donation to the cobalt surface and thus increase 1-olefin formation and inhibit isomerization (see Figure 6.9a and b).[46] Bromfield *et al.*[47] also reported that S^{2-} ions suppressed the formation of 2-olefin and promoted the formation of 1-olefin, which further supports the hypothesis. With an extended reaction time, sulfur compounds such as CoS or CoS_2 have been found to form on cobalt catalysts,[48,49] which promotes the isomerization of olefins and inhibits hydrogenation of 1-olefins.[50] This conclusion was in line with the unpublished gas phase results of our H_2S poisoning study, *i.e.* after 178 h of poisoning the 0.5%Pt–25%Co/Al_2O_3 catalyst using 1.0 ppm H_2S 2-C_4, the olefin content increased to 7.2% (from the initial value of 4.9%) and was accompanied by a decrease in 1-C_4 olefin content, *i.e.* from 41.8 to 31.8% (see Table 6.3). In a study on the effect of COS, adding 1.8 ppm COS

Table 6.3 Change in 1-C_4 and 2-C_4 olefin selectivity with time on a 0.5%Pt–25%Co/Al_2O_3 catalyst.[a]

		Olefin content (%)	
TOS (h)	X_{CO} (%)	1-C_4	2-C_4
6.3	49.2	41.8	4.9
178.3	46.5	31.5	7.2

[a]Reaction conditions: 220 °C, 2.0 MPa, H_2/CO = 2.0, 1.0 ppm H_2S.

for 170 h was also found to decrease the 1-olefin content.[49] In both cases, the sulfur compounds that formed likely increased the catalyst surface acidity and/or reduced the S back-charge-donation role and consequently enhanced isomerization of 1-olefins.

6.5 Conclusions

Generic models for the mean carbon number $\bar{n}$, mean H/C ratio β and chain growth probability α as a function of C_{5+} selectivity, have been developed in terms of material balance. The models can be used to predict these important parameters, *i.e.* $\bar{n}$, β and α, which are critically important in kinetic modeling and reactor design but are generally not available, especially when detailed oil and wax compositions are missing. The models were validated using the data for both iron and cobalt catalysts.

The overall hydrocarbon distribution of several iron and cobalt catalysts were studied. With the cobalt catalysts, the distribution followed the ASF distribution, except for high C_1 and low C_2. This trend has been attributed to several reasons, for example, multiple CH_4 formation paths, including methanation; multiple CH_4 formation sites and strong readsorption of ethylene. In the case of iron catalysts, the overall hydrocarbon distribution displayed two alpha models in the low carbon range of C_1–C_4 and the high carbon range of C_{5+}. The unique hydrocarbon distribution patterns reflect dissimilar chain growth mechanisms on these two types of catalysts. With the iron catalysts, it could be that both the CO insertion mechanism (responsible for C_{5+} chain growth) and the carbide mechanism (responsible for C_1–C_4 formation) were involved. With the cobalt catalyst, it is more likely that one type of chain growth monomer primarily participated in chain growth to the main products.

The distribution of individual 2–5 methyl paraffin on various iron and cobalt catalysts was also investigated. The results seen with both iron catalysts indicate the chain-size-dependent 2–5 methyl distribution. For example, the 2–5 methyl paraffins in the C_8–C_{15} range decreased with an increase in the carbon number on the iron catalysts, but less change was seen with the cobalt catalysts. These results could be due to a kinetic effect and greater steric hindrance for longer hydrocarbons. The formation of methyl paraffin was explained by the alkylidene mechanism, but the chain growth monomers or reaction pathways may vary between the iron and cobalt catalysts. In all cases, the fraction of total methyl paraffins on the iron catalysts were determined to be 15–22%, which is 5–7 times higher than that for the cobalt catalysts, *i.e.* 1–3%.

The fraction of 3-methyl branching was greater than that of 2-methyl and much greater than other methyls on both the cobalt and iron catalysts. These results might imply that ethylidene ($CH_2{=}CH{-}M$) was formed *via* reabsorption of ethene species and that it served as a chain growth monomer to promote the branching chain growth.

The effect of promoters K, Zr and Pt and the poisons CoS and HBr on iso-paraffin and 1-olefin formations were studied. Adding higher K loading to iron and 1 ppm COS and 5% Zr to cobalt promoted 1-olefin formation but suppressed the formation of methyl paraffins. Adding 20 ppm HBr enhanced iso-paraffin formation slightly. Contrarily, the Pt promoter decreased both the 1-olefin and the isoparaffin content. The CoS_x sulfur compounds formed during the reaction had an adverse effect on hydrocarbon formation: it decreased 1-olefin but improved methyl paraffin formation. A possible electronic effect of the promoters and a change in the surface acidity and kinetic effect were presumed to contribute to the promoter effect. For example, K, S^{2-} and Zr might provide back-donation electrons to CO on the catalyst, while H^+ might modify the surface acidity of the catalysts.

Author Contributions

Conceptualization, experimental and methodology, model development, data analysis, investigation, writing and validation – W. M.; GC/MS methodology and analysis for liquid products, investigation, writing and validation – W. S. and M. M.

Acknowledgements

This paper is dedicated to the late Professor Burtron H. Davis. His optimistic attitude to science and life inspires us. The authors are thankful for the financial support from the US-DOE (DE-FC26-08NT0006368) and the Commonwealth of Kentucky. Authors are very grateful to the English Editor of this special book, Juliet Gillies, for her spending time editing this paper (sentence and grammar), which I believe increased greatly the readability of this article. The authors are also grateful to Muthu K. Gnanamani for his helpful initial editing and suggestions for this book chapter.

References

1. H. H. Storch, N. Golumbic and R. B. Anderson, *The Fischer–Tropsch and Related Synthesis*, John Wiley & Sons, New York, 1951.
2. R. B. Anderson, *The Fischer–Tropsch Synthesis*, Academic Press, 1984.
3. B. H. Davis, in *Handbook of Heterogeneous Catalysis*, ed. G. Ertl, H. Knözinger and J. Weitkamp, John Wiley-VCH, New York, NY, 1997, pp. 13–35.
4. M. Dry, *Catal. Today*, 2002, **71**, 227.
5. M. Dry, *Chem. Technol.*, 1982, 744.
6. Y. W. Li, Bukur catalysis and engineering meeting at DoHa, December 2017.
7. F. G. Botes and J. W. Niemantsverdriet, *Catal. Today*, 2013, **215**, 112–120.
8. Z. Y. Liu, S. D. Shi and Y. W. Li, *Chem. Eng. Sci.*, 2010, **65**, 12–17.

9. X. D. Wen, Y. Yang, H. W. Xiang, H. J. Jiao and Y. W. Li, *Sci. Sin. Chim.*, 2017, **47**, 1298–1311.
10. J. Eilers, S. A. Posthuma and S. T. Sie, *Catal. Lett.*, 1990, 7, 253.
11. H. Schulz, *Appl. Catal.*, 1999, **186**, 3.
12. M. E. Dry, *Catal. Lett.*, 1990, 7, 241.
13. Significant coal-to-liquid project in production in Ningxia, http://www.china.org.cn/china/2016-12/29/content_40003229.htm.
14. S. Ledakowicz, H. Nettelhoff, R. Kokuun and W. D. Deckwer, *Ind. Eng. Chem. Process Des. Dev.*, 1985, **24**(4), 1043.
15. D. B. Bukur, B. Todic and N. Elbashir, *Catal. Today*, 2016, **275**, 66.
16. R. L. Zhang, J. Chang, Y. Y. Xu, L. R. Cao, Y. W. Li and J. L. Zhou, *Energy Fuels*, 2009, **23**, 4740.
17. T. J. Donnelly and C. N. Satterfield, *Appl. Catal.*, 1989, **52**, 93.
18. D. B. Bukur, M. Koranne, X. S. Lang, K. R. P. M. Rao and G. P. Huffman, *Appl. Catal.*, 1995, **126**, 85.
19. G. A. Huff and C. N. J. Satterfield, *Catal.*, 1984, **85**, 370.
20. R. S. Hurlbut, I. Puskas and D. J. Schumacher, *Energy Fuels*, 1996, **10**, 537.
21. E. Iglesia, *Appl. Catal., A*, 1997, **161**, 59.
22. H. Schulz and M. Claeys, *Appl. Catal.*, 1999, **186**, 91.
23. H. Schulz and M. Claeys, *Appl. Catal.*, 1999, **186**, 71.
24. B. C. Shi and B. H. Davis, *Appl. Catal.*, 2004, **277**, 61.
25. B. Todic, T. Bhatelia, G. F. Froment, W. Ma, G. Jacobs, B. H. Davis and D. B. Bukur, *Ind. Eng. Chem. Res.*, 2013, **52**, 669.
26. B. Todic, W. Ma, G. Jacobs, B. H. Davis and D. B. Bukur, *Catal. Today*, 2014, **228**, 32.
27. H. Schulz, *Topics Catal.*, 2003, **26**, 73.
28. H. Schulz, *Appl. Catal.*, 2020, **602**, 117695.
29. B. C. Shi, R. A. Keogh and B. H. Davis, *J. Mol. Catal.*, 2005, **234**, 85.
30. B. C. Shi, L. Wu, Y. X. Liao, C. F. Jin and A. Montavon, *Top. Catal.*, 2014, **57**, 451.
31. M. S. Luo, W. D. Shafer and B. H. Davis, *Catal. Lett.*, 2014, **144**, 1031.
32. W. Ma, W. D. Shafer and B. H. Davis, Branching products of iron catalysts, Unpublished data, 2013.
33. B. H. Davis, Technology development for iron Fischer–Tropsch catalysis, DOE final report, 1998.
34. B. H. Davis and G. Jacobs, Sensitivity of Fischer–Tropsch synthesis and water-gas shift catalysts to poisons from high-temperature high-pressure entrained-flow (EF) oxygen blown gasifier gasification of coal/biomass mixtures, DOE final report, 2011.
35. A. W. Weivearmp, H. S. Seelig, N. J. Bowman and W. E. Cady, *Ind. Eng. Chem.*, 1953, **45**, 343.
36. F. Fischer and H. Tropsch, *Brennst.-Chem.*, 1926, 7, 97–104.
37. J. T. Kummer and P. H. Emmett, *J. Am. Chem. Soc.*, 1953, **75**, 5177.
38. H. Pichler and H. Schul, *Chem. Ing. Technol.*, 1970, **42**, 1162.
39. B. Shi and C. Jin, *Appl. Catal., A*, 2011, **393**, 178–183.

40. J. Gaube and H. F. Klein, *J. Mol. Catal. A*, 2008, **283**, 60–68.
41. J. Patzlaff, Y. Liu, C. Graffmann and J. Gaube, *Appl. Catal., A*, 1999, **186**, 109.
42. R. A. Dictor and A. T. Bell, *J. Catal.*, 1986, **97**, 121.
43. W. Shafer, M. Gnanamani, U. Graham, J. Yang, C. Masuku, G. Jacobs and B. Davis, *Catalysts*, 2019, **9**, 259.
44. W. Ma, G. Jacobs, R. A. Keogh, D. B. Bukur and B. H. Davis, *Appl. Catal.*, 2012, **437–438**, 1.
45. W. Ma, G. Jacobs, P. Gao, T. Jermwongratanachai, W. D. Shafer, V. R. R. Pendyala, C. H. Yen, J. L. S. Klettlinger and B. H. Davis, *Appl. Catal.*, 2014, **475**, 314.
46. J. X. Xie, P. P. Paalanen, T. W. van Deelen, B. M. Weckhuysen, M. J. Louwerse and K. P. de Jong, *Nat. Commun.*, 2019, **10**, 67.
47. T. C. Bromfield and N. J. Coville, *Appl. Catal.*, 1999, **186**, 297.
48. W. P. Ma, G. Jacobs, W. D. Shafe, V. R. R. Pendyala, Q. F. Xiao, Y. F. Hu and B. H. Davis, *Catal. Lett.*, 2016, **146**, 1204.
49. V. R. R. Pendyala, G. Jacobs, W. P. Ma, W. D. Shafer, D. E. Sparks, A. MacLennan, Y. F. Hu and B. H. Davis, *Catal. Today*, 2018, **299**, 14.
50. C. Y. Tu, M. F. Li, H. F. Li, Y. Chu, F. Liu, H. Nie and D. D. Li, *RSC Adv.*, 2016, **6**, 33177.

CHAPTER 7

Fischer–Tropsch Synthesis Reactors

X. LU*[a] AND D. HILDEBRANDT[b]

[a] Wuhan Institute of Technology, 206 Guanggu 1st Road, Wuhan, China;
[b] University of the Witwatersrand, 1 Jan Smuts Avenue, Johannesburg, South Africa
*Email: xiaojunlu@wit.edu.cn

7.1 Introduction

Research and development on Fischer–Tropsch synthesis (FTS) reactors began in the 1930s. An efficient and reliable industrial-scale FTS reactor is crucial for the successful industrialization of coal/gas/biomass to liquid (CTL, GTL, BTL) technologies. Since the industrialization of FTS technology in Germany in the 1930s, the reactor design has gone through three main stages of development. The resulting new-generation industrial-scale FTS reactors led to corresponding developments and improvements in the process efficiency and economics.[1] Different combinations of catalysts and reactor designs were implemented during these three developmental periods driven by changes in market forces that led to different final products being targeted. In the first period (1930s–1940s), Germany developed fixed-bed reactors (FBR) and built nine facilities. This was the first industrialization of FTS. This was driven by the need to produce fuel during World War II, and the facilities were dismantled after the war. In the second period (1950s–1980s), Sasol implemented large-scale CTL projects using the extensive low-quality coal reserves in South Africa. A FBR and two different types of fluidized reactors were developed. In the third period (from the 1980s

Catalysis Series No. 44
Chemicals and Fuels from Biomass *via* Fischer–Tropsch Synthesis: A Route to Sustainability
Edited by Joshua Gorimbo, Xinying Liu, Yali Yao and Diane Hildebrandt

Published by the Royal Society of Chemistry, www.rsc.org

onward), new GTL and CTL projects were built in Malaysia, Qatar and China. FBRs, slurry-bed reactors (SBR), and fluidized-bed reactors were all used in projects in these countries, depending on the targeted product slate.

Three major types of FTS reactors are in operation at industrial scale: the FBR (in the form of a shell and tube reactor), the fluidized-bed reactor and the SBRs.[1,2] Although they were developed and applied at industrial scale during different periods in the past, the overall installed global capacities for the reactor types are of similar orders of magnitude, with approximately 155 000 bbl per day (Shell Pearl project, Shell Malaysia and projects in China) produced in FBRs, about 125 000 bbl per day produced in fluidized-bed reactors (Sasol II and III, and Yankuang in China), and about 183 000 bbl per day in slurry-bed reactors (Sasol Oryx project, and Shenhua, Yankuang, Lu'an and Yitai in China) (data as of the end of 2020).

This chapter cannot cover all the types of reactors that have been used commercially in the past or those currently in use, and it also cannot introduce all the new concepts for reactor designs that might have potential for commercial BTL applications. However, the most important types of reactors that have been used or that are currently being used for FTS processes are discussed, and some suggestions are made about which of these reactor types may be suitable for BTL.

7.2 Fixed-bed Reactors

A FBR was used in the first industrial application of FTS and is still one of the best reactors to use for either certain scales of application (including likely scales of operation for future BTL plants) or for specific product ranges. The development of FBR technology has focused mainly on improving heat transfer rates in the reactor, which leads to improvements in productivity per unit volume.

7.2.1 Development of FBRs for FTS

7.2.1.1 Reactors for FTS in Germany in the 1930s

In the 1930s, the FTS process in Germany was operated at normal pressure using low-pressure FBRs. The reactor was a box-shaped vessel comprising tubes and plates 5 m in length, 2.5 m in width and 1.5 m in height.[1–3] The catalyst was loaded between plates on the shell side, and cooling water passed through the tubes. Heat generated by the reaction was removed by water that circulated in the tubes (see Figure 7.1). Temperature control in the catalyst was achieved by adjusting the cooling water pressure based on the correlation between the pressure and the temperature of boiling water. The temperature of the water was controlled accurately in the range 180–200 °C, while the temperature in the catalyst bed was typically 5–8 °C above the temperature of the cooling water.[4,5] Although it was a low-pressure FBR, the cooling tubes could be operated up to 3 MPa, which was essential for allowing the water to

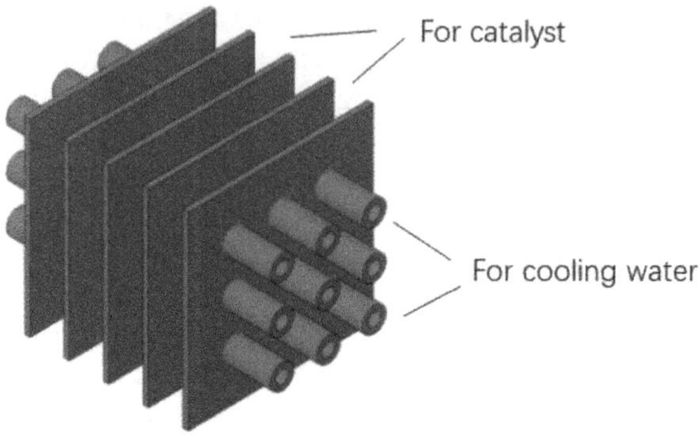

Figure 7.1 Internal structure of the normal pressure FT reactor.

reach the reaction temperature. It is obvious that the heat generated per volume of catalyst or, equivalently, per volume reactor was relatively low, due to both the low pressure and the low temperature of the reactor.

The catalyst used was the so-called standard catalyst developed by German scientists during World War II.[6] It was composed of Co, ThO_2, MgO and Kieselguhr.[3,7] The mean diameter of the catalyst particles used was 2–3 mm, which is the same dimension of the catalysts used in FTS FBRs today. Each reactor had 12 m^3 effective volume for the catalyst bed and cooling tubes, and about 900 kg cobalt metal was effectively loaded onto the catalyst.[3]

Instead of recycling the tail gas from the process, as is normally done today, this German FT process used 2–3 reactors in series to achieve a reasonable conversion of the syngas.[2,5] After leaving the reactor, the process stream was cooled to condense the liquid products and the water. The unconverted syngas with light hydrocarbons (incondensable light products) was fed to the next reactor in the series for further reaction. Water was separated in between the reactors, so that oxidation of the catalyst in the later reactor was mitigated, as the partial pressure of water vapor was reduced by the interstage water knockout.

7.2.1.2 Medium-pressure Reactor

From a reaction mechanism perspective, a higher pressure favors the production of long-chain hydrocarbons, which is desirable for processes that target fuel production. A medium pressure process and a new fixed-bed reactor were developed in Germany as a result of better understanding of the FTS.

The new design was different from previous designs for the FBR: it comprised a vertical pressure vessel consisting of 2044 annular tubes.[2,4,5] The catalyst was loaded into the annular space between the outer reactor tube and the inner cooling tube. The reactor was 4.5 m in height and 2.7 m in diameter. As illustrated in Figure 7.2, the diameter of the outer tube was 48×44 mm and that of the inner tube was 24×21 mm^2. This means the

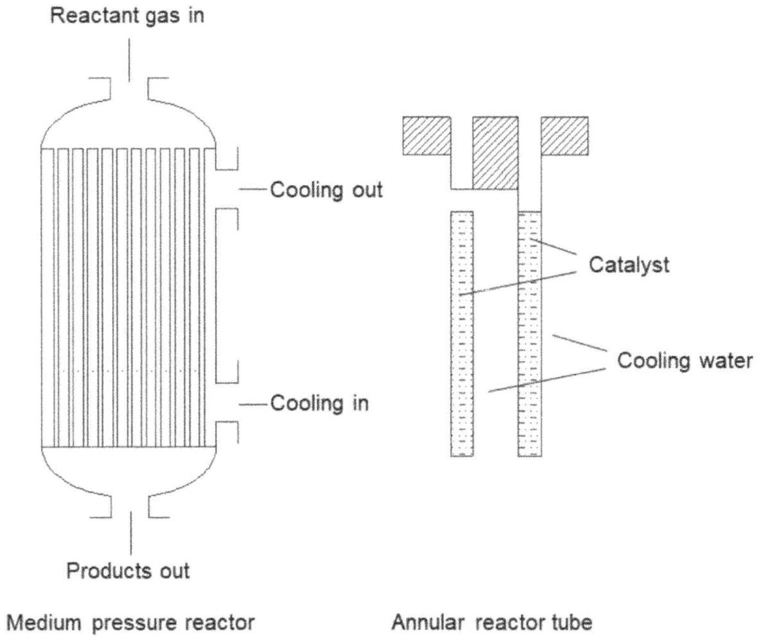

Figure 7.2 Medium pressure FBR and its internal structure.

annular gap was only 5 mm. Therefore, the catalyst pellets used were much smaller than those loaded in the FBR discussed in the previous section. Catalyst loading in this type of reactor was 10 m^3 only, as the space for charging of the catalyst was much smaller than with the previous types of FBRs. Saturated water was circulated in both the shell side and the inner reactor tube. The heat of the reaction was removed by boiling the water to keep the reactor at a constant temperature. The heat transfer efficiency was increased in this type of reactor compared to the previously discussed low-pressure reactor.

This reactor was operated at medium pressure (0.5–1.5 MPa), and the operating temperature was kept at 180–200 °C. The elevated pressure and low reaction temperature favored the formation of long-chain hydrocarbons and boosted the selectivity to heavy oil and wax. The increased pressure also resulted in a longer life span of the catalyst due to the solvent effect caused by condensation of some hydrocarbons on the catalyst at a higher pressure.[3]

The FBR design evolved from a rectangular to a cylindrical shape with domed ends, which was more suitable for the higher pressure used in the process. Most of the later reactor designs for FTS continued to use a cylindrical shell as the pressure vessel. The higher process pressure increased the reaction rate, and as a result better heat transfer was required – in both the catalyst bed and between the reactor tube and the cooling medium – to protect the catalyst. Placing the catalyst in the narrow gap in the annular tube addressed both these issues.

Manufacturing this type of reactor and loading the catalyst could, however, be difficult. A FBR used for highly exothermic reactions is normally composed of two streams, namely the process side (where the reaction takes place) and the utility side (where the cooling agent removes the reaction heat generated on the process side). Adding an additional side to the reactor makes the reactor hard to manufacture. Therefore, the scale-up and application of this type of reactor were challenging.

Each of these medium-pressure reactors was loaded with 10 m^3 catalyst containing approximately 1 ton of Co.[4] The elevated system pressure benefited the production rate per volume of catalyst. Similar to the reactors in series used for the low-pressure FTS process discussed in Section 7.2.1.1, three medium-pressure FTS reactors were connected in series in order to achieve a reasonably high conversion rate of the syngas feed. The GHSV increased after each stage, so that the GHSV of the process stream entering the last stage was three times that of the process stream flowing to the first stage.

As a result of the increased reactor pressure, short-chain carboxylic acid dissolved in the product water, which caused corrosion of the equipment, which was fabricated with carbon steel.[3] To prevent this corrosion, no condensation of water was allowed in the process in order to prevent an aqueous solution of carboxylic acid from forming. Thus the cooling stream used for the interstage product knockout between the reactors was preheated to 160 °C, and the process lines after each reactor were heated and well insulated. This design became standard practice for later FTS processes.

The conversion that can be achieved by a reactor is typically limited because of the production of water, which deactivates the catalyst at a high concentration. This is especially the case when using iron-based catalysts. There are two main strategies for achieving a high overall conversion. The original plants built in Germany in the 1940s used the reactors-in-series configuration with interstage knockout to achieve a high overall conversion in one pass. The newer designs use a single reactor that operates at a low- to medium-per-pass conversion, followed by the water knockout. The tail gas is recycled, and, in this way, a high overall conversion is achieved. However, start-up, operation and control of this system are more complicated. For smaller BTL plants, the reactor in series, single-pass configuration is probably the better arrangement due to its simplicity and robustness.

7.2.1.3 Sasol Arge Reactor

A new type of FBR, namely a shell and tube FBR, was developed by engineers from Germany and South Africa. This design became known as the Arge reactor. In 1955, five Arge reactors were constructed for commercial production with a design capacity of 550 bbl per day per reactor.[8–10] This scale was limited in comparison to that of the Sasol II and Sasol III process or of the Shell Oryx GTL and Shenhua CTL projects that were developed in later years. However, the Arge reactor, with its low operation temperature, was a milestone in the history of FTS, as many projects later used the same design concept for GTL,

CTL and BTL, although different types of catalysts were used in these processes. The Sasol Arge process was the only commercial-scale FBR that used co-precipitated iron as the catalyst for fuel and wax synthesis.

Each of the Arge reactors was approximately 17 m tall with 12 m long reactor tubes inside the reactor shell.[5,9,10] (See Figure 7.3 for the Arge reactor layout.) The diameter of the shell was approximately 3 m. The shell contained 2052 reactor tubes that were mostly filled with catalyst. The diameter of the reactor tube was 50 mm. About 40 m^3 catalyst was charged into each reactor. The catalyst used was the classic co-precipitated iron catalyst $Fe/Cu/K/SiO_2$.[5] The catalyst was prepared by dissolving iron and promoters in nitric acid and adding sodium carbonate to precipitate the metals. Potassium was added before the paste was screw-extruded into pellets with a dimension of ϕ 3×10 mm. The catalyst was prereduced before it was loaded into the reactor.

Saturated water was used in the shell side to remove the heat generated by the reaction. The bubbles of steam that formed on the outer surface of the reactor tubes resulted in a high-heat transfer rate. By adjusting the pressure in the boiler, the shell side temperature could be kept constant. The reactor was operated at 200–230 °C, and the reactor inlet pressure was set at 2.5–2.7 MPa. The H_2/CO ratio in the fresh syngas was 1.7–1.8, and the design GHSV was

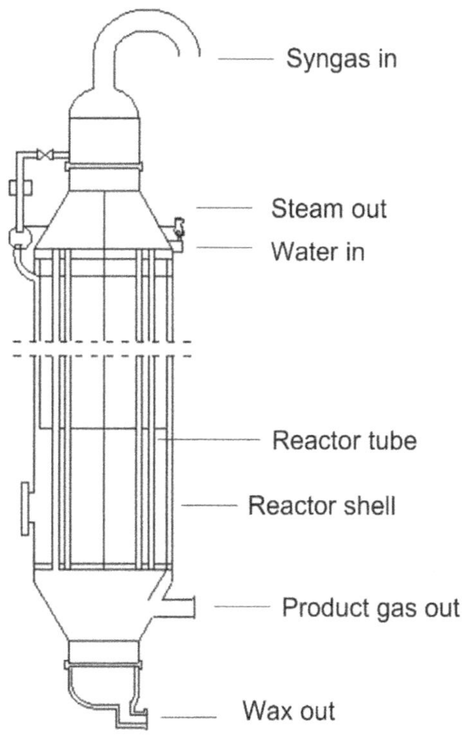

Figure 7.3 Sasol Arge reactor.

$500\,h^{-1}$. At the end of each operating cycle, the temperature in the reactor could be increased by 25–30 °C to compensate for the loss in catalyst activity. The life span of the catalyst was about one year.[5,10]

After about 60 years of running the Arge reactors, they were replaced by slurry reactors in the first decade of the 21st century. However, the Arge reactor set a standard reactor design for almost all later FTS projects that used FBRs.

7.2.1.4 Shell Middle Distillate Reactor

In 1993, the Shell GTL plant in Bintulu, Malaysia, used a similar shell and tube reactor for its project,[11,12] and later in 2011, the same reactor concept but with a larger size and capacity was used for the Shell Pearl project in Qatar. There were several reasons for deciding to use the multitubular FBR in these projects:[11,13]

- It is easy to scale up a multitubular FBR as one reactor tube has the same performance as all the tubes in the reactor.
- The catalyst can take up to 60% of the reactor volume so that the volume efficiency is high.
- There is no need for solid–liquid separation when using a FBR.

Four shell and tube reactors were installed in the Bintulu project. Each reactor had a diameter of about 7 m and contained about 26 000 reactor tubes.[14] Although the length of the reactor tube was the same as that used in the Arge reactor, the tube diameter was reduced from 2 inch to 1 inch for better heat transfer. However, for the same volume of catalyst loaded, the Shell reactor was much heavier than the Arge reactor due to the large increase in the number of reactor tubes. The reactor design capacity was 3000 bbl per day per reactor.[15] Cobalt ($Co/Zr/SiO_2$) catalyst[16] was used to boost the selectivity to long-chain hydrocarbons and prolong the catalyst life. When the reactor was operated at the typical conditions of 220 °C and 2.5 MPa, the selectivity to C_{5+} was higher than 85% with a chain growth factor, α, close to 0.9.[17,18] The catalyst needed to be regenerated every 9–12 months, and the overall life span of this cobalt catalyst was up to 5 years.[15] This significantly improved the economics of the project.

With the success of the Bintulu project, the middle distillate reactor and the GTL process were proven to be efficient and profitable. The same design was used for the Shell Pearl project in Qatar (see Figure 7.4), but the reactor diameter was further increased and comprised 29 000 reactor tubes. The height of the reactor was about 20 m, and the weight about 1200 tons. A new cobalt catalyst with higher activity was used in this project. Twenty-four reactors[19] were installed in the Pearl project with an overall capacity of 140 000 bbl per day, making it the largest single FTS project in the world.

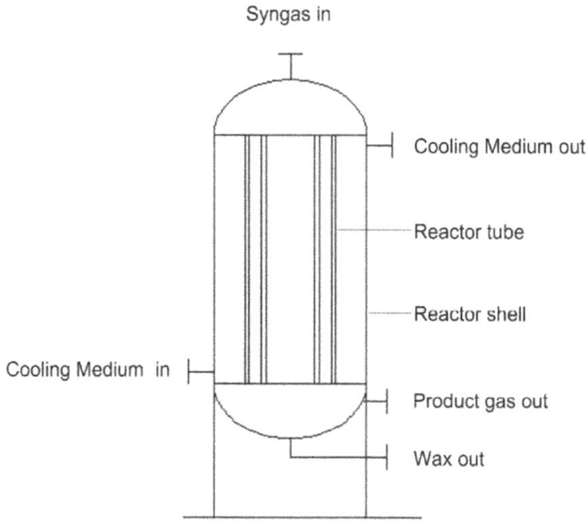

Figure 7.4 Illustrative drawing of a shell middle distillate reactor.

7.2.2 FBR Design

FBRs are suitable for a range of process scales: they are easy to scale up, the volume efficiency is high, and the reactor and auxiliary equipment are relatively simple and robust. Start-up is also reasonably straightforward, and the flow to the reactor can be varied over a fairly wide range. Thus FBRs are well suited to BTL applications.

Design methods for shell and tube reactors are reported in textbooks and the literature. Experience in the design of shell and tube heat exchangers can be adapted for the reactor design, as they are fundamentally similar. Indeed, the design of a commercial shell and tube reactor could be defined as a scale-up of a FBR. Several problems can occur in commercial reactors – such as temperature increase and blockage of the reactor tubes due to degradation of catalyst pellets – but they are not typically experienced when the catalyst is tested in microreactors in laboratories, where catalyst is used only in the order of grams. However, when tens of tons of catalyst are packed into thousands of reactor tubes, the heat generated by the reaction can be in the order of megawatts. Heat transfer and temperature control are therefore among the core issues that need to be addressed in terms of reactor design and operation.

The temperature profile in the reactor tubes is decided not only by the heat generated by the reaction but also by thermodynamics, mass transfer and kinetics. In this chapter, we focus on the logic of the design method and consider only these aspects, not the very detailed steps or even complex equations used during the design process.

7.2.2.1 Thermodynamics

The thermodynamics of the FTS reaction indicates whether the reaction is favored by feed and operation conditions and determines the heat of the reaction. The FTS reaction is extremely exothermic, with the standard heat of reaction for the formation of hydrocarbons being of the order of -156 kJ mol^{-1} CO converted. The formation of methane is even more exothermic, with a standard heat of reaction of -206 kJ mol^{-1}. This equates to an adiabatic temperature rise (that is, the increase in temperature that would occur if the reaction went to completion) of about 1500 K. Furthermore, the equilibrium constant, K_{eq}, for the FTS reaction is large: 300 °C $K_{eq} = 2 \times 10^8$ for hydrocarbons and $K_{eq} = 6 \times 10^{18}$ for methane. This implies that the reaction is effectively irreversible and that there is no equilibrium limit on the conversion of the syngas.

The design and operation of an FTS reactor are therefore much more complicated than for other exothermic but reversible reactions, such as methanol or ammonia synthesis, as the temperature can run away in an FTS shell and tube reactor, as there is no reaction equilibrium to limit the extent of the reaction. Thus if the temperature increases above the design temperature, the actual reaction rate in the reactor becomes higher than the reaction rate used in the reactor design. This in turn means that more heat is released per unit volume, and this results in an increase in the catalyst temperature. If the temperature rises to such an extent that the heat transfer rate becomes limiting, the temperature rise in the catalyst bed boosts the reaction rate, which in turn increases the rate of reaction further, which can lead to temperature runaway and loss of control of the reactor temperature.

As the temperature increases, the selectivity to methane increases, leading to increased heat release, as the formation of methane is more exothermic than the heat of reaction of longer-chain hydrocarbons. Therefore, it is very important to design the reactor so that in the event of interruption of feed gas to the reactor, the gas in the reactor is immediately released to prevent temperature runaway. Indeed, if at any stage the temperature in the reactor begins to increase uncontrollably, the gas in the reactor should be immediately removed, for example by purging with nitrogen or by dropping the pressure and letting the gas in the reactor escape.

A higher temperature results in carbon depositing on the catalyst, which leads to deactivation of the catalyst. The deposited carbon also creates forces in the porous catalyst that can cause the catalyst pellets to disintegrate into powder. This powder causes the pressure drop in the reactor tube to increase until the reactor is completely blocked. One of the fundamental reasons that shell and tube FTS reactors fail is that the thermodynamics of FTS and the design and operation of a reactor do not take this into account.

7.2.2.2 Mass Transfer

Typically, the catalyst is in pellet form with a mean diameter of 3–5 mm in a FTS FBR; the liquid phase products that are formed in the reactor fill the catalyst pores and cover the surface of the catalyst pellets. Therefore, mass transfer limitations must be taken into account when designing the reactor. The effect of mass transfer on the observed reaction rate can be evaluated

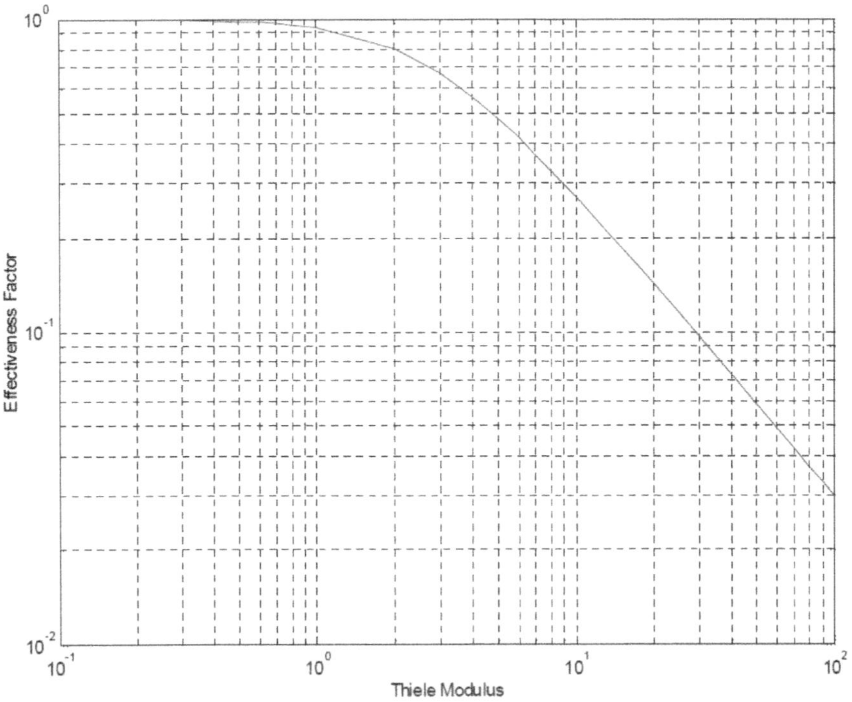

Figure 7.5 Effectiveness factor with Thiele modulus.

using the Thiele modulus, ϕ, when the catalyst dimension is larger than 1 mm. The relationship between the effectiveness and Thiele modulus is illustrated in Figure 7.5.

The Thiele modulus is the ratio of the particle diffusion time to the reaction time, with the general expression for the Thiele modulus given by:

$$\phi = \frac{R}{3}\left(\frac{k_V}{D_{eA}}\right)^{1/2} \tag{7.1}$$

where R is the characteristic radius of the pellet; k_V is the volumetric reaction rate constant; D_{eA} is the effective or intraparticle diffusion coefficient of species A in the catalyst.

The expression for the Thiele modulus can also be written as follows in the case of FTS.[20]

$$\phi = \frac{V_P}{A_{p,ext}}\sqrt{\frac{\rho_P c_{H_2,G} k_{m,H_2}}{D_{eff,H_2,L} c_{H_2,L}}} \tag{7.2}$$

where $V_p/A_{p,ext}$ is the ratio of particle volume to the external surface area; ρ_P is the density of the particle; $c_{H_2,G}$ is the concentration of hydrogen in the gas phase; k_{m,H_2} is the pseudo first-order rate of mass transfer of hydrogen;

$c_{H_2,L}$ is the concentration of hydrogen in liquid wax; $D_{eff,H_2,L}$ is the effective diffusion coefficient.

The Thiele modulus can be related to an effectiveness factor, η_{pore} by:[20]

$$\eta_{pore} = \frac{r_{m,H_2,eff}}{k_{m},\, H_2 c_{H_2,G}} = \frac{\tanh\phi}{\phi} \approx \frac{1}{\phi} \text{ (when } \phi > 2) \tag{7.3}$$

where $r_{m,H_2,eff}$ is the effective or observed reaction rate when taking internal mass transfer limitations into account. The actual reaction rate can be adjusted accordingly, by multiplying η_{pore} with the rate measured without mass transfer limitations, for example as measured using a powdered catalyst. Therefore, the observed rate of reaction for a catalyst pellet is described by:

$$r_{H_2,eff,FT} = \eta_{pore} r_{H_2,FT} \tag{7.4}$$

The effect of internal mass transfer limitations should be investigated during the catalyst development stage. Any inaccuracy in the reaction rate used in the design that is caused by inaccuracies in the estimated mass transfer limitations could be effectively mitigated if the kinetic model and its parameters are fitted under conditions that are close to the operating conditions in the commercial reactor.

In summary, mass transfer caused by the diffusion of reactants into the catalyst pores must be included in the calculations when designing a FBR. A simple way to estimate the effective rate of reaction is to determine the effectiveness factor and adjust the reaction rate accordingly.

7.2.2.3 Kinetics

Kinetics for FTS have been developed using either power law rate equations or equations making certain mechanistic assumptions. Both simple and complex forms of kinetic equations for iron and cobalt catalysts are reported in the literature. Kinetic equations developed from elementary reaction steps normally have a fairly complex form, with many parameters and rate constants; empirical kinetic equations normally have a relatively simple form and a very limited number of parameters. If we examine the kinetic equations in the literature, we find that most of them give us similar results at typical operating conditions. The reliability of the equations becomes less certain if the expressions are extrapolated to a concentration, temperature and pressure that are far from the conditions under which they were measured experimentally.

Typically, when we start designing a FBR for FTS, the catalyst has been chosen and the reaction kinetics have been measured and fitted using appropriate kinetic equations and values for the parameters. Rate equations with a limited number of parameters that have been fitted using data at operating conditions close enough to those to be used in the large-scale reactor should be applied. The measurements should preferably be done using the same size of catalyst pellet and gas velocity as that to be used in the large-scale reactor. If the kinetic equations have been developed from experimental data using a catalyst in a powdered form, the mass transfer effect should be taken into account during the design, using effectiveness factors, as discussed earlier.

Calculating the reaction rate allows us to estimate the heat generated by the reaction. In order to accurately determine the amount of heat generated at any point in the catalyst bed, detailed kinetic equations should be used for each product. Many kinetic equations that are reported in the literature describe the consumption rate of reactants, but very few rate equations describe the formation rates of all products. To simplify the calculations, the kinetic equations for CH_4, CO_2 are essential, and the rate of formation of the rest of the products can be based on the formation rate of $-CH_2-$ units. Once we know the amount of heat generated, we can calculate the required heat transfer rate in the reactor tube.

Examples of kinetic rate equations for FTS for a Fe catalyst are given in eqn (7.5)–(7.7):[21–23]

$$r_{\mathrm{m,H_2,FT}} = \frac{k_{\mathrm{m,H_2,eff,FT}}c_{\mathrm{H_2,G}}}{1 + 1.6\dfrac{c_{\mathrm{H_2,G}}}{c_{\mathrm{CO,G}}}} \tag{7.5}$$

where $k_{\mathrm{m,H_2,eff,FT}} = 5.1\,\mathrm{m^3kg^{-1}s^{-1}}\mathrm{e}^{\frac{-52000}{\mathrm{RT}}}$

$$r_{\mathrm{m,H_2,M}} = -\frac{\mathrm{d}n\mathrm{H_2}}{\mathrm{d}m_{\mathrm{cat}}} = k_{\mathrm{m,H_2,eff,M}}c_{\mathrm{H_2,G}} \tag{7.6}$$

where $k_{\mathrm{m,H_2,eff,M}} = 27.3\,\mathrm{m^3kg^{-1}s^{-1}}\,\mathrm{e}^{\frac{-70000}{\mathrm{RT}}}$

$$r_{\mathrm{m,H_2,S}} = k_{\mathrm{m,H_2,eff,S}}c_{\mathrm{H_2O,G}} \tag{7.7}$$

where $k_{\mathrm{m,H_2,eff,S}} = 155\,\mathrm{m^3kg^{-1}s^{-1}}\,\mathrm{e}^{\frac{-70000}{\mathrm{RT}}}$

where $r_{\mathrm{m,H_2,FT}}$ is the formation rate of $-CH_2-$ units; $r_{\mathrm{m,H_2,M}}$ is the formation rate of methane; $r_{\mathrm{m,H_2,S}}$ is the rate of the water-gas-shift reaction. These are typical empirical equations, with a limited number of parameters being used in the models. The temperature and mass transfer effects on the reaction rate are both included in the reaction rate constant.

In summary, no matter how complex the kinetic equations are, they need to describe the reaction rate in terms of the partial pressure of the reactants and the temperature and should include the internal mass transfer effects.

7.2.2.4 Heat Transfer

Heat transfer (including heat transfer in the catalyst bed and between the fluids and tube walls) is always the limiting factor in terms of reactor performance in a FBR, if the reaction is highly exothermic. With any reactor system, at steady state, the heat release and removal rates must balance. This means that for an exothermic reaction, all the heat generated must be removed by one or more methods. When the heat transfer coefficient is roughly fixed, as in a FBR, the only way to remove more heat from the catalyst bed is to elevate the temperature in the catalyst bed to increase ΔT across the heat transfer path. However, this also causes the reaction rate to increase, thereby increasing the rate of heat release and simultaneously accelerating the deactivation rate of the catalyst.

There are four sources of resistance to heat transfer in a shell and tube reactor: transfer across the catalyst bed; transfer across the boundary between the catalyst bed and the inner surface of the reactor tube; transfer through the wall of the reactor tube; transfer between the outer surface of the reactor tube and the cooling medium in the shell. The overall heat transfer coefficient U_{over} can be expressed as the sum of the four heat transfer coefficients, as described by:[20]

$$U_{\text{over}} = \left(\frac{d_{\text{R}}}{8\lambda_{\text{rad}}} + \frac{1}{\alpha_{\text{W,int}}} + \frac{d_{\text{wall}}}{\lambda_{\text{wall}}} + \frac{1}{\alpha_{\text{W,ex}}}\right)^{-1} \tag{7.8}$$

Typically, the heat transfer rate across the catalyst bed, λ_{rad}, limits the value of the overall heat transfer coefficient. The porosity of the catalyst causes the catalyst bed to have an effective thermal conductivity at least one order of magnitude lower than that of the reactor tube. Using a high thermal conductive material, such as silicon carbide, as the support can improve the effective thermal conductivity and thus improve the heat transfer in the reactor.

Estimating the value of λ_{rad} in a FTS reactor is not a straightforward process because the liquid products that are present in the catalyst particles and on the surface affect the conductivity of the solid. Furthermore, the amount of liquid and the composition of the liquid depend on the reaction operating temperature in a complex way.[24,25] The conductivity of the catalyst bed before the reaction is lower than that measured after the reaction has occurred for some time. Furthermore, if hot inert gas is passed over the spent catalyst bed, the thermal conductivity of the bed increases as FT liquid products evaporate out of the particles.[25] This has implications for the start-up of the reactor, which are discussed later in this chapter.

The highest temperature exists at the center of the catalyst bed in the radial dimension. The temperature rise can be expressed by eqn (7.9), and it is proportional to the square of the reactor radius R. This largely limits the diameter of the tube that can be used for shell and tube reactors for use in FTS.[24,25]

$$T_{\text{c}} - T_{\text{W}} = \frac{\Delta H_{\text{r}} r_{\text{a}} R^2}{4\lambda_{\text{eff}}} \tag{7.9}$$

where T_{c} is the temperature at the center of the catalyst bed; T_{w} is the temperature at the reactor wall; ΔH_{r} is reaction enthalpy; r_{a} is the rate of the reaction; λ_{eff} is the effective thermal conductivity in the radial direction.

At any point in the reactor, under steady state of operation, the temperature in the catalyst bed and the gas temperature are both constant;

therefore, the heat released by the reaction does not heat up either the catalyst or the gas, so it must be removed by conduction or convection. Eventually, the heat of reaction released in the catalyst bed must be removed by both the cooling medium in the reactor shell and the process gas that leaves the reactor. The energy balance can be written as a simple pseudo-homogeneous one-dimensional reactor model.[20]

$$\rho_G \sum \Delta_R H_{ir_{m,H_2},i} = \rho_{mol}\, c_P u_s \frac{dT}{dz} + \frac{4U_{over}(T - T_w)}{D_T} \tag{7.10}$$

where the term on the left is the amount of heat released by reaction(s) at position z in the bed; c_p is the heat capacity of the stream; u_s is the gas velocity; U_{over} is the overall heat transfer coefficient from the center of the bed to the tube wall; D_T is the reactor diameter.

In summary, the catalyst bed offers the highest resistance to heat transfer; thus the heat transfer rate through the catalyst bed essentially determines the amount of heat that can be removed in a reactor tube or bed. This sets the limit on the reaction rate that can be realized in a reactor tube and thus, in turn, the achievable catalyst activity. The temperature rise in the catalyst bed is proportional to the square of the tube diameter and thus is very sensitive to the choice of tube diameter. The heat generated by reaction in the tube matches the sum of the heat transferred out of the tube to the cooling medium and the heat carried out by the process stream by virtue of its temperature increase.

7.2.2.5 Reactor Model and Process Design

When we consider the FBR, there are gradients of the temperature and concentration of reactants and products in both the axial and radial dimensions. Four different types of models have been developed to describe a FBR: a one-dimensional pseudohomogeneous model; a two-dimensional pseudohomogeneous model; a one-dimensional heterogeneous model; a two-dimensional heterogeneous model. Among these, the first two are commonly used when the mass transfer (inter- and intraparticle) effects are included in the kinetic equations, and a lumped temperature is used for the gas and catalyst at any point of the bed.

The simulation results in the literature also show that the predicted temperature increase and the reaction rate data are similar for the one- and two-dimensional models. Therefore, a pseudohomogeneous one-dimensional model is good enough in most cases for simulating a FBR for FTS.

When we design a shell and tube reactor for FTS, we start by doing calculations or simulations for one reactor tube in order to determine its productivity. When we scale up, we assume that all tubes in the reactor behave in the same way, and the overall reactor productivity is simply determined by multiplying the productivity of a single tube and the number of tubes in the reactor. No matter whether the design is carried out numerically or by hand, the design must consider the interactions among the mass, the energy balance constraints and the rate equations. Although the equations in this section are written as continuous functions, the calculation process is discretized. The catalyst bed and the reactor tube are divided into small elements in both the radial and axial directions for 2-dimensional models and in the axial direction only for 1-dimensional models. Typically, the axial conduction, convection and diffusion terms are neglected in these models. For each small element, conversion of the reactants and formation of products are calculated by incorporating mass transfer and kinetics, and using the outlet temperature, pressure, flow rate and concentration (or partial pressure) of the previous element as the feed to the current element. Once the conversion of reactants and formation of products are calculated, the heat generated in the element can be determined. Using the energy balance described in Section 7.2.2.4, the amount of heat transferred radially and axially can be calculated, and the temperature of the reactant and products that leave the element can be calculated by difference. The calculation continues until either the desired conversion or the end of the tube is reached.

When calculating each element, if the temperature at the center of the catalyst bed increases to a value greater than the allowed maximum temperature, a new, smaller dimeter reactor tube needs to be selected. Using this new diameter, the same calculation needs to be carried out again, until the maximum temperature in the tube is below the allowable maximum temperature. Then the pressure drop under the design conditions should be calculated. If the result is within the acceptable level, the calculation and simulation for a single reactor tube is finished; if not, a reduced catalyst bed height should be used. In order to achieve the desired conversion with the same space velocity, the linear gas velocity needs to be reduced for a lowered bed height. An optimal pressure drop can be calculated by adjusting both the bed height and linear velocity to achieve the desired conversion over the catalyst bed. At this stage, the calculation and simulation for a single reactor tube is complete, and the scale for the reactor is done by multiplying the productivity of a single reactor tube by the number of tubes in the reactor.

7.2.3 Notes for Design and Operational Issues

7.2.3.1 Catalyst Loading

Catalyst loading is one of the key steps to ensuring the stable operation of a FBR, thereby protecting the catalyst. The key aspect during catalyst loading is that the pressure drop across all the tens of thousands reactor tubes needs to be roughly the same. In order to achieve this, the pressure drop across each reactor tube must be measured and adjusted during catalyst loading, as a standard procedure.

When the reactant gas is fed into the reactor, the distribution of the gas to the reactor tubes and the flow rate in each reactor tube depend on the difference in the pressure drop between all the reactor tubes. If some of the reactor tubes have higher pressure drops than others, they receive less reactant gas or even be bypassed. This reduces the productivity of the reactor and could lead to a higher temperature at the inlet of the tube, as the lower gas flow rate removes less heat from the zone that is exposed to the highest concentration of reactants and that therefore has the highest reaction rate. Furthermore, tubes with lower pressure drops receive more feed gas than they were designed for, which could result in a higher reaction rate and lead to a higher temperature and even catalyst damage.

7.2.3.2 Start-up of a Shell and Tube Reactor

Temperature runaway often takes place during the start-up of a FTS shell and tube reactor, which results in powdering of the catalyst pellets and subsequent failure of the reactor. Once these phenomena happen, they cannot easily be brought back under control. As explained earlier, temperature runaway needs to be addressed by either immediately dropping the reactor pressure or allowing the gas to escape or by flushing with an inert gas, such as nitrogen, in order to remove the feed gas from the reactor and thereby stop the reaction. This temperature runaway occurs very quickly, especially as the reaction is so exothermic, with an adiabatic temperature rise in the order of 1500 K. Thus even if the reactor is protected by immediate removal of the reacting gas from the tubes, the catalyst might still be damaged.

The catalyst needs to be reduced (activated) before it becomes active for Fischer–Tropsch (FT) reactions. A freshly reduced catalyst always has a higher activity compared to a catalyst operated under steady state for two main reasons: (1) the freshly reduced catalyst always has more active sites available to participate in reactions; (2) the pores of the catalyst pellets are not yet filled with liquid product; hence the mass transfer is much higher compared to a liquid-filled catalyst at steady state. Higher catalyst activity results in a higher reaction rate and subsequently more heat from the reaction being released. This extra heat cannot be removed by the cooling medium on the shell side, as the heat removal rate is designed based on the catalyst activity at steady state. Practical advice for operating a shell and tube reactor during start-up is that a lower partial pressure of syngas in the feed should be maintained to keep the reaction rate low, until liquid phase products are collected in the knockout vessels at the outlet of the reactor. This indicates that sufficient liquid products have been produced so as to fill the catalyst pores and cover the surface of the catalyst pellets. At this stage, the catalyst activity should be close to the steady state value used in the design, and the desired operating conditions can then be (slowly) changed so as to achieve full production.

These safeguards can be implemented fairly easily, either by automation or by relatively unskilled operators. This makes FBRs particularly suitable for BTL, as the small scale of the plant implies that the process must be designed so that fewer and relatively unskilled operators can run the plant.

7.3 Fluidized-bed Reactors

When catalyst is fluidized in the reactor vessel, one critical precondition among all the factors that result from the characteristics of the FTS reaction must always be satisfied to ensure the success of the industrialization of fluidized-bed reactors: no liquid phase products can be present in the reactor under operating conditions.[26] The presence of liquid causes agglomeration of the catalyst and thus inhibits fluidization of the catalyst, which results in failure of the operation of the fluidized-bed reactor. FTS is a process of synthesizing small molecules to large molecules. The reaction follows a chain growth mechanism that is similar to the polymerization process. A liquid phase does not form when the chain growth factor, α, is smaller than 0.7.[5,26] A higher temperature is required to achieve a lower α and thereby prevent the formation of heavy (liquid) hydrocarbons. An iron-based catalyst is used because of its robustness and low cost. A low α value means a much lower average molecular weight of the hydrocarbons, thereby pushing the selectivity to a higher gasoline fraction.

7.3.1 Reactor Development

7.3.1.1 Reactor Developed by Hydrocol

After World War II, relevant technologies for FTS and CTL were made public in the United States. Hydrocarbon Research Inc. developed an American version of the FTS process, called the Hydrocol Process.[27] A fluidized-bed reactor combined with an iron-based catalyst was developed for FTS. The pilot tests were conducted in Orion in New York, and an industrial-scale plant was built and run in Brownsville, Texas. This was the first industrial implementation of a high-temperature FTS process.

The Hydrocol process used two fixed fluidized-bed reactors for light hydrocarbon production. Each reactor was 5 m in diameter and 18 m in height.[9,28] The reactors held approximately 200 tons of iron catalyst each, which fluidized in the free space inside the reactor vessel. Some space inside the reactor was occupied by a heat exchanger, which was used to remove the heat generated by the reactions. The temperature inside the reactor was controlled by varying the pressure in the heat exchanger, following the boiling point curve of water. The reaction heat taken out by water was used to generate steam. Using a well-designed gas distributor and a balance between the dimension of the catalyst particles and the gas velocity, the height of the fluidized catalyst bed was maintained at a steady level.

The reactor was operated at a temperature ranging between 305 and 345 °C,[9,28] which is the typical temperature range for an iron-based high-temperature FT process. A sufficiently high temperature is a key parameter for successful operation of FTS in a fluidized-bed reactor. Although the chain growth probability α of the FT reaction depends on many factors, including the active metal, promoters, supports and partial pressure of the reactants, temperature remains the most important factor for tuning the α value and hence the product distribution. From a kinetic point of view, temperature is also a critical parameter for the reaction rate. This is why a high once-through conversion is normally easy to achieve in the fixed fluidized-bed reactor system, which eliminates the need for tail gas recycling. A syngas conversion of between 90 and 96% was achieved using a space velocity of 2000–3000 h^{-1} at 2.7 MPa feed pressure.[9]

The catalyst used in the Hydrocol reactor was fused iron – the same catalyst that had been used in ammonia synthesis. It contained 97% Fe_3O_4, 2.5% Al_2O_3 and 0.5% K_2O.[4] The fused iron catalyst was ground to powder with a mean diameter of 0.045 to 0.45 mm. The iron oxide was reduced to metallic iron using H_2 at 350–460 °C before it was sent to the reactor for fuel synthesis.[4] Later, this synthesized catalyst was replaced by natural magnetite that contained 0.5% K_2O. The performance of the natural magnetite was close to that of the synthesized iron catalyst.

One of the advantages of using a fluidized-bed reactor is that the catalyst can be loaded and unloaded online. This offers two important advantages: neither the reactor nor the process needs to shut down from time to time to replace the deactivated catalyst; an upstream sulfur removal unit is not necessary before the FT reactor, as the percentage of catalyst replacement can be increased when the catalyst has been deactivated by sulfur in the feed syngas.

7.3.1.2 Kellogg Circulating Fluidized-bed Reactor

After the Arge reactor was successfully commercialized in Sasolburg, South Africa, the demand for synthetic fuel was soon beyond the capacity of the installed FBRs. Sasol began the development of circulating fluidized-bed (CFB) reactors, named Synthol reactors, in 1955.[5,9] Two commercial CFB reactors, with a capacity of 2000 bbl per day each, were installed in Sasolburg for testing and demonstration. The diameter of these reactors was 2.3 m and the height 46 m.[2,26,28] The successful performance of these reactors led to 16 large-scale Synthol CFB reactors being constructed in the early 1980s for the Sasol Phase II and Phase III plants.

This reactor system is quite complicated and needs a complex support system to cope with the circulation of the catalyst. During operation, catalyst particles are carried by the feed gas from the bottom of the reactor into the reaction zone and are quickly transported out of the reactor and through the so-called goose neck. The catalyst is separated in cyclones, flows through a downpipe, and is picked up by the feed gas and carried back into the reactor. Because of the quick catalyst circulation in the CFB reactor, the equivalent amount of catalyst fluidized by the gas flow is about 4000 $t\,h^{-1}$, and the catalyst loss is about 0.0015%.[5,8]

The Synthol reactor needs to operate at higher temperatures than used in the Arge reactor, in order to avoid the formation of heavy hydrocarbons, which would result in catalyst agglomeration, which would, in turn, cause problems with fluidization of the catalyst. The operating temperature was in the range 290–340 °C, and the operating pressure was 1.9 MPa.[9] An 85% feed gas conversion was easily achieved due to the higher activity of the catalyst due to the high operating temperature. Some reaction heat was removed *in situ* using two sets of heat exchangers placed in the direction of the gas flow. However, these heat exchangers removed only about 30–40% of the heat of the reaction, and, as a result, the temperature of the exit gas stream was in the range 320–360 °C.

Fused iron prepared from magnetite ore was initially used for the catalyst. This catalyst did not have a high enough mechanical strength, and the turbulence caused by the high linear velocity in the reactor resulted in excess fines being formed, which both caused loss of catalyst and disturbed the fluidization behavior of the catalyst. Later, iron filings were used to replace the imported magnetite ore as the raw material for catalyst manufacture.[5] Promoters such as K_2O and structural promoters such as MgO and Al_2O_3 were added during the preparation of the catalyst. The prepared catalyst was reduced (activated) by H_2 for 2 days at a temperature range of 350–450 °C in a FBR, and the surface area increased from about 1 $m^2 g^{-1}$ to only about 30 $m^2 g^{-1}$.[29]

Fluidization of the catalyst in a CFB reactor depends on factors such as the selectivity and phase of the products, as well as on the operating conditions and properties of the catalyst. The operation of the Synthol reactor showed that it could deal with the fluctuations in syngas composition, flow, temperature and pressure, but the actual window of operation was small. One of the drawbacks of FTS operated at such a high temperature is the formation of coke during the reaction. This causes deactivation of the catalyst, but running at such a high temperature leads to a trade-off between catalyst lifetime and catalyst activity, in order to achieve a high rate of conversion in a single reactor.

Online replacement of the catalyst is one of the advantages of a CFB reactor, and fresh catalyst can, in principle, be added continuously without shutting down the reactor. However, this design objective was not fully realized, and, in practice, the Synthol reactors had to be shut down every 40–50 days to replace the spent catalyst.

Despite the advantages of this type of reactors including online catalyst replacement and a higher heat transfer rate, the increased complexity of operation of the reactor and the auxiliary equipment, as well as the small window of operation, probably limits its applicability in BTL processes.

7.3.1.3 Sasol Synthol Reactor

With the improved design and optimization of the Kellogg CFB reactor,[30] a larger version of the reactor was developed for the Sasol II and Sasol III expansion. Sixteen Synthol reactors were installed in Sasol II and Sasol III, with a 7500 barrel per day capacity per reactor.[29,31] The operating pressure was also increased to 2.5 MPa. Eight reactors were installed in each plant, and the overall productivity reached 120 000 bbl per day.[5] Although the design had been optimized, the concept and operation of the Sasol Synthol Reactor were the same as those of the Kellogg CFB reactor. A reactor layout for the Sasol Synthol Reactor is shown in Figure 7.6.

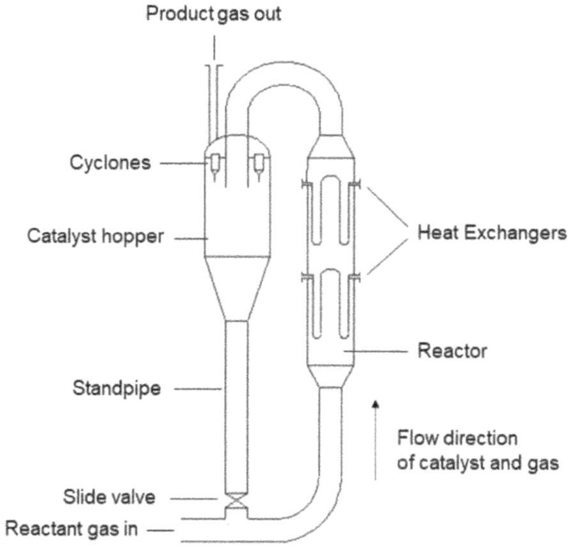

Figure 7.6 Simple illustration of the Sasol synthol reactor.

7.3.1.4 Sasol Advanced Synthol Reactor

The Sasol Advanced Synthol (SAS) reactor might be regarded as the most successful fluidized-bed reactor for high-temperature FTS. One of the advantages when using a CFB is that the reaction takes place in the reactor chamber on one side of the reactor and that regeneration of the catalyst takes place in another chamber on the other side of the reactor. This design is especially suitable for a reaction system in which the catalyst deactivates very quickly. However, deactivation of an Fe catalyst in FTS is not that fast.[5] The massive volume of catalyst circulating in the reactor also increases the operating cost. The low productivity per total reactor volume also increases the number of reactors needed to achieve production capacity. A fixed fluidizing-bed reactor, known as the SAS reactor, became the choice for the next generation of reactor at Sasol.[9,28,32] In principle, the concept of the SAS reactor is the same as that of the Hydrocol reactor.

In 1989, Sasol commissioned the first commercial-scale SAS reactor. The reactor had a diameter of 5 m and a height of 22 m with a capacity of 3500 bbl per day.[5,28] The success of the 5 m SAS reactor gave Sasol enough confidence to further scale up the fixed fluidized-bed reactor. From 1995 to 1999, Sasol installed four SAS reactors of 8 m in diameter and another four of 10.7 m in diameter at Sasol II and III to replace the 16 CFB reactors. The design capacity of the 8 m SAS reactor was 11 000 bbl per day and that of the 10.7 m reactor was 20 000 bbl per day. With the combined capacity of the new reactors, Sasol maintained its capacity at 120 000 bbl per day.[5]

7.3.2 Structure of the SAS Reactor

The SAS reactor is illustrated in Figure 7.7. The body of the reactor is a big empty drum/column. Inlet syngas is heated to 200 °C and sent to the reactor through a gas distributor at the bottom of the reactor. Although gas bubbles are invisible in a fluidized-bed reactor, the size of the bubbles leaving the gas distributor still plays an important role in fluidizing the catalyst in the reactor. The particle size of the catalyst used in the SAS reactor is about 60 μm. The volumetric flow rate of the gas in the reactor changes due to the reaction, and in order to keep the linear velocity constant, some of the reactor volume is taken up by internal parts. In the region where the cooling tubes are placed, up-flowing gas has the highest linear velocity; hence the catalyst particles have the best fluidization. When the gas and catalyst leave the cooling tube region, the gas velocity drops, and the catalyst particles start to settle. Most of the catalyst particles disengage from the gas and return to the well fluidized region. The remaining catalyst particles are separated out by the cyclones in the reactor. There is very little bypassing, and most of the reactant gas participates in the reaction in the cooling tube region, where the reaction heat is taken out through cooling tubes. The reactor is operated at the same temperature and pressure as the CFBs.

7.3.2.1 Gas Distributor

The gas distributor is one of the crucial internal parts of a fluidized-bed reactor. It provides support for the catalyst particles and allows even gas distribution in order to fluidize the catalyst particles. Many different types of gas distributors have been reported in patents and the open literature. The most commonly used ones are a group of tubes of different lengths in either a branched or parallel shape.

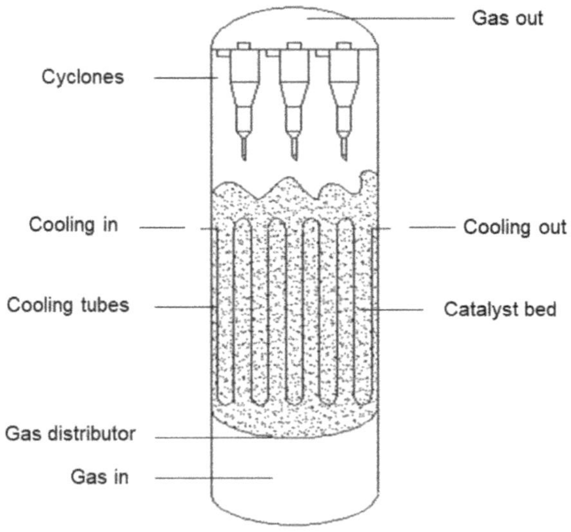

Figure 7.7 Simple illustration of the Sasol advanced synthol reactor.

7.3.2.2 Cooling Tubes

The reaction rate is high in a fluidized-bed reactor. A set of cooling tubes in either a U-shape or a coil shape is normally used to remove the heat. The part of the cooling tube in the direction of catalyst flow is round in shape to reduce the friction between the catalyst and the cooling tubes, thereby protecting them both. The high level of turbulence created by the gas enhances the heat transfer coefficient significantly. A high boiling point liquid is used in the cooling tubes in the reactor. The heat taken from the reactor by this cooling liquid is used to heat up water in a second-stage heat exchanger to produce medium-pressure steam.

7.3.2.3 Catalyst Separation Device

Separation of solid catalyst particles from the gas stream is less challenging compared to the solid–liquid separation that takes place in a SBR. The separation device only needs to separate solids and gas, unlike in slurry reactors, where fine solids need to be separated from the liquid at a high temperature and pressure. The catalyst size in a fluidized reactor for FTS is also bigger than that used in slurry reactors, which simplifies the solid separation further. Well-designed cyclones are the most commonly used devices to separate the catalyst.

The operation of a fluidized-bed reactor is more complicated than an FBR, as the gas velocity needs to be tightly controlled, in order to ensure satisfactory fluidization of the solid. Furthermore, a high-temperature FTS is best suited to the production of lighter products, which are more complicated to separate, store and transport than heavier products.

7.4 Slurry-bed Reactors for FTS

A slurry phase reactor is also called a bubble column reactor by scientists and engineers. To be consistent, we call it a slurry-bed reactor (SBR) in this chapter. The SBR is used because of the excellent heat and mass transfer performance and good interphase contact. It is an important technology that has excellent prospects for FTS applications for the production of liquid fuels.

Two major reasons for developing a new reactor to replace the fixed-bed and fluidized-bed reactors are:

1. The productivity per unit volume of a FBR is largely limited by the poor heat transfer characteristics of the FBR and the highly exothermic nature of the FTS reaction. Even if equal amounts of heat exchange area are used in both the fixed-bed and slurry-bed reactor, the achievable reaction rate in a FBR is still much lower than that in a slurry reactor. The effective thermal conductivity of the catalyst bed is always the limiting factor in determining the achievable catalyst activity in a FBR.

2. A fluidized-bed reactor produces a low boiling point product; however, there are economic drivers for producing long-chain hydrocarbons in FTS directly, as heavier hydrocarbons offer a richer choice of final product. The highest boiling point product in a fluidized-bed reactor must be such that no liquid is formed in the reactor. Thus, in order to keep the reactor operating in the gaseous state, the chain growth factor α is controlled to ensure it is lower than 0.7, and subsequently no (or few) long-chain hydrocarbons are produced in a fluidized-bed reactor. In contrast, an SBR can produce a product distribution of $\alpha > 0.9$ and is thus very versatile in producing long-chain hydrocarbon products.

7.4.1 Development of the SBR

Although the development and commercialization of the SBR was done by Sasol in the 1990 s, the slurry phase reactor was first proposed and tested at lab scale in Germany during World War II.[2] Later, in 1953, a semi-industrial-scale demonstration plant with a capacity of 11.5 tons per day of liquid hydrocarbon was built in Germany.[2] The operating results of this demonstration plant showed the technical characteristics and economic advantages of a SBR for FTS. In the 1980s, Sasol began to develop its own SBR for FTS, based on the German SBR design. In 1990, a 75 bbl per day pilot-scale SBR was built in Sasolburg, South Africa.[5,26]

In 1993, an industrial-scale reactor 5 m in diameter and 2500 bbl per day in capacity was built and put into operation,[33] which was the first commercialization of a SBR for a FTS application. A precipitated iron catalyst, like that used in the Arge reactor, was used. The typical operating conditions were a temperature of 245 °C and a pressure of 2.1 MPa. The syngas feed was maintained at a H_2/CO ratio of 1.55, which is suitable for an iron-based catalyst and is meant to suppress the H_2O partial pressure at the reactor outlet.[33,34] During operation, the catalyst was suspended in the liquid phase products (wax). Syngas was bubbled through the liquid phase and underwent reaction. The liquid product from the reactor went through a two-stage filtration system to separate the catalyst from the wax. Although the same recipe as that used for the Arge reactor was used to manufacture the catalyst, heavier products were produced by the SBR than by the Arge reactor. The chain growth factor, α, in a SBR is as high as 0.95.[5,34] The catalyst particle size used in a SBR is in the order of microns, which eliminates internal mass transfer resistance.

In 1997, the Oryx project in Qatar was initiated as a joint venture between Qatar Petroleum, Sasol and Philips Petroleum.[32] Sasol implemented the same low-temperature SBR technology as it had commercialized in South Africa, although the designed capacity was upscaled to 34 000 bbl per day. A different catalyst, namely a cobalt-based catalyst, was used for this project.[10] This was the first cobalt-based SBR system for commercial-scale FTS. The typical operating temperature is 230 °C and pressure is 2.5 MPa. According to a report, the newly developed Co catalyst experienced problems when it was used in the SBR and productivity did not reach the designed

Figure 7.8 Sasol Oryx slurry-bed reactor during installation. Reproduced from ref. 36 with permission from the Royal Society of Chemistry.

capacity for some time.[5] Two SBRs were installed in the Oryx project (see Figure 7.8). Each reactor weighs about 2100 tons, is 60 m in height and 10 m in diameter.[35] These remain the largest operating SBRs for FTS in the world, even though new reactors were constructed for later projects.[36]

In 2006 and 2007, three companies in China (Shenhua, Yitai and Lu'an) initiated three pilot-scale CTL projects, and all of them adopted the slurry phase reactor technology for FTS. The scale of each project was 3500 bbl per day (160 000 tons per annum). Precipitate iron was used in these projects and the operating temperature was set at 270–280 °C. This reaction temperature was adopted for at least two reasons: to compensate for the lower activity of iron at a low temperature; to produce fewer heavy hydrocarbons and so reduce the load on the catalyst separation unit. These three pilot-scale projects were an important step in the development of FTS in China. In 2013, Shenhua in China started its 88 000 bbl per day (4 megatons per year) CTL project. The same Fe-based medium-temperature FTS was implemented in this project. Eight SBR, each with a 11 000 bbl per day nominal capacity, were installed.

7.4.2 Reactor Structure

The SBR for FTS is simpler to operate, easier to manufacture and less expensive than shell and tube reactors. The structure is essentially a three-phase bubble column, as illustrated in Figure 7.9. The reactor is equipped

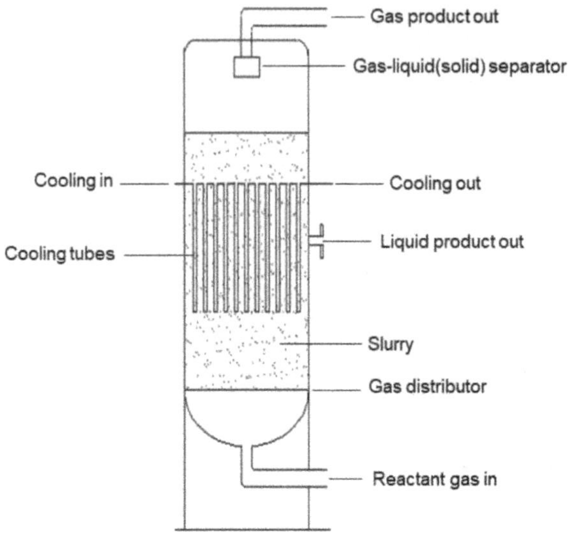

Figure 7.9 Simple illustration of a slurry reactor.

with a heat transfer coil, a gas distributor at the bottom, a gas–liquid separator at the top and a liquid level control device on the outside. The reactor is typically operated, as in the case of Sasol I, at a temperature of 250 °C and a pressure of 2.0–2.5 MPa. The preheated syngas enters the reactor from the bottom through the gas distributor and bubbles into the slurry, which consists of liquid paraffin wax and granular catalyst. Reactant gas makes contact with the catalyst pellets that are suspended in the liquid products and undergoes reaction, which generates hydrocarbon products and water. When the bubbles rise from the bottom of the reactor to the top, the majority of the reactants participate in the reaction and the bubbles shrink. The heat generated by the reaction is removed by built-in cooling tubes and produces steam. The liquid phase products are extracted from the reactor through a catalyst separator, which can be built-in or external. When the unreacted reactants and the gaseous phase products leave the slurry, the small droplets and catalyst fines carried by the gas are disengaged at the top of the reactor by a gas–liquid (solid) separator. The unreacted gas is recycled to the reactor to achieve a high rate of overall conversion.

The reactor can be regarded mainly as a big empty drum or column containing the necessary internal parts, including a gas distributor, cooling tubes, a catalyst separation device, and a gas–liquid (solid) separator. The internal parts are crucial for the stable operation of a slurry phase reactor.

7.4.2.1 Gas Distributor

In the SBR, the syngas and suspended catalyst in the slurry should stay in good contact to prevent channeling. A well-distributed gas flow can ensure good dispersion of the catalyst, as well as even temperature distribution in the slurry. Therefore, it is important that the gas distributor distributes the gas evenly. The gas distributor determines the size and frequency of the primary bubbles at the syngas inlet, and this, in turn, affects the area for mass transfer between gas and liquid. The size distribution and average size of the stable gas bubbles are closely related to the type, structure and nozzle size of the gas distributor. Several types of gas distributors have either been tried or are being used for slurry phase reactors, such as the porous plate, packed plate and sintered metal plate distributor, the nozzle distributor and the tube distributor. Patents have been filed for various gas distributor designs, among which the concentric ring tube distributor is commonly used. The selection and design of a gas distributor in a slurry phase reactor should meet the following requirements:

1. The gas distributor must distribute the inlet gas evenly with as little pressure drop as possible. Too high a pressure drop results in increased energy consumption. Under low-pressure-drop operating conditions, the gas distribution can be optimized by selecting the appropriate opening size and distributor/bed pressure drop ratio or by adopting certain predistribution methods.
2. The gas distributor gives good gas distribution in terms of both bubble size and bubble speed in the slurry to ensure good gas–liquid/solid contact near the distributor and thus reduce the possibility of a dead zone forming.

7.4.2.2 Heat Removal

We still need to remove the reaction heat effectively in the SBR. If heat cannot be removed effectively, the temperature rises, which results in carbon deposition, catalyst deactivation and even damage to the reactor. The heat generated by the reaction is removed by cooling tubes that are submerged in the slurry. An industrial-scale FT SBR could have several hundreds to thousands of cooling tubes. The diameter of the cooling tube is normally 3–5 inch[5] They can be arranged in one to three cooling tube bundles that are installed along the axial direction of the reactor. Cooling water flows through the cooling tubes, and the heat is removed by boiling water. The demineralized water is pumped into the cooling tubes, and part of it vaporizes to steam. The mixture of water and steam enters the steam drum for phase separation. The saturated water goes back to the cooling tubes, while the steam goes to the utilities. When the reactor is operated at 220–250 °C during low-temperature FTS, medium-pressure steam can be produced at 1.6 MPa. Different designs for cooling tubes can be found in patents and open publications.

One big advantage of a SBR over a FBR is that the heat transfer is much better in a SBR. Therefore, the productivity is much higher per unit reactor volume. However, the productivity of a SBR is still limited by heat transfer rather than by catalyst activity. Part of the reactor volume is taken by the cooling tubes, while the rest is occupied by slurry, which is where the reaction occurs. The activity of the catalyst, together with the slurry volume and catalyst fraction in the slurry, determines the heat duty of the reactor. On the other hand, the volume of the cooling tube bundle(s) (in terms of the number of cooling tubes and their total surface area) determines the amount of heat that can be removed. The balance of these two determines the productivity of a SBR. This means that the ultimate capacity of a SBR still depends on the heat exchange surface area that the cooling tubes in the reactor can provide. Once the maximum capacity of heat removal has been reached, higher catalyst activity cannot improve the productivity of the reactor but rather reduces the usage of catalyst.

7.4.2.3 Device for Gas–Liquid (Solid) Separation

It is essential to have a gas–liquid (solid) separation device above the liquid level inside the reactor. Gas enters at the bottom and leaves from the top of the reactor. When unreacted gas and gas products leave the slurry, waxy liquid droplets and catalyst fines are entrained. These need to be separated before the gas leaves the reactor chamber, in order to protect the downstream equipment. Such separation is normally achieved by the collision of droplets and fines on baffles arranged inside the separator chamber. The baffles alter the flow direction of the gas and trap the droplets and fines. The droplets and catalyst fines accumulate on the baffle surface and flow back into the reactor. A high-efficiency separator can trap liquid droplets that are larger than 8 μm and catalyst fines that are larger than 5 μm.

7.4.2.4 Catalyst Separation from Liquid Product

The catalyst material needs to stay in the reactor in order for the reaction to occur, while the liquid phase product must be removed continuously from the reactor in order to maintain the liquid level in the reactor. Thus the catalyst needs to be separated from the liquid products as the slurry is extracted from the reactor. This separation should give a good quality wax for further upgrading and for preventing problems in downstream equipment due to entrained catalyst in the liquid product. Furthermore, the separation needs to either retain the catalyst or return it to the reactor. This separation must also occur at the reaction temperature and pressure and thus can be quite challenging.

Various devices with different mechanisms have been proposed for separating the catalyst, including settling due to gravity, centrifugal or magnetic forces, and filtration using pressure differences. Among these techniques, filtration is the most commonly used method for separating the catalyst from the slurry. A two-stage filtration system is used in the reactors in South Africa, Qatar and China. In such a filtration system, the first stage filter is placed inside the reactor to separate most of the catalyst, while a second stage filter is placed externally for further separation. The designs and drawings for these separation devices, including the filters, can be found in the patents.

The filtration of the catalyst is achieved by building up a filtration cake formed by the catalyst itself. The filter itself only provides the support to form the catalyst cake for filtration. The filter is normally assembled from wedge-shaped parts and the V-shaped gap creates the path for the wax to flow. The catalyst particles bridge at the very end of the V-shaped gap and trap catalyst particles to form a cake. Liquid wax squeezes through the gaps in the catalyst cake and flows out of the reactor. When the catalyst cake is thick enough to create too high a pressure drop, the catalyst cake is broken and flushed back into the reactor by reverse flow of either the liquid products or the reactant gas.

7.4.3 Design of a SBR for FTS

Design and scale-up methods for SBRs have been reported in the literature by many researchers. In this section, we do not intend to explain all the details or describe all the different equations that have been developed and reported by researchers; rather, we narrate the process for reactor design, similar to what was done in the section on FBR design.

In each step of the design process, one can find various empirical equations to use for the calculations. Again, the results derived using different empirical equations could be similar if the values of the operating parameters fall within the range used when developing the different empirical equations. An engineer who wants to design a SBR might feel lost in a forest of hundreds of different empirical equations, and because of the many approaches and uncertainty, it is hard to validate the simulation results or calculations. Therefore, in this section, we first discuss the important aspects of SBR design and then describe the design process.

7.4.3.1 Mass Transfer in a SBR

Unlike the case of a FBR, the size of the catalyst particles used in SBRs are in the order of microns. The mass transfer limitations due to diffusion of reactants into the catalyst particles are very low and hence can be neglected. However, the diffusion of reactant gas into the slurry becomes the dominant mass transfer resistance.

At steady state, the overall mass transfer J for reactant in either the gas phase (G), liquid phase (L) or solid phase (S) can be expressed by eqn (7.11)–(7.13):[37]

$$J_G = k_G a\ (P_G - P^*) \tag{7.11}$$

$$J_L = k_L a\ (C^* - C_L) \tag{7.12}$$

$$J_S = k_S a_P\ (C_L - C_S) \tag{7.13}$$

where P_G is the partial pressure of reactant in gaseous phase; P^* is the partial pressure of the reactant at vapor–liquid equilibrium; C^* is the concentration of reactant in liquid phase at vapor–liquid equilibrium; C_L and C_S are the concentration of reactant in bulk liquid and in the catalyst, respectively; a is the surface area of the bubbles in the slurry; a_p is the external surface area of the catalyst.

P^* and C^* can be related by Henry's law:

$$P^* = C^* H_e \tag{7.14}$$

where H_e is Henry's constant for reactant dissolved in the slurry.

As previously explained, in the case of a SBR, J_G and J_S can be overlooked, as gas mixing and diffusion of gas into small catalyst particles occur much faster than the mass transfer between the bulk gaseous phase and the liquid.[37,38] Therefore, we should mainly focus on the parameter J_L.[39] In order to estimate the value of J_L, we need to estimate the mass transfer coefficients k_L or k_La for the reactants.

Measurement of the mass transfer coefficient, k_L and the volumetric mass transfer coefficient, k_La have been reported by many researchers for different simulated compositions of gas, liquid and solid under various operation conditions.[40,41] Here, we use one particular example to estimate the value of k_L. Deckwer *et al.* proposed that eqn (7.15) could be used to calculate k_L in an FTS reactor:[41]

$$k_L \mathrm{Sc}^{2/3} = A \left[\frac{\mu_L (\rho_L - \rho_G) g}{\rho_L} \right]^{1/3} \tag{7.15}$$

where A is an empirical factor, and its value depends on the average diameter of the bubbles; Schmidt numbers $\mathrm{Sc} = \mu_L/\rho_L D_L$ and D_L are the diffusion coefficient of the reactant gas in the liquid. Among the gas, liquid and solid phases in the reactor, the liquid and solid phases can be regarded as one continuous phase; therefore, ρ_L and μ_L in eqn (7.15) can be written as ρ_{SL} and μ_{SL}:

$$\rho_{SL} = \varepsilon'_S \rho_S + \varepsilon'_L \rho_L = \varepsilon'_S \rho_S + (1 - \varepsilon'_S) \rho_L \tag{7.16}$$

where ε'_S is the volumetric fraction of solid in the slurry without gas; ε'_L is the volumetric fraction of liquid in the slurry without gas.

When the solid fraction in the slurry is low (when $\varepsilon'_S < 0.16$), the viscosity of the slurry is also low and can be expressed as:[41]

$$\mu_{SL} = \mu_L (1 + 4.5 \varepsilon'_S) \tag{7.17}$$

When the solid fraction is high (when $0.16 < \varepsilon'_S < 0.60$), μ_{SL} can be expressed as:

$$\mu_{SL} = \mu_L \left[1.25 \varepsilon'_S + 10.05 \varepsilon'^2_S + 2.73 \times 10^{-3} \exp(16.6 \varepsilon'_S) \right] \tag{7.18}$$

In contrast, k_La is more dependent on the structure of the internal parts of the reactor, the properties of the slurry, the operational conditions (T, P, ν), the size of the catalyst particles and the fraction of catalyst particles in the slurry.[42–44]

This is only one example of the correlations used for calculating the mass transfer coefficients k_L and k_La. Many other empirical equations are reported in the literature. Before selecting empirical equations reported in the literature for use in calculations, one should understand the characteristics of the reactor system and compare these to the conditions for which the equations were derived. Equations developed for situations that are similar to those in the reactor system that is being designed should be used. Although the calculation of the mass transfer coefficients k_L and k_La can be very complex and are dependent on operating conditions and the properties of the gas, liquid and solid in the reactor, the purpose of the calculations is quite straightforward, *i.e.* we need to find out how much reactant is available to participate in the reaction at any time or in any position in the reactor. Once we know this value, it can be used in the next step of the calculation, which is to estimate the reaction rate.

In summary, the mass transfer coefficient and mass transfer model give us a way to calculate the amount of gas that is available to participate in the reaction.

7.4.3.2 Kinetics

We have already explained that the mass transfer in the catalyst can be neglected in a SBR; therefore, the kinetics can be incorporated in the calculation/simulation. Kinetics are key in describing the consumption of the reactants and the formation of the products. However, kinetics are not as complex as other aspects in designing a SBR as the kinetics developed in the laboratory can be used in a fairly straightforward way. Empirical equations should be used with a limited number of parameters that have been developed for the same catalyst under the operating conditions used in the full-scale reactor. The importance of kinetics for the rate of reaction of CO and for the products should be emphasized when designing a SBR for two reasons: (1) to determine the heat generated by the reaction accurately, which is necessary for the design of all FT reactors; (2) to determine the amount of liquid phase product that is produced in order to control the liquid level in the reactor.

In summary, kinetics gives us the consumption rate of the reactants and the formation rate of the products. It also gives us the amount of heat generated by the reaction.

7.4.3.3 Heat Transfer

The heat transfer in a FBR limits the achievable productivity per unit volume of reactor. With the presence of a liquid phase in the SBR, the temperature in the liquid is close to isothermal. Due to the large heat capacity of the liquid, temperature fluctuation in the reactor is largely damped, and no temperature runaway is experienced. Heat transfer in the SBR reactor is limited by the rate of heat exchange between the slurry and the cooling tubes submerged in the slurry, which allows a much higher heat transfer rate in an SBR than in a FBR. The rate of heat removal from a SBR is described by:

$$Q = UA_{\mathrm{suf}}\left(T_{\mathrm{slurry}} - T_{\mathrm{cooling}}\right) \tag{7.19}$$

where U is the overall heat transfer coefficient or the heat exchanged in the SBR.

The overall heat transfer coefficient U depends on the heat transfer coefficients between the boundary of the slurry and cooling tube, the cooling tube wall and the boundary of the cooling tube and cooling medium. The heat transfer coefficient at the boundary of the slurry and cooling tube depends on the properties of the catalyst used and the operating conditions, including temperature, pressure, gas space velocity, size of gas bubbles and catalyst fraction in the slurry.[45,46] For example, Saxena, *etc.*[47] reported the following empirical equation for the heat transfer coefficient between the slurry and the tube wall:

$$U = 0.12\left(\frac{g^2\rho_{\mathrm{L}}}{\mu_{\mathrm{L}}}\right)^{1/6}\left(\frac{\rho_{\mathrm{L}} - \rho_{\mathrm{G}}}{\rho_{\mathrm{L}}}\right)^{1/3}\left(k_{\mathrm{L}}\rho_{\mathrm{L}}c_{\mathrm{pL}}\right)^{1/2} \tag{7.20}$$

where g is the gravitation constant; c_{pL} is the heat capacity of the slurry.

The terms of the empirical equation also depend on the characteristics of the reactor system, and we need to verify the similarity between a system reported in the literature and the proposed design in order to ensure that the equation is applicable. It has been reported that U increases with an increase in slurry temperature, superficial velocity of gas and size of catalyst particle and decreases with an increase in viscosity of slurry and catalyst density. Contradictory results have been reported regarding the effect of the solid fraction of the slurry on the U value.

Due to good mixing in the slurry, the temperature in the reactor is fairly constant, especially in the case of small L/D_{T} ratios. The temperature difference between the gas inlet and outlet is normally within 5 °C.

In summary, the heat transfer coefficient is used in the heat transfer calculation to determine the required surface area of the cooling tubes. After determining the reaction rate and thus the heat released by the reaction, the energy balance can be used to determine the cooling area and thus the number of cooling tubes required.

7.4.3.4 Gas Holdup in a SBR

The slurry bed is composed of liquid product, solid catalyst and gas bubbles. The overall volume and the height of the slurry bed depend on the volume fraction of the gas bubbles in the slurry. The bubble content also affects the gas–liquid contact surface area and subsequently impacts the mass and heat transfer characteristics.[48–50] The relationship between the volumetric density of the catalyst and the gas holdup can be expressed as:

$$C_{cat} = \frac{W_{cat}}{V_R(1 - \varepsilon_{internal})} = \frac{(1 - \varepsilon_G)\varepsilon_P}{\rho_P} \tag{7.21}$$

where C_{cat} is the volumetric density of the catalyst; V_R is the reactor volume; $\varepsilon_{internal}$ is the volume fraction of the internal parts; ε_G is the gas holdup.

The gas holdup can be affected by many factors in the reactor, including the dimensions of the reactor (height and diameter), the gas distributor, the properties of the slurry (including density and viscosity) and the operating conditions (including the superficial velocity of the gas, temperature and pressure). The gas fraction increases with increasing gas superficial velocity and liquid viscosity, while it decreases with increasing liquid velocity and surface tension. Research[49] also found that there is a linear relationship between the gas content and the mass transfer coefficient k_L.

In summary, we need to calculate the gas fraction of the slurry. It affects the volume of the slurry, the height of the slurry bed and the mass transfer from the bulk gas to the slurry. It is used to calculate the gas conversion, product selectivity and mass transfer coefficient.

7.4.3.5 Distribution of Catalyst Particles

As the catalyst particles are suspended in the slurry, there are two possible distributions for the catalyst particles: an even suspension and a distributed suspension.[51,52] Fluidization of catalyst particles in the slurry occurs because of the gas bubbles rising from the bottom of the slurry to the top. Therefore, the evenness of the catalyst distribution depends mainly on the catalyst particle size, the catalyst fraction in the slurry and the gas superficial velocity.[51]

A minimum gas velocity is required to ensure suspension of the catalyst in the slurry. The minimum gas velocity also determines the lowest space velocity of operation. Eqn (7.22) and (7.23) correlate the minimum gas velocity u_{ms} to the solid fraction ε_S.[51,52]

When $Re_T = \dfrac{D_T u_{ms} \rho_G}{\mu_G} < 500$:

$$\frac{\varepsilon_S \rho_P}{\rho_{SL}} = 6.84 \times 10^{-4} C_\mu \left(\frac{D_T u_{ms} \rho_G}{\mu_G}\right) \left(\frac{\sigma_L \varepsilon_G}{u_{ms} \mu_L}\right)^{-0.23} \left(\frac{u_t \varepsilon_G}{u_{ms}}\right)^{-0.18} \eta^{-3} \tag{7.22}$$

when $Re_T = \dfrac{D_T u_{ms} \rho_G}{\mu_G} > 500$:

$$\frac{\varepsilon_S \rho_P}{\rho_{SL}} = 1.072 \times 10^{-1} C_\mu \left(\frac{D_T u_{ms} \rho_G}{\mu_G}\right) \left(\frac{\sigma_L \varepsilon_G}{u_{ms} \mu_L}\right)^{-0.23} \left(\frac{u_t \varepsilon_G}{u_{ms}}\right)^{-0.18} \eta^{-3} \tag{7.23}$$

where Re_T is the Renault's number; ρ_P is the catalyst density; ρ_{SL} is the density of the slurry; ρ_G is the gas density; D_T is the diameter of the reactor;

σ_L is the liquid surface tension; μ_G is the gas viscosity; μ_L is the liquid viscosity; η is the wettability factor of the solid; C_μ is the viscosity correctional factor.

The distribution of catalyst particles in the slurry bed can be described by sedimentation models, and eqn (7.24) is one example.[53] Among all the parameters, the catalyst diameter has the biggest influence on the catalyst distribution in the slurry.

$$D_S \frac{d^2 C_S}{dx^2} + \left(u_{SS} - \frac{u_L}{(1-\varepsilon_G)} \frac{dC_S}{dx} \right) = 0 \tag{7.24}$$

where D_S is the distribution factor in the axial direction; C_S is the catalyst fraction in the slurry; u_{SS} is the sedimentation speed of the catalyst; u_L is the liquid velocity.

u_{SS} and D_S can be calculated using the following correlations:

$$u_{SS} = 1.2 u_t (u_G/u_t)^{0.25} \left[(1 - V_{cat}) / (1 - V_{cat}^*) \right]^{2.5} \tag{7.25}$$

$$D_S = u_G D_T (1 + 8Fr^{0.85})/13Fr \tag{7.26}$$

where u_t is the sedimentation speed of the catalyst; V_{cat} is the volumetric fraction of the catalyst in the slurry; V_{cat}^* is the volumetric fraction of the catalyst particles in the slurry at equilibrium.

In summary, there is a catalyst distribution mainly in the axial direction of the reactor, which can be described by a sedimentation model. With the right model, we can calculate the catalyst fraction at different positions in the slurry bed. The result is used in the reaction rate calculation.

7.4.3.6 Reactor Model and Design Process

Once we understand the preceding aspects, the SBR can be designed by using appropriate models. Designing a SBR is much more complicated than designing a shell and tube reactor or FBR. Even with advanced tools, such as computational fluid dynamics (CFD), it is still difficult to validate the simulation results. This is due to the number of equations used and the complexity of these equations, the number of parameters incorporated into the correlations and the simplifying assumptions made during design.

At least three types of flow or reactor models have been proposed in the literature: (1) plug flow, for both gas and liquid; (3) plug flow for gas and continuously stirred tank reactor (CSTR) for liquid (including solid); (3) CSTR for both gas and liquid (including solids). Obviously, a CSTR model is more suitable for a reactor with a low L/D_T ratio, such as in the case of large-scale commercial reactors; a plug flow model better describes situations with a high L/D_T ratio, such as a laboratory reactor. When we consider the characteristics of commercial-scale SBRs, we can consider that the slurry is well mixed (*i.e.* behaves as a CSTR) and that the gas can either be flowing plug flow, which we will call the "Plug Flow Model", or that the gas is also well mixed, which we will call the "CSTR model".

Due to the mathematical differences, the results derived using either a CSTR or a plug flow model represent the two extremes of mixing in a SBR. For simple kinetics, these cases could also represent the extremes in the average reaction rates in the reactor. The performance of an actual SBR typically falls between these two extremes. This means that a single model assumption in the design is often not good enough to give accurate results. With this in mind, we could divide the reactor into two separate zones along the axial dimension during simulation and design. The lower part could be regarded as one or more fully mixed CSTRs, and the upper part of the bed could be regarded as one plug flow zone. With higher turbulence in the lower part of the reactor, catalyst distribution is close to homogeneous, and the slurry exhibits better mixing with more gas holdup. As the gas flows upward, the gas bubbles coagulate. The axial position in the reactor, which is where we assume the pseudo plug flow behavior starts, is where the gas bubbles start to stabilize. Therefore, during CFD simulation, instead of regarding the whole reactor volume as a single reactor type with homogeneous mixing, it could be divided into several different reactor types for simulation purposes.

The number of fully mixed CSTRs used in the modeling depends on the sedimentation behavior of the catalyst and the mixing of the gas with the slurry. If the catalyst settles slowly when the mixing of the gas in the slurry is good, a model with fewer CSTRs or perhaps only a single CSTR could be adopted.

The accuracy of this reactor model depends strongly on factors such as sedimentation of the catalyst and the stability of the gas bubbles. A faster linear flow rate of inlet gas, either along the wall or at the center of the reactor, could cause deviation from the predicted reactor performance.

When doing simulations, the reactor can be divided into reactor cells, and each cell can be regarded as a CSTR. Under conditions of plug flow, for example, the output of the previous cell is the input for the next cell. The final design result is the output of the final cell. The kinetics, mass transfer and catalyst concentration, as well as the temperature, pressure and gas space velocity, can be used to calculate the rate of gas consumption and the rate of product formation. Once the overall formation rate of the products is known, the amount of heat generated by the reaction can be determined, and the heat exchange area and number of cooling tubes can be estimated. The reactor volume can then be adjusted to include the volume of the cooling tubes, and the calculation process can be iterated until the volume of the cooling tubes is distributed so as to ensure heat balance across the cooling tube.

7.4.4 Design and Operational Notes

As discussed in Section 7.4.2.4, filtration is typically used to separate the catalyst particles and the liquid products from the slurry. The filter cake is formed by the catalyst particles, and, as the filter cake grows, the pressure drop across the filter increases until the filter cake needs to be dislodged by back-flushing either gas or liquid through the filter. The filtration occurs at the reactor temperature and pressure, and this separation has led to problems during operation. In a demonstration project, after operating for some time with cycles of normal operation and back-flushing, the pressure drop across the filter was observed to increase with time. In order to extract the liquid products from the reactor, the operating pressure drop across the filter had to be increased, until a point was reached at which the filter burst. (This incident was not openly reported, but it did happen.)

Carbon deposits are found on the surface of the filter. In FBRs, carbon is deposited because of excess heat being generated and poor heat transfer in some regions. It is abnormal to experience carbon deposits in a SBR because of the much greater heat capacity and good heat transfer provided by the slurry. When carbon deposits are found, it suggests that some hydrocarbons have undergone thermal cracking due to localized high temperature. This in turn suggests that localized high temperature has occurred, even though the overall slurry is controlled at the reaction temperature.

The appearance of localized high temperature is believed to be caused by a higher reaction rate and, in particular, a higher CH_4 formation rate. As the back-flushing in the case mentioned earlier was accomplished by flowing H_2 into the reactor, localized areas of high partial pressure of H_2 were formed. Therefore, the reaction rate, especially the CH_4 formation rate, was promoted, and the excess heat caused a localized higher temperature, resulting in carbon formation.

Temperature management in FTS reactors is one of the most important aspects in both the design and operation of such systems, even when a reactor system is believed to have efficient heat transfer. It must be borne in mind that problems can still be experienced due to the interaction between the reaction rate and heat transfer. Heat management for an FTS reactor should be considered not only during the reactor design phase but also during the process and/or reactor operation phase.

In terms of application of SBRs to BTL, online separation of the catalyst from the slurry remains critical, and it may not be simple to either automate the process or for it to be run by semiskilled operators. Furthermore the scale-up and operation of SBRs is not straightforward or even that well understood. Thus the reactor may not be able to run satisfactorily with either varying gas flow rates or compositions. The feed to a BTL plant is likely to be variable, and therefore the tight control needed for the successful operation of an SBR may not be achievable in BTL.

7.5 New Types of FBRs

The development of catalysts has followed two main routes: (1) pursuing a higher turnover frequency with the same amount of catalyst or active metal loading; (2) changing the product selectivity, in order to produce a different product spectrum that better matches market demand or that has a higher market value. While researchers are developing new and better catalysts for

FTS, the development of a better reactor is also crucial for the success of commercial projects. Based on market trends, long-chain hydrocarbon products and α-olefins in the carbon range of C_2 to C_{15} are still the major products that should be produced *via* FTS in the coming years.

Planning for the production of high value-added liquid phase products is especially crucial for BTL processes in order to improve the economics. In the meantime, the limited scale of BTL projects suggests that FBRs are a good option for BTL. The product selectivity in FTS is sensitive to operating conditions, especially temperature. Thus FTS should optimally be operated under isothermal conditions. In very exothermic reactions, such as FTS, the use of catalysts with higher activity is always limited by the heat transfer in the reactor.

In this section, two new types of FBRs that show better heat transfer performance are discussed.

7.5.1 Microchannel Reactors

One way to enhance heat transfer is to use a plate heat exchanger as the reactor rather than a shell and tube heat exchanger. As introduced in Section 7.2.2.4, the heat transfer resistance in a FBR is mainly in the catalyst bed. Thus the dimensions of the catalyst bed in the direction of heat transfer to the cooling medium need to be reduced so that overall heat transfer can be improved. By using a plate heat exchanger, the heat transfer dimension of the catalyst bed can be reduced, thereby improving the heat transfer rate in the reactor.

The size of the catalyst must also be reduced because of the smaller dimension between the plates – typically of the order of 0.1 mm, compared to a catalyst particle size of about 3 mm in a conventional shell and tube reactor. This improves the catalyst activity, as the mass transfer resistance becomes smaller or even negligible. With this better reactor design, the activity of the cobalt-based catalyst used in the microchannel reactor can be 5–10 times higher than that used in the standard FBR design. This increases the productivity per unit volume of the reactor, and hence a much small reactor volume can be used to achieve the same production rate. The distance between adjacent plates can be in the order of millimeters, and a U-shaped heat conductive plate can be placed between two plates to achieve a very high heat exchange coefficient. The available heat exchange area per unit volume of catalyst is much larger compared to that of a conventional shell and tube reactor.

The cooling medium flows in the cooling channels next to the catalyst chambers in cross flow. This allows a good match between improved catalyst activity and increased heat transfer, which set the dimension of the microchannel-type FTS reactor.[54–56] However, as the catalyst particle size is much smaller in this reactor, there is a dramatic increase in pressure drop for the same length of bed. The microchannel reactor from Velocys has a catalyst bed designed for 60 cm in height.[57] Figures 7.10 and 7.11 show the dimensions and configuration for the microchannel reactor from Velocys. This reactor has been applied in small-scale commercial plants for synthesizing fuel using FTS.

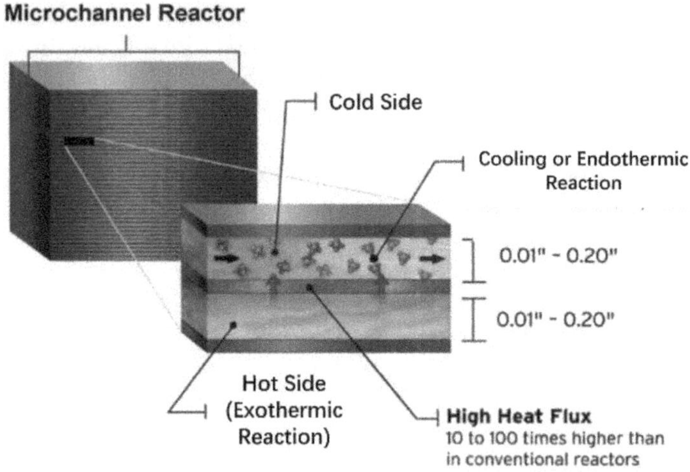

Figure 7.10 Microchannel reactor for FTS. Reproduced from ref. 58 with permission from Elsevier, Copyright 2010.

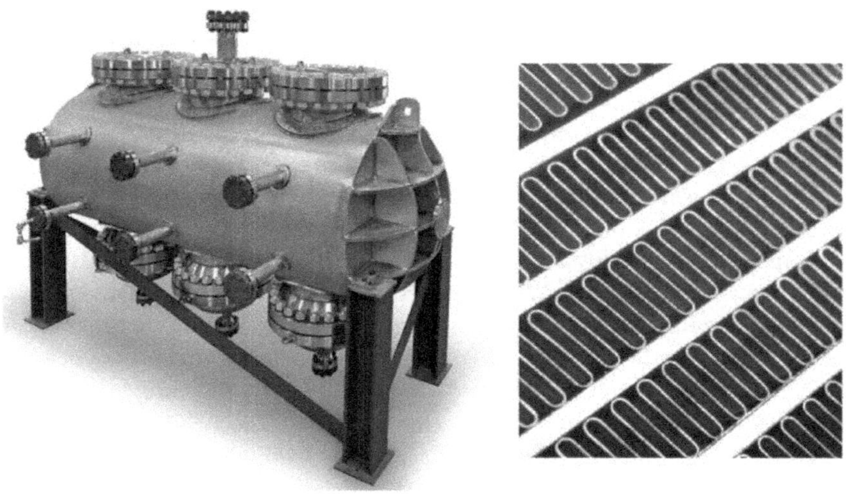

Figure 7.11 Microchannel reactor (left) and catalyst loading space (right). Reproduced from ref. 59 with permission from Edward T. Dodge, Copyright 2014.

The microchannel reactors for FTS are designed and manufactured in modules and are rectangular in shape, as shown in Figure 7.10. Reactor modules are then assembled into a horizontal vessel in parallel, as shown in Figure 7.11. The number of reactor modules can be adjusted to match the design capacity. The reactor modules are welded onto the inner wall of the reactor vessel to create an inlet gas manifold at the top and a liquid product collector at the bottom. The two volumes at the sides of the vessel created by the welding are manifolds for the cooling medium to enter and leave the reactor. Because of the rectangular shape, the reactors only take a fraction of the volume of the reactor shell. If more reactor modules are arranged in parallel in the radial direction, less vessel volume will be used for the reactor and catalyst. This results in a much smaller volume usage for the reactor vessel when compared to conventional shell and tube reactors.[58]

Loading and unloading of the catalyst can also be challenging. Once the reactor modules are fixed inside the reactor vessel, loading and unloading of the catalyst can only be done through the many manholes (or handholes) on both the top and bottom of the reactor vessel, as shown in Figure 7.11.[59]

As a new design for a FBR, the microchannel reactor is an important step in reactor development for FTS. This new design is based on a deep understanding of both the behavior of the catalyst and the characteristics of a FBR. The microchannel reactor has advantages, such as higher productivity per volume of catalyst and reactor and better temperature management but also has disadvantages, such as a higher pressure drop and a higher manufacturing cost. One probably cannot expect to obtain advantages with a new reactor design without some trade-offs, and this should be expected when science and engineering are incorporated into a reactor design, as pros always come with cons, due to the limits set by either the science or the engineering.

7.5.2 Heat Pipe Reactor

Heat is removed from the reactor by a cooling medium, which undergoes a full cycle of phase change. The liquid phase coolant is heated by the heat of the reaction and is vaporized. The gaseous phase coolant is then cooled by a cooling source external to the reactor and converts back into the liquid phase. By undergoing this phase change cycle, the heat generated by the reaction is taken out of the reactor, so that the temperature in the catalyst bed can be managed. Reexamination of a shell and tube reactor for FTS reveals a typical axial temperature profile, as shown in Figure 7.12. It is best to ensure an isothermal temperature profile in the reactor; however, it is not achievable in

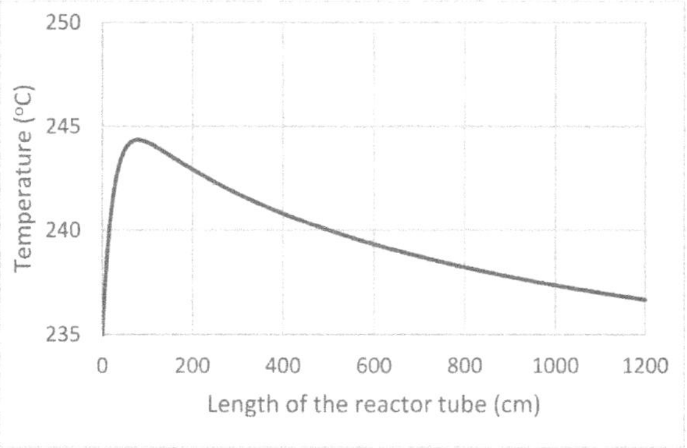

Figure 7.12 Typical temperature profile at the center of the catalyst bed in a shell and tube reactor.

practice. Under realistic reactor operating conditions, a reasonable temperature rise is normally allowed for, but an elevated temperature causes damage to the catalyst and hence shortens the catalyst life and perhaps triggers a temperature runaway. Therefore, it should be strictly avoided. When we look again at the temperature profile in a tube, it is not the overall temperature rise that we should reduce or eliminate but rather the highest temperature along the profile. If a secondary heat removal mechanism is introduced axially at the center of the catalyst bed in the radial dimension, this results in a much lower maximum temperature in the radial direction.

We know that for any FBR that allows for internal heat exchange, there are normally two compartments, namely a process side and a utility side. An additional compartment in the reactor could cause a challenge in terms of manufacturing the reactor, given that a secondary utility side would need flow connections with the process streams in and out of the reactor. This means that a secondary utility side should be able to remove reaction heat from the center of the catalyst bed, but it should not need material exchange or connections with the process or utility streams flowing into or out of the reactor.

A piece of pipe containing either a single component liquid or a liquid mixture and with both ends sealed – called a heat pipe- can be used in the center of the catalyst bed as the secondary heat removal mechanism, without adding much complexity to the reactor manufacturing or catalyst loading.[60] Heat pipes are one of the most efficient ways to move heat, or thermal energy, from one point to another. The most common heat pipe is cylindrical in cross section, with a wick inside that runs axially. Cool working fluid moves through the wick from the colder side (condenser) to the hotter side (evaporator), where it vaporizes. This vapor then moves back to the condenser's heat sink, carrying thermal energy along with it. The working fluid condenses, releasing its latent heat in the condenser, and then repeats the cycle to continuously transfer heat from the condenser end of the pipe to the evaporator end of the pipe.

A shell and tube reactor with heat pipes is demonstrated in Figure 7.13. Heat is taken out of the catalyst bed by turning the cooling medium from liquid to vapor. This heat is then transported by the heat pipe outside the tubes, where it heats up the feed gas, thereby condensing the cooling medium back to liquid phase. The liquid formed by condensation flows downward *via* the grooves on the inner wall of the heat pipe. Heated by the heat from the catalyst bed, the feed gas can have a lower feed temperature before entering the reactor, which also reduces the heat duty of the syngas preheater.

By adding heat pipes along the center of the catalyst bed, the heat transfer distance is largely reduced without reducing the reactor tube diameter. With the enhanced heat transfer coefficient of the heat pipe, the diameter of the heat pipe needs to be only a fraction of the diameter of the reactor tube. If, for example, the diameter of the heat pipe takes up 1/4 of the reactor tube diameter, it only reduces the catalyst volume by 1/16, which has a small impact on the amount of catalyst loaded, but this can suppress the temperature rise at the center of the catalyst bed. As the maximum temperature difference in the bed is reduced by the heat pipe, the reactor can be run at a

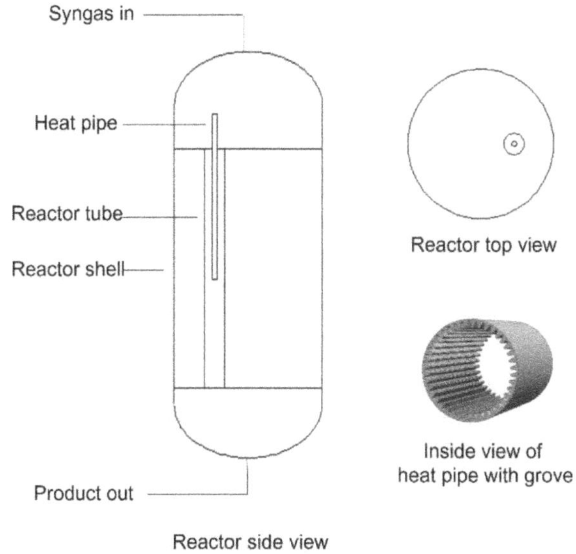

Figure 7.13 Shell and tube reactor with heat pipes used as the secondary heat removal mechanism.

higher overall temperature, which means that the average reaction rate can be significantly higher than without the heat pipe. The benefit of this type of reactor is that it has the advantages of a conventional shell and tube reactor – including low manufacturing and maintenance costs – while improving the heat management aspect of the reactor, thereby improving the stability of the reactor operation and prolonging the catalyst life. The disadvantages of this reactor are that it has an additional requirement for reactor manufacturing and that assembling the reactor can also be challenging. At the end of the day, selecting a suitable reactor for highly exothermic reactions, such as FTS, requires ensuring a balance between reliability when scaling up, manufacturing cost, stability of operation and maintenance requirements.

7.6 Reactor Selection for BTL

In principle, all the different types of reactors that have been introduced in this chapter can be used in indirect conversion of BTL products *via* the FTS route. However, the choice of a suitable reactor for a BTL process needs to take some key aspects into account, which are linked to the properties of the biomass itself, as well as the market trends for the products. The advantages and disadvantages of the different types of reactors are not the only factors that we need to consider when selecting a reactor for a BTL application. Some other strategic aspects also need to be considered in the scope of assessment, including the products and their workup, availability of catalysts and readiness of downstream processes and technologies.

7.6.1 Scale

The production of oil or chemicals from biomass *via* FTS is much smaller in scale than a process that uses coal or natural gas as feedstock. When considering a reasonable transport distance, the scale of a BTL is normally below 1000 bbl per day, even in areas rich in biomass resources. Instead of scaling up reactors (and processes) from past demonstration reactors for commercialization of CTL and GTL, scaling down is the path for the development of BTL reactors (and processes). Some points of view believe that the fixed-bed FT reactors (Shell) and slurry FT reactors (Sasol) are operable only at the scale of the Pearl or Oryx plants. Any attempt to scale these reactors down may result in problems in control, operation or meeting the design production capacity. With smaller scale of BTL plants, a more reliable reactor with less complexity of operation is preferred.

7.6.2 Plan for Products

When FTS is used for fuel production, all three types of reactors can be used, and the corresponding catalyst and downstream processes are well developed. However, we cannot ignore the fact that there is a trend worldwide to replace gasoline and diesel vehicles with electrical ones. If this trend escalates, then future applications of FTS in biomass conversion should be aimed at the production of chemical products rather than fuel replacement products. From this point of view, the production of long-chain hydrocarbons can meet both the current needs for oil production, as well as the long-term need for chemical products such as wax products, base oils for lubricants, *etc.* Therefore, reactors that are suitable for producing high-boiling-point products using a low operating temperature are probably preferable.

As some countries still require fuel production, using bifunctional catalysts to produce mainly petrol and diesel fractions from syngas in one synthesis reactor is a good option. The bifunctional catalyst is an FTS catalyst that is combined with an acid molecular sieve composite at a microdimension, which simultaneously assists the FTS reactions and catalytic cracking of long-chain hydrocarbons. The advantage of using this type of catalyst is to save on the downstream oil refining process and hence shorten the whole process. This can reduce the complexity of the process and improve the economy of the plant.

Bifunctional catalysts are currently in use at the laboratory to small-scale demonstration plant scale. At present, the structure of bifunctional catalysts is more suitable for use in FBRs. It may be less attractive to use such catalysts in FBRs for large-scale GTL or CTL projects, but they might be a suitable option for BTL projects. After development and engineering, these catalysts may be suitable for use in SBRs as well, but they are difficult to use in fluidized-bed reactors due to the small molecular weight of the synthesized products and excessive cracking of products at a high operating temperature. Given the difficulties in design and operation, the lower value of the primary products and the complexity of the downstream processes after the reactor, fluidized-bed reactors need to be carefully assessed before being chosen for BTL projects.

7.6.3 Economic Feasibility

As the project scale is increased, the advantages of economy of scale can become significant. A higher feed conversion and a lower consumption of feed for the same amount of product are always the optimal targets for large-scale projects, even when elevating the CAPEX (capital expenditure) as a trade-off. This is consistent with what we often see in projects where efforts are made to increase feedstock conversion or product yield by even 2–3%. This is not a big improvement in relative value, but it is obvious in the improvement of absolute output for a large-scale project. However, in small and medium-sized projects, such as the BTL projects discussed in this book, the scale effect is much smaller than that of large projects, so the fixed investment per unit product produced is much higher. Therefore, the sensitive parameter in feasibility assessment can shift from conversion to investment and operating costs. Rather than seeking extremely high raw material utilization, it is better to aim for the minimum fixed investment and operating cost per unit of production. Such a decision strategy also determines the choice of reactor. In small and medium-sized projects, it is more sensible to choose a reactor that is stable and reliable and that has a lower operating cost, but that is not necessarily the most efficient option. At the same time, when selecting the reactor, we should consider the reactor and all the necessary ancillary equipment used during reactor operation and start-up, as well as the overall investment and operating cost of the whole process for the different reactor types.

A higher time online for the reactor – and therefore for the process as a whole – also improves the economic feasibility. A stable reactor that uses a catalyst with a long life span, such as a cobalt-based catalyst, could be the best choice for a BTL project.

7.7 Summary

The development of FT reactors has gone through different stages, with the development of reactor designs being driven by market demand for the products and catalyst innovation. The success of FTS technologies and their industrial application largely relies on successful interaction between the reactor and the catalyst. With the better scientific understanding and experience that has accumulated in the field of FTS, BTL is unquestionably feasible from a technical perspective. A FBR, preferably with higher heat transfer characteristics, better temperature control and a simpler process, is appropriate for a BTL project.

At the end of this chapter, the authors would like to clearly state that, in terms of the reactor and the process for BTL, we surely need to develop better reactors with higher efficiency and better heat management. However, identification and planning for the production and marketing of the desired products and the development of an innovative process should be given consideration first. With innovative processes and better initial planning for production and marketing of the product, choosing a suitable reactor type with a better design will improve the economic feasibility of the BTL project. The reactor is crucial for the success of a BTL (and a CTL and GTL) project as it converts the feed to products; however, it is not the decisive element in terms of the feasibility or success of a BTL project. The reactor only uses a small portion of the CAPEX of the whole process (normally 5–10%, depending on the raw material used for syngas production and the complexity of the overall process) and even less of the OPEX (operating expenses). Often, it is not the reactor choice and design that prevent the implementation of a BTL project but the process of identifying and planning the best product slate, as well as the lack of an innovative process design to reduce CAPEX.

Abbreviations

BTL	Biomass to liquid
CAPEX	Capital expenditure
CFB	Circulating fluidized bed
CFD	Computational fluid dynamics
CSTR	Continuously stirred tank reactor
CTL	Coal to liquid
FBR	Fixed-bed reactor
FT	Fischer–Tropsch
FTS	Fischer–Tropsch synthesis
GTL	Gas to liquid
OPEX	Operating expenses
SAS	Sasol advanced synthol reactor
SBR	Slurry-bed reactor

Symbols

a	Surface area of the bubbles in the slurry
a_p	External surface area of the catalyst
$A_{p,ext}$	Particle external surface area
A_{suf}	Surface area
$c_{H_2,G}$	Concentration of hydrogen in gas
$c_{H_2,L}$	Concentration of hydrogen in liquid wax
c_p	Heat capacity
c_{pL}	Heat capacity of the slurry
C^*	Concentration of reactant in liquid at vapor–liquid equilibrium

C_L	Concentration of reactant in bulk liquid
C_S	Concentration of reactant in the catalyst
C_μ	Viscosity correctional factor
C_{cat}	Volumetric density of catalyst in slurry
d_{wall}	Reactor wall (tube) thickness
D_{eA}	Intraparticle diffusion coefficient of species A
$D_{eff,H_2,L}$	Effective diffusion coefficient of hydrogen
D_T	Reactor diameter
D_S	Distribution factor of catalyst in axial direction
H_e	Henry's constant
ΔH_r	Reaction enthalpy
g	Gravitation constant
J_G	Mass transferred in gas
J_L	Mass transferred in liquid
J_S	Mass transferred in solid catalyst
k_L	Mass transfer coefficient
k_La	Volumetric mass transfer coefficient
k_{m,H_2}	Pseudo-first-order rate of mass transfer of hydrogen
$k_{m,H_2,eff,FT}$	FT reaction rate constant
$k_{m,H_2,eff,M}$	Methanation reaction rate constant
$k_{m,H_2,eff,S}$	Water-gas-shift reaction rate constant
k_V	Volumetric reaction rate constant
K_{eq}	Equilibrium constant
L	Length of the reactor
P_G	Partial pressure of reactant in gaseous phase
P^*	Partial pressure of reactant at vapor–liquid equilibrium
r_a	Reaction rate
$r_{H_2,FT}$	FT reaction rate
$r_{H_2,eff,FT}$	FT effective reaction rate
$r_{m,H_2,M}$	Methane reaction rate
$r_{m,H_2,S}$	Water-gas-shift reaction rate
R	Radius
Re_T	Renault's number
Sc	Schmit number
T	Temperature
T_c	Temperature at the center of the catalyst bed
T_w	Temperature at the reactor wall
ΔT	Temperature difference
u_{ms}	Minimum gas velocity
u_s	Gas velocity
u_{SS}	Sedimentary speed of catalyst in slurry under operation
u_t	Sedimentary speed of catalyst
U	Heat transfer coefficient
U_{over}	Overall heat transfer coefficient
V_{cat}	Volumetric fraction of catalyst in slurry
V^*_{ast}	Volumetric fraction of catalyst in slurry at equilibrium

V_P	Particle volume
V_R	Reactor volume
W_{cat}	Mass of catalyst
z	Height or axial direction
α	Chain growth factor of the FTS
$\alpha_{W,int}$	Heat transfer coefficient at the reactor internal wall
$\alpha_{W,ex}$	Heat transfer coefficient at the reactor external wall
ε'_L	Volumetric fraction of liquid in slurry
ε'_s	Volumetric fraction of solid in slurry
ε_G	Gas fraction in slurry
ε_P	Particle fraction in slurry
$\varepsilon_{internal}$	Volumetric fraction of the reactor internal parts
η	Wettability factor of solid
η_{pore}	Effectiveness factor for reaction rate
λ	Thermal conductivity
λ_{rad}	Radial thermal conductivity
λ_{wall}	Thermal conductivity of the reactor wall
λ_{eff}	Effective thermal conductivity
μ_L	Liquid viscosity
μ_G	Gas viscosity
μ_{SL}	Slurry viscosity
ρ	Density
ρ_P	Density of the particle
ρ_{mol}	Gas molar density
ρ_{SL}	Density of slurry
ρ_G	Gas density
ρ_L	Liquid density
σ_L	Liquid surface tension
ϕ	Thiele modulus

References

1. A. Steynberg and M. Dry, *Fischer–Tropsch Technology, in Studies in Surface Science and Catalysis*, Elsevier, New York, 2004.
2. B. H. Davis, *Top. Catal.*, 2005, **32**, 143.
3. H. H. Storch, N. Golumbic and R. B. Anderson, *The Fischer–Tropsch and Related Synthesis*, John Wiley & Sons, New York, 1951.
4. F. Asinger, *Paraffins Chemistry and Technology*, Pergamon, Oxford, 1968.
5. A. de Klerk, *Fischer–Tropsch Refining*, John Wiley & Sons, New York, 2011.
6. A. N. Stranges, *Stud. Surf. Sci. Catal.*, 2007, **163**, 1.
7. J. L. Casci, C. M. Lok and M. D. Shannon, *Catal. Today*, 2009, **145**, 38.
8. J. C. Hoogendoorn, Presented in part at Clean Fuels and Coal Symposium, 1973.
9. M. E. Dry, *Catalysis Science and Technology*, Springer, Berlin, 1981, vol. 1.
10. R. L. Espinoza, A. P. Steynberg, B. Jager and A. C. Vosloo, *Appl. Catal., A*, 1999, **186**, 13.

11. S. T. Sie, M. M. G. Senden and H. M. W. van Wechem, *Catal. Today*, 1991, **8**, 371.
12. R. Overtoom, N. Fabricius, W. Leenhouts, *Presented in part at Annual Gas Processing Symposium*, Amsterdam, 2009.
13. S. T. Sie and R. Krishna, *Appl. Catal., A*, 1999, **186**, 55.
14. M. Senden and M. McEwan, Presented in part at World Petroleum Congress, Calgary, Canada, 2000.
15. F. J. M. Schrauwen, *Handbook of Petroleum Refining Processes*, McGraw-Hill, New York, 2004.
16. D. J. Moodley, J. van de Loosdrecht, A. M. Asib and J. W. Niemantsverdriet, *Advances in Fischer–Tropsch Synthesis, Catalysts and Catalysis*, Taylor & Francis, Boca Raton, 2009.
17. J. Eilers, S. A. Posthuma and S. T. Sie, *Catal. Lett.*, 1990, **7**, 253.
18. S. T. Sie, *Rev. Chem. Eng.*, 1998, **14**, 109.
19. http://dejong.nl/project/shell-pearl-gtl/.
20. A. Jess and C. Kern, *Chem. Eng. Technol.*, 2009, **32**, 1164.
21. A. Jess, R. Popp and K. Hedden, *Appl. Catal., A*, 1999, **186**, 321.
22. T. Kuntze, PhD Thesis, University Karlsruhe, Germany, 1991.
23. H. Raak, PhD Thesis, University Karlsruhe, Germany, 1995.
24. X. Zhu, X. Lu, X. Liu, D. Hildebrandt and D. Glasser, *Ind. Eng. Chem. Res.*, 2010, **49**, 10682.
25. X. Zhu, X. Lu, X. Liu, D. Hildebrandt and D. Glasser, *Chem. Eng. J.*, 2014, **247**, 75.
26. M. E. Dry, *Today*, 2002, **71**, 227.
27. P. C. Keith, *Oil Gas J.*, 1946, **45**, 102.
28. A. P. Steynberg, M. E. Dry, B. H. Davis and B. B. Breman, *Stud. Surf. Sci. Catal.*, 2004, **152**, 64.
29. A. P. Steynberg, R. L. Espinoza, B. Jager and A. C. Vosloo, *Appl. Catal., A*, 1999, **186**, 41.
30. W. C. A. Holtkamp, F. T. Kelly and T. Shingles, *ChemSA*, 1977, **3**, 44.
31. M. E. Dry, *Appl. Catal., A*, 1996, **138**, 319.
32. J. Collings, *Mind Over Matter: The Sasol Story*, A half-Century of Technology Innovation, Sasol, Johannesburg, 2002.
33. A. P. Steynberg, *Stud. Surf. Sci. Catal.*, 2004, **152**, 1.
34. B. Jager and R. Espinoza, *Catal. Today*, 1995, **23**, 17.
35. A. de Klerk and E. Furimsky, *Catalysis in the Refining of Fischer–Tropsch Syncrude*, Royal Society of Chemistry, Cambridge, 2010.
36. Y. Lin and G. W. Huber, *Energy Environ. Sci.*, 2009, **2**, 68.
37. W. K. Lewis and W. G. Whitman, *Ind. Eng. Chem.*, 1924, **16**, 1215.
38. S. H. Fogler, *Elements of Chemical Reaction Engineering*, Prentice Hall PTR, New Jersey, 1992.
39. C. N. Satterfield, *Mass Transfer in Heterogeneous Catalysis*, MIT Press, Cambridge, 1970.
40. A. Behkish, PhD Thesis, University of Pittsburgh, 2004.
41. W. D. Deckwer, Y. Louisi, A. Zaidi and M. Ralek, *Ind. Eng. Chem. Process Des. Dev.*, 1980, **19**, 699.

42. R. Krishna and J. M. van Baten, *Catal. Today*, 2003, **79/80**, 67.
43. H. Letzel, J. C. Shouten, R. Krishna and C. M. van den Bleek, *Chem. Eng. Sci.*, 1999, **54**, 2237.
44. A. Behkish, Z. Men, J. R. Inga and B. I. Morsi, *Chem. Eng. Sci.*, 2002, **57**, 3307.
45. S. Kumar, K. Kusakabe, K. Raghuathan and L.-S. Fan, *AIChE J.*, 1992, **38**, 733.
46. W. Chen, T. Hasegawa, A. Tsutsumi, K. Otawara and Y. Shigaki, *Chem. Eng. J.*, 2003, **96**, 37.
47. S. C. Saxena, N. S. Rao and A. C. Saxena, *Chem. Eng. Commun.*, 1990, **96**, 31.
48. D. J. Vermeer and R. Krishna, *Ind. Eng. Chem. Process Des. Dev.*, 1981, **20**, 475.
49. W. D. Deckwer, *Bubble Column Reactor*, John Wiley & Sons, Chichester, 1992.
50. P. M. Wilkinson, A. P. Spek and L. L. van Dierendonck, *AIChE J.*, 1992, **38**, 544.
51. N. K. Roy, D. K. Guha and M. N. Rao, *Chem. Eng. Sci.*, 1964, **19**, 215.
52. K. Koide, T. Yasuda and S. Iwamoto, *J. Chem. Eng. Jpn.*, 1983, **16**, 7.
53. W. D. Deckwer, Y. Serpemen, M. Ralek and B. Schmidt, *Ind. Eng. Chem. Process Des. Dev.*, 1982, **21**, 231.
54. L. C. Almeida, O. Sanz, J. D'Olhaberriague, S. Yunes and M. Montes, *Fuel*, 2013, **110**, 171.
55. C. Cao, J. Hu, S. Li, W. Wilcox and Y. Wang, *Catal. Today*, 2009, **140**, 149.
56. S. R. Deshmukh, A. L. Y. Tonkovich, K. T. Jarosch, L. Schrader, S. P. Fitzgerald, D. R. Kilanowski, J. J. Lerou and T. J. Mazanec, *Ind. Eng. Chem. Res.*, 2010, **49**, 10883.
57. R. Lipski, *Smaller-Scale GTL Enters the Mainstream*, Velocys, 2012.
58. J. J. Lerou, A. L. Tonkovich, L. Silva, S. Perry and J. McDaniel, *Chem. Eng. Sci.*, 2010, **65**, 380.
59. https://www.edwardtdodge.com/2014/10/31/velocys-small-scale-gas-to-liquids/.
60. D. Glasser, X. Lu and D. Hildebrandt, *Fixed Bed Reactor*, *US Pat.*, 10751683B2, 2020.

CHAPTER 8

Fischer–Tropsch Catalysts

H. BECKER, K. WEIN AND R. GÜTTEL*

Institute of Chemical Engineering, Ulm University, Albert-Einstein-Allee 11, 89081 Ulm, Germany
*Email: robert.guettel@uni-ulm.de

8.1 Introduction

During the exothermic Fischer–Tropsch (FT) reaction, CO is hydrogenated by means of a solid catalyst forming a broad mixture of aliphatic hydrocarbons and the co-product water, according to eqn (8.1), which represents the stoichiometry for one characteristic chain element in paraffin and olefin formation. In order to utilise CO_2 in hydrocarbon production, a two-step approach is frequently applied, consisting of the reverse water-gas-shift (RWGS) reaction (eqn (8.2)), followed by CO hydrogenation (eqn (8.1)). This leads to the overall stoichiometry provided in eqn (8.3). Obviously, the hydrogenation of CO_2 requires larger amounts of H_2 and also leads to twice the amount of water formed stoichiometrically per carbon atom (compared to CO hydrogenation). In principle, certain aliphatic alcohols are formed as well (eqn (8.4)), depending on the catalyst used. The overall stoichiometry for the conversion of biomass into hydrocarbons is depicted in eqn (8.5) using the example of sugar (fructose) building blocks.

$$n\mathrm{CO} + 2n\mathrm{H_2} \rightarrow (\mathrm{CH_2})_n + n\mathrm{H_2O} \quad \Delta_\mathrm{R}H^0_{\mathrm{C_{20}H_{42}}} = -20 \cdot 154\ \mathrm{kJ\ mol^{-1}} \tag{8.1}$$

$$\mathrm{CO_2} + \mathrm{H_2} \rightleftharpoons \mathrm{CO} + \mathrm{H_2O} \quad \Delta_\mathrm{R}H^0 = 41\ \mathrm{kJ\ mol^{-1}} \tag{8.2}$$

$$n\mathrm{CO_2} + 3n\mathrm{H_2} \rightarrow (\mathrm{CH_2})_n + 2n\mathrm{H_2O} \quad \Delta_\mathrm{R}H^0_{\mathrm{C_{20}H_{42}}} = -20 \cdot 113\ \mathrm{kJ\ mol^{-1}} \tag{8.3}$$

Catalysis Series No. 44
Chemicals and Fuels from Biomass *via* Fischer–Tropsch Synthesis: A Route to Sustainability
Edited by Joshua Gorimbo, Xinying Liu, Yali Yao and Diane Hildebrandt

Published by the Royal Society of Chemistry, www.rsc.org

$$nCO + 2nH_2 \rightarrow H(CH_2)_nOH + (n-1)H_2O \quad \Delta_R H^0_{C_{10}H_{21}OH} = -10 \cdot 147\,kJ\,mol^{-1} \tag{8.4}$$

$$(CH_2O)_n + nH_2 \rightarrow (CH_2)_n + nH_2O \quad \Delta_R H^0_{C_{20}H_{42}} = -20 \cdot 54\,kJ\,mol^{-1} \tag{8.5}$$

The product spectrum of the polymerisation-like FT reaction follows the Anderson–Schulz–Flory (ASF) distribution and can be described by the chain growth probability α. In particular, the molar fraction x_n of a hydrocarbon with a chain length n in the product mixture can be calculated according to eqn (8.6). Förtsch *et al.*[1] introduced an extended model to account for deviations typically observed for the CH_4, C_2 and C_{10+} selectivities.

$$x_n = (1-\alpha)\alpha^{n-1} \tag{8.6}$$

$$\alpha = \left\{1 + k_\alpha \left[\frac{p_{H_2}}{p_{CO}}\right]^\beta \exp\left[\frac{\Delta E_{A,\alpha}}{R}\left(\frac{1}{T_{ref}} - \frac{1}{T}\right)\right]\right\}^{-1} \tag{8.7}$$

Importantly, the chain growth probability depends on the operating conditions, as expressed by eqn (8.7), which was introduced by Vervloet *et al.*,[2] based on experimental data for a Re-promoted Co/Al_2O_3 catalyst. In this equation, p is the partial pressure, T is the temperature, R is the universal gas constant. The apparent activation energy $\Delta E_{A,\alpha}$, as well as k_α and β, are model parameters. Recently, Becker *et al.*[3] did an extensive simulation study and reported changes of the conditions inside the porous catalyst and the reactor volume. This leads to chain growth probability, which depends on the spatial location that affects the product distribution obtained at the reactor outlet. The raw FT syncrude requires further processing and upgrading, since the product spectrum obtained from FT synthesis is not directly compatible with the demands of the subsequent value chain, *i.e.* for utilisation as fuel or as chemical building blocks.

Depending on the source of biomass and the gasification technology used, the raw synthesis gas consists of H_2, CO, CO_2 and CH_4 as main components and may also contain further impurities, such as H_2S, NH_3 and HCl.[4,5] Therefore, deactivation is a major issue to be considered and was recently reviewed comprehensively.[6] Besides carbon formation *via* the Bouduard reaction (eqn (8.8)) or thermal and chemical sintering, the oxidation of the active species at high water partial pressures associated with high conversions plays an important role. In addition, poisoning by impurities is a major issue and requires extensive efforts in synthesis gas cleaning.

$$2CO \rightleftharpoons CO_2 + C_{(s)} \quad \Delta_R H^0 = -172\ kJ\,mol^{-1} \tag{8.8}$$

The synthesis gas composition may also differ from the stoichiometrically favourable values with implications on the choice of suitable catalyst materials and conditioning measures. For instance, the presence of either CO or CO_2 affects the required and formed amount of H_2 and H_2O, respectively.

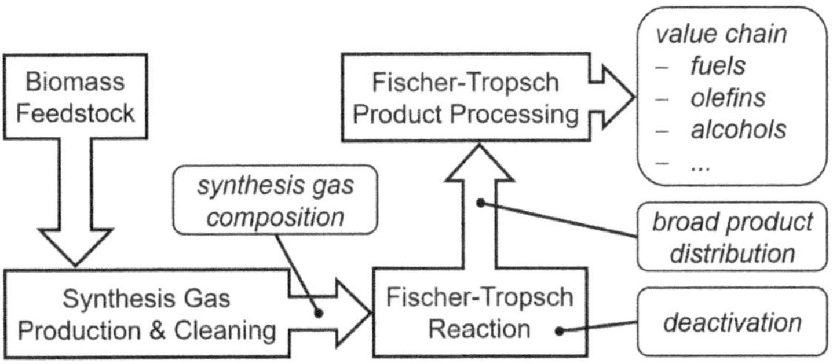

Figure 8.1 Challenges for FT catalysts in biomass-to-liquids (BtL) processes.

Furthermore, CH_4 passes through the FT reactor rather inert, not contributing to the formation of the desired hydrocarbons.

The main resulting challenges for catalyst materials applied in biomass-to-liquids (BtL) *via* the FT process route are illustrated in Figure 8.1. The requirements for catalyst design arising from those challenges are the focus of this chapter. We in particular emphasise approaches in order to improve the robustness and tolerance of the activity and product distribution. As per Güttel and Turek,[7] the impact of active and support material is distinguished from that of the catalyst structure and geometry in this chapter in separate discussions of the micro- and mesoscale, respectively.

8.2 Microscale: Catalyst Materials

8.2.1 Overview and Active Phases

At the microscale, the intrinsic activity, selectivity and stability of FT catalysts are governed by particle size and type of active material, support material and the chemical and structural promotors used. The most frequently applied active materials are Co, Fe and Ru. These active materials are doped by chemical and structural promotors to increase dispersion and reducibility, as well as to inhibit deactivation. The active compound is usually supported on porous substrates, which provide a sufficient surface area for immobilisation of small active metal nanoparticles. An overview of typical support materials, chemical and structural promotors, and the advantages and disadvantages of the most commonly used catalyst materials is provided in Table 8.1. Recently, Shaver *et al.*[8] did an extensive experimental study on various Co-, Fe- and Ru-based catalyst materials and emphasised the effect of promotors and supports on product selectivity.

The relevance of the active materials for biomass-derived syngas depends on their activity, selectivity and stability, as well as on the flexibility towards syngas composition. In particular, the capability of hydrogenating both CO and CO_2 is important, and this also depends on the active phases present under FT conditions. In this regard, cobalt exhibits a low WGS (see eqn (8.2)) activity[9] and is therefore suitable for CO hydrogenation into liquids *via* FT

Table 8.1 Features of catalyst materials typically used in FT synthesis. See comprehensive reviews in ref. 11, 14, 37, 129 and 130 as well.

Applications	Promotors and supports	Advantages/disadvantages
Co-Based catalysts[10,21,22,34]		
Hydrogenation of CO into aliphatic hydrocarbons: long-chain paraffins and olefins	Promotors:[20,131] Re, Ru, Pt, Pd, *etc.* Supports:[21] oxides (Al_2O_3, SiO_2, TiO_2), carbonaceous,[105,132–134] zeolitic,[135] and mesoporous[136] materials, *etc.*	Advantages: high specific reaction rate Disadvantages: no direct hydrogenation of CO_2
Fe-Based catalysts[12,13,23,24,34]		
Hydrogenation of CO and CO_2 into aliphatic hydrocarbons: short-chain paraffins and olefins	Promotors:[23,24] K, Cu, Al, Zn, *etc.* Supports:[24] unsupported, oxides (Al_2O_3, SiO_2, TiO_2), carbonaceous, zeolitic materials, *etc.*	Advantages: low cost Disadvantages: high CO_2 selectivity
Ru-Based catalysts[15,19,137]		
Hydrogenation of CO into aliphatic hydrocarbons and of CO_2 into methane	Promotors:[137] not required, but Mn[27] or Cs,[31] reported Supports:[32] oxides (Al_2O_3, SiO_2, TiO_2),[27–29] carbonaceous,[16,31,32] and zeolitic[30] materials, *etc.*	Advantages: active at moderate temperatures, long lifetime Disadvantages: high cost

reaction, with a preferred stoichiometric H_2/CO ratio of 2. The catalyst precursor species obtained after calcination is cobalt oxide, which requires activation by reduction usually performed under an H_2 atmosphere, to yield metallic cobalt that is postulated to be the active phase.[10,11] Besides the metallic species, oxides and carbides are also present under FT conditions and might be active under FT conditions as well.[10,11] However, cobalt carbides are preferably formed under CO-rich conditions and need to be stabilised,[10] which makes them less likely for probably CO lean, biomass-derived syngas.

In contrast, iron-based catalysts provide the capability for hydrogenation of CO and CO_2, due to their activity with the WGS (eqn (8.2)),[11] which is beneficial for biomass-derived syngas.[12] In particular, a low H_2/CO ratio can be compensated by CO/CO_2 interconversion *via* WGS.[12] It is generally accepted that iron is present under FT conditions in the form of oxides, carbides and the metallic state, but the specific iron phases that cause the catalyst activity are still under debate.[11,13] During activation, the iron oxide precursors undergo reduction towards metallic iron, which is subsequently carburised in the presence of CO.[13] Importantly, an induction phase is observed with iron-based catalysts under FT conditions, which probably originates from carburisation. De Smit and Weckhuysen assign this to the lower activation energy required for carbon diffusion in Fe (43.9–69.0 kJ mol^{-1}) compared to that of the FT reaction (89.1 ± 3.8 kJ mol^{-1}), which causes the rate for carbon diffusion to be higher than that for the FT reaction.[13] Therefore, iron-based catalysts require a specific activation procedure under syngas after reduction of the iron oxide precursors.

Ruthenium-based catalysts are considered most active for FT synthesis and provide a high chain growth probability that exceeds that even of iron and cobalt, and they also exhibit improved stability.[11,14] Despite these advantages, Ru is studied far less, probably due to the high costs and thus a lower relevance for industrial applications. It has to be emphasised that Ru is reported to be active for CO hydrogenation towards liquid hydrocarbons as well as for CO_2 methanation,[15] which gives rise to a certain amount of flexibility regarding the CO/CO_2 content in the syngas mixture. The active phase is metallic Ru, which is derived by reduction of the RuO_2 precursor with H_2, and it remains stable under FT conditions.[11] In contrast to Fe and Co, the application of Ru-based catalysts in aqueous and organic phases, as well as in biphasic systems, has been reported.[16] In particular, the biphasic approach was shown to combine the advantages of the single-phase operation, namely high activity and chain growth probability. Therefore, the hydrophobic catalyst was localised in the organic phase. Despite the high price, the unique properties and advantages render Ru an attractive candidate for the conversion of biomass-derived syngas.

8.2.2 Activity and Promotors

The intrinsic rate of CO conversion in FT reaction depends on the turnover frequency (TOF) and the number of available active sites. To minimise the

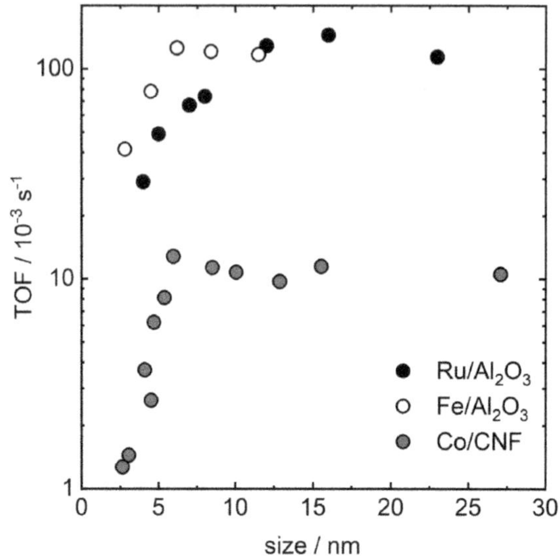

Figure 8.2 Particle size effect on activity for Co/CNF (220 °C, $H_2/CO = 2$, 1 bar; data from Bezemer *et al.*[40]), Fe/Al_2O_3 (280 °C, $H_2/CO = 2$, 9.8 bar; data from Park *et al.*[39]) and Ru/Al_2O_3 (250 °C, $H_2/CO = 10$, 1.85 bar; data from Carballo *et al.*[38]). (The TOF values depend on operating conditions and are thus not directly comparable.)

amount of active material required for a certain CO conversion, the number of active sites must be increased. Assuming a constant density of active sites at the surface of the metal nanoparticles, this can be realised by reducing the particle size and thereby increasing the specific metal surface area. For structure insensitive reactions, the TOF is expected to be independent of particle size;[17] therefore a minimal particle size is desired. For the FT reaction with Co-, Fe- and Ru-based catalysts, however, the TOF depends on the size of the active metal nanoparticles, as illustrated in Figure 8.2. Although the operating conditions and support materials are not directly comparable, a similar trend becomes apparent with the most relevant active metals: below *ca.* 10 nm, the TOF increases with size, while it is constant above that value. Consequently, the FT reaction can be considered structure sensitive for small metal nanoparticles, while it is insensitive for larger ones, which agrees with the analyses done by Liu *et al.*,[18] and Ma and Dalai.[19]

Various noble metal promotors are reported to improve reducibility and thereby the activity with cobalt-based catalysts. The most intensively studied are alumina-, silica- or titania-supported materials promoted by Pt, Re or Ru with a mass fraction of less than *ca.* 1 wt% in most cases (see Table 8.1). Ma *et al.*[20] compared an unpromoted $Co/\gamma\text{-}Al_2O_3$ catalyst with its counterpart promoted with Pd, Pt, Re and Ru. The authors reported an increase in the degree of reduction, which they attributed to improved H_2 dissociation facilitated by the noble metals and electronic effects induced by alloying. Furthermore, the Co dispersion was found to decrease with the type of promotor in the order Re > Ru > Pt > unpromoted. Given the manifold studies on the effect of the support, the reader is referred to the recent comprehensive reviews by Munirathinam *et al.*[21] and Gholami *et al.*[22] Munirathinam *et al.* point out that the metal–support interactions play a pivotal role in the dispersion and reducibility of cobalt, which affects the activity observed. In particular, oxide supports exhibit medium to strong interactions, while carbonaceous materials or SiC provide less pronounced interactions with cobalt. Textural properties, such as pore size distribution, are also important. For instance, mesoporous SiO_2 materials lead to the formation of smaller cobalt nanoparticles with a narrow distribution range and therefore affect the activity. This is discussed in Section 8.3.

Iron does not necessarily need to be supported on porous substrates, but is usually promoted by K, Mn, Cu, Al, Si, Zn or other elements (see Table 8.1). In terms of unsupported, precipitated iron catalysts, Chun *et al.*[23] recently summarised the promotor effects comprehensively, as follows. The authors point out that Cu is considered the best promotor for facilitating the reduction of Fe_2O_3 *via* Fe_3O_4 into metallic Fe. In a first step, copper oxide is reduced to metallic copper, which enables H_2 spillover to the iron oxides and thereby their reduction. At the same time, alkali promotors, K in particular, are used to improve selectivity towards the desired products but also to facilitate carburisation during the activation phase. Furthermore, structural promotors, such as SiO_2 and Al_2O_3, are employed to improve catalyst stability in terms of thermal sintering and attrition. Importantly, the authors state that an optimal content of each promotor can be used to exploit the maximum benefit, while the impact of the promotors affect each other, which gives rise to a complex interaction, as illustrated in Figure 8.3.

Torres Galvis and De Jong[24] point out that the role of the support with iron-based catalysts is rather complex. On the one hand, strong interactions between iron and oxidic support materials, such as Al_2O_3 or SiO_2, lead to the formation of mixed iron oxides, which are inactive in the FT reaction. On the other hand, supports that provide weak interactions facilitate the activation of iron, but at the same time these materials are less stable against loss of active surface under reaction conditions. Similar to cobalt, the pore structure of the

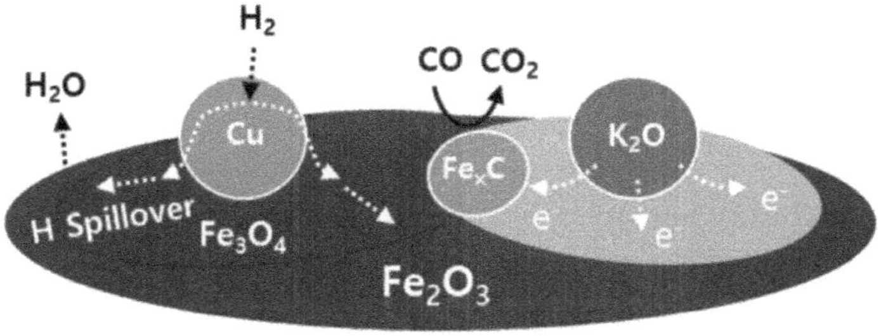

Figure 8.3 Scheme illustrating the effects of reduction and alkali promotors on the activation of precipitated iron catalysts. Reproduced from ref. 23 with permission from Springer Nature, Copyright 2020.

support affects the activity of iron-based catalysts as well by confining the size of the iron carbide particles formed under reaction conditions.[25] Supported iron catalysts are promoted by the same compounds reported for their unsupported counterparts. Recently, the effect of various alkali promotors – such as Rb, Cs and K – has been studied systematically by Shafer *et al.*[8]

Ruthenium-based catalysts are either used as unsupported nanoparticles[26] or supported on oxides (*e.g.* SiO_2, Al_2O_3, TiO_2),[27–29] zeolites[30] and carbonaceous[16,31,32] materials. In some recent studies, promotors, such as Mn[27] or Cs,[31] were compared directly to unpromoted but supported ruthenium. Surprisingly, the results revealed that the compounds even have a negative impact on the activity. However, due to a lack of comprehensive studies, general conclusions cannot be drawn yet.

8.2.3 Selectivity

In principle, the product selectivity of the FT reaction follows the ASF distribution (see eqn (8.6)), irrespective of the active metal being Co, Fe or Ru. However, the chain growth probability depends on the active material, the support material and the employed promotors, as well as the operating conditions. The effects of the support materials and promotors on the selectivity are manifold and are briefly introduced in terms of the industrially relevant Fe- and Co-based catalysts. As a general trend, the product spectrum of cobalt catalysts is characterised by aliphatic paraffins with marginal olefin formation and a rather high chain growth probability, while iron catalysts allow a broader spectrum of products to be produced, which also contain olefins and oxygenates to a greater extent (Table 8.1).[12,33]

This gives rise to the use of iron-based FT synthesis in order to substitute steam cracking as a major source for lower olefin production in the C_2–C_4 range *via* the FT-to-olefin (FTO) process.[24] In order to tailor the product spectrum and to favour long-chain hydrocarbon formation,[34] and thereby suppress unwanted CH_4, K or Na are applied as promotors. The addition of S to K- or Na-promoted Fe catalysts supported on SiO_2 or SiC is reported to increase the selectivity in C_2–C_4 olefins.[35] Higher alcohols represent another class of attractive products accessible *via* CO hydrogenation using Fe-based catalysts, as shown by Luk *et al.*[36] with zeolite-supported FeCu compounds. In contrast, for cobalt-based catalysts, high C_{5+} and low CH_4 selectivity are desired. However, the impact of noble metal promotors and support materials on the product distribution has not been clearly revealed yet. While a certain effect was reported in the review done by Gholami *et al.*,[22] the limited comparability of the catalysts mentioned has to be considered. Nevertheless, some transition metal oxides (such as ZrO_2 and MnO_x) are reported to increase C_{5+} selectivity and the olefin-to-paraffin ratio, probably due to the lowered hydrogenation ability.[37]

Apart from the effect of the chemical composition of the catalyst materials, an effect caused by the active metal particle size has also been reported. With Ru/Al_2O_3 catalysts, CH_4 selectivity appears to be rather constant at *ca.* 75% (calculated by relating the TOF for CH_4 formation to that of

CO conversion) for particle sizes of 4 to 23 nm.[38] In contrast, Fe- and Co-based catalysts exhibit a decline in CH_4 selectivity with an increase in metal particle size before reaching constant values for *ca.* 10 nm and above.[39,40] Accordingly, C_{5+} selectivity also increases with particle size and reaches constant values above 10 nm. In combination with the effect of size on activity, the optimal active metal nanoparticle size is in the order of 10 nm or slightly below, since it combines the highest activity and selectivity towards higher hydrocarbons with a large specific active surface area.

The main challenges with FT selectivity are the undesired formation of CO_2 and CH_4, as well as tailoring the product spectrum. CO_2 formation is mainly caused by the Bouduard (eqn (8.8)) and water-gas-shift (WGS) (eqn (8.2)) reactions, depending on the presence of water and the respective catalyst activity. CH_4 selectivity can be minimised to below 10% with the appropriate choice of reaction conditions, active metal particle size and catalyst material composition, which is usually associated with an increase in chain growth probability, and thus results in the formation of long-chain hydrocarbons. Therefore, tailoring of the product spectrum requires an additional degree of freedom in catalyst design, which is provided by multifunctional catalysts. These materials consist of the FT active compound supported on a solid acid material, such as various zeolites,[12,41,42] which allows tuning of the formation and cracking/isomerisation rates of hydrocarbons separately from each other. If the metal nanoparticle is completely encapsulated within the microporous zeolite shell, even shape selectivity can be exploited to narrow the product distribution by impeding large molecules during diffusion through the pore system. Hence, this approach provides the ability to break the ASF distribution, and it can be applied to different scales when designing FT catalysts and reactors.[7,43] The concept was pioneered by Tsubaki *et al.*[44] and subsequently applied to iron[45–47] and cobalt[48–50] catalysts by various research groups. Figure 8.4 shows the impact of the microporous zeolite support on the carbon number distribution in comparison to Co/SiO_2 catalysts.[48] Hierarchical zeolites that provide additional mesopores are reported to provide an even higher CO conversion and lower C_{12+} selectivity than microporous zeolites.[49,51]

8.2.4 Stability and Deactivation

Among the deactivation mechanisms summarised by Bartholomew,[52] particle growth, carburisation, reoxidation and poisoning are the most frequently discussed in terms of FT synthesis using Fe[13] and Co[6,53] catalysts (see Figure 8.5). *Operando* characterisations are powerful tools for understanding the causes of deactivation and the relationship between catalyst structure and performance during operation, as recently reviewed by Fischer and Claeys.[54] For instance, in a recent *operando* study based on X-ray absorption spectroscopy (XAS) and diffraction (XRD), three activity regimes were revealed for a commercial Co–Ni–Re/γ-Al_2O_3 catalyst.[55] The experiments were performed under industrially relevant conditions, *i.e.* at 250 °C

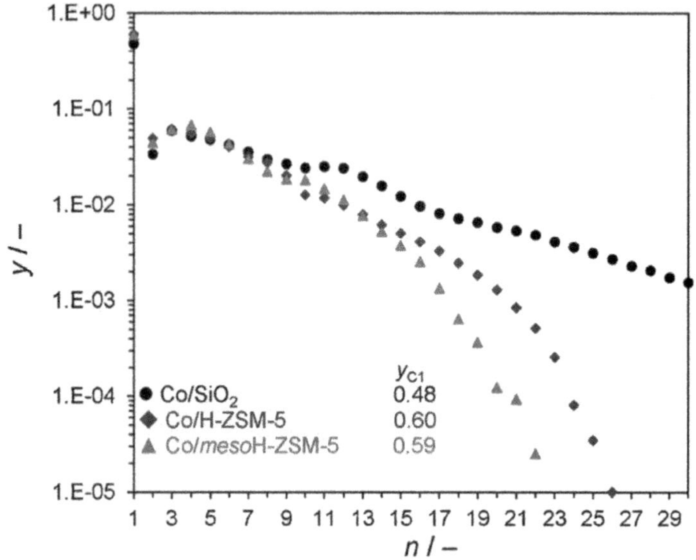

Figure 8.4 Carbon number distribution for conventional and bifunctional cobalt catalysts during FT synthesis (513 K, 15 bar, $H_2/CO = 2$). Reproduced from ref. 48 with permission from Elsevier, Copyright 2013.

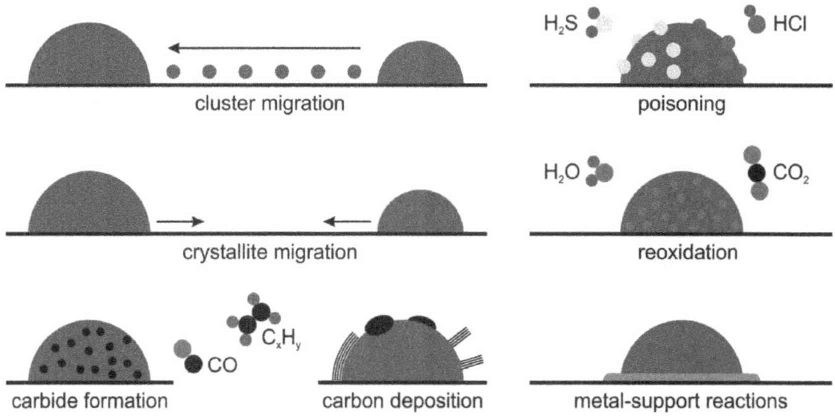

Figure 8.5 Mechanisms of catalyst deactivation during conversion of biomass-derived syngas into hydrocarbons *via* the FT synthesis.

and 30 bar for 310 h on stream. However, the authors point out that the formation of carbon depositions on the Co surface and of $CoAl_2O_4$ species by solid-state reactions are responsible for apparent activity loss being superimposed by pore filling with liquid hydrocarbons.

The particle growth causes a loss of active surface and is induced by chemical and thermal sintering *via* cluster (Ostwald ripening) or crystallite migration and coalescence.[6] While the former might be facilitated by subcarbonyl formation in the presence of CO,[56] the latter becomes more pronounced at higher temperatures that exceed the Hüttig temperature (30% of the absolute melting point). Since the melting points differ among the oxidic, carbidic and metallic species of the active metal, the relevance of thermal sintering depends on the state of the catalyst during operation. Furthermore, the melting point of nanoparticles decreases significantly below that of bulk material;[57] therefore thermal sintering becomes more important. In order to prevent sintering, the choice of the synthesis concept, the support material and promotors, and the activation procedure is very important.[58]

The research group of De Jong extensively studied the impact of synthesis and activation procedure for Co-based catalysts. In particular, they developed the melt impregnation method,[59] elaborated the importance of the drying conditions after impregnation with a nitrate precursor solution[60] and studied the impact of the calcination atmosphere.[61] Lower temperatures required for reduction and calcination also affect the solid-state reactions of the active material with the support, which are responsible for the formation of inactive mixed species and strong metal–support interactions. Sintering of active nanoparticles can also be prevented by encapsulating the nanoparticles inside a microporous matrix, as recently revealed by *in situ* electron microscopy studies done by Straß-Eifert *et al.*[62] Therefore, the authors compared Co nanoparticles at the external surface of zeolite crystals with those encapsulated and monitored their evolution for 1 h at 700 °C in a syngas atmosphere.

The high water partial pressure associated with high conversion is known to induce reoxidation of the active material, which facilitates the formation of inactive metal–support compounds, as studied comprehensively by Wolf *et al.* for Co catalysts.[63,64] The authors deposited Co nanoparticles on alumina, silica and titania and exposed the materials to different H_2/H_2O ratios at 220 °C, in order to simulate high conversion. The observed reoxidation is even accelerated in the presence of CO.[65] Similar observations were reported by Tsakoumis *et al.* for CoRe/γ-Al_2O_3 catalysts studied *in situ* under typical FT conditions.[66] Apart from deactivation, the cobalt oxidation also leads to an increase in CH_4 selectivity.[67] While sintering of the Co nanoparticles occurs in the early stage of FT operations, carbon deposition and surface oxidation become the main deactivation modes with a longer time on stream (TOS).[68]

The deactivation by carbon deposit formation under FT conditions has been exemplified for Fe-based catalysts. Interestingly, carbonaceous iron species formed by carburisation during activation are proposed as being active, but this is still under debate.[11,13] Several mechanisms for deactivation associated with the iron carbide species are discussed in the literature according to De Smit and Weckhuysen.[13] For instance, active iron carbide can be oxidised towards inactive iron oxides or interconverted into an inactive carbide. Furthermore, carbon deposits can be formed that block the active surface. This process seems to be facilitated by basic promotors, such as Na_2O or K_2O. Pendyala *et al.*[69] compared the FT performance of unpromoted and K-promoted Fe catalysts in a continuous slurry stirred tank reactor and observed similar deactivation rates for both during *ca.* 300 h on stream. Mössbauer analyses performed *ex situ* revealed that the unpromoted catalyst suffers from reoxidation, while deactivation of the promoted counterparts originates from a carbon layer formation at the catalyst surface.[69]

8.3 Mesoscale: Pore Structure and Catalyst Geometry

As fixed-bed reactors are most likely to be utilised for BtL processes, gradients in concentration and temperature fields inevitably occur. These can have a significant influence on the overall reaction rate and selectivity[3,70] and also depend heavily on the operating conditions, especially the H_2/CO syngas ratio but also the temperature. High syngas ratios favour a high reaction rate but lower the average chain length of the products significantly. In contrast, at low syngas ratios, the selectivity improves, yielding mainly long-chained hydrocarbons, whereas the reaction rate is hampered significantly.[71] Similarly, the temperature has a positive effect on the reaction rate, but when the temperature is too high, the selectivity drops sharply towards products with a low carbon number.[2,70,72] Consequently, it is crucial to limit gradients of concentration and temperature to an extent that is manageable for optimal operations. Temperature gradients can be influenced by the catalyst's size and shape, but the size of the reactor is often much more relevant;[73] therefore it is not considered in this chapter, which focuses on the catalyst. The catalyst's size and shape affect the internal gradients within the pores significantly due to diffusion limitations. Hence, structuring the pore system can help to improve its accessibility and thereby the performance of the catalyst.[74]

8.3.1 Transport Effects

The typical stoichiometric syngas ratio chosen – about 2 – avoids gradients of this ratio over the axial direction of the reactor, which is governed by convection.[75] However, as larger catalyst pellets or catalyst layers are often desired, an inclination for a rather large catalyst diffusion length is prevalent, due to allowing a higher catalyst mass in the reactor or reducing the pressure drop.[76] This means that diffusion limitations occur, and since diffusion coefficients are different for hydrogen and carbon monoxide – typically by a factor of about 2.5 – hydrogen is transported faster than it is consumed in comparison to CO.[77] In turn, this leads to an increase in the syngas ratio when diffusion limitations occur,[2,78] which has consequential effects on rate and selectivity.

Overall, diffusion limitations can be seen as a key design challenge, and papers that deal with reducing their influence by modifying the pore structure of the catalyst to improve the accessibility for the reactants are seen in a large number of publications.[79–89] Nonetheless, attempting to avoid concentration gradients entirely is not only futile but also not sensible, as some performance gain is observed with moderate diffusion limitations.[72] So the goal cannot be to just improve the diffusion within the catalyst:

it should also be to obtain an accurately defined quantity of ideally oriented pores that allow sufficient access to a large number of active sites.

8.3.2 Pore Structure

As explained, a well-performing catalyst requires a large number of active metal nanoparticles of the right size in a volume as small as possible, while still being sufficiently accessible. To obtain such a catalyst, the porous support material must fulfil two roles. First, it needs to stabilise the active metal by immobilising the nanoparticles within its pores and preventing them from agglomeration and sintering. Secondly, the pores themselves provide a volume in which the reactants and products can be transported through diffusion. For stabilisation of the catalyst in an optimum size range of about 5 to 10 nm, the catalyst pore size needs to be in the same range, as a correlation is often found between pore size and the size of the active metal nanoparticles.[90–93] Hence, the catalysts should ideally have pores exclusively this size and be available in a large total volume. With this as the objective, different catalyst support types have been used for the preparation of catalysts and have been characterised and tested.

In this chapter, three basic support types are considered, although other categorisations are possible. An illustration of the general shape of these support materials is given in Figure 8.6. The first type (Figure 8.6a) comprises mostly commercially available supports, which may be synthesised *via* precipitation, pyrolysis or spray drying.[90,94] The porous structure of these supports is relatively unordered, as it is a result of a random process. The second type of support material (Figure 8.6b) features a highly ordered pore structure with uniform and evenly spaced cylindrical pores that were generated using a template. The last of the basic support types (Figure 8.6c) is based on a collection of fibres that allow for large void spaces between the individual fibres, which may or may not be connected to each other.

The basic shape of the pore structure of these materials is illustrated in Figure 8.6, and several implications can be drawn regarding catalysts

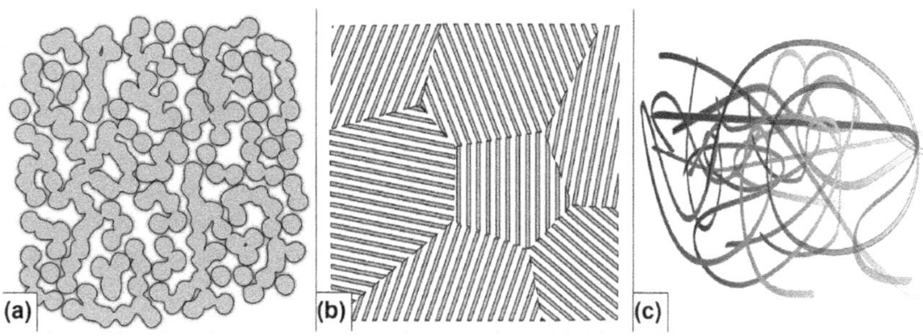

Figure 8.6 Schematic illustration of different types of general support pore structures: (a) conventional support with random pore structure; (b) highly ordered cylindrical pore structure; (c) agglomerates of fibres.

obtained from them. The conventional support, with its random pore structure features pores of nonuniform size, results in the active metal nanoparticles also exhibiting a nonuniform size distribution. This could cause some active particles to be too small for an effective TOF, whereas larger ones would exhibit a low dispersion, thereby lowering the efficiency of the utilisation of the active metal. Nevertheless, the random nature of the structure also results in interconnected pores, which are less prone to blocking. However, this blocking can be seen as a drawback for the ordered porous supports,[95] as a single pore might even get blocked by a single metal nanoparticle upon loading with active metal, making a considerable part of that pore inaccessible. Nonetheless, the uniform pores of the ordered supports can be utilised to yield and stabilise active metal nanoparticles with a better defined size.[90,96] In contrast to this very defined yet restricted pore structure, fibres offer a highly accessible pore space but without volume to confine and stabilise the active metal nanoparticles. This is also visible in Figure 8.7a, which depicts the total pore volume over the average pore size for various support types and materials.

In terms of carbon fibres, very few articles report an actual average pore size,[97,98] albeit many have reported on total pore volume.[40,99–103] This may indicate a lack of suitable mesopores, which can stabilise nanoparticles by confinement. However, the hollow core of the carbon nanotubes (CNTs) can provide mesopore volume, but this is only accessible when the CNTs are functionalised, which introduces defects and opens the tubes.[97,98] In contrast, fibres based on alumina are not hollow, but they provide appropriately sized mesopores[80,104] with a total volume that is much higher than that of their conventional support type counterparts. These conventional alumina-based supports can have a pore size ranging from about 5 to 50 nm and a pore volume of 0.25 to just above 1 $mL\,g^{-1}$. This seems low in comparison to silica, which exhibits a similar pore size, but a pore volume of about 0.7 to 2.4 $mL\,g^{-1}$, when the conventional support type is considered. The ordered silicas extend the pore size range to the lowest observed values down to about 2 nm and also exhibit the highest and lowest pore volumes for silica, *i.e.* 2.5 and 0.4 $mL\,g^{-1}$, respectively. Similarly, the ordered carbon support material exhibits a pore size of 3 to 8 nm and a pore volume of 0.7 to 1.4 $mL\,g^{-1}$.

Since all sources, included in this evaluation used ordered mesoporous silica as a template for the ordered carbon materials, a certain correlation between the two materials is likely. This could also mean that these two materials are the most promising for obtaining the best performing catalysts, as they offer the highest pore volume in the pore size range of 5–10 nm. However, as the pore volume is based on the support mass, the quantity alone does not suffice to assess how much pore volume can be integrated into a reactor with a given support material, as its volume also contributes to the effective volume of the support. Since the density of alumina is about twice as high as that of either silica or carbon, the pore volume of alumina might be comparable to that of silica and carbon and still close to the ideal

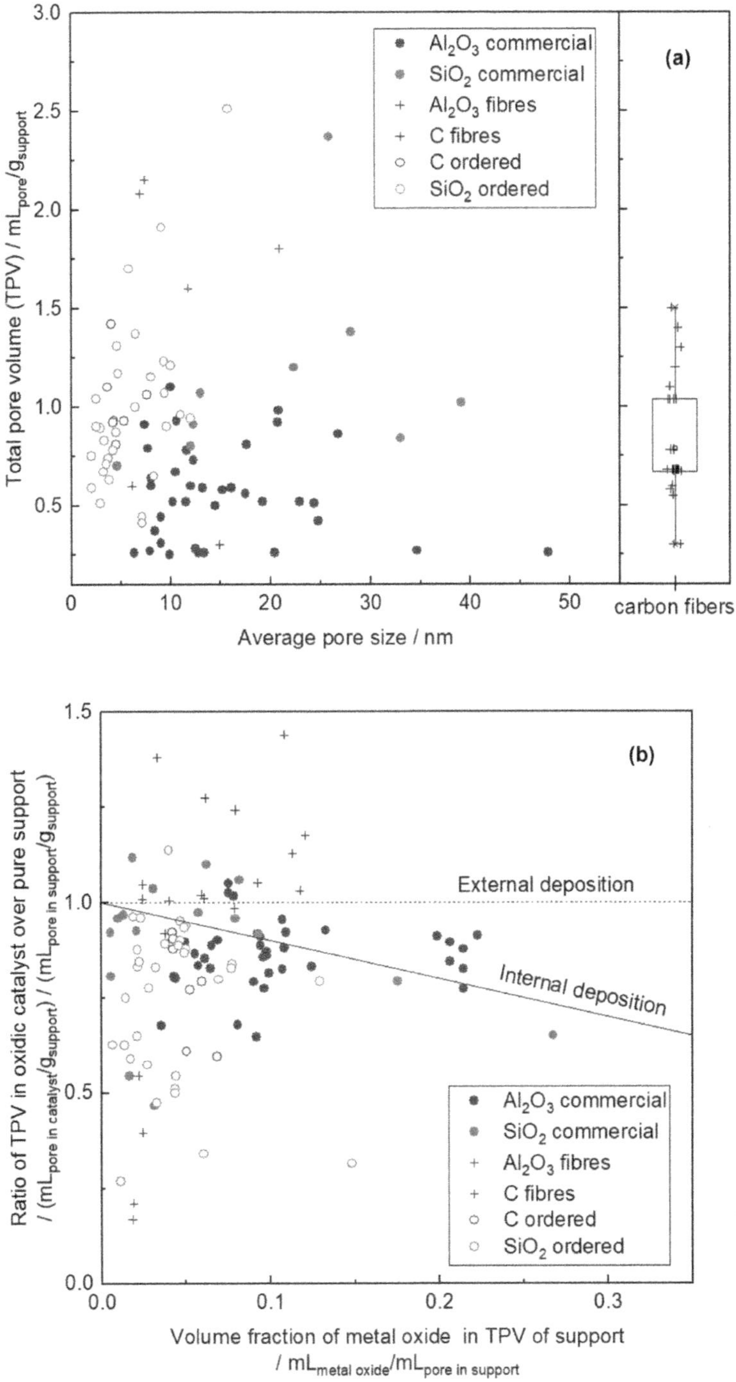

Figure 8.7 (a) Total pore volume over average pore size for various support materials and structures; (b) fraction of pore volume after impregnation and calcination as a function of volume fraction of active metal oxide. Data evaluated from various sources.[40,59,80,90–93,95–104,117–127]

pore size of about 8 nm. Hence a closer look at the characteristics of the catalysts made from the different materials is required.

Although various methods for catalyst preparation exist,[105] most catalysts reported on in this investigation use an impregnation method. During preparation, the active metal or its oxide can be deposited on the catalyst support, either within the pores or on the outer surface of the support. While deposition inside the pores is favourable – due to stabilisation of the nanoparticles – it also reduces the available pore volume. Deposition on the outer surface only provides a marginal active surface area but does not reduce the volume. Figure 8.7b shows the remaining fraction of pores after impregnation plotted over the calculated volume fraction of metal oxide within the pores. When only external deposition occurs, the pore volume remains constant at unity, as illustrated by the horizontal line. In the case of internal deposition, the remaining fraction of pores is reduced by the volume of the active metal oxide. Hence, a linear drop from one to zero would be expected in that case – also illustrated by a line. Ideally, all points ought to be between these two lines, but this is only approximately true for conventional alumina and silica supports. When the actual value is below the line for internal deposition, this indicates pore blocking, which is largely the case for silica- and carbon-based ordered supports. The fibrous materials can be found either quite close to the external deposition line, as would be expected, or well above (carbon) or below (alumina) the two lines. For the carbon supports, this means that, upon preparation, the quantifiable pore volume increases, which would contribute to better diffusive accessibility. This increase might be due to a rearrangement of the fibres, and thus the pore system, which is possibly caused by the surface tension of the liquid used for impregnation and by the decomposition of the active metal precursor. Such a rearrangement, but with an opposite tendency, is also likely to have occurred with the alumina fibres, which exhibit an excessive drop in pore volume, despite the low fraction of metal oxide deposited.

Although investigating the success of metal loading of various supports is vital for choosing the right support material, eventually, the efficacy of a catalyst is assessed by the volume-specific surface area of the active metal, which largely defines the observed rate, given that this active surface area is effectively accessible. This accessibility is strongly correlated to the porosity; hence a comparison of porosity and active surface area for various supports can aid in finding a suitable catalyst. Figure 8.8 depicts the active surface area of cobalt catalysts over catalyst porosity, assuming spherical cobalt nanoparticles. The cobalt nanoparticle size is either based on XRD measurements or obtained from hydrogen chemisorption, with the latter method being affected by blocked pores, while the former is not.

While the XRD results suggest that a high surface area is obtainable when less porosity is present – either because more metal is loaded into the support or small pores, with a corresponding lower pore volume being utilised – this is not the case in the results obtained from chemisorption. Here, only medium porosities achieve the highest surface areas, and this is especially visible with

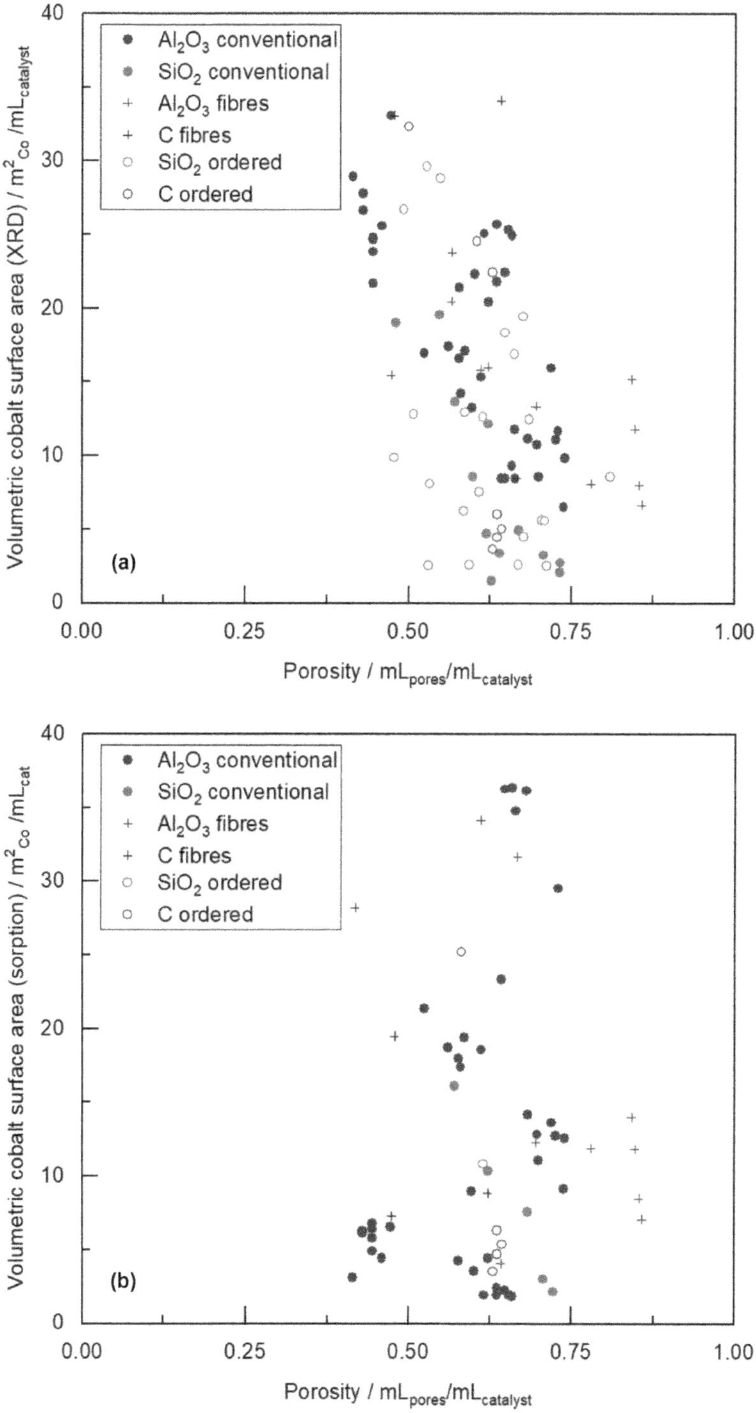

Figure 8.8 Volumetric surface area of cobalt (a) determined from XRD or (b) *via* chemisorption as a function of calculated porosity of the reduced catalyst. Data evaluated from various sources.[80,90–93,96–98,101,103,104,117–122,124–126,128]

alumina. For low cobalt loadings, a high porosity can be obtained but without a high surface area; for a lower porosity, the surface area is also lowered, which can be attributed to restricted access to the cobalt sites. This indicates that a trade-off between porosity and activity is required when effective utilisation of the active surface area is the objective. Although other factors also need to be accounted for, and the chemisorption results are merely an indication of the imposed mass transport restrictions during synthesis, structuring the pores still seems to be a viable option to improve the catalysts.

With fixed-bed reactors, larger catalyst pellets allow the lowering of the pressure drop,[106] while with monolithic[86] channels, foams[88] and microchannels,[84] a thicker catalyst coating typically increases the catalyst holdup. In both cases, an increase in the diffusion length would improve the reactor economics. But to avoid diffusion limitations when the diffusion length is increased, the catalyst pore system needs to be structured, which can be achieved *via* several routes. The catalyst preparation can start with nanoparticle synthesis of the active metal, and this core can subsequently be encapsulated to yield a core–shell catalyst.[107,108] With this method, the correlation between pore size and active metal particle size is nullified, and the pore volume can be tuned independently of the metal loading. Thus high metal loading without pore blockage can be achieved, together with sufficiently small pores to stabilise the active metal.[109] Another possibility is to organise the nanofibres into aligned arrays and use this as the support. This offers advantages due to the orientation of the pores towards the fluid phase, and it can improve the mechanical stability of the support, as the fibres are attached to a surface. Despite several examples for the synthesis of these structures made of carbon[110] or titania[111] having been reported, no case was found in which these structures were used as catalyst support for FT synthesis. An inverted structure of the fibres is available from anodic oxidation of titanium[112] or aluminium,[113] which yields layers with fairly large, aligned pores and which can be altered by hydrothermal treatment.[114]

8.3.3 Hierarchical Structure

Aside from direct structuring of the support pores, hierarchical networks can also be used to employ smaller mesopores in stabilising the active metal and larger transport pores. By doing so, the inhibiting effect of small pores on the mass transport is also reduced. Simulations done on generic catalysts have shown the importance of finding optimal values for volume fraction and size of the transport pores, compared to diffusion inside the mesopores to maximise the yield of a catalyst pellet.[74,115,116] This optimisation is also required for FT catalysts, as introducing transport pores is only viable when the diffusion length is large enough to impose restrictions and when transport pores provide an adequate diffusivity, since only then is the removal of the active phase compensated for by increased efficiency.[72] However, practical examples often lack this optimisation, as it is challenging to generate porous supports with a freely adjustable diffusion length and transport pore volume fraction.

By impregnating a conventional silica material with high porosity and large pores, with a sol of titania or silica, a bimodal porous support and catalyst were prepared, which exhibited better performance than a catalyst with only either transport pores or mesopores.[82,83] However, whether this advantage was caused by a better suited cobalt surface area for the bimodal catalyst or by improved mass transport could not be demonstrated. Another catalyst tested in the FT reaction used hollow mesoporous silica spheres as a support, which proved superior to a catalyst based on regular ordered mesoporous silica.[79] The catalyst pellet size was not reported, so the difference cannot be attributed to improved mass transport. Another attempt at employing catalyst layers of different thickness and with and without transport pores could not show a significant improvement, probably because of the irregularly shaped transport pores.[85] An investigation done using catalyst pellets of different sizes could at least show that when transport pores are present, the selectivity improves when the diffusion-restricted large pellets are used, while no difference could be found for small pellets.[81]

8.4 Summary and Conclusions

The BtL application has demanding requirements in terms of the design of FT catalysts, which need to be active, highly selective towards the desired product fraction and stable for long-term operation. Furthermore, the syngas feedstock may not provide the ideal composition and contains CO_2, CO, H_2 and impurities. Finally, the catalyst design should also provide the ability to implement compact reactors in remote production units.

The chapter focused on the most promising active compounds for FT synthesis and summarised typical promotor and support materials. The impact of the catalyst properties on activity and selectivity was discussed, together with aspects related to stability. The link between the active compound and reactor design was provided by elaborating on the impact that pore structure and geometry have on catalyst performance. Hence, the chapter provided an overview of the main degrees of freedom to be considered when designing FT catalysts for particular BtL applications, as well as of the most promising parameters to be chosen.

Abbreviations

ASF	Anderson–Schulz–Flory
BtL	Biomass-to-liquids
CNTs	Carbon nanotubes
FT	Fischer–Tropsch
RWGS	Reverse water-gas-shift
TOF	Turnover frequency
TOS	Time on stream
WGS	Water-gas-shift
XAS	X-ray absorption spectroscopy
XRD	X-ray diffraction

References

1. D. Förtsch, K. Pabst and E. Groß-Hardt, *Chem. Eng. Sci.*, 2015, **138**(3), 333.
2. D. Vervloet, F. Kapteijn, J. Nijenhuis and J. R. van Ommen, *Catal.: Sci. Technol.*, 2012, **2**(6), 1221.
3. H. Becker, R. Güttel and T. Turek, *Catal.: Sci. Technol.*, 2019, **9**(9), 2180.
4. G. Schaub and R. Edzang, *Chem. Ing. Technol.*, 2011, **83**(11), 1912.
5. R. Zennaro, M. Ricci, L. Bua, C. Querci, L. Carnelli and A. d'Arminio Monforte, Syngas: The Basis of Fischer-Tropsch, in *Greener Fischer-Tropsch Processes for Fuels and Feedstocks*, ed. P. M. Maitlis and A. De Klerk, Wiley-VCH Verlag GmbH & Co. KGaA, Weinheim, Germany, 2013, pp. 17–51.
6. M. Rahmati, M.-S. Safdari, T. H. Fletcher, M. D. Argyle and C. H. Bartholomew, *Chem. Rev.*, 2020, **120**(10), 4455.
7. R. Güttel and T. Turek, *Energy Technol.*, 2016, **4**(1), 44.
8. W. Shafer, M. Gnanamani, U. Graham, J. Yang, C. Masuku, G. Jacobs and B. Davis, *Catalysts*, 2019, **9**(3), 259.
9. G. P. van der Laan and A. A. C. M. Beenackers, *Catal. Rev.*, 1999, **41**(3–4), 255.
10. I. C. ten Have and B. M. Weckhuysen, *Chem. Catal*, 2021, **154**, 162.
11. A. V. Puga, *Catal.: Sci. Technol.*, 2018, **8**(22), 5681.
12. S. Abelló and D. Montané, *ChemSusChem*, 2011, **4**(11), 1538.
13. E. de Smit and B. M. Weckhuysen, *Chem. Soc. Rev.*, 2008, **37**(12), 2758.
14. H. Jahangiri, J. Bennett, P. Mahjoubi, K. Wilson and S. Gu, *Catal.: Sci. Technol.*, 2014, **4**(8), 2210.
15. M. R. Axet and K. Philippot, *Chem. Rev.*, 2020, **120**(2), 1085.
16. V. V. Ordomsky, A. Y. Khodakov, B. Legras and C. Lancelot, *Catal.: Sci. Technol.*, 2014, **4**(9), 2896.
17. R. A. van Santen, *Acc. Chem. Res.*, 2009, **42**(1), 57.
18. J.-X. Liu, P. Wang, W. Xu and E. J. Hensen, *Engineering*, 2017, **3**(4), 467.
19. W. Ma and A. K. Dalai, *Reactions*, 2021, **2**(1), 62.
20. W. Ma, G. Jacobs, R. A. Keogh, D. B. Bukur and B. H. Davis, *Appl. Catal., A*, 2012, **437–438**, 1.
21. R. Munirathinam, D. Pham Minh and A. Nzihou, *Ind. Eng. Chem. Res.*, 2018, **57**(48), 16137.
22. Z. Gholami, Z. Tišler and V. Rubáš, *Catal. Rev.*, 2020, **134**(8), 1.
23. D. H. Chun, G. B. Rhim, M. H. Youn, D. Deviana, J. E. Lee, J. C. Park and H. Jeong, *Top. Catal.*, 2020, **63**(9–10), 793.
24. H. M. Torres Galvis and K. P. de Jong, *ACS Catal.*, 2013, **3**(9), 2130.
25. E. C. Ra, K. Y. Kim, E. H. Kim, H. Lee, K. An and J. S. Lee, *ACS Catal.*, 2020, **10**(19), 11318.
26. C. Xiao, Z. Cai, T. Wang, Y. Kou and N. Yan, *Angew. Chem., Int. Ed.*, 2008, **47**(4), 746.
27. M. Nurunnabi, K. Murata, K. Okabe, M. Inaba and I. Takahara, *Appl. Catal., A*, 2008, **340**(2), 203.

28. J. M. González Carballo, E. Finocchio, S. García, S. Rojas, M. Ojeda, G. Busca and J. L. G. Fierro, *Catal.: Sci. Technol.*, 2011, **1**(6), 1013.
29. Y. Zhang, X. Yang, X. Yang, H. Duan, H. Qi, Y. Su, B. Liang, H. Tao, B. Liu, D. Chen, X. Su, Y. Huang and T. Zhang, *Nat. Commun.*, 2020, **11**(1), 3185.
30. J. Sun, X. Li, A. Taguchi, T. Abe, W. Niu, P. Lu, Y. Yoneyama and N. Tsubaki, *ACS Catal.*, 2014, **4**(1), 1.
31. J. L. Eslava, X. Sun, J. Gascon, F. Kapteijn and I. Rodríguez-Ramos, *Catal.: Sci. Technol.*, 2017, 7(5), 1235.
32. J. Kang, S. Zhang, Q. Zhang and Y. Wang, *Angew. Chem., Int. Ed.*, 2009, **48**(14), 2565.
33. S. S. Ail and S. Dasappa, *Renewable Sustainable Energy Rev.*, 2016, **58**(16), 267.
34. J. Horáček, *Monatsh. Chem.*, 2020, **151**(5), 649.
35. F. Jiang, M. Zhang, B. Liu, Y. Xu and X. Liu, *Catal.: Sci. Technol.*, 2017, 7(5), 1245.
36. H. T. Luk, C. Mondelli, S. Mitchell, D. Curulla Ferré, J. A. Stewart and J. Pérez–Ramírez, *J. Catal.*, 2019, **371**, 116.
37. Q. Zhang, J. Kang and Y. Wang, *ChemCatChem*, 2010, **2**(9), 1030.
38. J. M. G. Carballo, J. Yang, A. Holmen, S. García-Rodríguez, S. Rojas, M. Ojeda and J. L. G. Fierro, *J. Catal.*, 2011, **284**(1), 102.
39. J.-Y. Park, Y.-J. Lee, P. K. Khanna, K.-W. Jun, J. W. Bae and Y. H. Kim, *J. Mol. Catal. A: Chem.*, 2010, **323**(1–2), 84.
40. G. L. Bezemer, J. H. Bitter, H. P. C. E. Kuipers, H. Oosterbeek, J. E. Holewijn, X. Xu, F. Kapteijn, A. J. van Dillen and K. P. de Jong, *J. Am. Chem. Soc.*, 2006, **128**(12), 3956.
41. D. X. Martínez-Vargas, L. Sandoval-Rangel, O. Campuzano-Calderon, M. Romero-Flores, F. J. Lozano, K. D. P. Nigam, A. Mendoza and A. Montesinos-Castellanos, *Ind. Eng. Chem. Res.*, 2019, **58**(35), 15872.
42. W. Zhou, K. Cheng, J. Kang, C. Zhou, V. Subramanian, Q. Zhang and Y. Wang, *Chem. Soc. Rev.*, 2019, **48**(12), 3193.
43. S. Sartipi, M. Makkee, F. Kapteijn and J. Gascon, *Catal.: Sci. Technol.*, 2014, **4**(4), 893.
44. J. Bao, J. He, Y. Zhang, Y. Yoneyama and N. Tsubaki, *Angew. Chem., Int. Ed.*, 2008, **47**(2), 353.
45. C. C. Amoo, M. Li, A. Noreen, Y. Fu, E. Maturura, C. Du, R. Yang, X. Gao, C. Xing and N. Tsubaki, *ACS Appl. Nano Mater.*, 2020, **3**(8), 8096.
46. B. Gu, C. Zhou, S. He, S. Moldovan, P. A. Chernavskii, V. V. Ordomsky and A. Y. Khodakov, *Catal. Today*, 2020, **357**, 203.
47. C. Zhu, M. Zhang, C. Huang, Y. Han and K. Fang, *ACS Appl. Mater. Interfaces*, 2020, **12**(52), 57950.
48. S. Sartipi, J. E. van Dijk, J. Gascon and F. Kapteijn, *Appl. Catal., A*, 2013, **456**, 11.
49. A. Straß-Eifert, L. I. Wal, C. Hernández Mejía, L. J. Weber, H. Yoshida, J. Zečević, K. P. Jong and R. Güttel, *ChemCatChem*, 2021, **13**(11), 2726.

50. A. Carvalho, M. Marinova, N. Batalha, N. R. Marcilio, A. Y. Khodakov and V. V. Ordomsky, *Catal.: Sci. Technol.*, 2017, 7(21), 5019.
51. S. Sartipi, K. Parashar, M. J. Valero-Romero, V. P. Santos, B. van der Linden, M. Makkee, F. Kapteijn and J. Gascon, *J. Catal.*, 2013, **305**, 179.
52. C. H. Bartholomew, *Appl. Catal., A*, 2001, **212**(1–2), 17.
53. N. E. Tsakoumis, M. Rønning, Ø. Borg, E. Rytter and A. Holmen, *Catal. Today*, 2010, **154**(3–4), 162.
54. N. Fischer and M. Claeys, *Catal. Today*, 2020, **53**(29), 293001.
55. M. Loewert, M.-A. Serrer, T. Carambia, M. Stehle, A. Zimina, K. F. Kalz, H. Lichtenberg, E. Saraçi, P. Pfeifer and J.-D. Grunwaldt, *React. Chem. Eng.*, 2020, **5**(6), 1071.
56. M. Claeys, M. E. Dry, E. van Steen, P. J. van Berge, S. Booyens, R. Crous, P. van Helden, J. Labuschagne, D. J. Moodley and A. M. Saib, *ACS Catal.*, 2015, **5**(2), 841.
57. F. Gao and Z. Gu, Melting Temperature of Metallic Nanoparticles, in *Handbook of Nanoparticles*, ed. M. Aliofkhazraei, Springer Reference, Springer, Cham, Switzerland, 2016, pp. 661–90.
58. C. H. Bartholomew, M. Rahmati and M. A. Reynolds, *Appl. Catal., A*, 2020, **602**, 117609.
59. T. M. Eggenhuisen, J. P. den Breejen, D. Verdoes, P. E. de Jongh and K. P. de Jong, *J. Am. Chem. Soc.*, 2010, **132**(51), 18318.
60. P. Munnik, P. E. de Jongh and K. P. de Jong, *J. Am. Chem. Soc.*, 2014, **136**(20), 7333.
61. M. Wolters, L. J. van Grotel, T. M. Eggenhuisen, J. R. Sietsma, K. P. de Jong and P. E. de Jongh, *Catal. Today*, 2011, **163**(1), 27.
62. A. Straß-Eifert, T. L. Sheppard, C. D. Damsgaard, J.-D. Grunwaldt and R. Güttel, *ChemCatChem*, 2021, **13**(2), 718.
63. M. Wolf, E. K. Gibson, E. J. Olivier, J. H. Neethling, C. R. A. Catlow, N. Fischer and M. Claeys, *ACS Catal.*, 2019, **9**(6), 4902.
64. M. Wolf, N. Fischer and M. Claeys, *Chem. Catal.*, 2021, **1**(5), 1014.
65. M. Wolf, B. K. Mutuma, N. J. Coville, N. Fischer and M. Claeys, *ACS Catal.*, 2018, **8**(5), 3985.
66. N. E. Tsakoumis, J. C. Walmsley, M. Rønning, W. van Beek, E. Rytter and A. Holmen, *J. Am. Chem. Soc.*, 2017, **139**(10), 3706.
67. C. Lancelot, V. V. Ordomsky, O. Stéphan, M. Sadeqzadeh, H. Karaca, M. Lacroix, D. Curulla-Ferré, F. Luck, P. Fongarland, A. Griboval-Constant and A. Y. Khodakov, *ACS Catal.*, 2014, **4**(12), 4510.
68. M. Sadeqzadeh, S. Chambrey, J. Hong, P. Fongarland, F. Luck, D. Curulla-Ferré, D. Schweich, J. Bousquet and A. Y. Khodakov, *Ind. Eng. Chem. Res.*, 2014, **53**(17), 6913.
69. V. R. R. Pendyala, U. M. Graham, G. Jacobs, H. H. Hamdeh and B. H. Davis, *ChemCatChem*, 2014, **6**(7), 1952.
70. G. Chabot, R. Guilet, P. Cognet and C. Gourdon, *Chem. Eng. Sci.*, 2015, **127**, 72.
71. M. Lualdi, S. Lögdberg, M. Boutonnet and S. Järås, *Catal. Today*, 2013, **214**, 25.

72. H. Becker, R. Güttel and T. Turek, *Catal.: Sci. Technol.*, 2016, **6**(1), 275.
73. A. Jess and C. Kern, *Chem. Eng. Technol.*, 2012, **35**(2), 369.
74. S. Gheorghiu and M.-O. Coppens, *AIChE J.*, 2004, **50**(4), 812.
75. M. Ostadi, E. Rytter and M. Hillestad, *Chem. Eng. Res. Des.*, 2016, **114**, 236.
76. C. Knobloch, R. Güttel and T. Turek, *Chem. Ing. Techn.*, 2013, **85**(4), 455.
77. Z. A. Makrodimitri, D. J. M. Unruh and I. G. Economou, *J. Phys. Chem. B*, 2011, **115**(6), 1429.
78. D. B. Bukur, M. Mandić, B. Todić and N. Nikačević, *Catal. Today*, 2020, **343**, 146.
79. J. Li, Y. Xu, D. Wu and Y. Sun, *Catal. Today*, 2009, **148**(1–2), 148.
80. C. Liu, J. Li, Y. Zhang, S. Chen, J. Zhu and K. Liew, *J. Mol. Catal. A: Chem.*, 2012, **363–364**, 335.
81. D. Merino, I. Pérez-Miqueo, O. Sanz and M. Montes, *Top. Catal.*, 2016, **59**(2–4), 207.
82. M. Shinoda, Y. Zhang, Y. Yoneyama, K. Hasegawa and N. Tsubaki, *Fuel Process. Technol.*, 2004, **86**(1), 73.
83. N. Tsubaki, Y. Zhang, S. Sun, H. Mori, Y. Yoneyama, X. Li and K. Fujimoto, *Catal. Commun.*, 2001, **2**(10), 311.
84. L. C. Almeida, F. J. Echave, O. Sanz, M. A. Centeno, G. Arzamendi, L. M. Gandía, E. F. Sousa-Aguiar, J. A. Odriozola and M. Montes, *Chem. Eng. J.*, 2011, **167**(2–3), 536.
85. H. Becker, R. Güttel and T. Turek, *Catal. Today*, 2016, **275**, 155.
86. R. Guettel, J. Knochen, U. Kunz, M. Kassing and T. Turek, *Ind. Eng. Chem. Res.*, 2008, **47**(17), 6589.
87. F. Kapteijn, R. M. de Deugd and J. A. Moulijn, *Catal. Today*, 2005, **105**(3–4), 350.
88. A. Egaña, O. Sanz, D. Merino, X. Moriones and M. Montes, *Ind. Eng. Chem. Res.*, 2018, **57**(31), 10187.
89. H. Li, J. Wang, C. Chen, L. Jia, B. Hou and D. Li, *RSC Adv.*, 2017, 7(16), 9436.
90. A. Y. Khodakov, A. Griboval-Constant, R. Bechara and F. Villain, *J. Phys. Chem. B*, 2001, **105**(40), 9805.
91. Ø. Borg, S. Eri, E. Blekkan, S. Stosater, H. Wigum, E. Rytter and A. Holmen, *J. Catal.*, 2007, **248**(1), 89.
92. H. Xiong, Y. Zhang, K. Liew and J. Li, *J. Mol. Catal. A: Chem.*, 2008, **295**(1–2), 68.
93. I. T. Ghampson, C. Newman, L. Kong, E. Pier, K. D. Hurley, R. A. Pollock, B. R. Walsh, B. Goundie, J. Wright, M. C. Wheeler, R. W. Meulenberg, W. J. DeSisto, B. G. Frederick and R. N. Austin, *Appl. Catal., A*, 2010, **388**(1–2), 57.
94. E. Rytter and A. Holmen, *Catal. Today*, 2016, **275**(17), 11.
95. T. M. Eggenhuisen, J. Zečević, H. Talsma, K. P. de Jong and P. E. de Jongh, *J. Phys. Chem. C*, 2012, **116**(13), 7480.
96. A. Y. Khodakov, R. Bechara and A. Griboval-Constant, *Appl. Catal., A*, 2003, **254**(2), 273.

97. A. Tavasoli, M. Trépanier, A. K. Dalai and N. Abatzoglou, *J. Chem. Eng. Data*, 2010, **55**(8), 2757.
98. H. Xiong, M. A. Motchelaho, M. Moyo, L. L. Jewell and N. J. Coville, *J. Catal.*, 2011, **278**(1), 26.
99. R. M. M. Abbaslou, J. Soltan and A. K. Dalai, *Appl. Catal., A*, 2010, **379**(1–2), 129.
100. T. W. van Deelen, H. Yoshida, R. Oord, J. Zečević, B. M. Weckhuysen and K. P. de Jong, *Appl. Catal., A*, 2020, **593**, 117441.
101. A. Tavasoli, K. Sadagiani, F. Khorashe, A. A. Seifkordi, A. A. Rohani and A. Nakhaeipour, *Fuel Process. Technol.*, 2008, **89**(5), 491.
102. T. O. Eschemann, W. S. Lamme, R. L. Manchester, T. E. Parmentier, A. Cognigni, M. Rønning and K. P. de Jong, *J. Catal.*, 2015, **328**, 130.
103. H. Almkhelfe, X. Li, P. Thapa, K. L. Hohn and P. B. Amama, *J. Catal.*, 2018, **361**(1), 278.
104. V. Vosoughi, S. Badoga, A. K. Dalai and N. Abatzoglou, *Fuel Process. Technol.*, 2017, **162**(279), 55.
105. A. C. Ghogia, A. Nzihou, P. Serp, K. Soulantica and D. Pham Minh, *Appl. Catal., A*, 2021, **609**, 117906.
106. J. Knochen, R. Güttel, C. Knobloch and T. Turek, *Chem. Eng. Process.*, 2010, **49**(9), 958.
107. N. Kruse, A. G. Machoke, W. Schwieger and R. Güttel, *ChemCatChem*, 2015, **7**(6), 1018.
108. R. Xie, D. Li, B. Hou, J. Wang, L. Jia and Y. Sun, *Catal. Commun.*, 2011, **12**(5), 380.
109. J. Yang, X. Fang, Y. Xu and X. Liu, *Catal.: Sci. Technol.*, 2020, **10**(4), 1182.
110. H. Almkhelfe, J. Carpena-Núñez, T. C. Back and P. B. Amama, *Nanoscale*, 2016, **8**(27), 13476.
111. H. Eskandarloo, M. Hashempour, A. Vicenzo, S. Franz, A. Badiei, M. A. Behnajady and M. Bestetti, *Appl. Catal., B*, 2016, **185**, 119.
112. H. Li, G. Wang, J. Niu, E. Wang, G. Niu and C. Xie, *Results Phys.*, 2019, **14**(6), 102499.
113. L. Zhang, H. Chu, H. Qu, Q. Zhang, H. Xu, J. Cao, Z. Tang and J. Xuan, *Int. J. Hydrogen Energy*, 2018, **43**(6), 3077.
114. C. Zhang, Y. Pu, F. Wang, H. Ren, H. Ma and Y. Guo, *Chin. J. Chem. Eng.*, 2020, **28**(5), 1311.
115. G. Wang, E. Johannessen, C. R. Kleijn, S. W. de Leeuw and M.-O. Coppens, *Chem. Eng. Sci.*, 2007, **62**(18–20), 5110.
116. E. Johannessen, G. Wang and M.-O. Coppens, *Ind. Eng. Chem. Res.*, 2007, **46**(12), 4245.
117. A. Y. Khodakov, A. Griboval-Constant, R. Bechara and V. L. Zholobenko, *J. Catal.*, 2002, **206**(2), 230.
118. S. Mohammadnasabomran, A. Tavasoli and Y. Zamani, *React. Kinet., Mech. Catal.*, 2019, **128**(1), 217.
119. H. Xiong, Y. Zhang, S. Wang and J. Li, *Catal. Commun.*, 2005, **6**(8), 512.

120. W. Ma, G. Jacobs, D. E. Sparks, M. K. Gnanamani, V. R. R. Pendyala, C. H. Yen, J. L. Klettlinger, T. M. Tomsik and B. H. Davis, *Fuel*, 2011, **90**(2), 756.
121. C. Xing, G. Yang, P. Lu, W. Shen, X. Gai, L. Tan, J. Mao, T. Wang, R. Yang and N. Tsubaki, *Microporous Mesoporous Mater.*, 2016, **233**, 62.
122. K. Shimura, T. Miyazawa, T. Hanaoka and S. Hirata, *J. Mol. Catal. A: Chem.*, 2014, **394**, 22.
123. D. Yin, W. Li, W. Yang, H. Xiang, Y. Sun, B. Zhong and S. Peng, *Microporous Mesoporous Mater.*, 2001, **47**(1), 15.
124. X. Li, M. U. Nisa, Y. Chen and Z. Li, *Ind. Eng. Chem. Res.*, 2019, **58**(8), 3459.
125. Y. Zhao, S. Huang, L. Wei, Y. Zhang, A. Lin, C. Liu and J. Li, *Ind. Eng. Chem. Res.*, 2020, **59**(7), 3279.
126. Y. Yang, L. Jia, B. Hou, D. Li, J. Wang and Y. Sun, *Catal.: Sci. Technol.*, 2014, **4**(3), 717.
127. H. Tang and J. Li, *J. Fuel Chem. Technol.*, 2011, **39**(8), 615.
128. S. Storsæter, B. Tøtdal, J. C. Walmsley, B. Tanem and A. Holmen, *J. Catal.*, 2005, **236**(1), 139.
129. A. Y. Khodakov, W. Chu and P. Fongarland, *Chem. Rev.*, 2007, **107**(5), 1692.
130. E. van Steen and M. Claeys, *Chem. Eng. Technol.*, 2008, **31**(5), 655.
131. F. Diehl and A. Y. Khodakov, *Oil Gas Sci. Technol.*, 2009, **64**(1), 11.
132. Y. Chen, J. Wei, M. S. Duyar, V. V. Ordomsky, A. Y. Khodakov and J. Liu, *Chem. Soc. Rev.*, 2021, **50**(4), 2337.
133. T. Fu and Z. Li, *Chem. Eng. Sci.*, 2015, **135**(1–2), 3.
134. M. J. Valero-Romero, M. Á. Rodríguez-Cano, J. Palomo, J. Rodríguez-Mirasol and T. Cordero, *Front. Mater.*, 2021, **7**, 258.
135. A. A. Adeleke, X. Liu, X. Lu, M. Moyo and D. Hildebrandt, *Rev. Chem. Eng.*, 2020, **36**(4), 437.
136. L. U. Okonye, Y. Yao, D. Hildebrandt and R. Meijboom, *Sustainable Energy Fuels*, 2021, **5**(1), 79.
137. Z. Gholami, N. Asmawati Mohd Zabidi, F. Gholami, O. B. Ayodele and M. Vakili, *Rev. Chem. Eng.*, 2017, **33**(4), 138.

CHAPTER 9

BTL-FT Products Refining Downstream Options and Processes

Y. ZHAO AND C. M. MASUKU*

Davidson School of Chemical Engineering, Purdue University, West Lafayette, IN 47907, USA
*Email: cmasuku@purdue.edu

9.1 Introduction

Fossil fuels, especially petroleum, are the primary energy sources used to meet current fuel demands. However, they are nonrenewable and pollute the environment during combustion.[1] There is a strong push for alternative energy sources that could be converted into fuel. Fischer–Tropsch synthesis (FTS) converts synthesis gas (syngas) into a wide range of hydrocarbons, and the reaction effluents are further refined into many kinds of chemicals and fuels.[2] Syngas is a mixture of predominantly CO and H_2 that can be produced through a number of processes, such as natural gas reforming, coal gasification and biomass gasification.[3]

Biomass as a renewable resource has captured some attention recently,[4] as it has the advantage of reducing greenhouse gas (GHG) emissions during photosynthesis. Thus research efforts have focused on investigating the biomass-to-liquid process *via* FT reaction (BTL-FT), including economic analysis,[5] process design[6] and refining.[7] Refinery designs remain a challenge due to various Fischer–Tropsch (FT) effluents containing a wide

Catalysis Series No. 44
Chemicals and Fuels from Biomass *via* Fischer–Tropsch Synthesis: A Route to Sustainability
Edited by Joshua Gorimbo, Xinying Liu, Yali Yao and Diane Hildebrandt

Published by the Royal Society of Chemistry, www.rsc.org

spectrum of hydrocarbons, multiple refining options, various products and changing specifications for desired liquid fuels.[8] This results because combining the variety of refinery pathways and the various compositions of product fuel into an integrated process increases the complexity of BTL-FT syntheses due to the large number of configurations that need to be considered.

9.2 Biomass-to-liquids Process *via* Fischer–Tropsch Reaction

9.2.1 Fischer–Tropsch Reaction Mechanism

FTS was initially developed at the beginning of the 20th century in order to investigate technology that could convert alternative materials into fuel instead of petroleum. Later in the 20th century, industrial chemical plants used to produce fuels from FTS were constructed in Germany, Japan, the United States and South Africa. FTS is still a popular research field, since it is a potential alternative method that could be used to solve the shortage of liquid fuel used in the transportation industry. In terms of the plants that have been built, syngas was obtained mainly using the methane reforming or coal gasification methods. Biomass is a renewable material and therefore a promising direction for research; therefore, the installation of additional plants is under consideration.

The main FTS reactions are hydrogenation of carbon monoxide to alkanes and alkenes in the presence of a catalyst. This could be expressed as per eqn (9.1) and (9.2):

$$n\mathrm{CO} + (2n+1)\mathrm{H_2} \rightarrow \mathrm{C}_n\mathrm{H}_{2n+2} + n\mathrm{H_2O} \tag{9.1}$$

$$n\mathrm{CO} + (2n\mathrm{H_2}) \rightarrow \mathrm{C}_n\mathrm{H}_{2n} + n\mathrm{H_2O} \tag{9.2}$$

Nickel, iron, cobalt and ruthenium are the main metals that catalyze CO hydrogenation activity. Nickel has very high activity, and it was used as a catalyst by an industrial FT plant in Germany in the 20th century. However, there was a problem with forming volatile carbonyl, which led to the continuous loss of metal. The other issue is high selectivity to methane, which reduces the liquid fuel yield. Ruthenium is prohibitively expensive, which has limited its use in industrial applications. Thus iron and cobalt are the most commonly used catalysts.

Iron has the advantage of low price. In addition to the alkanes and alkenes, iron catalysts also produce oxygenates and promote the water-gas-shift (WGS) reaction, which converts carbon monoxide and water into hydrogen and carbon dioxide. Thus the iron-based catalyst process is less sensitive to the H_2/CO ratio. When a cobalt catalyst is applied, extra WGS should be considered to adjust the H_2/CO ratio prior to the FT reaction. Cobalt catalyst produces less oxygenated products in comparison to iron, prompting simplification of the process with fewer oxygenate separation units.

Catalysts for FT synthesis are sensitive to contaminants from biomass, like sulfur and nitrogen. Thus the thorough cleaning of syngas before sending it to the FT reactor is necessary. FT synthesis is a highly exothermic reaction, and the release of heat should be incorporated to control temperature.

9.2.2 Fischer–Tropsch Product Distribution

The FT reaction product has similarities to crude oil and is sometimes referred to as synthetic crude or syncrude. The design of the FT refinery process depends on the composition of this syncrude. FT syncrude has a composition distributed over a wide range of hydrocarbons (from C_1 to C_{30+}) and oxygenates. The distribution depends on the selectivity of the FT reaction, which is influenced by the reactor conditions, including temperature, catalyst type, pressure and H_2/CO ratio. Significant effort has been made to investigate the chain growth mechanism and the mathematical models for the carbon number distribution.

According to Anderson *et al.*,[9] the product distribution of the hydrocarbons can be described by the Anderson–Schulz–Flory (ASF) equation:

$$m_n = (1-\alpha)\alpha^{n-1} \tag{9.3}$$

where m_n is the mole fraction of a hydrocarbon with chain length n; α is the chain growth probability factor that is independent of n; α determines the total carbon number distribution of the FT products.

The logarithm of eqn (9.3) gives:

$$\log(m_n) = \log\left(\frac{1-\alpha}{\alpha}\right) + n\log\alpha \tag{9.4}$$

Hence a plot of the logarithm of the molar concentration of hydrocarbons of chain length n, m_n, *vs.* the carbon number n would produce a straight-line plot, the slope of which is $\log(\alpha)$ and the intercept $\log\left(\frac{1-\alpha}{\alpha}\right)$. Most studies assume that the ASF distribution is a model of the product selectivity.[10] However, significant deviations have been reported in the literature; the most prevalent deviation is the two-alpha positive deviation product distribution model.[11]

9.2.3 BTL-FT Syncrude Composition

FT plants that produce liquid fuel from syngas can be characterized according to three categories of feedstock: coal-to-liquid process, gas-to-liquid process, biomass-to-liquid process. Synthetic gas is a precursor for the FT reaction and should be produced first and then converted into syncrude. The most widely used treatment method used for coal and biomass is gasification, while for natural gas it is reforming. However, the feedstock does not determine syncrude composition but influences the treatment method used to convert feedstock into syngas. The raw syngas needs to be cleaned to remove the components containing heteroatom from the feedstock, while the gas cleaning technology used determines the quality of the synthesis gas rather than the feedstock used as the carbon source. In other words, syncrude composition from the BTL-FT process can be regarded similar to general FT synthesis.

Syncrude composition, which is described by chain-growth probability α, is dependent on the reaction operating conditions, including operating temperature, type of catalyst, H_2/CO ratio and pressure. The parameters of ASF distribution, especially α, can be correlated with these operating conditions. These factors influence the refining design. It can be deduced that, for alkenes and alcohols, the consumption ratio of H_2/CO is fixed at 2; for alkanes, the ratio decreases with the increase in carbon number. The ratio is 3 when undesired methane is produced, but it approaches 2 as the products become heavier. The most suitable ratio for a cobalt catalyst is 2.05–2.015, since the WGS reaction is not promoted. For the iron catalyst, the ratio is approximately 1.65. Thus once the feed has been converted to syngas, the H_2/CO ratio must be modified to the desired ratio of FT reaction, in order to deduce the composition of undesired light hydrocarbons in the syncrude, *e.g.*, methane.

9.2.4 Generic Composition of FT Syncrude

Understanding the syncrude composition could help with designing the refining process and permit efficient utilization of materials. The main components of FT syncrude are hydrocarbons and oxygenates. Hydrocarbons comprise paraffins, olefins and aromatic classes, while oxygenates are composed of alcohols, acids, aldehydes, ketones and esters. A minor composition of metal carboxylates is also observed. It is deduced that, due to the formation of carboxylic acids, the metal contained in equipment or catalysts may be eroded, which leads to the formation of metal carboxylates. However, the content is too minor to be considered.

The generic composition syncrude corresponds to three typical industrial operating conditions. These are detailed here for purposes of comparison and providing a reference for refinery design.

9.2.4.1 Paraffins

Alkanes are saturated hydrocarbons with a repeating unit $(-CH_2-)_n$. Paraffins produced by FT synthesis can be characterized as the linear type, the branched isomer type and the cycloparaffin type. Most chain branches are methyl branches that are randomly distributed. Low-temperature FT (LTFT) gives highly linear hydrocarbons, while high-temperature FT (HTFT) products contain more branched isomers compared to LTFT products. Also, the utilization of iron-based catalyst contributes to an increased number of branched hydrocarbons. The majority of branched hydrocarbons are of the

Table 9.1 Cetane numbers of various hydrocarbons compounds. Reproduced from ref. 12 with permission from John Wiley & Sons, Copyright © 2011 Wiley-VCH Verlag GmbH & Co. KGaA.

Class	Compounds	Cetane number
Linear paraffin	*n*-Decane	77
	n-Dodecane	87
	n-Tetradecane	95
	n-Hexadecane	100
	n-Octadecane	106
Branched paraffin	2,2,4,4,6,8,8,-Heptamethylnonane	15
	7,8-Dimethyltetradecane	40
	5-Butyldodecane	45

single methyl-branched paraffin type. The probability of chain branching decreases with an increase in carbon number. The extent of branching is a crucial factor in terms of the quality of liquid fuel.

The cetane number is an important specification for diesel, which represents the level of compression ignition delay behavior. A higher cetane number represents a short delay. As shown in Table 9.1, paraffins with the same carbon number but different structures have different cetane numbers. Linear paraffins show a higher number and better behavior. However, branched paraffin is beneficial in promoting the property of gasoline. The octane number (ON) is a measure of a fuel's ability to resist compression before ignition, and this rises with an increase in the degree of branching and a reduction in the carbon number. Thus when a liquid fuel like gasoline or diesel is produced as the required product, it is essential to differentiate between a range of hydrocarbons.

9.2.4.2 *Olefins*

Olefins refer to unsaturated hydrocarbons containing at least one double bond. FT product hydrocarbons are mainly composed of mono-olefins and some diolefins. The three primary components of the olefins produced are 1-olefins, *trans*-2-olefins and *cis*-2-olefins. About 85% of the double bonds in mono-olefins are at the terminal position.[10] Secondary conversions like isomerization contribute to the observation of internal olefins. Double bonds are shifted from the terminal position to the internal position. Branched and cyclic olefins are also found in the products. The use of iron-based catalyst produces more olefin compared with cobalt. The number of olefins increases with the loading of promoter. Alkali metal promoted iron catalyst has the highest selectivity toward olefins compared with other promoters.[10]

9.2.4.3 *Aromatics*

Aromatics comprise a small portion, about 6% (weight fraction), of hydrocarbons obtained from HTFT synthesis. As is illustrated in Table 9.2, the

Table 9.2 Hydrocarbon distribution by carbon number and class. Reproduced from ref. 13 with permission from American Chemical Society, Copyright 1953.

Carbon number	Yield wt%	Yield Mole%	Hydrocarbon class, vol% Paraffins	Hydrocarbon class, vol% Aromatics	Hydrocarbon class, vol% Olefins	Ratio of *trans*-internal to terminal double bonds
1	8.9	28.7	100	0	0	—
2	7.8	13.7	78.6	0	23.2	—
3	15.9	19.4	18.5	0	81.5	—
4	13.9	12.7	13.4	0	86.6	0.032
5	11.5	8.47	12.7	0	87.4	0.022
6	8.6	5.26	12.8	0.3	86.9	—
7	6.6	3.46	12.4	3.7	83.9	0.048
8	5.0	2.30	13.2	5.4	81.4	0.051
9	3.9	1.60	15.0	7.5	77.5	0.049
10	3.0	1.11	15.1	12.2	72.7	0.056
11	2.4	0.81	15.4	11.0	73.6	0.061
12	1.8	0.55	16.2	15.6	68.2	0.078
13	1.4	0.40	15.4	21.7	62.9	0.094
14	1.1	0.29	16.5	19.9	63.6	0.11
15	1.0	0.25	17.4	—	—	—
16	0.9	0.21	—	—	—	0.15
30	6.3	0.79	—	—	—	—
Total	100.0	100.00	—	—	—	—

amount of aromatic hydrocarbons is approximately 0, with a carbon number around 5, an increase in carbon number leads to increased content of aromatic hydrocarbons. Monocyclic, dicyclic and tricyclic structures are all observed in aromatic hydrocarbons. Nonaromatic rings were found attached. For benzene series, monoalkyl and *ortho*-dialkyl are the main components.

9.2.4.4 Carbonyl Compounds

Carbonyl compounds such as ketones and aldehydes have been reported as FT products.[10] Aldehydes are the primary product of FTS, while ketones are considered a secondary product formed during carboxylic acid decomposition.[14] The distribution of carbonyl compounds is similar to the distribution of acids and alcohols; however, the branching pattern of carbonyl compounds resembles that of hydrocarbons.

9.3 Fischer–Tropsch Refining Technologies

FTS was developed in response to the demand to find alternatives for petroleum in fuel production process and transportation – fuels such as gasoline, jet fuel and diesel. Gasoline is composed of hydrocarbons mainly in the C_5–C_{11} range, while jet fuel is in the C_8–C_{16} range, and diesel is in the C_{11}–C_{22} range. As fuel specifications have become increasingly stringent, the potential to extract chemicals from the process has been explored, and

Table 9.3 Composition of chemicals in FT aqueous phase. Reproduced from ref. 15 with permission from American Chemical Society, Copyright 1987.

Category	Compound	Mass (%)
Nonacid	Aldehydes	5
	Alcohols	
	CH_3OH	1
	C_2H_5OH	55
	C_3H_7OH	16
	C_4H_9OH	7
	C_{5+} alcohols	2
	Ketones	14
Acids	Acetic	70
	Propionic	16
	Butyric	9

opportunities other than fuel production have been investigated to satisfy economic requirements.

Chemicals other than fuels were first recovered from the process in the 1960s.[2] Ethylene recovered from FTS and nitrogen from an air separation plant were used in the production of plastic and ammonia. Later, more chemicals were targeted for extraction,[2] including ethylene and propylene. Liquefied petroleum gas is mainly composed of propane and butane, which has a high heating value and can be used as fuel gas. These components are extracted from the short-chain hydrocarbons obtained from FT syncrude. Refining technology like isomerization and etherification are introduced. Ethers like tertiary amyl methyl ether (TAME) can be produced, which are used as an octane booster for gasoline.[2] Linear alpha-olefins can be recovered for use in the polymer industry, or they can be converted to linear aldehydes and alcohols by hydroformylation for use in the production of detergents.

Low-molecular-weight oxygenates appear predominantly in the aqueous phase, with their typical composition as indicated in Table 9.3. They comprise mainly alcohols, ketones and acids, which can be separated *via* distillation. The aldehydes, ketones and alcohols are obtained at the top, while carboxylic acids are retained in the bottom.[15] Carboxylic acids can be recovered for use in the production of soap and grease. High-value chemicals can be considered to be produced so that the reliance on high fuel prices to satisfy economic performance can be relieved.[2]

9.3.1 Skeletal Isomerization

The gasoline octane number is dependent on the skeletal structure of its components, which is improved with an increase in the degree of branching. Syncrudes produced by FTS are mostly linear and have a low degree of branching, which leads to insufficient ON.[16] Certain reactions are introduced to produce clean gasoline with high ON. Skeletal isomerization is a reaction that can change the structure of both paraffins and olefins to produce isomers and increase the branching. Skeletal isomerization of paraffin follows a three-step process:

- Dehydrogenation of *n*-paraffin into *n*-olefin.
- Isomerization of *n*-olefin into iso-olefin, resulting in a branched hydrocarbon.
- Hydrogenation of iso-olefin into iso-paraffin.[17]

The prevailing mechanism of isomerization can be separated into three routes: monomolecular route, bimolecular route and a combination of these two routes.[18] The monomolecular route reveals protonated cyclopropane branching, while the bimolecular route includes the dimerization of two pronated carbonium ions and further cracking into the ions of the desired iso-structure or by-products. For example, with isomerization of butane, C_8 intermediates are achieved and cracked into *n*-butane, i-butane or the by-products C_3 and C_5. It was observed that the bimolecular route is more favorable for isomerization of C_4; however, for hydrocarbons with a carbon number higher than 4, the monomolecular route is prevalent.[19] However, the dominant mechanism still depends on the reaction conditions.[18]

Isomerization of olefins is also applied in preparation for etherification with alcohols to produce fuel ethers such as methyl tertiary butyl ether (MTBE) and TAME.[12] Isomerization of paraffins can be separated into two main routes: isomerization of butane and isomerization of light naphtha hydrocarbons (C_5 and C_6). Production of isobutane is the preparation step for a further alkylation reaction that converts isobutane and butene into trimethylpentanes with an ON of 100. Light naphtha hydrocarbons consisting mainly of linear pentane and hexane have an ON of about 70. After isomerization, linear paraffins are converted into isomers, and the research ON (RON) can be increased to a value of 84 to 92.[17]

As fuel regulations have become more stringent, the use of aromatics and olefins to produce gasoline with high ON is restricted due to environmental concerns. Skeletal isomerization of paraffins has proved to be a simple yet cost-effective method that has attracted attention. In contrast, the isomerization of olefins (C_4, C_5) is energy intensive and is thus not regarded as important refining technology.

9.3.2 Double-bond Isomerization

FTS syncrudes are rich in olefins, especially for the HTFT process. Double-bond isomerization can increase ON by converting linear α-olefins into linear internal olefins. This is a technology worth considering, particularly for naphtha range hydrocarbons, which contain a large number of linear α-olefins. For example, with pentene, ON increases from 84 to 87 by converting 1-pentene to *trans*-2-pentene; however, more significant enhancement was observed with an increase in the olefin chain length, from 69.9 for 1-hexene to

86.8 for *trans*-2-hexene and from 52.6 for 1-heptene to 71.1 for *trans*-2-heptene, while the ON of the isomerized internal olefins is still moderate. Thus double-bond isomerization is not the preferred choice to improve the ON.

9.3.3 Alkylation

Alkylation is a reaction that introduces an alkyl group to an organic molecule. It was developed when there was demand to take full advantage of the by-products of naphtha cracking, such as ethylene, propylene, butene and isobutane, which are produced by the cracking of heavy hydrocarbons.[12] Alkylation in the refinery process can be separated into aliphatic alkylation and aromatic alkylation. Aromatic alkylation happens between benzene and an olefin and is used to reduce the benzene content in gasoline while retaining the ON.[12] It can also be used for chemical production.

Aliphatic alkylation occurs between a branched paraffin and an olefin and produces a highly branched alkylate.[20] Alkylates have a high ON, high heat of combustion, low vapor pressure and an ideal boiling range. The most common alkylation is done using isobutane and light olefins in the refining process, which aims at improving the ON of the gasoline product. The most widely used catalysts are hydrofluoric acid (HF) and sulfuric acid. The use of liquid acids prompts the reaction but has a disadvantage in that the acidic catalysts are not environmentally friendly. Although aluminum chloride is also used, it is difficult to manage, and regeneration means that commercial acceptance is limited. The majority of the industrial plants prefer sulfuric acid as a catalyst because of its lower price, efficient utilization of isobutane and low degree of self-alkylation. When the HF alkylation process is applied, a recovery system is needed due to the high cost of the acid. The processes are operated at a low temperature in the 0–16 °C range. The isobutane-to-olefin ratio has a significant influence in that, while the ratio is increased, the yield of both alkylate and ON is improved.[21]

Aliphatic alkylation requires isobutane and light olefins as reactants. Since FTS syncrudes are mostly linear, little isobutane exists. Skeletal isomerization is usually incorporated in the process to convert *n*-butane into isomers for alkylation. However, the isobutane content cannot be too high; otherwise undesired side reactions are triggered, *e.g.*, oligomerization and self-alkylation. The type of olefin used influences the effect of alkylation, and the use of propylene and pentene result in a higher acid consumption and a lower ON in the alkylate product[21] (see Table 9.4).

The alkylation reaction is more suitable for HTFT than for LTFT because with LTFT neither C_2–C_4 olefins nor isobutene is present in significant quantities. Also, the amount of propylene and ethylene produced is not sufficient to carry out the aromatic alkylation process.

9.3.4 Aromatization

Aromatization can convert both paraffins and olefins into aromatic hydrocarbons, which improves the availability of aromatics as chemicals and

Table 9.4 Influence of the olefin feed on isobutane and acid consumption and alkylate ON. Reproduced from ref. 21 with permission from American Chemical Society, Copyright 1971.

Catalyst	H_2SO_4		
Olefin feed	$C_3^{=}$	$C_4^{=}$	$C_5^{=}$
Isobutane consumption, vol% olefin	115–135	100–125	90–125
Acid consumption, lg per gal alkylate	1.0–3.0	0.4–1.2	0.7–2.0
Alkylate octane	87–91	91–97	88–92
RON	85–88	88–95	86–90
MON			

increases the ON of gasoline products. Aromatization can be classified according to its reactants into LPG (C_2–C_4) aromatization and light naphtha components (C_{5+}) aromatization.[22] ZSM-5 zeolite is the most widely used catalyst. It can convert short-chain alkanes into aromatic hydrocarbons, such as benzene, toluene, ethylbenzenes and xylenes. Light paraffins are first dehydrogenated into corresponding olefins; then oligomerization and cracking are involved, and larger alkenes are formed. Naphthenes are formed through cyclization and further converted into benzenic hydrocarbons by hydrogen transfer.[23] Metal can be added to produce a bifunctional catalyst.

9.3.5 Reforming

Catalytic reforming is a process that converts low-grade naphtha range hydrocarbons (C_6–C_{10}) into cycloalkanes and aromatics with a high ON. Hydrogen is produced during the dehydrogenation process and can be provided to the hydroprocessing units like hydrocracking. Reforming is considered the optimal technology to refine naphtha because it has a superior ability to convert paraffins to aromatics and hydrogen. The reaction is operated at a high temperature and high pressure, which is considered energy intensive. With an increase in temperature, the ON of the products increases. Catalytic reforming can be separated into three types, according to the catalyst used: Pt/Al_2O_3 reforming, bimetallic reforming, Pt/l-zeolite reforming.

The Pt/Al_2O_3-based catalyst is bifunctional and has acidic sites. The reactions involved are dehydrogenation of the naphthenes into aromatics, dehydrogenation followed by isomerization of cycloalkanes into aromatics and dehydrocyclization. Dehydrogenation–hydrogenation and hydrogenolysis take place on metals. Skeletal isomerization and hydrocracking are catalyzed on metallic and acid sites. Alkylation and disproportionation are promoted by acidic sites. Cycloparaffins react more rapidly. Since FTS syncrude is mostly linear, this makes the Pt/Al_2O_3-based catalyst an inappropriate choice, as a low aromatics yield may be obtained.

Bimetallic catalysts consist of Pt and another metal (usually rhenium); tin, iridium and others and the different elements have different positive properties. For example, when iridium is used as the second metal, a significant effect on activity can be observed.

The Pt/l-zeolite-based catalyst is monofunctional, and no acid site exists. It has high selectivity for aromatization. However, the catalysts are very sensitive to sulfur concentration, and precautions must be taken to remove sulfur in the feed. However, this factor can be ignored since the FTS-derived feed is already sulfur free. Coke is formed during the process, which can cause deactivation of the catalyst; thus a regeneration process is required.

9.3.6 Oligomerization

Since the alkylation process is not environmentally friendly due to the use of acidic catalysts, olefin oligomerization is another method worth considering to convert short-chain olefins to liquid fuel.[24] This is an area of commercial importance. Gaseous hydrocarbons in syncrudes have a relatively high ON content: ethane has a RON of 111.4, while ethylene has a RON of 100.4. Although short-chain hydrocarbons can be dissolved in the oil phase, their existence is limited by the vapor pressure specification of fuel products. Oligomerization is introduced to convert the gaseous olefins into liquid hydrocarbons and can be classified into two categories according to the mechanistic routes: acid-catalyzed carbonium chemistry and metal-catalyzed 1,2-insertion.

Acidic catalysts include solid phosphoric acid (SPA), acidic resin, amorphous silica alumina (ASA) and zeolites. This method has been applied for more than 80 years to oligomerize light olefins into fuels or precursors for petrochemicals. The mechanism is the carbonium route,[24] which is illustrated in Figure 9.1. Note that the first olefin inserted is black, the second is blue and the third is red.

Olefin oligomerization was not commercialized until the development of the SPA catalyst. It has a high selectivity toward gasoline and is also cost-effective and environmentally friendly. However, the challenges are that only a limited amount of water and oxygenates can be tolerated in the feed, and it has a limited lifetime. Acidic resin selectively oligomerizes the highly reactive olefins, like branched olefins isobutene. SPA catalysts have limited applicability for FTS syncrudes because of the low degree of branching.

ASA has proved to be more stable at a high temperature than SPA or acidic resin, and ASA works well with FTS syncrudes. Although high-quality gasoline with a high ON can be achieved by olefin oligomerization, the hydrogenated distillate has a low cetane number.

Metal-based catalysts use a Cosse–Arlman 1,2-insertion mechanism, followed by β-hydride elimination to yield products,[24] as illustrated in Figure 9.2. The most frequently used metal is nickel.

9.3.7 Etherification

Oxygenates can increase the octane rating and combustion quality, with alcohols and ethers being the two main choices. Ethers are preferred

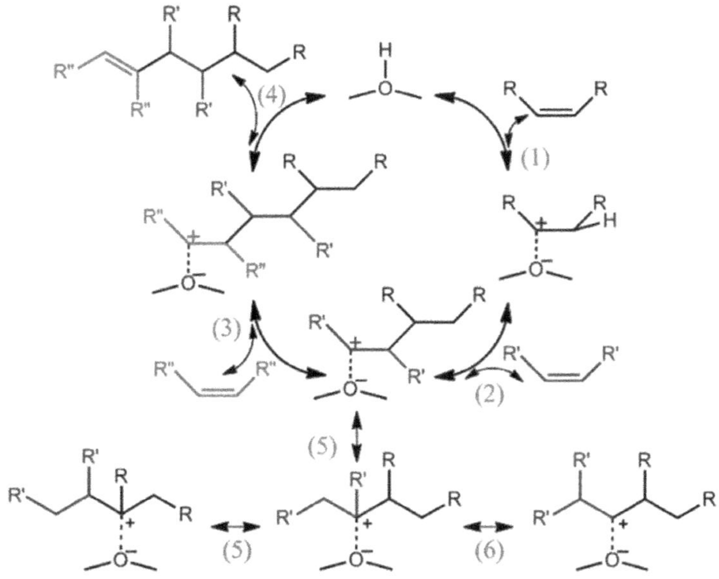

Figure 9.1 Carbenium ion mechanism utilized in acid catalyzed oligomerization reactions. Reproduced from ref. 24 with permission from Elsevier, Copyright 2017.

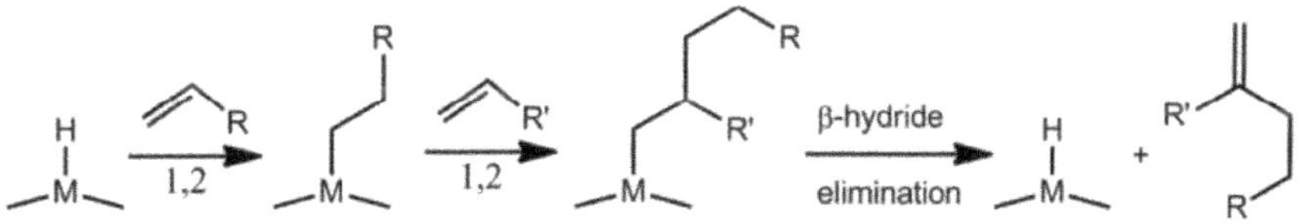

Figure 9.2 Mechanism of oligomerization while using metal catalyst. Reproduced from ref. 24 with permission from Elsevier, Copyright 2017.

because they have low vapor pressure for a similar carbon number, which can reduce the vapor pressure of gasoline.[12] Etherification takes place between an iso-olefin with a double bond on a tertiary carbon and an alcohol. It was traditionally used for the purification of linear olefins to remove close-boiling isomers.[25] For example, the formation of ethyl *tert*-butyl ether (ETBE) can be obtained from isobutene and ethanol as reactants.[26] The reaction is illustrated in eqn (9.5).

$$(CH_3)_2C{=}CH_2 + HO\text{-}C_2H_5/(H^+) \rightarrow (CH_3)_3\text{-}O\text{-}C_2H_5 \tag{9.5}$$

ETBE is used as an octane booster for gasoline. The commercial catalyst is Amberlyst-15, which is a macroreticular sulfonic resin. Increased acidity and increased pore size can lead to improved performance of the catalyst.[25] Alcohol is usually in excess to prevent the oligomerization of olefins. The catalysts pronate alcohols instead of olefins. Oxygenates have been shown to inhibit etherification selectivity,[27] and the reaction is reversible and exothermic. Thus a low operating temperature is preferred from an equilibrium perspective. Etherification is not energy intensive.[12]

9.3.8 Hydroprocessing

9.3.8.1 Hydrotreating

Hydrotreating is a process that can remove heteroatom-like oxygen through hydrodeoxygenation (HDO), sulfur through hydrodesulfurization (HDS), nitrogen through hydrodenitrogenation (HDN) and metals through hydrodemetallization (HDM) in organic compounds, as well as removing impurities like olefins and aromatics in hydrocarbons under high pressure. The presence of sulfur and nitrogen can lead to deactivation of the catalyst. Olefins can result in the formation of coke under high temperature, while metal in hydrocarbons can damage the reforming catalyst.[28] Thus they need to be removed. O, S and N are removed in the form of water, hydrogen sulfide and ammonia, respectively. Olefins are dehydrogenated, while metals are reposited upon hydrotreating the catalyst. The principal types of catalyst used for hydrotreating are bimetallic or trimetallic. For example, Co/Mo is the general catalyst used for HDS; Ni/Mo is used for HDN; Co/Mo/other metals on neutral support can prevent the polymerization of olefins, and this is used for olefin hydrotreating.

However, these processes are designed for crude oil refining, and differences need to be considered in terms of the FTS refining process. Firstly, the syncrudes are sulfur and nitrogen free. The use of sulfide base metal hydroprocessing catalysts requires the continuous addition of sulfur-containing compounds. Second, oxygenate content is different, especially when biomass is used as the feedstock. Hydroprocessing of an oxygenate-rich stream is more challenging. Third, extra attention needs to be paid to heat management of the hydrotreater design. Last but not least is industrial practice to load the HDM catalyst as the top layer for hydrotreating. Carboxylic acids are produced during FTS, and their existence leads to the corrosion of equipment and leaching of catalysts. However, treatment of corresponding metal carboxylates cannot be achieved by HDM but can be achieved through a thermal decomposition pathway.[29]

9.3.8.2 Hydrocracking

Hydrocracking is a process that is mainly used to crack heavy hydrocarbons (like waxes) in syncrude into lighter materials, especially the naphtha- and distillate-range products. It can also reduce the content of polynuclear aromatics, which helps to satisfy the specification. Diesel fuel products produced from FTS syncrude *via* catalytic hydrocracking exhibit better properties than crude oil, such as a high cetane number and no sulfur or nitrogen heteroatoms. Catalysts used for hydrocracking are bifunctional with a hydrogenation–dehydrogenation function and an acidic function. The mechanism for hydrocracking is illustrated in Figure 9.3.

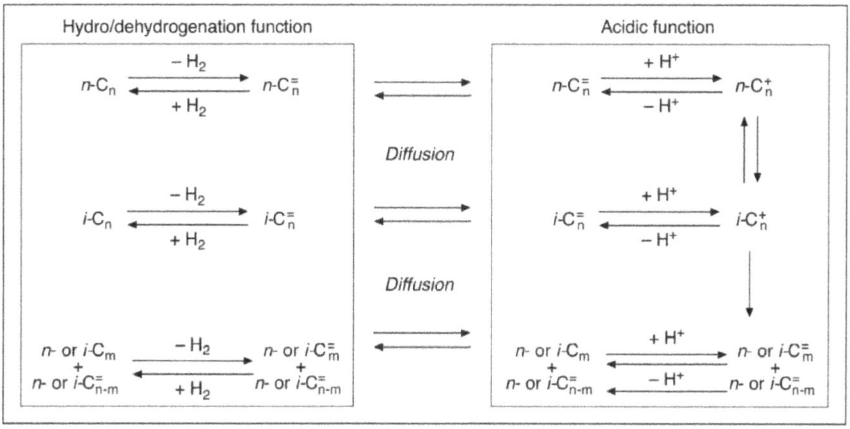

Figure 9.3 Hydrocracking mechanism. Reproduced from ref. 30, https://doi.org/10.2516/ogst/2008047, under the terms of the CC BY 4.0 license https://creativecommons.org/licenses/by/4.0/.

The formation of cracked products follows two steps: a hydroisomerization step and a hydrocracking step. The *n*-paraffin is first dehydrogenated into an olefin, which diffuses to the acidic site on the catalyst, and the protonated molecule is achieved. The carbocation is then isomerized and cracked by β-scission to produce a lighter olefin and a light carbocation. Ideal hydrocracking includes features that enable the conversion of heavy hydrocarbons into light distillate and naphtha, the minimization of the cracking of light hydrocarbons, and high selectivity toward isomerization to improve the cold flow properties. The ideal hydrocracking is related to the choice of operating conditions and the formation of catalysts. Under such ideal conditions, the production of middle distillates can be maximized. The oxygenates contained in syncrude may affect the activity and selectivity of hydrocracking, but they can be used as a modifier.

9.3.8.3 *Hydroisomerization*

Hydroisomerization is used to increase the degree of branching of paraffins. It can be separated according to the carbon number range of feed material. The cracking probability for C_6 and lighter aliphatic compounds is low, and they cannot crack to produce stable carbocation intermediates. Thus the hydroisomerization process of light hydrocarbons (C_4 and C_5/C_6) does not include hydrocracking, and it is analogous to skeletal isomerization. With C_{6+} hydrocarbons, cracking can proceed. Besides catalytic hydrocracking, waxes in syncrude can also be hydroisomerized to produce lubricant base oil. Also, with the existence of high-molecular-weight linear paraffins, the wax shows a high pour point, which leads to coagulation at a low temperature. Thus it is necessary to remove the heavy linear paraffins. Catalytic dewaxing *via* hydroisomerization is economical. Hydroisomerization of heavy hydrocarbons is always accompanied by hydrocracking those heavy hydrocarbons into light materials.

The hydroisomerization process also requires bifunctional catalysts, which is similar to hydrocracking. Hydrogenation–dehydrogenation occurs on metallic sites and isomerization cracking occurs on acidic sites. The different formation of catalysts exhibits different isomerization selectivity. Two mechanisms rule the hydroisomerization step. For Type A isomerization, the branching degree of the carbocation is not modified, but the branching undergoes positional change *via* alkyl and hydride shift. For Type B isomerization, the branching degree of carbocation is increased or decreased *via* the formation of cyclic carbonium ion intermediates.[30]

Type A:

Type B:

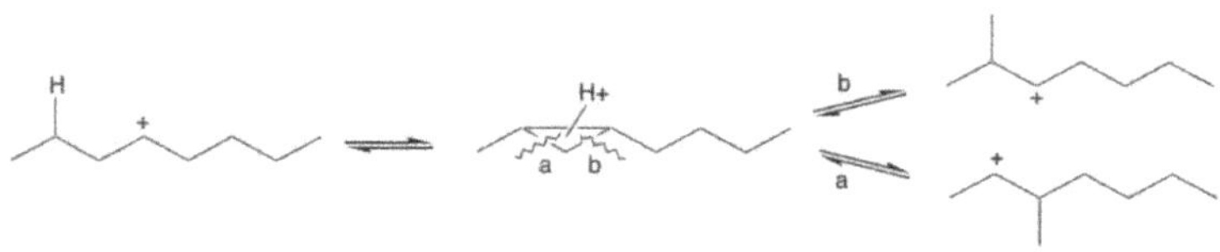

9.4 Downstream Options for Syncrudes

9.4.1 C_2 Hydrocarbons

Ethylene content is high in HTFT syncrude, and research has been done on converting ethane into higher-range hydrocarbons. ZSM-5 zeolite is used as the catalyst. Aromatization occurs at a relatively high temperature above 250 °C. At a low temperature, olefins oligomerize and then crack. The major product is dimer. With an increase in the amount of alumina in the catalyst, cracking is facilitated. Furthermore, aromatic alkylation can be considered to convert ethane into a range of gasoline products. However, ethane is commonly used as a fuel gas instead of being refined into liquid fuel.

9.4.2 C_3 Hydrocarbons

HTFT has the highest C_3 hydrocarbon content, which reaches 15% in syncrude. Propylene has a higher content than propane. Both propylene and propane can be sold directly as chemicals after purification. Propane is a kind of fuel that is the main component of liquified petroleum gas (LPG). Propylene can be polymerized to produce plastics. Except for being used as material for chemicals, C_3 hydrocarbons can also be further treated into components of liquid fuels.

Both aromatic alkylation and aliphatic alkylation are available for propylene. Benzene and propylene can be alkylated into cumene, which has a high ON and the potential to be used as a chemical. Though alkylation of propylene with isobutane is feasible, it is less desirable than alkylation between butene.

Aromatization can convert both paraffin and olefin into aromatics. Thus aromatics can be produced from propylene and propane *via* aromatization. Propylene can be oligomerized into heavier hydrocarbons to serve as components of liquid fuel. It is preferable to produce diesel range product to improve cold flow properties.

9.4.3 C_4 Hydrocarbons

Butane can be blended into gasoline fuels or used directly as a component of LPG after separation from butene. 1-Butene obtained *via* etherification to remove isomers like 2-butene can be used for polymerization into plastics, while most C_4 hydrocarbons need to be refined. The upgrading technologies used for C_4 hydrocarbons may be varied on account of the olefin-to-paraffin ratio. With HTFT, olefins account for most of the composition of syncrude. Oligomerization of butene is a good choice. 1-Butane produces better product than 2-butane.[31] Aliphatic alkylation following isomerization is also a process worth considering. The existence of isomerization is to convert the linear FTS paraffins into isomers. When butane content is too low compared with butene, even when all the butane is isomerized, there is still butene left which cannot be alkylated.

9.4.4 C_5/C_6 Hydrocarbons

The C_5/C_6 stream has high volatility and a low octane number. Isomerization is a vital process to satisfy the octane number demand by increasing the branching degree. The catalysts used are bifunctional, with both metallic and acidic sites. Pt-chlorinated alumina, sulfated zirconia and zeolite catalysts are most widely used in industrial refiners. The reaction activity of chlorinated alumina is the highest.[32] There is growing interest in oligomerization/dimerization of C_5/C_6 light alkenes into diesel and gasoline blending components. It is exothermic, and the low temperature and high pressure are beneficial to the equilibrium.[32] The structure of oligomers is influenced by the porosity of the zeolite. With increasing porosity, the degree of branching increases.[33] Isopentene can be used as the feed for etherification to produce ethers like tertiary amyl methyl ether.[2] Since FTS syncrudes are mostly linear. Straight olefins need to be isomerized before the etherification process. C_6 can also be used as the reactant for aromatization to produce benzene as chemical; however, it is not a desired component in fuel and can cause environmental pollution.[32]

9.4.5 C_7–C_{11} Hydrocarbons

This stream cannot be used for direct blending because it contains mostly linear olefins and paraffins and has a low octane number. Hydroisomerization can be applied to increase the degree of branching. However, for hydrocarbons with a carbon number higher than 7, hydrocracking is obvious during the hydroisomerization process. There is reduction in ON due to the side reaction.[16] Catalytic reforming can also be considered. Straight hydrocarbons are converted into aromatics with a high octane number. However, FTS syncrude is a poor feedstock for reforming. The larger $(N+2A)$ is (N represents percentage of naphthenes; A refers to percentage of aromatics), the less severe the operating conditions are. Because of the low $(N+2A)$ number, FTS syncrudes are poor materials for catalytic reforming. The Pt/l-zeolite is monofunctional and without acidic sites and so is more efficient.[12]

9.4.6 C_{10}–C_{15} Hydrocarbons

The kerosene range hydrocarbons can be hydrotreated to produce jet fuel products. If the refining process is aimed at achieving maximum gasoline yield, the stream can also be hydrocracked to produce naphtha-range hydrocarbons.[8]

9.4.7 C_{15}–C_{20} Hydrocarbons

Diesel-range hydrocarbons follow a similar refining process to the kerosene stream. The hydrotreated product shows a good cetane number, which satisfies the specifications for diesel fuel. This stream can also be hydrocracked to increase the yield of lighter hydrocarbons.[8]

9.4.8 $C_{20}+$ Wax

Hydrocracking is applied to reduce the heavy residue into naphtha, kerosene- and diesel-range hydrocarbons. Hydroisomerization and olefin saturation can synergistically work with hydrocracking. Cracking breaks bonds, and the formed olefin products are consequently hydrogenated into paraffinic products.[8,34]

9.5 Conclusion

Biomass-to-liquids processes *via* the Fischer–Tropsch reaction produces a range of hydrocarbons that offers a variety of refining opportunities. This process can produce high-quality diesel or gasoline alongside other commodity chemicals. The biomass feedstock can integrate well with fossil fuel sources to reduce the overall carbon dioxide emissions of currently applied processes.

Abbreviations

ASA	Amorphous silica alumina
ASF	Anderson–Schulz–Flory
ETBE	Ethyl *tert*-butyl ether
FT	Fischer–Tropsch
FTS	Fischer–Tropsch synthesis
GHG	Greenhouse gas
HDM	Hydrodemetallization
HDN	Hydrodenitrogenation
HDO	Hydrodeoxygenation
HDS	Hydrodesulfurization
HTFT	High-temperature FT
LPG	Liquified petroleum gas
LTFT	Low-temperature FT
MTBE	Methyl tertiary butyl ether
ON	Octane number
RON	Research octane number
SPA	Solid phosphoric acid
TAME	Tertiary amyl methyl ether
WGS	Water-gas-shift

References

1. T. Seadira, G. Sadanandam, T. A. Ntho, X. Lu, C. M. Masuku and M. Scurrell, *Rev. Chem. Eng.*, 2018, **34**, 695.
2. L. P. Dancuart, R. de Haan and A. de Klerk, *Stud. Surf. Sci. Catal.*, 2004, **152**, 482.
3. R. C. Baliban, J. A. Elia and C. A. Floudas, *Ind. Eng. Chem. Res.*, 2010, **49**, 7343.
4. E. van Steen and M. Claeys, *Chem. Eng. Technol.*, 2008, **31**, 655.
5. I. S. Tagomori, P. R. R. Rochedo and A. Szklo, *Biomass Bioenergy*, 2019, **123**, 134.
6. R. C. Baliban, J. A. Elia, R. Misener and C. A. Floudas, *Comput. Chem. Eng.*, 2012, **42**, 64.
7. A. P. Steynberg, W. U. Nel and M. A. Desmet, *Stud. Surf. Sci. Catal.*, 2004, **147**, 37.
8. A. de Klerk, *Energy Environ. Sci.*, 2011, **4**, 1177.
9. R. B. Anderson, H. Friedel and H. H. Scorch, *J. Chem. Phys.*, 1951, **19**, 313.
10. W. D. Shafer, M. K. Gnanamani, U. M. Graham, J. Yang, C. M. Masuku, G. Jacobs and B. H. Davis, *Catalysts*, 2019, **9**, 259.
11. A. A. Muleja, J. Gorimbo and C. M. Masuku, *Catalysts*, 2019, **9**, 746.
12. A. de Klerk, *Fischer–Tropsch Refining*, John Wiley & Sons, 2011.
13. A. W. Weitkamp, H. S. Seelig and N. J. Bowman, *Ind. Eng. Chem.*, 1953, **45**, 343.

14. D. G. Cain, A. W. Weitkamp and N. J. Bowman, *Ind. Eng. Chem.*, 1953, **45**, 359.
15. M. E. Dry, *ACS Symp. Ser.*, 1987, **328**, 18.
16. Y. N. Yang, S. Y. Peng and B. Zhong, *Catal. Lett.*, 1992, **16**, 351.
17. N. V. Chekantsev, M. S. Gyngazova and E. D. Ivanchina, *Chem. Eng. J.*, 2014, **238**, 120.
18. V. Nieminen, M. Kangas and T. Salmi, *Ind. Eng. Chem. Res.*, 2005, **44**, 471.
19. K. Surla, H. Vleeming and D. Guillaume, *Chem. Eng. Sci.*, 2004, **59**, 4773.
20. A. de Klerk and P. L. de Vaal, *Ind. Eng. Chem. Res.*, 2008, **47**, 6870.
21. W. L. Lafferty, Jr and R. W. Stokeld, Origin and Refining of Petroleum-alkylation and isomerization, *Adv. Chem.*, 1971, **103**, 130.
22. X. Guo, G. Liu and E. D. Larson, *Ind. Eng. Chem. Res.*, 2011, **50**, 9743.
23. M. Guisnet, N. S. Gnep and D. Aittaleb, *Appl. Catal., A*, 1992, **87**, 255.
24. C. P. Nicholas, *Appl. Catal., A*, 2017, **543**, 82–97.
25. A. de Klerk, *Ind. Eng. Chem. Res.*, 2004, **43**, 6349.
26. J. Hidalgo, M. Zbuzek and R. Černý, *Open Chem.*, 2014, **12**, 1.
27. D. Smook and A. de Klerk, *Ind. Eng. Chem. Res.*, 2006, **45**, 467.
28. S. Parkash and D. Hydrotreating, *Refining Processes Handbook*, Elsevier, 2003.
29. A. de Klerk, *Catal. Today*, 2008, **130**, 439.
30. C. Bouchy, G. Hastoy and E. Guillon, *Oil Gas Sci. Technol. Revue de l'IFP*, 2009, **64**, 91.
31. M. Golombok and J. de Bruijn, *Ind. Eng. Chem. Res.*, 2000, **39**, 267.
32. A. Aitani, M. N. Akhtar and S. Al-Khattaf, *Energy Fuels*, 2019, **33**, 3828.
33. G. Bellussi, F. Mizia and V. Calemma, *Microporous Mesoporous Mater.*, 2012, **164**, 127.
34. M. Fahim, T. Al-Sahhaf and A. Elkilani, Hydroconversion, in *Fundamentals of Petroleum Refining*, Elsevier, 2010.

CHAPTER 10

BTL-FT to Transportation Fuels

Y. ZHAO AND C. M. MASUKU*

Davidson School of Chemical Engineering, Purdue University, West Lafayette IN 47907, USA
*Email: cmasuku@purdue.edu

10.1 Introduction

Fossil fuel (petroleum, natural gas and coal) is the main energy source used worldwide and contributes about 80% of total energy consumption.[1] More than half of the refined petroleum and other liquid fuels are used in the transportation sector, which is the largest consumer of liquid fuels.[2] Although still the primary source of energy, fossil fuel is nonrenewable, and stocks are finite. Concern that global reserves will eventually run out has persisted for decades.[3] According to BP's Statistical Review of World Energy 2016, global coal stores will provide some 114 more years of production, while the estimate for petroleum and natural gas is roughly 50 years.[4] However, before this process of exhausting fossil fuels reserves becomes a problem, the climate change crisis may make itself felt. Combusting fossil fuel for energy and transportation causes a large amount of CO_2 emission, which is the biggest driver of global climate change. In order to meet the global climate change target of keeping average warming of the planet below 2 °C, about two-thirds of known fossil fuel stores need to be left untouched.[3] Thus it is necessary to investigate alternative methods of producing fuel, especially liquid fuel used in the transportation industry. One way to address these challenges is to replace fossil fuel with renewable resources. Currently,

Catalysis Series No. 44
Chemicals and Fuels from Biomass *via* Fischer–Tropsch Synthesis: A Route to Sustainability
Edited by Joshua Gorimbo, Xinying Liu, Yali Yao and Diane Hildebrandt

Published by the Royal Society of Chemistry, www.rsc.org

renewables account for merely 15% of total energy consumption worldwide. However, according to the U.S. Energy Information Administration (EIA), its share will continue to increase in the next three decades, and it will eventually become the primary source of energy worldwide and will overtake petroleum in terms of size of share.[1]

Renewable energy is a collective term for a number of different energy sources, including hydropower, solar, wind and biomass. Of these, biomass is the only carbon-based resource that can be used as a substitute for fossil fuel, as liquid fuel can be produced from it. Biomass refers to any material of biological origin, including organic material produced by photosynthesis, living microorganisms and all dead and waste biomass such as animal residue and agricultural residue. These can be obtained continuously from a wide range of sources within a short period of time.[5,6] However, in order to avoid a potential effect on the food market, the use of non-food-based biomass for energy is necessary. Liquid fuel derived from biomass is called biofuel, with various types being produced. EIA estimates that the percentage of biofuel used compared to gasoline, diesel and jet fuel will continue to increase in the future.[1] If biofuel is used, CO_2 emission can be reduced. The utilization of biomass not only satisfies the demand for sustainable development but also helps to reduce greenhouse gas (GHG) emissions due to the photosynthesis effect.

Biomass is a promising carbon-based feedstock for replacing fossil fuel. It is estimated that the terrestrial biomass carbon worldwide is approximately 50 times that of fossil-fuel-related carbon.[6] Green liquid fuel can be produced from biomass *via* a series of biological and thermochemical routes. The biological route relies on microbial production of fuel from biomass saccharides.[7] Thermochemical routes include biomass pyrolysis, gasification and liquefaction, which can convert biomass into syngas or bio-oil.[8] These thermochemical routes have certain advantages over the biological routes. For instance, no chemical pretreatment is required, and processing time is more efficient. The thermochemical methods used include the well-known biomass-to-liquid *via* Fischer–Tropsch (BTL-FT) process that can convert carbonaceous feedstock into liquid fuel through gasification of biomass followed by Fischer–Tropsch synthesis (FTS).[9]

FTS is a well established reaction that has increasingly been incorporated into a sustainable process to produce synthetic liquid fuel from synthetic gas (syngas), which is composed of CO and H_2. Syngas can be obtained *via* natural/shale gas reforming, coal or biomass gasification;[10] however, a metallic catalyst must be used. The most widely used options are cobalt and iron. FTS can be operated at both low- and high-temperature conditions, and different distributions of hydrocarbons are generated. Liquid transportation fuel like gasoline, diesel and jet fuel can be produced *via* further refining of the FTS syncrude. FT liquid fuel products not only have similar characteristics to conventional petroleum fuel but also have special advantages, in that FT fuel is free of sulfur and consists of very few aromatics and nitrogen compounds, which leads to less emission of pollutants.[11] Thus the production of a substitute synthetic liquid fuel *via* the BTL-FT process is one approach that could be considered to shift the current dependence on petroleum.[10]

10.2 Fuel Specifications

There are both technical and political concerns about fuel specifications.[12] The specifications vary around the world and are country dependent, including in areas like the European Union, North America, India and Japan.[13] The three main types of transportation fuel produced are motor-gasoline, diesel fuel and jet fuel, and the property specifications for these are detailed next.

10.2.1 Motor-gasoline

Gasoline is a transportation fuel used in ignition engines. It is composed of hydrocarbons that mainly range from C_5 to C_{11}. The guidelines that are widely accepted for motor-gasoline are the specifications developed by European Committee for Standardization (CEN) and American Society for Testing and Material (ASTM).[12,13] Examples are ASTM D4814 (Standard Specification for Automotive Spark-ignition Engine Fuel); EN228 (Automotive Fuels–Unleaded Petrol–Requirements and Test Methods). Specifications are provided for the following properties in ASTM D4814: volatility, lead content, corrosion by copper and silver, gum content, sulfur content, oxidation stability.[14] Other properties mentioned in EN228 are octane number (ON) (research octane number (RON) and motor octane number (MON)), density, olefins and aromatics content, benzene and oxygenate content.[15] The requirements and test methods are summarized in Table 10.1.

10.2.1.1 Octane Number

Certain molecules in gasoline tend to trigger self-ignition and detonation under high temperature and high pressure, before reaching the ignition spark, which can lead to low efficiency of the engine. ON is a measure of the gasoline's antiknock property, and this is one of the most important properties of gasoline. The higher the number, the better the fuel's capability to withstand compression.

There are two kinds of octane ratings: RON and MON. RON is used to evaluate fuel performance for city driving, when speed and temperature are low; MON represents fuel performance for highway driving, when a faster engine speed and higher fuel temperature are required. The two rating methods use different conditions.[16] ASTM D2699 (Research Octane Number of Spark-Ignition Engine Fuel) and ASTM D2700 (Motor Octane Number of Spark-Ignition Engine Fuel) are specifications for testing.

The value of the octane rating is dependent on compound class, chain length and the structure of the components in the gasoline. For hydrocarbons

Table 10.1 Requirements and test methods for Automotive Fuels-Unleaded Petrol in EN 228. Reproduced from ref. 15 with permission from John Wiley & Sons, Copyright © 2016 Wiley-VCH Verlag GmbH & Co. KGaA.

Properties	Units	Min	Max	Test method
RON	—	95.0	—	EN ISO 5164/ ASTM D2699/IP237
MON	—	85.0	—	EN ISO 5163/ ASTM D2700/IP236
Lead content	mg l^{-1}	—	5.0	EN 237/ASTM D3341
Density	kg m^{-3}	720.0	775.0	EN ISO 3675/D1298/ IP 160 EN ISO 12185/D4052
Sulfur content	mg kg^{-3}	—	10.0	EN ISO 13032/ ISO13032 EN ISO 20846/ ASTM D5453
Oxidation stability	Minutes	360	—	EN ISO 7536/ ASTM D525/IP 40 ASTM D 7525
Existent gum content	mg per 100 ml^3	—	5	EN ISO 6246/ ASTM D381
Copper strip corrosion (3 h at 50 °C)	Rating	Class 1		EN ISO 2160/ ASTM D130/IP 154
Olefins content	% (v/v)	—	18.0	EN 15553/ ASTM D1319/IP156
Aromatics content		—	35.0	EN ISO 22854/ ASTM D6839
Oxygenate content:	% (v/v)			EN 13132/
-Methanol		—	3.0	ASTM D4815
-Ethanol		—	10.0	EN ISO 22854/
-Iso-propyl alcohol		—	12.0	ASTM D6839
-Iso-butyl alcohol		—	15.0	
-*Tert*-butyl alcohol		—	15.0	
-Ethers (5 or more C atoms)		—	22.0	
-Other oxygenates		—	15.0	

with the same carbon number, aromatics have the highest ON, followed by oxygenates and naphthenes.[12] Linear paraffins have the lowest ON. Most olefins have a higher ON than paraffins when their structures are similar, but this is also dependent on the skeletal structure. With an increase in the degree of branching and deduction in the carbon number, ON increases.[12]

It is important to obtain the ON of a specific gasoline fuel accurately. Much research has been carried out to correlate the ON with the gasoline constitution. Ghosh *et al.*[16] developed a detailed gasoline composition-based octane model that can be used to predict the RON and MON of gasoline fuel products, regardless of origin. The standard error between model prediction and measurement value was within 1. Correa-Gonzalez *et al.*[17] compared the potential of various machine learning (ML) algorithms to predict the RON and concluded that the nonlinear support vector machines could achieve the best result (absolute error of 0.9).

Each hydrocarbon component contributes to the octane number assigned to the gasoline. FT fuel has a low ON, which is unfavorable for gasoline production; therefore, efforts should be made to increase the ON. Processes like alkylation, aromatization, catalytic reforming and etherification are considered to increase the degree of branching and the concentration of aromatics and oxygenates.

10.2.1.2 Volatility

Volatility is a measure of the evaporation characteristics of a gasoline product, which is considered the key characteristic for good drivability. It is a function of temperature and pressure; therefore, adjustments must be made in terms of altitude and the seasonal temperature of the location in which it will be used. Volatility is measured in terms of three properties: vapor pressure, distillation profile, vapor/liquid ratio.[18,19]

Fuels with high volatility – which also means high vapor pressure – are volatile. This can trigger the vapor lock phenomenon, which is caused by the liquid fuel changing into gas while still in the fuel delivery system. The gaseous phase occurs inside the closed conduit, resulting in a loss of power and engine stalling.[20] Gasoline volatility influences a vehicle's drivability as well as its volatile organic compounds (VOC) emissions. The evaporative emissions from gasoline can contribute to ground-level ozone and ozone-related health and environmental problems. Thus specifications have been developed by the Environmental Protection Agency (EPA) to restrict this.[18] Reid vapor pressure (RVP) is the vapor pressure at 100 °F, as measured by the ASTM D323 (Standard Test Method for Vapor Pressure of Petroleum Products (Reid Method)).[12] It is a common measure of gasoline volatility. Since vapor pressure increases with an increase in the temperature of liquid, the EPA regulates the vapor pressure of gasoline during summertime to reduce gasoline emissions of VOC.[18] This means that producing gasoline with a low volatility is desired; however, although there is no minimum limit specified for vapor pressure in ASTM D4814, a vapor pressure that is too low is not desirable because it may cause cold start problems.[21]

Gasoline is a mixture of a range of hydrocarbons with different boiling points. The distillation profile represents the increasing temperatures at which gasoline evaporates for a fixed series of increasing volume percentages or *vice versa*. It is measured by ASTM D86 (Distillation of Petroleum Products at Atmospheric Pressure). ASTM D4814 sets the distillation temperatures at the 10%, 50% and 90% evaporation points. Each set point is related to specific gasoline performance categories. Gasoline with good front-end volatility has low evaporation and easy starting characteristics and is free from vapor lock. Gasoline with a good midrange volatility is good in terms of short-trip economy, delivers rapid warm-up and good acceleration and protection against icing and hot stalling. Gasoline with good tail-end volatility is free of engine deposits and has minimal VOC exhaust emission. Temperature at a vapor/liquid ratio of 20 is the factor most correlated with vapor lock and other hot fuel handling problems, and gasoline products with a higher ratio number have better resistance against vapor lock.[19]

10.2.1.3 Oxygenate Content

Oxygenated organic compounds are added to gasoline to increase the octane number or to extend gasoline supplies.[23] The oxygenates most commonly added are ethers and alcohols, especially methyl tertiary butyl ether (MTBE), tertiary amyl methyl ether (TAME) and ethyl tertiary butyl ether (ETBE) with an ON of 100–114, and ethanol with an ON of 113. Ethers are preferred due to various reasons. Ethanol has a higher heat evaporation temperature than ether. It may result in reduced drivability since additional heat is required to vaporize the fuel.[19] The use of ethanol-blended gasoline may also affect evaporative emissions because ethanol can raise volatility.[23] Gasoline that contains ethanol can make the permeability problem worse.[19] The use of MTBE is forbidden by some states in the United States. Ethanol and heavier ethers are the main additives considered. It is worth mentioning that MTBE use has not been banned in Europe and some other countries.[12]

Proper oxygenating helps to ensure good performance of gasolines. It can increase the antiknock capability of gasoline. The oxygenated gasoline tends to have a lower ozone reactivity than nonoxygenated fuels. Moreover, oxygenated gasoline can induce a lean shift, which can reduce CO emissions, even though its performance is dependent on the operating system.[23]

With the BTL-FT process, a certain amount of oxygenate (mainly alcohols) exists in the FT hydrocarbon. However, most of the short-chain oxygenates are in the aqueous phase instead of the oil phase. Reaction like etherification can be considered in the refining process to increase the content of oxygenates in gasolines.

10.2.1.4 Olefin Content

Olefins are hydrocarbons that contain at least one carbon double bond. They are a major component of gasoline. The existence of olefins can promote the reactivity of gasolines during combustion and improve the octane number. With the development of modern vehicles, olefin content currently has no significant effect on emissions, including HC, NO_x, CO, benzene and so on, except for emissions of 1,3-butadiene. However, this may eventually be reduced. An olefin content that is too high can result in engine deposits and gum formation, especially those caused by dienes. Also, the evaporation of olefins contributes to ozone formation, and the combustion products form toxic dienes. Thus olefin content is restricted as per the specifications. This can be tested using standard methods that involve various technologies, including gas chromatography (ASTM D4815), infrared spectroscopy (ASTM D5845) and so on.

As illustrated in Table 10.2, with the BTL-FT process, olefins are major components in the gasoline materials produced by FTS. Their treatment

Table 10.2 Composition of gasoline-range materials produced by FTS on a mass basis. Reproduced from ref. 13 with permission from the Royal Society of Chemistry.

	Syncrude composition (mass%)		
Compound	Fe-HTFT (%)	Fe-LTFT (%)	Co-LTFT (%)
C_5–C_{10}			
Olefins	77.25	62.60	39.00
Paraffins	12.87	26.83	60.00
Aromatics	5.09	0.00	0.00
Oxygenates	4.79	10.57	1.00

needs to be designed meticulously, and reactions like alkylation and aromatization can be included in the refining process.

10.2.1.5 Aromatic Content

Aromatics in gasoline are hydrocarbons that contain at least one benzene ring. Aromatic components rarely occur in BTL-FT gasoline, but the HTFT process that uses an iron catalyst may contain a small quantity. The addition of aromatics in gasoline fuel can boost the octane number and increase the antiknock capability. In addition, aromatics are high-energy, dense fuel molecules. However, their existence also triggers problems. Aromatics, especially heavy aromatics, can increase the tendency to form engine deposits that can increase tailpipe emissions, while the combustion of aromatics contributes to the formation of benzene,[23] which is a toxic carcinogen. Under climate conditions with high temperatures, the benzene evaporates into the atmosphere and causes exposure to attendants at fueling stations.[24] A linear relationship between CO_2 emission and aromatic content also exists.[23]

10.2.1.6 Sulfur Content

Sulfur is a heteroatom that exists naturally in crude oil, and the use of fuel containing sulfur can result in many disadvantages. First, the combustion of these types of fuel produces sulfur dioxide (SO_2), which is the major precursor of acid rain.[19] Second, the existence of sulfur reduces the efficiency of the catalyst of exhaust gas catalytic converters, which is used to change harmful substances (like carbon monoxide (CO), hydrocarbons (HC) and NO_x in exhaust gas) into less harmful components like carbon dioxide (CO_2), water and nitrogen (N_2). Sulfur increases vehicle emissions, and it is also reported that sulfur can increase ozone formation and adversely affect exhaust gas oxygen sensors.[23]

With the BTL-FT process, the sulfur-containing compounds are also poisonous to FTS catalysts. Sulfur is introduced in the form of H_2S and $(CH_3)_2S$ as typical biomass-derived gas impurities. The compounds have a negative impact on reaction activity, even though they do not influence the product distribution.[22] There is no positive effect reported for sulfur existing in gasoline; thus strict restrictions are placed on sulfur content. However, with the BTL-FT process, the hydrocarbons produced are sulfur-free because they are removed through the Claus process before the FTS reaction occurs.[12]

10.2.1.7 Oxidation Stability

Oxidation stability shows how stable hydrocarbons, especially unsaturated ones, are able to resist oxidation reactions in the presence of air, *i.e.* because these reactions initialize the formation of undesired glutinous resin.[25] Oxidation stability is evaluated by determining the tendency to develop gum in storage by standard methods like ISO 7536[26] (Petroleum Products–Determination of Oxidation Stability of Gasoline–Induction Period Method) and ASTM D525[27] (Oxidation Stability of Gasoline (Induction Period Method)).

10.2.1.8 Gum Formation

The slow oxidation of unstable hydrocarbons like olefins and diolefins in storage at ambient temperature contributes to the formation of oxidized solids with a high molecular weight that can become insoluble and sticky. The oxidation products are called "gums" or "deposits". They can stick to the vehicle system and have an adverse effect on engine performance, emission and durability,[28] ASTM D381[29] (Gum Content in Fuels by Jet Evaporation) is used to measure the existing gum content by evaporating gasoline and weighing the residue. The summation of soluble gum that is measured by ASTM D381 and the insoluble gum[30] that is collected by filtration is referred to as potential gum.[12,19]

10.2.1.9 Corrosion

To determine the corrosiveness of the fuel, two main types of ratings are used: Copper Strip Corrosion and Silver Strip Corrosion. They are tested respectively by ASTM D130 (Detection of Copper Corrosion from Petroleum Products by the Copper Strip Tarnish Test) and ASTM D4814 Annex A1 (Corrosiveness of Silver from Petroleum Products by the Silver Strip Test). The Copper Strip Corrosion test is not sensitive to conventional gasoline because the sulfur species in the gasoline can provide a protection to copper. As more stringent specifications are issued to lower the sulfur content, this method will become more practical for modern gasoline products.[19]

10.2.2 Diesel Fuel

Diesel is a mixture of hydrocarbons mainly within the C_{11}–C_{22} range. It is the fuel used for compression ignition engines.[12] Thus good compression ignition capability is required. The most commonly used specifications for diesel fuels are ASTM D975 (Standard Specification for Diesel Fuel Oils)[32] for the United States and EN 59033 for the European Union, while other standards are used in other countries. The minimum cetane number required by CEN is higher than that required by ASTM. The following properties are dealt with in the regulations: cetane number, density, viscosity, sulfur content, aromatics content, flash point, ash content, distillation characteristics, corrosion and lubricity.[33] Table 10.3 lists the requirements and test methods for diesel fuel[33] and jet A fuel.[34]

Diesel fuel produced using the BtL-FT process is different from conventional petroleum-based diesel. FT syncrude distribution is dependent on operating conditions, especially temperature. LTFT syncrude contains more straight-run distillates than HTFT, making the process beneficial for the production of diesel.[13] Since FT syncrude is mostly linear, the diesel fuel

Table 10.3 Detailed requirements and test methods of diesel fuel in EN 590 and Jet A aviation fuel in ASTM .

Properties	Units	Min.	Max.	Test method	Ref.
Specifications for EN 590 diesel fuel					33
Cetane number		51	—	EN ISO 5165	
Cetane index		46	—	EN ISO 4264	
Density at 15 °C	kg m^{-3}	820	845	EN ISO 3675 EN ISO 12185	
Polycyclic aromatic hydrocarbons	% (m / m^{-1})	—	11	EN 12916	
Sulfur content	mg kg^{-1}	—	50	EN ISO 20846 EN ISO 20847 EN ISO 20884	
			10	EN ISO 20846 EN ISO 20884	
Flash point	°C	>55	—	EN ISO 2719	
Oxidation stability	g m^{-3}	—	25	EN ISO 12205	
Copper strip corrosion (3 h at 50 °C)	Rating	Class 1		EN ISO 2160	
Lubricity, WSD at 60 °C	μm	—	460	EN 2156-1	
Viscosity at 40 °C	mm^2 s^{-1}	2	4.5	EN ISO 3104	
Distillation % (v/v) recovered at				EN ISO 3405	
250 °C	% (V/V)	85	<65		
350 °C	% (V/V)				
0.95% (v/v) recovered at	°C		360		
Specifications for Jet A					34
Aromatics	% vol	—	25	ASTM D 1319	
Sulfur	% mass	—	0.3	ASTM D 3227	
Distillation:	°C			ASTM D 86	
10% recovered		—	205		
50% recovered		—	Report		
90% recovered		—	Report		
Final boiling point		—	300		
Flash point	°C	38	—	ASTM D 56	
Density at 15 °C	kg m^{-3}	775	840	ASTM D 1298	
Freezing point	°C	—	− 40	ASTM D 2386	
Viscosity at −20 °C	mm^2 s^{-1}	—	8	ASTM D 445	
Net heat of combustion	MJ kg^{-1}	42.8	—	ASTM D 4529	

produced has outstanding ignition quality. The fuel is sulfur-free.[31] However, the FT-derived distillates are reported to have a low density, which affects consumption.[12]

10.2.2.1 Cetane Number

Since diesel engines are compression ignition engines that ignite fuel at a high temperature, this results in compressed gases. The cetane number (CN) is a critical property used to evaluate the ignition and combustion ability of the diesel.[12] The standard test method is ASTM D613 (Standard Test Method for Cetane Number of Diesel Fuel Oil).[35] It measures the ignition delay. The higher the number, the better the quality of the fuel. Specifications have minimum limits for CN, so as to restrict the performance of the fuel. The minimum cetane number requirement set by CEN is 51;[33] with ASTM, it is 40.[32]

Diesel fuel with a high CN has a short delay period, which is an indication of good autoignition capability. The fuel starts to combust shortly after being injected into the compression cylinder. Engines may benefit from a higher CN when starting in cold weather. Also, a high CN can mean a reduction in emissions. NO_x is diminished in all diesel engines with high CN. Particulate matter (PM) emissions are engine dependent because prompt ignition may cause the fuel to burn in a diffusion-controlled phase, which increases PM emissions.[35] However, an increase in CN is not a limitless benefit: if ignition is too rapid, it may result in improper mixing of fuel and air, which can reduce engine performance and increase emissions.[12]

CN is dependent on the hydrocarbon structure. Linear paraffins have a high CN, and CN rises with an increase in carbon number. The CN of iso-paraffins is related to the skeletal structure: multiple short side chains lead to a low cetane number, while few side chains containing four or more carbons result in a high CN. Aromatics and olefins follow the same rules: the fewer and the longer the side chains, the higher the CN. Hydrocarbons with more than one aromatic ring have a CN lower than 20.[36]

Since the BtL-FT process produces mostly linear paraffins and very low aromatic content, the derived diesel fuel generally exceeds the CN specifications.[13]

10.2.2.2 Flash Point

Flash point is measured using the standard test method ASTM D93 (Flash-point by Pensky–Martens Closed Cup Tester). It is the lowest temperature at which sufficient vapor above the fuel liquid can be ignited.[36] The lower the temperature, the easier for the fuel to ignite. However, too low a CN can make the fuel dangerous to handle. Flash point is regarded as an assessment for safety in handling and use, like fire risk in storage and not an engine performance property.[12,36]

10.2.2.3 *Distillation Characteristics*

The distillation profile indicates the amount of fuel that evaporates at a given temperature. It is measured by ASTM D86 (Distillation of Petroleum Products).[37] The fuel is heated, and the temperature at a certain recovered volume is recorded.[36] The profile can be divided into three parts: light end, midrange and heavy end. Each is related to different properties. The light end affects starting ability; midrange affects viscosity and density; heavy end relates to boiling point. Too much fuel in the heavy end causes an increase in the PM emission due to coking.[23]

10.2.2.4 *Sulfur Content*

The existence of sulfur in fuels is disadvantageous in many respects. The combustion of fuels containing sulfur can cause the formation of SO_2, which is the primary sulfur compound emitted from an engine. It may damage the environment by triggering acid rain. Sulfur in diesel also results in higher PM emissions. SO_2 can be further oxidized into SO_4, when the sulfate is combined with water molecules and aerosols are formed.[23] Sulfate can also poison the catalyst in an exhaust emission control system and lead to higher exhaust emissions, as well as to corrosion and wear of the engine.[12] Thus specifications for sulfur content are becoming more and more stringent. Various standards apply for certain diesel fuels to measure the sulfur contents using X-ray spectrometry, ultraviolet fluorescence and so on.[36] However, with the BtL-FT process, the fuel produced is sulfur-free, so it is not impacted by the strict regulations.

10.2.2.5 *Aromatics*

Aromatics refer to hydrocarbons containing at least one benzene ring. Its content in diesel fuel is measured using ASTM D5168 (Determination of Aromatic Content of Diesel Fuels by Supercritical Fluid Chromatography), with monoaromatics and polycyclic aromatics being separated by supercritical carbon dioxide.[38] ASTM D1319 (Hydrocarbon Types in Liquid Petroleum Products by Fluorescent Indicator Adsorption)[36] can also be applied. The aromatics content in diesel fuel has an impact on many aspects of a fuel, including CN, density, combustion and exhaust emissions. The content is limited by the minimum CN requirement, since aromatics, especially polyaromatics, have a low cetane number. However, BtL-FT-derived fuel generally meets the CN regulations. Aromatics in synthetic fuel are considered a density booster,[13] as more aromatics contribute to a higher flame temperature, which results in an increase in NO_x emissions. Aromatics also lead to the formation of polycyclic aromatic hydrocarbons (PAH), and the emission of PM and the evaporation of PAH are both related to the PAH content.[23]

10.2.2.6 Cold Flow Properties

Diesel fuel needs good ignition properties, which require a high content of *n*-paraffins with a high CN. Although diesel quality is improved with the inclusion of linear paraffins, another problem is caused: *n*-paraffins have limited solubility in fuels and precipitate into wax crystal at a low temperature, which can cause clogging of the fuel system.[23] Thus cold flow properties can be improved by introducing branching. This is a critical measurement of fuel performance. The cold flow property is related to climatic and seasonal conditions; hence, the specifications are dependent on region.[12]

The cold flow property is defined by wax-related tests. ASTM D2500 measures the cloud point (CP) – also called the wax appearance point. It is the temperature at which the heaviest paraffins start to precipitate. EN116 is a test for the cold filter plugging point (CFPP). It determines the lowest temperature at which fuel can pass through a standard filter. CFPP is theoretically lower than CP because small wax crystals can still satisfy the test after precipitating at CP. ASTM D4539 (Low Temperature Flow Test) is a test developed to mimic the slow constant cooling process when a diesel engine is left overnight.[23,36] Additives that interact with wax crystals can be added to improve the cold flow property. They mitigate the effect of wax crystals by modifying their size, shape and even the degree of agglomeration.[36]

10.2.2.7 Density

Diesel fuel is used in compression ignition engines. The fuel is injected on a volume basis. A higher heating value per volume is proportional to density;[23] thus, the lower the fuel density, the higher the consumption of fuels and the lower the power output. From the perspective of fuel economy and engine performance, a high density is preferred.[39] However, increased density can increase PM emissions and NO_x emissions by conventional diesel engines, since modern engines tend to have been subjected to an exhaust emission treatment.[23] Hence specifications for density have both a lower and an upper boundary.

Density increases with an increase in the carbon number. It is also related to molecule structure. For hydrocarbons with the same carbon number, aromatics have the highest density, followed by naphthenes, with paraffins ranking last.[36] BtL-FT diesel fuel has a low density, but it can be compensated for a little by the aromatics in the fuel.

10.2.2.8 Lubricity and Viscosity

Viscosity has an impact on engine performance. Very-low-viscosity causes engine wear and fuel leakage; very high viscosity results in incomplete fueling due to the reduced flow rate.[23] Thus viscosity specifications have both a lower and an upper boundary. It is measured by ASTM D455

(Kinematic Viscosity of Transparent and Opaque Liquids).[40] The time needed for the fuel sample to flow through the capillary is recorded, and it is proportional to the kinematic viscosity of fuel. Viscosity is largely dependent on chain length and not as much on hydrocarbon structure. It increases with an increase in the CN.[36]

Lubricity is an indication of a fuel's ability to resist the friction between surfaces that are in relative motion. It can be separated into hydrodynamic lubrication and boundary lubrication. Hydrodynamic lubrication is related to viscosity because the friction happens between the layers of fuel liquid. The higher the viscosity, the better the hydrodynamic lubrication. Boundary lubrication provides a protection by the fuel adhering to the solid surface.[36] Certain content of oxygenates, sulfates, aromatics and nitrogen-containing compounds can contribute to a good lubrication performance. They follow a trend of oxygenates having the best enhancing performance, followed by N-containing compounds, with sulfates ranking last.[12] The heavy hydrotreating introduced due to the stringent specifications for sulfur content leads to a reduction of natural lubricants and causes a problem.[39] BtL-FT fuels obtained after hydrotreating can have the same problem. Low lubricity can cause engine wear. Additives can be used to increase lubricity.[12]

There are multiple methods used to evaluate lubricity, including ASTM D6079 (Evaluating Lubricity of Diesel Fuels by the High-frequency Reciprocating Rig (HFRR)) and ASTM D6078 (Evaluating Lubricity of Diesel Fuels by the Scuffing Load Ball-on-cylinder Lubricity Evaluator (SLBOCLE)). HFRR measures the wear scar of a steel ball that simulates the engine and has proved to have a good correlation with actual wear.[36]

10.2.3 Jet Fuel

Aviation fuel products can be categorized into aviation turbine fuel (also called jet fuel) and aviation gasoline, according to the type of engine used. Jet fuel is used in turbine engines, and aviation gasoline is used in piston engines. Jet fuel is a combination of a wide range of hydrocarbons in C_8–C_{16}. There are several grades of jet fuel, including Jet A-1, Jet A, Jet B, TS-1, JP-5 and so on. These grades are region and usage or purpose dependent. For example, Jet A-1 and Jet A are the two main kerosene fuels.[34] Jet A-1 is suitable for areas outside the United States (especially Europe), while Jet A is normally only available in the United States. Jet A-1 has a lower freeze point, while Jet B contains a wider range of hydrocarbons, including both naphtha and kerosene fractions. It is more flammable, and there is demand in very cold climates, like the Arctic regions. TS-1 is the fuel available in Russia. JP-5 is used for military purposes. The specifications for each grade are also factor dependent. The main specifications for Jet A-1 class are the UK's DEF STAN 91-91 (Jet A-1) and the ASTM D1655 (Jet A-1). The Jet A specifications are defined in ASTM D1655 (Jet A). The basic properties of Jet A in the ASTM D1655 specification are listed in Table 10.3, *e.g.* heat of combustion, sulfur and aromatics content, flash point, freeze point, density and viscosity.[34] Jet fuel is required to have good combustion characteristics and a high energy content because it is used to power aircraft, and these are the critical performance properties.

10.2.3.1 Combustion Characteristics

Turbine engines operate continuously. In the early stage of combustion, carbonaceous particles are formed, especially in terms of high aromatics and naphthenes content. They can be consumed by flame; however, their presence under high temperature and pressure flame conditions produces more heat, which can cause the formation of hot spots on the combustor wall. Cracks and premature engine failure can occur. If the particles are not totally consumed by the flame, they cause erosion of turbine blades and stators. The combustor can become plugged, and air is then not supplied. Particles are also responsible for the emission of smoke. Thus there are specifications for the aromatic content of jet fuel.

10.2.3.2 Energy Content

The energy content of jet fuel is expressed by the net heat of combustion. It refers to the amount of heat released when a unit of fuel is completely burned to form vapor water and carbon dioxide.[12] The net heat can be categorized into two types according to units: energy per unit mass of fuel and energy per unit volume of fuel. The net energy content is related to fuel composition. The higher the hydrogen content, the higher the gravimetric energy content. Hydrocarbons with a high hydrogen content (like paraffins) have a low density, which is also the characteristic of BtL-FT-derived fuels. Hydrocarbons with a low hydrogen content (like aromatics) have a high density, which results in a high volumetric energy content.[34,41] Since fuels are sold on a volumetric basis, a high energy per unit volume of fuel is generally preferred.[12] The net heat of combustion is measured by ASTM D4809 (Heat of Combustion of Liquid Hydrocarbon Fuels by Bomb Calorimeter). There are also two estimation methods: ASTM D3338 and ASTM D4529.[34]

10.2.3.3 Stability

Fuel with good stability can stay unchanged when exposed to disadvantageous conditions. Stability can be classified into two categories: storage stability and thermal stability. Instability during storage mainly happens because of oxidation reactions. This is influenced by the content of reactive compounds in the fuel. However, storage stability should not be an issue most of the time.[34]

Thermal stability is a critical property because jet fuel is used as the primary energy source for aircraft. The formation reactions of gum and particulates speed up with heating, which can cause problems like fuel fouling and coking, and this is harmful to the engine system. Thermal stability is measured by ASTM D324 (Thermal Oxidation Stability of Aviation Turbine Fuels).[34] It can be enhanced by means of additives.[42] FT-derived fuels have a similar thermal stability to conventional fuel types.[43]

10.2.3.4 Viscosity and Freezing Point

Unlike gasoline and diesel fuel, which is mainly used on the ground, jet fuel is used in aviation. The fuel is exposed to very low temperatures at altitude; thus the specifications for stringent low-temperature properties include the viscosity and freezing point.

Viscosity refers to the fluid's resistance to flow.[34] It increases with a decrease in temperature.[43] If the viscosity of jet fuel is too high, it may result in injection problems, since the fuel is designed to spray in the form of little droplets in order for it to evaporate quickly. High viscosity can also lead to a high pressure drop in the pipelines, which requires more driving force from the pumps. Thus the upper limit for viscosity has been determined. It is measured by ASTM D445 (Kinematic Viscosity of Transparent and Opaque Liquids).[34] FT-derived jet fuel products have an identical viscosity to that of conventional fuel types.[43]

Freezing point is the temperature at which the last wax crystal completely disappears. The higher the freezing point, the easier it is for the fuel to precipitate. Thus specifications have upper boundaries. The standard testing method is ASTM D2386 (Freezing Point of Aviation Fuels).[34]

10.3 Biomass-to-liquid-based Fischer–Tropsch (BtL-FT) Process

10.3.1 Process Description

Biomass-to-liquid (BtL) synthesis is a multistep process that includes syngas generation *via* biomass gasification, syngas conversion to hydrocarbons *via* FTS, and separation and refining of FT syncrude. Figure 10.1 depicts a general BtL process flow sheet.

Biomass contains a high level of moisture. In order to maintain efficiency of gasification, the moisture concentration needs to be controlled to be below a certain threshold. Thus the raw biomass should be dried and milled first.[11] The treated biomass is then gasified. Gasification technologies can be separated into the direct and indirect types. Directly heated gasifiers take advantage of the heat obtained *via* reactions between oxygen and organic compounds. High-purity oxygen is required for this process. This is restricted by the high capital cost of air separation units required to produce pure oxygen. Indirect gasifiers are heated through a sequence of steps. The primary reaction during indirect gasification is devolatilization of biomass to gas. Condensable vapors in the gas are converted into tar, and char is produced as reaction products. Char can be burned to produce heat. Tar is cracked into CO and H_2 using a cracker. Feedstock composition, gasifier

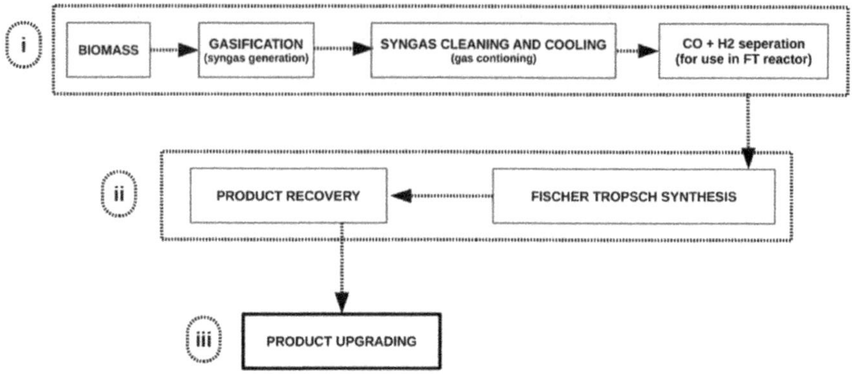

Figure 10.1 BtL-FT flowsheet. Reproduced from ref. 11 with permission from Elsevier, Copyright 2016.

type, residence time, operating parameters and the existence of the water-gas-shift (WGS) reaction all affect syngas composition.[44]

The main elements contained in biomass are C, O, H and a small percentage of N, Cl and S, which contribute to the contaminants produced in syngas, *e.g.* CO_2, H_2S, NH_3 and HCl. They must be removed before being sent to the FT reactor to prevent deactivation of the catalyst. H_2S can be converted to sulfur products through the Claus process. CO_2 can be recovered using a Rectisol unit[70] and NH_3 and HCl *via* adsorption.

The desired reactions during FTS are paraffins and olefins production with an H_2/CO ratio of $(2n + 1/n)$ and 2. Syngas composition within this range is ideal because undesired reactions may be triggered if the H_2/CO ratio is out of range. Iron-based catalysts are less sensitive to the H_2/CO ratio because it can promote the WGS reaction, which converts CO and H_2O into H_2 and CO_2. When a cobalt catalyst is applied, extra WGS should be considered to adjust the H_2/CO ratio prior to FTS.[12]

The operating conditions of FTS can be characterized into three categories by temperature and catalyst type: high-temperature FT (HTFT) catalyzed by an iron catalyst; low-temperature FT (LTFT) catalyzed by an iron catalyst or LTFT catalyzed by a cobalt catalyst.[13] The use of a cobalt catalyst at a high temperature is restricted because high light gas selectivity is observed under these conditions.[71]

Understanding syncrude composition could aid in designing the refining process and permit efficient use of materials. The main components of FT syncrude are hydrocarbons (including paraffins, olefins and aromatics) and oxygenates (mainly composed of alcohols, aldehydes, with the existence of acids, ketones and esters with a carbon number ranging from C_1 to C_{30+}). The generic composition of syncrude corresponds to three typical types of operating conditions, as listed in Table 10.4. It can be concluded that the distribution of syncrude produced by HTFT mainly lies in the C_1–C_5 range. A low content of heavier products like wax is achieved. HTFT syncrude also contains a certain number of aromatics, while the olefins to paraffins ratio is

Table 10.4 Generic FT syncrude composition (mass basis). Reproduced from ref. 13 with permission from the Royal Society of Chemistry.

	Syncrude composition (mass%)		
Compound	Fe-HTFT	Fe-LTFT	Co-LTFT
Gaseous product (C_1–C_4)			
Methane	12.7	4.3	5.6
Ethylene	5.6	1.0	0.1
Ethane	4.5	1.0	1.0
C_3–C_4 olefins	21.2	6.0	3.4
C_3–C_4 paraffins	3.0	1.8	1.8
Naphtha (C_5–C_{10})			
Olefins	25.8	7.7	7.8
Paraffins	4.3	3.3	12.0
Aromatics	1.7	0	0
Oxygenates	1.6	1.3	0.2
Distillate (C_{11}–C_{22})			
Olefins	4.8	5.7	1.1
Paraffins	0.9	13.5	20.8
Aromatics	0.8	0	0
Oxygenates	0.5	0.3	0
Residue/wax (C_{22+})			
Olefins	1.6	0.7	0
Paraffins	0.4	49.2	44.6
Aromatics	0.7	0	0
Oxygenates	0.2	0	0
Aqueous product			
Alcohols	4.5	3.9	1.4
Carbonyls	3.9	0	0
Carboxylic acids	1.3	0.3	0.2

high. With FTS catalyzed by an iron catalyst under low temperature, about 50% of the syncrude mass is C_{22+} hydrocarbons, *e.g.* wax. The paraffin-to-olefin ratio decreases with an increase in the carbon number, which means that the products become more paraffinic for longer chain length. However, a certain number of oxygenated compounds still exist, especially short-chain oxygenates in the aqueous phase. The content of aromatic compounds approaches zero. With cobalt, the products are more paraffinic, and less oxygenates are produced compared to the iron catalysis process.

Although it depends on the operating conditions, the components of hydrocarbons produced by the BtL-FT process follow some common rules, *i.e.*:

- FT effluents are mostly linear, with a low degree of branching.
- FT fuel is free of sulfur.
- FT effluent contains a certain amount of oxygenates, and most of the short-chain oxygenates are in the aqueous phase.
- FT effluent is poor in terms of aromatics and naphthenes.

Raw FT syncrude cannot be used directly: it needs to be upgraded and refined into useful products. BtL-FT can produce various chemicals, such as ethylene, propene for plastic production, refined wax for daily use chemicals like candles, oxygenates like ethanol and so on.[53] However, its main function is to produce transportation fuel products – mainly motor-gasoline, diesel fuel and jet fuel.[13]

10.3.2 Refining Transportation Fuel

As illustrated already, converting biomass into liquid transportation fuel is a multistep process that includes syngas generation *via* biomass gasification, syngas conversion to hydrocarbons *via* FTS, and separation and refining of FT syncrude. The product obtained depends on both conversion and also the upgrading process.[7] It is assumed that the processes used for the production of transportation fuel use similar routes prior to FTS. The distribution of FT syncrude (as a result of FTS operating conditions) and the design of the refining process produces different products.

10.3.2.1 Motor-gasoline

The distribution of FT hydrocarbons is largely dependent on operating conditions such as temperature and catalyst type, which affects the design of the refining process. A higher temperature favors hydrocarbons with a low boiling point. A cobalt catalyst produces less oxygenated products than an iron catalyst. According to the aforementioned contents, HTFT is beneficial in producing gasoline products, with the main distribution being in the low-boiling-point range.

It should be noted that even for HTFT with the ideal chain growth probability to produce mostly low-boiling-point hydrocarbons, heavy residues also occur in the FT syncrude. Since it is not economical to convert the hydrocarbon already in the range of a specific fuels into others, the refining process with the aim of only producing one specific fuel is not practical. However, the yield of the fuel can be maximized.

The Hydrocol process operated industrially in the 1950s. It is a gas-to-liquid (GtL) process that is catalyzed with iron under a high-temperature condition. The hydrocarbon distribution obtained is illustrated in Table 10.5.[45] It can be concluded that the syncrude from the Hydrocol process is mainly composed of light gases and naphtha-range hydrocarbons, predominantly straight-chain olefins with double bonds that are mostly in the 1-position.[46] The refining process should aim at further treating the effluent already in the gasoline range and converting the syncrude out of range as much as possible. The steps in the refining treatment include but is not restricted to:

- Applying oligomerization to convert the light gases into gasoline-range products; and
- Getting rid of the undesired oxygenates and obtaining useful oxygenates like alcohols.

The straight-run syncrude is composed of a large number of linear hydrocarbons, which leads to a low ON. The ON increases in Hydrocol mainly through two processes:[12] first, double-bond isomerization is triggered in the bauxite treatment; second, the liquid hydrocarbons are obtained *via* oligomerization

However, this process was designed to meet past fuel regulations, and it might be too simplistic to meet the requirements of the more stringent specifications. Also, it is restricted to the HTFT refining process. De Klerk[13] developed a refining process to produce maximum motor-gasoline, which satisfies EN228:2004 specifications. It is shown in Figure 10.2.

The aqueous phase contains most of the short-chain oxygenates. The carboxylic acids are harmful to the FTS process since they can corrode the equipment and introduce metal heteroatoms. Thus carboxylic acids can be hydrogenated into alcohols. Short-chain alcohols are ideal boosters in terms of the ON. They are used to blend the motor-gasoline. Alcohol dehydration is

Table 10.5 Composition of C_3 and heavier gas and oil fractions from the Hydrocol process. Reproduced from ref. 15 with permission from John Wiley & Sons, Copyright © 2011 Wiley-VCH Verlag GmbH & Co. KGaA.

Product	Boiling range (°C)	Total (mass%)	Olefins (mass%)
Liquid petroleum gas	C_3–C_4	32	82
Naphtha	C_5–204	56	85–90
Distillate	>204	8	75–85
Residue	—	4	—

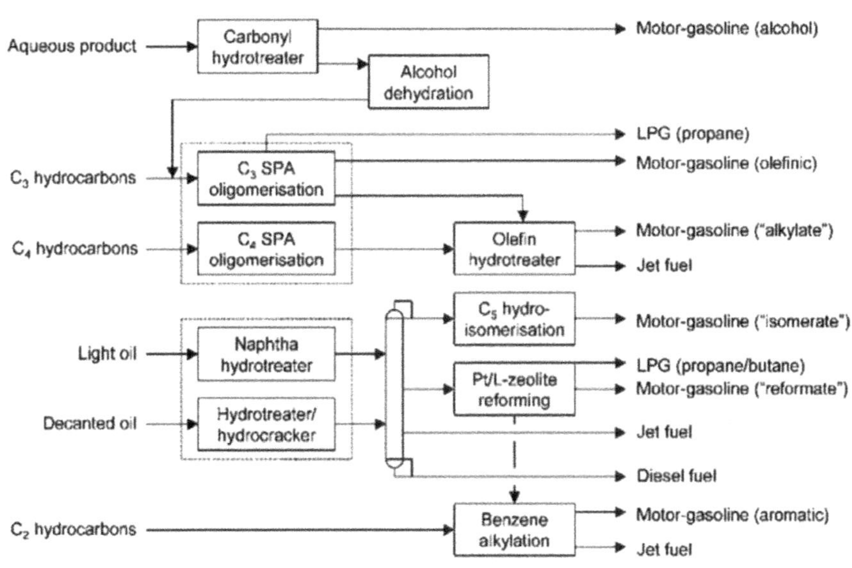

Figure 10.2 HTFT refinery design maximizing motor-gasoline production. Reproduced from ref. 13 with permission from the Royal Society of Chemistry.

a selective reaction that can produce olefins. C_3 and C_4 are oligomerized to form gasoline-range hydrocarbons. The olefins obtained from C_3 are not hydrotreated to prompt the separation of propane that has a high market value, while C_4 products are hydrotreated to paraffins. Naphtha-range syncrude is hydrotreated to increase the degree of saturation and remove minor heteroatoms, and then hydroisomerized to enhance branching. C_6 can follow the same route, while it is also suitable for reforming to produce benzene for the aromatic alkylation reaction. The straight-run kerosene and distillate-range hydrocarbons can be used as jet fuel and diesel fuel that satisfy the specifications following hydrogenation. Heavy residues are hydrocracked into a range of different products. While the goal is to maximize gasoline yield, a certain form of cracking is required. In the end, the motor-gasoline product accounts for about 70% (by mass) of the total liquid fuel obtained.

The differences between HTFT and LTFT determine the different refining processes. With LTFT, more heavy hydrocarbons are produced, with less olefins and no aromatic content, especially with cobalt-based LTFT. The refining process already illustrated can also be used for LTFT, except that the heavy residues are sent for fluid catalytic cracking instead of hydrocracking, so as to produce more aromatics and olefins that can increase the ON, as illustrated in Figure 10.3.[13]

EIA states that, in 2020, biofuels accounted for 5% of total energy consumption by the transportation industry; the ethanol share was about 4%, while the combined share of biodiesel and renewable diesel was about 1%.[47] It can be deduced that the production of renewable gasoline is not often considered.[48]

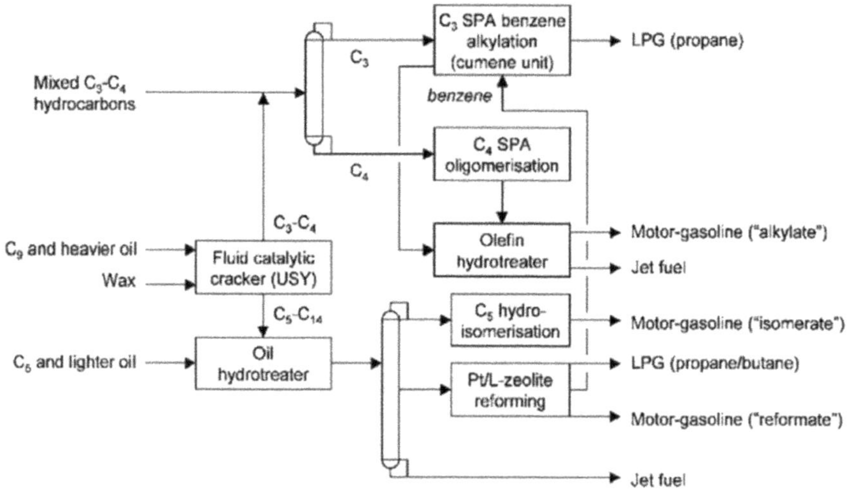

Figure 10.3 LTFT refinery design maximizing motor-gasoline production. Reproduced from ref. 13 with permission from the Royal Society of Chemistry.

10.3.3 Refining Jet Fuel

Aviation is a fast-growing field, with the demand for air travel increasing every year. Since aviation needs to be environmentally sustainable, reducing CO_2 emissions has become the focus of the International Civil Aviation Organization (ICAO), airline companies and academic institutions worldwide. The Carbon Offsetting and Reduction Scheme for International Aviation (CORSIA) has been implemented by ICAO, and starting in 2019, all aircraft operators with emissions greater than 10 000 tons of CO_2 must report on an annual basis.[50] Additionally, they are required to purchase eligible emission credits to offset emissions.[52] The aviation industry is being urged to reduce emissions, and Delta Air Lines announced its goal of achieving carbon neutrality and zero-impact aviation in 2020. The airline is investing in innovative solutions, including carbon capture and storage and sustainable aviation fuel (SAF). SAF has the potential to reduce CO_2 emissions by 80% compared to conventional fuels and is seen as the long-term solution to reducing GHG emissions. Delta plans to replace 10% of its fuel with SAF by 2030,[49] which prompts the need to investigate producing SAF.

SAF can be categorized into two types: biofuel and synthetic fuel. Synthetic SAF can be produced *via* the BtL-FT process. The Northwest Advanced Bio-Fuels (NWABF) project is developing these types of refineries to produce cellulosic renewable jet fuel. They use the gasification of woody biomass to produce syngas for FTS.[51] The project aims to reduce the carbon footprint of aviation by using synthetic SAF. Other companies, including Fulcrum Bio-Energy and NESTE, are also dedicated to the development of SAF. Bio-jet fuel produced *via* FTS is regulated by ASTM International Method D7566.[52]

A typical BtL-FT process that produces jet fuel is provided in Figure 10.4. The first two parts are similar to the process indicated in Section 10.4.1. Biomass feedstock is dried and then gasified into raw syngas. The WGS reaction is introduced into the system if it is catalyzed by cobalt. After the cleaning treatment, syngas is processed by FTS to produce liquid fuels. The reaction can be carried out at either a low or a high temperature. LTFT produces more linear waxes than HTFT. Part of the unconverted syngas is recycled, and another part is sent to the power generation unit. Heavy residue from FT syncrude is hydrocracked into lighter products composed of jet fuel and diesel. Naphtha is obtained after hydrogenation.

However, this may not be the most ideal process for maximizing SAF production. According to De Klerk,[13] the jet fuel yield can be maximized by converting light gases, naphtha, distillate and residue. Since FT syncrude has a low density, the inclusion of cyclic hydrocarbons can satisfy the lower density limit. Branching is also required to lower the maximum freezing point. Thus reactions like oligomerization, aromatization, catalytic reforming and (hydro)isomerization are also vital for jet fuel refining process.

Figure 10.5 shows a refining process aiming at maximizing the production of jet fuel for HTFT. The treatment of light gases and naphtha-range hydrocarbons is critical. Partitioned C_3–C_8 streams are oligomerized in a

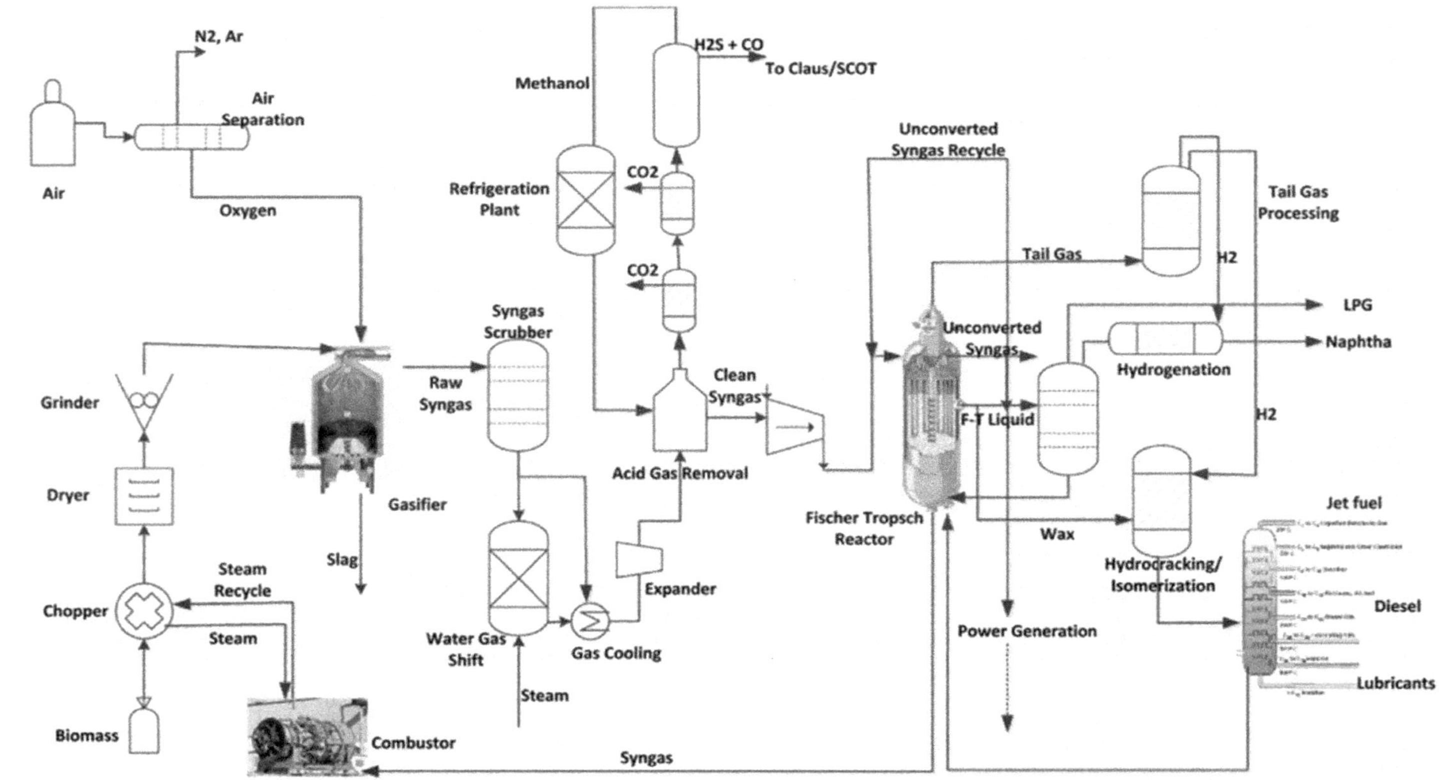

Figure 10.4 BtL-FT process that produces jet fuel. Reproduced from ref. 54 with permission from Elsevier, Copyright 2016.

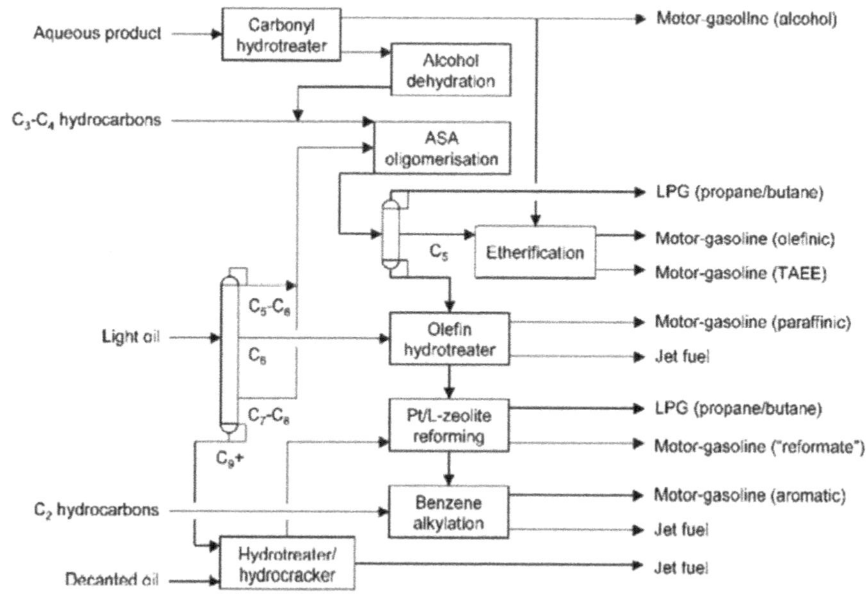

Figure 10.5 HTFT refinery design maximizing jet fuel production. Reproduced from ref. 13 with permission from the Royal Society of Chemistry.

combined unit for maximum kerosene production. Since olefins result in instability of jet fuel, an olefin hydrotreatment is introduced. The straight-run kerosene-range hydrocarbons do not meet jet fuel specifications; therefore, they need to be reformed or isomerized. Although the distillate range hydrocarbons can be sold as diesel fuel after mild hydroprocessing, with the goal of achieving maximum yield of kerosene, they are hydrocracked along with the heavy residue. The severity of hydrocracking is high, so that no diesel fuel is produced. The jet fuel that is finally obtained accounts for 52.2% (by mass) of all the products.

With LTFT, less olefins are produced, which requires less hydrotreatment compared to HTFT. C_3–C_5 hydrocarbons are oligomerized into the motor gasoline and jet fuel range. The oligomerization unit is combined with aromatic alkylation. Since hydrocracking of wax produces light hydrocarbons, after aromatization, the C_6–C_7 aromatics produced are alkylated with C_3–C_5 olefines. ZSM-5 is used as the catalyst for aromatization since it was developed for LPG conversion into aromatics. Hydrocracking is accompanied by hydroisomerization, and the degree of branching can be increased to meet the freezing point specification, as shown in Figure 10.6. The jet fuel that is finally obtained accounts for 62.7% (by mass) of all the products.

Although the refining process designed by De Klerk is theoretically feasible, it has not been validated by large-scale production. The Sierra BioFuels Plant of Fulcrum BioEnergy Company is the first commercial plant in the United States that can convert solid waste into renewable transportation

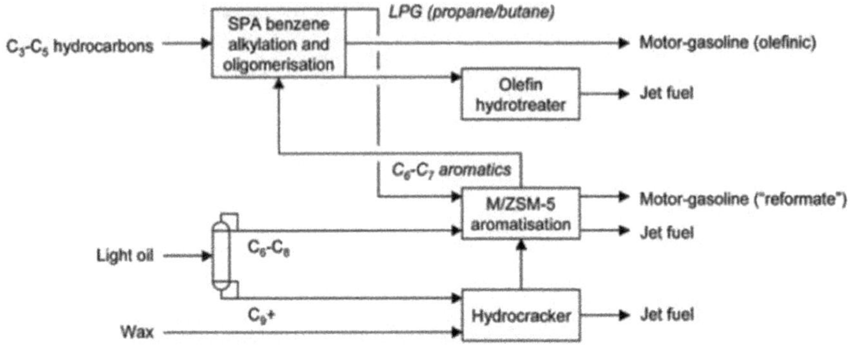

Figure 10.6 LTFT refinery design maximizing jet fuel production. Reproduced from ref. 13 with permission from the Royal Society of Chemistry.

fuel, including jet fuel and diesel, *via* the FT process.[55] According to a report on this plant, a fixed-bed tubular reactor is used. The purified syngas reacts with a proprietary catalyst in the reactors to form three intermediate FT products: a heavy-fraction FT liquid (HFTL) product, a medium-fraction FT liquid (MFTL) product, and a light-fraction FT liquid (LFTL) product, commonly called naphtha. In this process, naphtha is recycled to the partial oxidation unit with the remaining tail gas to be reformed to hydrogen and carbon monoxide, without further upgrading into motor-gasoline products. The hydrotreatment, hydrocracking and hydroisomerization upgrading steps are used to upgrade the combined HFTL and MFTL products into jet fuel.[56]

Various institutions have made considerable effort to develop SAF; however, there are still challenges to be solved. The cost of SAF remains a challenge, especially in terms of synthetic jet fuel.[52] The pretreatment and gasification of biomass and the syngas cleaning sections account for 75% of the total cost.[54] The carbon conversion efficiency also needs to be improved. The hybrid natural gas and biomass process concept has been proposed based on a combination of carbonization, gasification, steam reforming and FT, in order to reduce the high cost of the BtL-FT process.[57] Most of the technologies are still in the early stages, and SAF needs more research and development to achieve cost competitiveness.[52]

10.3.4 Refining Diesel Fuel

In order to meet the goals of relieving environmental pressure and sustainable development, renewable diesel has become a green transportation fuel that is attracting a lot of attention worldwide. The EIA has said that the United States' renewable diesel capacity could continue to increase, given recent announcements and developing projects. For example, the Clean Fuel Program (CFP) was enacted by Oregon State Legislature in 2009 to reduce the carbon intensity (CI) of transportation fuels in the state. It aims to reduce the

CI of fuel by 10% by 2025; the baseline in 2015 was 5% biodiesel blended with conventional diesel.[58] This program aims to encourage the use of cleaner fuel, such as biofuel, renewable fuel, ethanol and electricity by providing incentives and the requirement to create demand for cleaner fuel.[60] This has brought tremendous opportunities in the field of renewable clean fuel.[58] The Low Carbon Fuel Standard (LCFS) is a similar program created by the California Air Resources Board in 2011 to reduce GHG emissions throughout the state by 20% by 2030 and by 80% by 2050. One of the methods that has been applied is the use of low-carbon fuel.[59] This has also promoted the development of renewable transportation fuel products. The U.S. renewable diesel production capacity in 2020 totaled nearly 0.6 billion gallons per year (gal per year), and the capacity may increase to 5.1 billion gal per year in 2024.[61] Companies like Marathon Petroleum Corporation are constructing plants to produce renewable diesel in California according to state policies. The facility is expected to start production in 2022.[62]

The BtL-FT process is especially suitable for the production of renewable diesel because straight-run FT syncrude has a high cetane number, which shows good ignition and combustion ability. On the contrary, undesired low octane number is obtained for gasoline. In addition, FT diesel is free of sulfur and has a low aromatic content, which results in clean combustion. It also has high oxidation stability and a high heating value after hydroprocessing, which reduces the need for antioxidant additives and makes it an attractive substitute for conventional diesel fuel.[63] The cold weather behavior of renewable diesel is better than that of biodiesel;[64] however, FT syncrude still faces disadvantages, such as low density, which needs to be solved *via* a further refining process.

The flow sheet of the whole process can be separated into two figures, as illustrated in Figures 10.7 and 10.8. Figure 10.7 represents the biomass conversion section that is similar in most BtL-FT processes: gasification of biomass converts biomass into combustible syngas; gas cleaning aims at removing contaminants in raw syngas and/or adjusting the H_2/CO ratio by including WGS and FT. Carbon capture and storage (CCS) is incorporated, as well as pressure swing adsorption (PSA), which is used to separate H_2 for further use.[65]

Figure 10.8 shows an industrial upgrading process for HTFT syncrude obtained from natural gas. The production of diesel *via* FTS has been applied industrially for several decades. However, among the industrial chemical plants established in Germany, Japan, the United States and South Africa since the early 1900s, natural gas and coal are the main carbon-based feedstock types. Although the refining process was not designed for a biomass-based process, it has compatibility for different feedstocks. The refinery feed composition is dependent on the FT catalyst and the operating conditions, not on feedstock type.[66]

Both HTFT and LTFT can be used to produce diesel fuel. HTFT syncrude is lighter and more branched, olefinic and aromatic compared to LTFT.

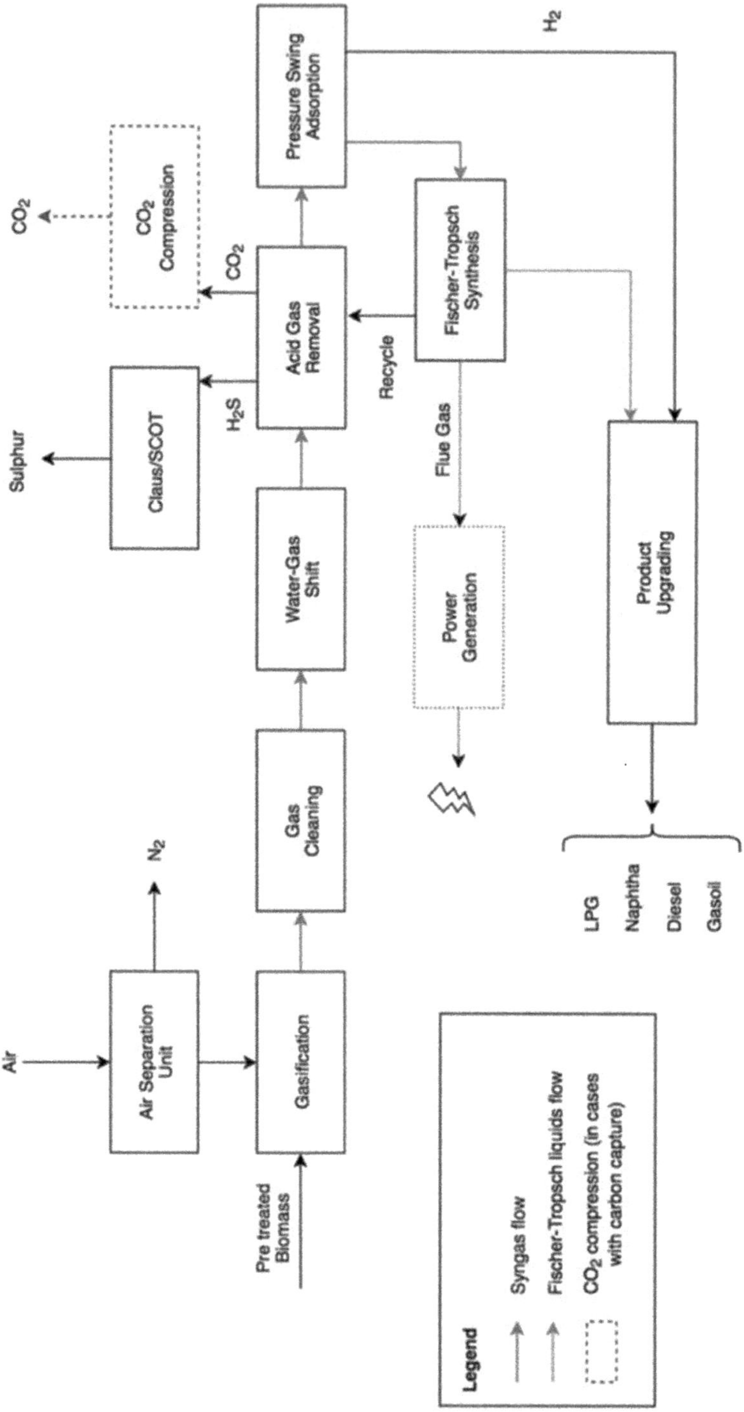

Figure 10.7 Flowchart of the biomass conversion process. Reproduced from ref. 65 with permission from Elsevier, Copyright 2019.

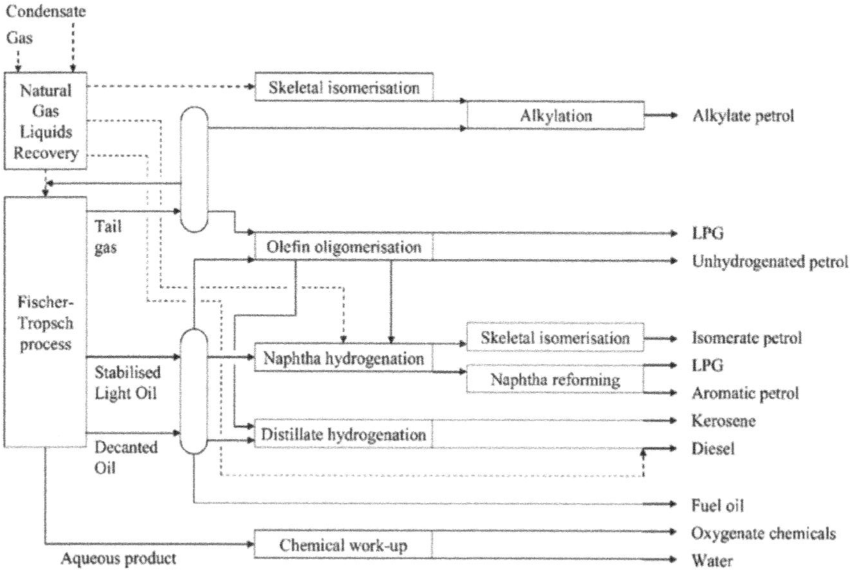

Figure 10.8 Flowchart of the HTFT refinery of PetroSA at Mossel Bay. Reproduced from ref. 66 with permission from American Chemical Society, Copyright 2009.

Although a low degree of branching leads to a high cetane number, it also results in poor cold flow properties. In addition, the low aromatic content and major linear paraffinic content contribute to a low density diesel.[66] Figure 10.7 shows a refinery flow sheet for the HTFT refinery (PetroSA) at Mossel Bay. The highly olefinic tail gas is alkylated with iso-paraffins into alkylate gasoline or oligomerized with light hydrocarbons into gasoline. The naphtha-range syncrude is hydrogenated to saturate the large number of olefins that exist. The saturated product is then skeletally isomerized and catalytically reformed into aromatics to increase the ON of the gasoline products. The distillate-range products are hydrogenated into kerosene and diesel.[67]

It should be noted that hydrocracking is not included in the process depicted in Figure 10.8 due to the small fraction of heavy hydrocarbons. While hydrocracking is an important reaction in the process of refining, especially for LTFT when heavy residue can account for nearly 50% (by mass) of the total syncrude. It is reported by Sasol and UOP that for total wax conversion in LTFT, the middle-distillate yield may reach 80% (by mass). Since FT syncrude is free of sulfur and nitrogen, both metallic and acidic catalysts can be used without being poisoned.[68] Hydrocracking serves two purposes: it converts a high boiling range (syn)crude to middle distillates and increases the degree of branching, which also improves the cold flow properties by hydroisomerization.

Dancuart *et al.*[67] reported a possible flow sheet of the upgrading process of LTFT products. From their research,[67] only LPG, naphtha and diesel are obtained as products. Mild hydrocracking of wax can yield about 80% diesel,

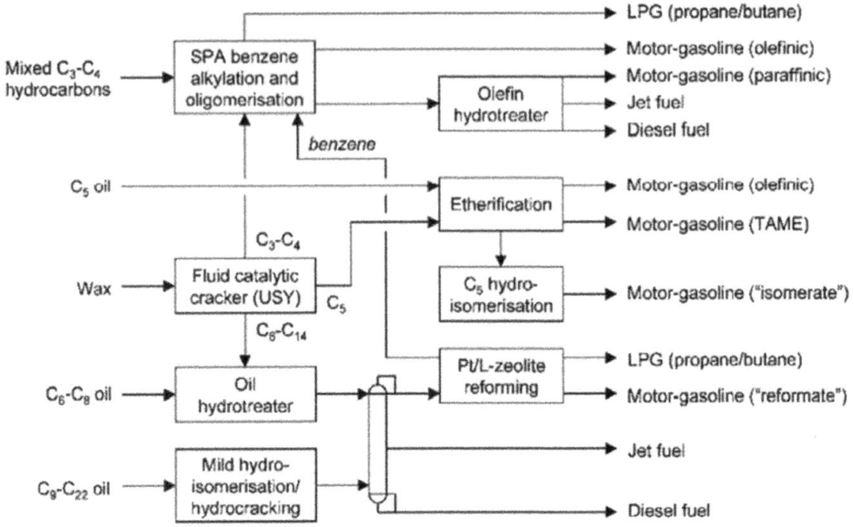

Figure 10.9 LTFT to maximize the yield of diesel with production of fuels on specifications. Reproduced from ref. 13 with permission from the Royal Society of Chemistry.

15% naphtha and 5% C_1–C_4 gases. The naphtha is an excellent feed for the production of lower olefins, including ethylene and propylene by steam cracking. The ethylene yield can be maximized when cracking at the highest severity. The highly linear, paraffinic characteristics suggest that the synthetic naphtha is not suitable for use as gasoline blends.[67] Other products can also be achieved *via* different refining process, *e.g.* jet fuel and gasoline.

A LTFT refinery was developed by De Klerk to maximize the yield of diesel.[13] The LTFT syncrude can easily yield distillate-range hydrocarbons, while the challenge is to produce diesel that satisfies specifications. The FT diesel density is between 0.72 and 0.82 $g\,ml^{-1}$,[63] while EN 590 has a general requirement of 0.82–0.845 $g\,ml^{-1}$.[69] The density of the distillate needs to be increased, as well as the cold flow property. Figure 10.9 shows that the fluid catalytic cracker is applied for wax cracking because it can produce more olefins for alkylation and oligomerization. The straight-run distillate and kerosene streams are mildly hydroisomerized to increase the degree of branching, so that the cold flow properties are improved. C_6–C_8 are catalytically reformed into aromatics. The benzene produced is used for aromatic alkylation. The aromatics obtained can increase the ON of gasoline, as well as improve the density of diesel. Jet fuel is coproduced to improve density without much loss in cetane number.[13] However, only 24.8% (by mass) of diesel can be achieved to specification.

10.4 Conclusion

The specific refinery design is dependent on many changing requirements, like specifications for the products, refinery location, size, feedstock cost, product pricing and so on. Thus there is no one best configuration. The decision regarding the various refinery configurations is a trade-off. For example, the treatment of short-chain hydrocarbon is optional. Also, ethylene can itself be sold as a product that has great value. However, it can also be aromatically alkylated with benzene to increase the ON for gasoline or oligomerized into heavier hydrocarbons. The design may be varied according to the desired result of the process.

Abbreviations

ASTM	American Society for Testing and Material
BtL	Biomass-to-liquids
CN	Cetane number
BTL-FT	Biomass-to-liquid *via* Fischer–Tropsch
CEN	European Committee for Standardization
CFP	Clean Fuel Program
CFPP	Cold filter plugging point
CI	Carbon intensity
CO	Carbon monoxide
CO_2	Carbon dioxide
CORSIA	Carbon Offsetting and Reduction Scheme for International Aviation
CP	Cloud point
CN	Cetane number
EI	U.S. Energy Information Administration
EPA	Environmental Protection Agency
ETBE	Ethyl tertiary butyl ether
FT	Fischer–Tropsch
FTS	Fischer–Tropsch synthesis
GHG	Greenhouse gas
GtL	Gas-to-liquid
HFRR	High-frequency reciprocating rig
HTFT	High-temperature Fischer–Tropsch
ICAO	International Civil Aviation Organization
LCFS	Low-carbon fuel standard
ML	Machine learning
MON	Motor octane number
MTBE	Methyl tertiary butyl ether
MTFL	Medium-fraction FT liquid
N_2	Nitrogen
NWABF	Northwest Advanced Bio-Fuels
ON	Octane number
PAH	Polycyclic aromatic hydrocarbons
PM	Particulate matter
PSA	Pressure swing adsorption
RON	Research octane number

RVP	Reid vapor pressure
SAF	Sustainable aviation fuel
SLBOCLE	Scuffing Load Ball-on-cylinder Lubricity Evaluator
SO_2	Sulfur dioxide
TAME	Tertiary amyl methyl ether
VOC	Volatile organic compounds

References

1. U.S. Energy Information Administration. Available from: https://www.eia.gov/outlooks/ieo/ [accessed 31 July 2021].
2. U.S. Energy Information Administration. Available from: https://www.eia.gov/outlooks/ieo/pdf/ieo2019.pdf [accessed 31 July 2021].
3. Our World in Data. Available from: https://ourworldindata.org/energy [accessed 31 July 2021].
4. BP. Available from: http://oilproduction.net/files/especial-BP/bp-statistical-review-of-world-energy-2016-full-report.pdf [accessed 31 July 2021].
5. D. Santos, R. Gonçalves and A. C. Alencar, *Int. J. Hydrogen Energy*, 2020, **45**, 18114.
6. R. S. Dhillon and G. V. Wuehlisch, *Biomass Bioenergy*, 2013, **48**, 75.
7. D. Klein-Marcuschamer and H. W. Blanch, *AIChE J.*, 2015, **61**, 2689.
8. P. Ibarra-Gonzalez and B. G. Rong, *Ind. Eng. Chem. Res.*, 2020, **59**, 14881.
9. H. Mahmoudi, M. Mahmoudi and O. Doustdar, *Biofuels Eng.*, 2017, **2**, 11.
10. C. A. Floudas, J. A. Elia and R. C. Baliban, *Comput. Chem. Eng.*, 2012, **41**, 24.
11. S. S Ail and S. Dasappa, *Renewable Sustainable Energy Rev.*, 2016, **58**, 267.
12. A. de Klerk, *Fischer–Tropsch Refining*, John Wiley & Sons, 2012.
13. A. de Klerk, *Energy, Environ. Sci.*, 2011, **4**, 1177.
14. ASTM D 4814, Standard specification for automotive spark-ignition engine fuel.
15. EN228, Automotive fuels – Unleaded petrol – Requirements and test methods.
16. P. Ghosh, K. J. Hickey and S. B. Jaffe, *Ind. Eng. Chem. Res.*, 2006, **45**, 337.
17. S. Correa-Gonzalez, Y. Kroyan and T. Sarjovaara, *Energy Fuels*, 2021, **35**, 9332.
18. United States Environmental Protection Agency. Available from: https://www.epa.gov/gasoline-standards/gasoline-reid-vapor-pressure [accessed 31 July 2021].
19. Chevron. Available from: https://www.chevron.com/-/media/chevron/operations/documents/motor-gas-tech-review.pdf [accessed 31 July 2021].
20. C. F. Taylor, *Internal Combustion Engine in Theory and Practice*, The MIT Press, 2nd edn, 1985.
21. S. A Shirazi, B. Abdollahipoor and J. Martinson, *Fuel*, 2019, **252**, 542.
22. Ø. Borg, N. Hammer and B. C. Enger, *J. Catal.*, 2011, **279**, 163.

23. OICA. Available from: https://www.oica.net/worldwide-fuels-charter/ [accessed 31 July 2021].
24. J. F. Periago, A. Zambudio and C. Prado, *J. Chromatogr. A*, 1997, **778**, 263.
25. A. Andziulis, *Transport*, 2006, **21**, 218.
26. ISO 7536, Petroleum products – Determination of oxidation stability of gasoline-Induction period method.
27. ASTM D 525, Standard Test Method for Oxidation Stability of Gasoline.
28. F. Pradelle, S. L. Braga and A. R. F. A. Martins, *Energy Fuels*, 2015, **29**, 7753.
29. ASTM D 381, Standard Test Method for Gum Content in Fuels by Jet Evaporation.
30. ASTM D 873, Standard Test Method for Oxidation Stability of Aviation Fuels.
31. D. Leckel, *Energy Fuels*, 2009, **23**, 2342.
32. ASTM D 975, Standard Specification for Diesel Fuel.
33. EN 590, Specifications for Automotive Diesel Fuel.
34. ASTM D 1655, Standard Specification for Aviation Turbine Fuels.
35. ASTM D 613, Standard Test Method for Cetane Number of Diesel Fuel.
36. G. Marketing Available from: https://www.onsitepoweradvisor.com/wp-content/uploads/2011/12/Chevron-DFTR.pdf [accessed 31 July 2021].
37. ASTM D 86, Standard Test Method for Distillation of Petroleum Products and Liquid Fuels at Atmospheric Pressure.
38. ASTM D 5168-20, Standard Test Method for Determination of Aromatic Content of Diesel Fuels by Supercritical Fluid Chromatography.
39. Wiley Online Library, Available from: https://onlinelibrary.wiley.com/doi/abs/10.1002/14356007.a16_719.pub2 [accessed 31 July 2021].
40. ASTM D 455, Standard Test Method for Kinematic Viscosity of Transparent and Opaque Liquids.
41. Air Force Research Laboratory. Available from: https://apps.dtic.mil/sti/pdfs/AD1093317.pdf [accessed 31 July 2021].
42. S. P. Heneghan, S. Zabarnick and D. R. Ballal, *J. Energy Resour. Technol.*, 1996, **118**, 170.
43. C. A. Moses and P. N. J. Roets, *J. Eng. Gas Turbines Power*, 2009, **131**, 431.
44. National Renewable Energy Laboratory. Available from: https://www.nrel.gov/docs/fy16osti/66291.pdf [accessed 31 July 2021].
45. B. H. Davis and L. O. Mario, *Advances in Fischer–Tropsch Synthesis, Catalysts, and Catalysis*, CRC Press, 2009.
46. H. Pichler, *Adv. Catal.*, 1952, **4**, 271.
47. U.S Energy Information Administration, Available from: https://www.eia.gov/energyexplained/use-of-energy/transportation.php [accessed 31 July 2021].
48. T. G. Kreutz, E. D. Larson and G. Liu, *25th Annual International Pittsburgh Coal Conference*, Pittsburgh, Pennsylvania, 2008.
49. DELTA, Available from: https://news.delta.com/delta-spotlights-ambitious-carbon-neutrality-plan-path-zero-impact-aviation-earth-month [accessed 31 July 2021].

50. Northwest Advanced Bio-Fuels, Available from: https://nwabiofuels.com/corsia-mandate-timeline/ [accessed 31 July 2021].
51. Northwest Advanced Bio-Fuels, Available from: https://nwabiofuels.com/ [accessed 31 July 2021].
52. L. Zhang, T. L. Butler and B. Yang, *Green Energy to Sustainability: Strategies for Global Industries*, 2020, pp. 85–110.
53. M. E Dry, *Industrial Chemicals via C1 Processes*, 1987, vol. 328, p. 18.
54. W. C. Wang and L. Tao, *Renewable Sustainable Energy Rev.*, 2016, **53**, 801.
55. Fulcrum BIOENERGY, Available from: https://fulcrum-bioenergy.com/technology/our-process/ [accessed 31 July 2021].
56. Green Car Congress, https://www.greencarcongress.com/2018/05/20180517-fulcrum.html [accessed 31 July 2021].
57. A. H. Tanzil, K. Brandt and M. Wolcott, *Biomass Bioenergy*, 2021, **145**, 105942.
58. U.S. Gain, Available from: https://www.usgain.com/blog/get-to-know-oregons-clean-fuel-program/ [accessed 31 July 2021].
59. U.S. Gain, Available from: https://www.usgain.com/blog/understanding-the-california-low-carbon-fuel-standard-lcfs/ [accessed 31 July 2021].
60. Oregon Clean Fuel Program, Available from: https://www.oregon.gov/deq/ghgp/cfp/Pages/default.aspx [accessed 31 July 2021].
61. U.S Energy Information Administration, Available from: https://www.eia.gov/todayinenergy/detail.php?id=48916 [accessed 31 July 2021].
62. BIOMASS Magazine, Available from: http://biomassmagazine.com/articles/17790/marathon-proceeds-with-renewables-conversion-at-martinez-refinery [accessed 31 July 2021].
63. S. Bezergianni and A. Dimitriadis, *Renewable Sustainable Energy Rev.*, 2013, **21**, 110.
64. S. L. Douvartzides, N. D. Charisiou and K. N. Papageridis, *Energies*, 2019, **12**, 809.
65. I. S. Tagomori, P. R. R. Rochedo and A. Szklo, *Biomass Bioenergy*, 2019, **123**, 134.
66. B. I. Kamara and J. Coetzee, *Energy Fuels*, 2009, **23**, 2242.
67. L. P. Dancuart, R. de Haan and A. de Klerk, *Stud. Surf. Sci. Catal.*, 2004, **152**, 482.
68. W. Böhringer, A. Kotsiopoulos and M. de Boer, *Stud. Surf. Sci. Catal.*, 2007, 163, 345.
69. EN 590. Standard for diesel fuel.
70. L. R. Matthews, A. M. Niziolek and O. Onel, *Ind. Eng. Chem. Res.*, 2016, **55**, 3203.
71. S. S. Ail and S. Dasappa, *Renewable Sustainable Energy Rev.*, 2016, **58**, 267.

CHAPTER 11

Sustainable Aviation Fuel from Biomass via *Gasification and Fischer–Tropsch Synthesis*

M. DOSSOW,* V. DIETERICH, A. HANEL AND S. FENDT*

Technical University Munich, Chair of Energy Systems, Boltzmannstr. 15, 85748 Garching, Germany
*Emails: marcel.dossow@tum.de; sebastian.fendt@tum.de

11.1 Civil Aviation Fuel and the Regulatory Framework

Crude oil-based conventional aviation fuel (CAF) typically consists of hydrocarbons in the range C_8–C_{16}, with approximately 20% paraffinic, 40% iso-paraffinic, 20% naphthenic and 20% aromatic content.[1] CAF production is conducted in large-scale refineries by atmospheric distillation of selected crude oil fractions, followed by further processing.[2] In the refinery process, jet fuels are typically primary blends produced from straight-run distillates or include hydrocracked components.[3]

11.1.1 International Standards for Aviation Fuel

Worldwide, there are several standards for civil and military jet fuels. The standardized aviation fuels used in civil aviation are Jet A in the USA, Jet B in Canada, TS-1 in Russia, and Jet A-1 in the rest of the world.[2] To minimize the differences among the fuel standards, the major jet fuel suppliers have

Catalysis Series No. 44
Chemicals and Fuels from Biomass *via* Fischer–Tropsch Synthesis: A Route to Sustainability
Edited by Joshua Gorimbo, Xinying Liu, Yali Yao and Diane Hildebrandt

Published by the Royal Society of Chemistry, www.rsc.org

agreed on the Aviation Fuel Quality Requirements for Jointly Operated Systems, which combines the most restrictive requirements of the Jet A-1 fuel specifications and the International Air Transport Association (IATA) guidelines.[2] Therefore, Jet A-1 has been chosen as the single international standard to define aviation turbine fuel. Jet A-1 used in existing aviation fleets needs to meet the requirements laid out in ASTM D1655.[4] In this chapter, CAF refers exclusively to Jet A-1 derived from fossil resources as the international standard for civil aviation fuel.

CAF is not specified by composition, molecular structure or purity: it is limited by the boundary conditions specified by the respective standards, including standardized test methods.[2] The imposed inspection tests are either fundamental or empirical, and any fuel that meets all requirements is acceptable for aircraft operations.[2] Table 11.1 shows some of the most important property requirements and associated ASTM test methods for Jet A-1 fuel. Due to the potential risk to flight operations, changes to the specifications in ASTM D1655 are difficult.[2]

Sustainable aviation fuel (SAF) is defined as jet fuel derived from biomass or other waste and used as a blending component to meet the specifications for use in aircraft.[5] SAF requires technical certification to ensure that the fuel complies with the requirements for use in existing aircraft and sustainability certification to ensure that the SAF complies with the defined sustainability criteria.[6] The temperature, viscosity and opacity (clear/clean) requirements for aviation fuel are the most difficult challenge.[7] The international standardized pathways for synthetic fuels are summarized in Table 11.2.

Currently, SAF has to be blended with CAF to meet the specified requirements.[5,8] The blending ratios are provided in Table 11.2. Blending of FT-SPK, SPK/A, HEFA-SPK, ATJ-SPK and CH-SK is allowed up to 50% with CAF, while HHC-SPK and SIP-SPK can only be blended up to 10%. Thus blends that are in accordance with the standardization specifications can be considered equivalent to certified CAF.[5,8,9]

Table 11.1 Selected Jet A-1 property requirements and test methods specified by ASTM D1655. Reproduced from ref. 39 with permission from Elsevier, Copyright 2016.

Fuel property requirement	Value	ASTM test methods
Maximum aromatic content by volume	26.5%	D6379
Maximum total sulfur by mass	0.3%	D1266, D2622, D4294, D5453
Minimum flash point	38 °C	D56, D93, D3828
Density minimum-maximum range at 15 °C	775–840 kg m^{-3}	D1298, D4052
Maximum freezing point	− 47 °C	D5972, D7153, D7154, D2386
Maximum viscosity at 220 °C	8.0 mm^2 s^{-1}	D445, D7042
Minimum net heat of combustion	42.8 MJ kg^{-1}	D4529, D3338, D4809
Minimum smoke point	18 mm	D1322
Maximum naphthalene content by volume	3.0%	D1840

Table 11.2 Synthetic jet fuels approved for use according to ASTM D7655.[8]

Pathway	Certification	Product		Blend (%)
Syngas-to-aviation fuels	2009 Annex A1	Fischer–Tropsch hydroprocessed synthetic paraffinic kerosene[a]	FT-SPK	50
	2015 Annex A4	Fischer–Tropsch synthetic paraffinic kerosene with aromatics	FT-SPK/A	50
Oil-to-aviation fuels	2011 Annex A2	Hydroprocessed esters and fatty acids synthetic paraffinic kerosene	HEFA-SPK	50
	2020 Annex A7	Hydroprocessed hydrocarbons, esters and fatty acids	HHC-SPK[b]	10
Direct sugars-to-aviation fuels	2014 Annex A3	Hydroprocessed fermented sugars to synthetic iso-paraffins[c]	HFS-SIP	10
Alcohols-to-jet fuels	2016/2018 Annex A5	Alcohol-to-jet synthetic paraffinic kerosene	ATJ-SPK	50
Catalytic hydrothermolysis jet[30]	2020 Annex A6	Catalytic hydrothermolysis synthetic kerosene	CH-SK	50

[a]Synfuel consisting mainly of *n*-alkanes.
[b]Also HC-HEFA-SPK.
[c]Synfuel consisting mainly of iso-alkanes.

11.1.2 Greenhouse Gas Emissions from Aviation and International Emission Control

International air traffic is contributing massively to global warming, since aviation uses kerosene almost exclusively from fossil sources.[3] The life cycle greenhouse gas (GHG) emissions from the production and combustion of CAF are about 87.5 g CO_2 equivalents per MJ, with direct GHG emissions of about 73.2 g_{CO_2} per MJ.[10,11] With a lower heating value (LHV) of 43.15 MJ per kg,[12] the direct emission factor of CAF is 3.159 g_{CO_2} per g_{CAF}.

With an average annual increase of 3%, CAF consumption rose from about 122 Mt in 1986 to about 316 Mt in 2018.[13,14] All the main aviation stakeholders predict a massive increase in revenue passenger-kilometers of 3.9% (IATA) to 4.9% per year (Airbus) over the next few decades.[11,15] According to IATA, global civil-aviation-related carbon emissions rose from 860 Mt_{CO_2} in 2017 to about 905 Mt_{CO_2} in 2018 – an increase of 5.2% in a single year.[16] The International Council on Clean Transportation estimates that global commercial aviation emitted about 2.4% of the 37.9 Gt_{CO_2} emitted globally from fossil fuel use in 2018[17] Within the EU, aviation is accountable for 3.6% of the GHG emissions and for 13.4% of the emissions from the transport industry, which means aviation is the second largest source of GHG emissions from the transport industry (after road traffic).[18]

Apart from direct CO_2 emissions, air traffic fuel emissions affect the radiative balance of the atmosphere through various mechanisms. These include direct emissions of nitrogen oxides (NO_x), which influence atmospheric chemistry and result in changes in the amount of ozone (O_3) and methane (CH_4).[11] Exhaust emissions of CO, unburned hydrocarbons and nitrogen also contribute to air traffic emissions, although they are related to combustion conditions rather than to fuel properties.[3]

Today, air transport emissions are subject to the permitted levels of emissions of NO_x, CO, unburned hydrocarbons and smoke, which is controlled by the United Nations agency for aviation, the International Civil Aviation Organization (ICAO). However, these emission values only relate to the landing-and-takeoff cycle, and exhaust emissions at altitude are not restricted.[2] On a European level, the European Union Aviation Safety Agency (EASA) is responsible for regulating GHG emissions from air transportation.[18] However, there is no single objective for emission control and regulation regarding CO_2 emissions in-flight and between countries.

11.1.3 International Emission Reduction Goals for Aviation

Due to the projected worldwide growth in passenger air traffic over the next few decades, it is estimated that international aviation fuel consumption will triple from 2017 to 2050.[19] This indicates the need for infrastructural, operational and technological improvements to avoid a proportional increase in emissions. To address its climate impact and to participate in a united approach to reducing emissions, the aviation industry body, IATA, set three global goals in 2009:[20]

1. Improving fuel efficiency by an average of 1.5% annually from 2009 to 2020.
2. Capping net emissions through carbon-neutral growth from 2020 (CNG2020).
3. Halving net emissions by 2050 from 2005 levels.

These targets are similar to the ICAO targets of 2010, consisting of a global annual average fuel efficiency improvement of 2% and CNG2020.[21] ICAO member states submitted first action plans in 2013, addressing policies and actions to limit or reduce the impact of aviation on the global climate, with updates in 2015 and 2018.

Taking into account technological improvements due to improved fuel efficiency and advanced air traffic management (ATM) – such as lower fuel consumption per passenger-kilometer – it is estimated that international aviation fuel consumption will increase by 80–300% from 2017 to 2050.[19,22] Even in the aspirational 2% fuel efficiency increase scenario, which is unlikely to be met by 2050, global fuel consumption is projected to almost double compared to 2014.[19,22]

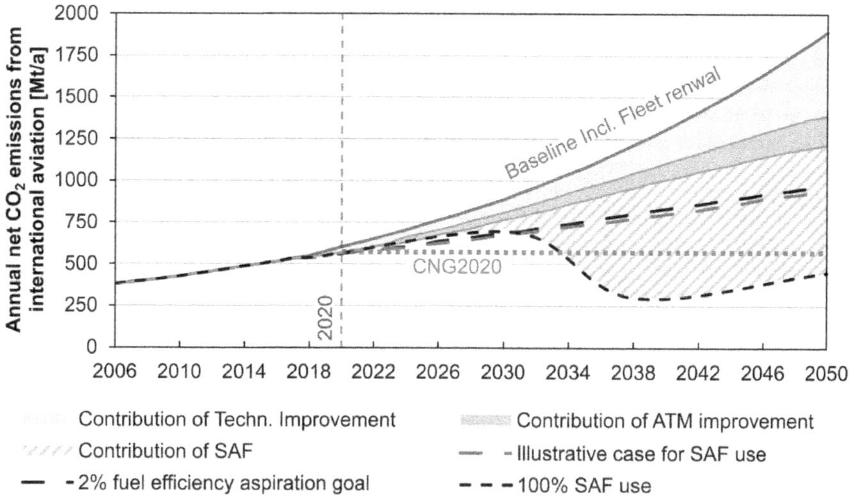

Figure 11.1 Annual net CO_2 emissions from international aviation until 2019 and future estimations until 2050.[22,23,106]

Figure 11.1 shows ICAO's latest projections in terms of CO_2 emissions from international aviation, due to the expected rise in CAF consumption. Different scenarios for annual net CO_2 emissions caused by air traffic have been developed, depending on the fuel consumption increase.[15,21,23]

ICAO's baseline scenario shows that aviation emissions are expected to more than triple by 2050,[†] at which time aircraft might account for 25% of the global carbon budget[17,19] Even under the assumption of massive contributions from improved technology and ATM, CO_2 emissions from international aviation will more than double.[17,19] Other effective approaches are required to meet the CNG2020 target,[15] as increased efficiency through new technologies and infrastructure will only slow down the predicted growth in GHG emissions from aviation (but not stop the continued growth).[19,22]

Since aviation requires high-energy-density fuels and electric commercial long-distance aircraft are unlikely to be used before 2040, there are few realistic alternatives to kerosene and jet fuel.[24–26] Current renewable fuels, such as ethanol, higher alcohols, pyrotechnic oils, biodiesel, methane and hydrogen, cannot replace kerosene because they do not meet the requirements for use in aviation, such as the physical and combustion properties, and the storage and sustainability requirements.[7] The use of sustainable,

[†]SARS-CoV-2 impact on air traffic and political measures: ICAO reported a decrease in global passenger numbers of −60% for 2020 compared to 2019.[24] In a most likely scenario (based on widespread vaccination campaigns), Eurocontrol estimates that global passenger traffic will recover to 95% of the pre-Corona 2019 levels by late 2024.[25] This estimate reflects the latest operational data but is subject to substantial changes with the situation evolving. Nonetheless, it must be assumed that air traffic will eventually recover and that the estimations of future net CO_2 emissions will merely be delayed.

low-emission fuels in aviation could fill the gap to carbon-neutral growth.[25,27]

In 2016, ICAO called for coordinated policies to accelerate the development and use of SAFs and for the industry to adopt the so-called Carbon Offsetting and Reduction Scheme for International Aviation (CORSIA). In order to help the aviation sector achieve its climate targets in the short and medium terms, CORSIA covers approximately 80% of the growth in CO_2 emissions between 2021 and 2035.[27,28] In practice, airlines will be required to compensate for their growing GHG emissions by buying carbon offsets. If an aviation company finances a reduction in emissions elsewhere – for example through the implementation of carbon reduction projects in developing countries – this amount will be considered an emission reduction by the company.[29] Aircraft operators will also receive carbon credits for using SAFs and fossil-based "lower carbon aviation fuels".[18,28] These alternative fuels must comply with sustainability criteria and must be certified by an approved certification scheme.[18,28] To be counted as an eligible emissions unit, an alternative fuel must deliver at least 10% in GHG savings and must not be made from biomass obtained from land with a high carbon stock.[28] Within the EU, however, GHG emission reductions of at least 65% are required for SAFs produced in installations that start operations after 2021.[18]

11.2 Sustainable Aviation Fuel from Biomass and Fischer–Tropsch Synthesis

As advanced SAFs are based on renewable energy sources and residues, they are a realistic option to reduce CO_2 emissions.[18,26] Due to the almost identical properties,[3] the use of such drop-in SAFs allows existing aircraft to be operated without any technical modifications, thus significantly reducing the cost associated with large-scale use of SAFs at the system level.[3,15]

11.2.1 Comparison of Sustainable Aviation Fuel Production Pathways

The seven approved SAF pathways (see Table 11.2) result in five conversion technology families:

(1) Syngas-to-aviation fuels *via* the BtL pathway, including gasification and Fischer–Tropsch synthesis (FTS), to produce hydroprocessed synthetic paraffinic kerosene with or without aromatics.
(2) Oil-to-aviation fuels by producing hydroprocessed esters and fatty acids (HEFA) or hydroprocessed hydrocarbons, esters and fatty acids from biogenic lipids.[30]
(3) Direct sugars-to-aviation fuels by processing hydroprocessed fermented sugars into synthetic isoparaffins.

(4) Alcohols-to-jet fuels to produce synthetic paraffinic kerosene.
(5) Catalytic hydrothermolysis jet by producing catalytic hydrothermolysis synthetic kerosene.

Fuel production from biomass generally involves a broad range of environmental, social and economic consequences. Any SAF production pathway needs to show high emission reduction potential, in addition to a renewable and sustainable feedstock. To avoid land use change, advanced SAF production processes must therefore use biomass residues or biogenic waste exclusively as feedstock. In particular, lignocellulosic residues show great potential as sustainable biomass feedstock in sufficiently large amounts to meet at least part of the aviation industry's demand for fuel.[15] Since HEFA is exclusively derived from plant and animal fats, (waste) oils and grease, the oil-to-aviation pathway faces the potential danger of a shortage of sustainably available raw materials. However, all other pathways can theoretically, use lignocellulosic biomass as feedstock for enzymatic hydrolysis or syngas fermentation. PtL processes theoretically have the advantage of being an unlimited fuel, if water, renewable energy and a CO_2 source are sustainably available.[21]

With fuel costs accounting for about 27% of an airline's total operating costs in 2015, economic profitability is the major hurdle to be overcome for the deployment of SAFs.[27] The fuel price is primarily related to feedstock cost.[27] Hamelinck *et al.* estimated that feedstock costs 45–58% of total fuel production costs.[31] For SAFs, the availability of an appropriate biomass feedstock – which can vary considerably with type, season and country – mainly determines its price. Thus different selling prices are achieved for fuels from different feedstock and conversion pathways.[31] SAF production pathways should further reduce costs by focusing on feedstock productivity, optimal reaction conditions (cheap catalyst), factory scale, and recovery and utilization of by-products.[27]

Figure 11.2A shows the specific GHG emissions, and Figure 11.2B the estimated minimum selling price (MSP) for the five SAF production pathways. Based on this assessment of sustainability and economic criteria, the BtL-FT pathway is the most promising candidate for civil SAF production in the short and medium terms. It is one of the most sophisticated technologies, with large-scale facilities already built,[3] and it also enables the highest GHG emission savings.

11.2.2 Status of Commercialization of the BtL-FT Process for the Production of Sustainable Aviation Fuel

Aviation fuel produced *via* the BtL-FT pathway was approved several years ago, and Sasol is producing FT-SPK from coal at a large, commercial scale. The Secunda plant in South Africa produces 160 000 bbl_{fuel} per day,[3] and no technical problems with its use have been reported.[32] SAF from natural gas *via* FTS has also been commercially available at a large scale from the Pearl plant in Qatar since 2012.[32] Shell routinely uses the produced SAF in blends

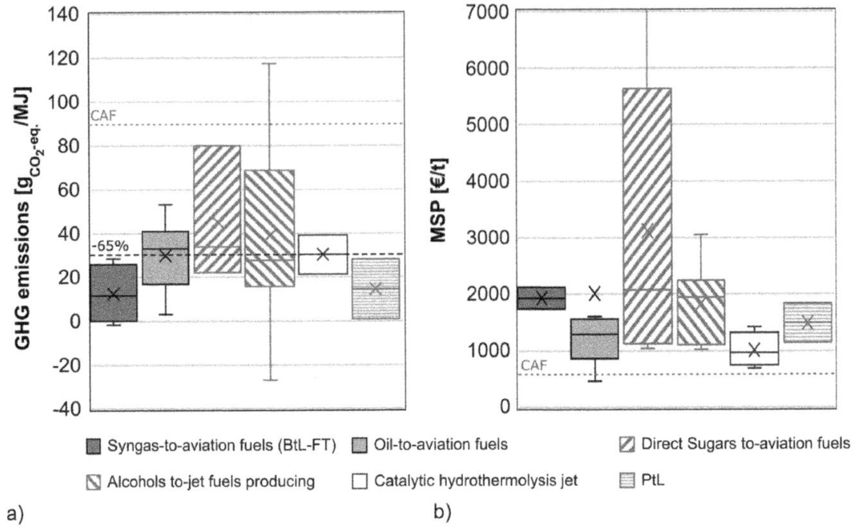

Figure 11.2 (A) Sustainability criteria assessment in terms of GHG emissions.[18,27] (B) Economic criteria assessment in terms of minimum selling price (MSP)[27,33,107] for the five SAF production technologies. (The price of CAF in 2019 averaged about 575 € per t.[108]).

(up to 25%) with conventional kerosene.[32] In addition, several industrial commercialization projects for the use of biomass as a feedstock have been initiated and started in recent years – most notably by the U.S.-based companies Red Rock Biofuels and Fulcrum Bioenergy and the French BioTFuel project.[15] With SAF from biomass being close to starting up the first commercial facilities, a Fuel Readiness Level (FRL) between 7 and 9 is estimated. The BtL-FT process route reaches product yields between 0.026 t_{fuel} per $t_{biomass,dry}$ and 0.258 t_{fuel} per $t_{biomass,dry}$,[33] and the resulting MSP is between 1700 € and 2120 €.[15,34]

ASTM D7566 explicitly names syngas as feedstock for the process certified in Annex A1.[30] Therefore, the process is not restricted to certain types of feedstock but is valid for all FT-based pathways, as long as the fuel is "wholly derived from syngas *via* the FT process using Iron or Cobalt catalyst" and is subsequently refined to meet the defined specifications.[35,36] As all biomass components, including lignin, are recovered and residues can be used as raw material, the FT route poses no sustainability risks.[37] EASA reported GHG emissions of 5.3–28.5 $g_{CO_2\text{-eq}}$ per MJ for FT-SPK.[18] Compared to CAF, these values correspond to GHG reductions of 67–94%.

11.2.3 Fuel Properties of BtL-FT Derived Sustainable Aviation Fuel

SAF produced *via* the BtL-FT pathway is similar in performance to CAF. It has excellent low-temperature properties, improved thermal stability and higher specific energy than conventional fuel, and it is a clean burning fuel, which results in an increase in the life of the combustor and turbine.[3,10] The biggest challenge with FT-derived SAF is the very low sulfur and aromatics content of less than 0.5 wt%.[38] The blending requirements are mainly due to the required aromatic content: a minimum of about 8 vol% of aromatics is required for synthetic fuel in the final blend. This is to ensure adequate seal swell within the fuel system, so as to prevent fuel leaks.[8,39] Elastomers used as seals – particularly those used in diaphragms, fuel tanks, fuel tank coatings and sealants, linings and a wide variety of static and dynamic seals in all parts of the fuel system – expand in the presence of aromatic compounds, which are found in conventional petroleum-derived jet fuel.[2,10,39]

The ASTM D1655 and ASTM D7566 standards both specify that the aromatic content of jet fuel must not exceed 26.5% by volume, as a high aromatic content generally leads to smoke formation and soot deposition.[39] With naphthalenes or bicyclic aromatics, which are particularly likely to form more soot, the limit is less than 3%.[4,39] It is important to note that ASTM D7655 (Annexes A1 and A3) explicitly limits the aromatics content of SAF from FTS to less than 0.5% prior to blending.[8] This implies that CAF must have an aromatics content of at least 16.8% to allow 50% blending.[32,39] In reality, much of the conventional kerosene produced in Europe has an aromatics content below this level, which means that a 50% blend might not be possible in practice.[32]

11.3 The BtL-FT Pathway for Sustainable Aviation Fuel Production

The BtL-FT process allows for the production of SAF based on carbon-neutral biomass residues and is also certified for use in civil aviation. The process is considered the most advanced of several thermochemical conversion pathways[15] and consists of seven major steps:

1. Pretreatment of biomass feedstock.
2. Syngas generation through gasification.
3. Quenching of syngas.
4. Sour water-gas shift (WGS).
5. Acid gas removal (AGR).
6. FT synthesis to produce synthetic, liquid crude (FT crude) from syngas.
7. Upgrading and refining the crude product into SAF.

The biomass feedstock is converted to syngas through thermal decomposition, which is the gasification step. Gasification takes place at high temperatures (1200–1600 °C) and deconstructs the feedstock into CO, H_2, and CO_2 primarily, as well as some ash.[30] The gas mixture is then separated, conditioned and cleaned to produce pure syngas. In the subsequent FTS, syngas is converted into long hydrocarbons and water.[26] The Fischer–Tropsch (FT) process step can be considered a polymerization reaction of CO employing iron-based or cobalt catalysts, depending on the selected temperature and desired products.[3,6] The final product is then cracked, hydrotreated, isomerized and fractionated to obtain liquid aviation fuels.[30] Typically, the FT process does not produce the cycloparaffins and aromatic compounds found in petroleum-based jet fuel.[30] Since FT crude has different properties than crude oil, the refining process to produce on-specification aviation fuels is also different than that used in conventional refineries.[35] The main steps of the BtL-FT pathway for SAF production are explained in detail in the following sections.

11.3.1 Pretreatment

The three main conversion steps in the pretreatment process are drying, torrefaction and grinding of the biomass. The moisture in the feedstock is reduced to 15% to decrease the energy requirements for the subsequent torrefaction step, as well as to reduce the requirements for control parameters.[40,41] To eliminate the initial moisture, biomass is exposed to elevated temperatures to evaporate the water without changing the chemical composition of the biomass. This is termed "nonreactive drying". Typical parameters for nonreactive drying are 50–150 °C, with heating at below 50 °C per min at ambient pressure for 30–120 min. Mass yields are between 90 and 95%.[42] Typically, wet biomass is dried to a final moisture content in a range of 6–15% for further processing and storage.[43]

Torrefaction is a mild thermochemical pretreatment technology, defined as isothermal pyrolysis, that is used to decompose woody biomass.[44,45] The product has a high energy density, is hydrophobic, compactable and grindable, and has a low oxygen-to-carbon (O/C) ratio, which makes it suitable for use in gasification.[42] Since biomass retains most of its energy and simultaneously loses its hygroscopic properties, torrefaction is typically used to upgrade lignocellulosic biomass to a higher-quality fuel for a subsequent conversion process.[42,46]

11.3.2 Entrained Flow Gasification

In the presented BtL-FT process, the pulverized torrefied biomass is converted to syngas through gasification. The most characteristic feature of biomass gasification processes is the choice of reactor type, which has the largest impact on product gas composition and efficiency.[47] Slagging, pressurized oxygen-blown entrained flow gasification (EFG), was selected for this study to produce the highest-quality syngas with a low methane content and almost no tars present.[48] When using oxygen as the gasification agent, high temperatures and high levels of H_2 and CO can be reached in the syngas, which increases the calorific value of the fuel. For BtL-FT processes, the oxygen required for EFG is typically produced in a cryogenic air separation unit (ASU).[47]

In EFG, the syngas composition, the H_2/CO ratio, in particular, can be controlled within certain limits, by adjusting the gasification parameters to approximate the compositions needed for FTS. The amount of oxygen provided to the gasification chamber is the most important parameter, with the equivalence ratio λ defined as the molar ratio of the total oxygen content to the stoichiometric oxygen content required for complete combustion.[49] In general, the gasification temperature increases with increasing values of λ, while the H_2/CO ratio decreases.[50,51] In EFG, the minimum gasification temperature is limited by the ash properties and specifically a melting temperature of about 1260 °C. In the case of pressurized, oxygen-blown EFG of pulverized biomass, a further limitation is the energetic loss due to increasing CH_4 and C_6H_6 formation at lower temperatures. Thus λ for biomass EFG is typically about 0.4.[49,51] The upper gasification temperature limit could require a so-called moderator to regulate the temperature within the gasification chamber, due to material limitations. The most common moderator is steam.[48] Steam used with oxygen as the gasification agent leads to more syngas and less CH_4 at higher temperatures, as both steam gasification and steam methane reforming are endothermic. The WGS reaction and the steam methane reforming process determine the bulk gas composition inside the gasifier.[52] By adding more steam, the reduction of CO in favor of H_2 is promoted by the WGS reaction.[50] Thus the H_2/CO ratio can be increased by the addition of steam.[50,52] Since more steam being added also results in lower gasification temperatures, there is an operating point at which a maximum amount of syngas is obtained.[48] To avoid a high CH_4 yield, temperatures above 1400 °C are recommended.[52]

The gasification pressure is generally based on the requirements of processes downstream of the gasifier. Gasification at elevated pressures has a number of advantages, such as savings in terms of compression energy and a reduction of equipment size, so that practically all modern processes are operated at a pressure of at least 10 bar and up to as high as 100 bar.[31,48]

11.3.3 Gas Conditioning

Typical impurities in the syngas after gasification of biomass include particles (dust, char, ash), halide compounds, nitrogen and sulfur compounds. These contaminants can lead to erosion, corrosion and deposits on downstream process components. In addition, impurities can often lead to poisoning of the catalyst in downstream processes. The required FT feed gas specifications for different contaminants (particles, halide compounds, nitrogen and sulfur) are given in Table 11.3. Possible syngas conditioning processes include quenching with a subsequent scrubber and a sour WGS reactor upstream of the AGR unit. The specific steps involved are explained next.

11.3.3.1 Quenching

Since there is always a need to cool the syngas, all entrained ash particles inevitably pass through the critical temperature range at which the ash becomes sticky[48] The best way to avoid this problem is to quench the gas as

Table 11.3 Provisional FT feed gas phase specifications for different contaminants.[31,105]

Contaminant	Target
Particles (dust, char, ash)	Complete removal
Halide compounds	<10 ppb
Total nitrogen	<1 ppm
NH_3	1 ppm[59]
HCN	10 ppb[59]
Total sulfur	<1 ppm
H_2S	<10 ppb[59]

quickly as possible to a temperature at which the ash becomes solid – typically at about 900 °C.[48] Thus for slagging gasifiers, a quenching step is employed to cool the syngas rapidly after gasification and simultaneously remove most of the particles and impurities in the syngas.[53]

Quenching is reliable and inexpensive compared to cooling by heat exchangers, but it is also less efficient because it degrades the thermal energy in the syngas to a medium-low temperature.[54] With EFG, quenching can be accomplished using either a gas, chemical or water quench.[48]

When the ash enters the water quenching area, the sudden drop in temperature causes it to solidify.[48] The amount and quality of the slag produced depend on the feedstock used.[55] A distinction can be made between a partial quench, with just enough water being evaporated to reduce the gas to a specific temperature; a full quench, with sufficient water being evaporated to saturate the gas with water vapor.[48] In the full quench design, raw syngas is exposed to both a full quench and a syngas scrubber.[54] As previously mentioned, chloride compounds might cause fouling, create deposits and cause poisoning of the catalyst, while WGS and AGR units can experience problems related to ammonia. Wet scrubbers remove the remaining particulate matter and water-soluble contaminants (primarily HCN, NH_3, and chlorides) by spraying the syngas with water.[54,55] Consequently, a full quench step serves as a first step in the gas treatment process to ensure safe and low-maintenance operation.

11.3.3.2 Sour Water-gas Shift

Since the syngas produced *via* biomass gasification typically has a H_2/CO ratio of below 1, a WGS reactor is employed to increase the value to more than 2, as required for the FT process.[56] The WGS reaction takes place according to eqn (11.1):

$$H_2O_{(g)} + CO \rightleftharpoons H_2 + CO_2, \quad \Delta H_R^0 = -41.2 \text{ kJ mol}^{-1} \tag{11.1}$$

Depending on the reaction temperature, H_2 (forward WGS at 250–400 °C) or CO (reverse WGS at 400–500 °C) can be concentrated,[48,57] whereas pressure has no influence on the reaction equilibrium.

The WGS reactor uses a sour catalyst, which is capable of handling H_2S impurities. Sour WGS reactors utilize catalysts to enhance the reaction rate.[54] The catalysts are capable of handling or even need sulfur, and they operate at temperatures between 300–500 °C, which is why the WGS is located upstream of the AGR reactor.[48] Contrarily, sweet WGS can only operate in the absence of S-compounds and are thus located downstream of AGR.[48] In general, high CO conversion rates of approximately 96–98% are possible.[58]

The amount of steam in the feedstream to the WGS reactor can also influence the reaction equilibrium. Depending on the desired H_2/CO ratio in the product stream, an H_2O/CO ratio between 1.65 and 2 might have a positive effect on the WGS conversion.[55,58] Besides the WGS reaction, most of the COS is converted to H_2S and HCN to NH_3 in the sour WGS reactor.[54,55]

11.3.3.3 Acid Gas Removal

An important component of the gas treatment step is acid gas treatment. In the syngas environment, AGR includes desulfurization and CO_2 removal,[48,55] which is especially important in BtL-FT processes, due to the WGS step upstream of the AGR, which would otherwise lead to unsuitably high CO_2 concentrations for the downstream FTS[59] Furthermore, deep sulfur removal is needed, as downstream FTS is very sensitive to sulfur poisoning[55] Commercially available AGR technologies mainly use a cold gas cleaning process, since they operate at ambient temperatures or lower.[48] Using a ZnO guard bed for chemical adsorption of trace H_2S is the prominent exception with a hot gas cleaning process.[48]

Various AGR technologies are available for gasification plants that most commonly use ad- or absorption technologies.[58] The solvent-based AGR plants comprise two sections: one for the absorption of impurities into the solvent and one for solvent regeneration.[60] The solvent can be a physical solvent (such as methanol) or a chemical solvent (such as amines).[60] In addition to the gas purity required for the subsequent synthesis and the raw gas composition and the selectivity of the AGR process, issues related to corrosion and economic boundary conditions are other important selection criteria for the selection of a suitable AGR technology.

H_2S is typically dissolved in a solvent from which it can be recovered as elemental sulfur in a Claus or similar type of unit and then be used in a variety of processes.[55,61] To ensure a total sulfur content of less than 1 ppm in the product gas stream, and a high CO_2 concentration in the raw syngas, only ZnO, Rectisol™, Purisol® and Selexol® are suitable process technologies. It should be noted that H_2 and CO losses during AGR must be minimized to maintain the desired H_2/CO ratio for FTS.[26,62]

Selexol® was chosen as the solvent for the AGR unit. The advantages of the Selexol® system for gasification applications are that no chemical reactions take place and the acidic gases have a favorable solubility compared to other light gases[63] Especially HCN and NH_3 are highly soluble in Selexol®.[48]

Typically, a two-stage Selexol® pressure swing process configuration is used.[64–66]

11.3.4 Fischer–Tropsch Synthesis

FTS is the central conversion step to produce liquid fuel. The process was developed for liquefying coal by Franz Fischer and Hans Tropsch in 1925.[67] The product of FTS is synthetic crude oil, or syncrude, which mainly consists of a mixture of linear long-chain hydrocarbons. In the following sections, FT reactions, the influence of process parameters and catalysts, and different reactor concepts are explained.

11.3.4.1 Process Conditions, Catalysts and Reactor Technologies Used in the Production of Sustainable Aviation Fuel

The main FT reactions are based on the hydrogenation of CO over metallic catalysts. Highly exothermic polymerization or chain extension reactions form a mixture of organic compounds by chain growth through the gradual addition of C species,[68] as per eqn (11.2):

$$2nH_2 + nCO \rightarrow [-CH_2-]_n + nH_2O, \quad \Delta H_R^0 = -152 \text{ kJ (mol n)}^{-1} \qquad (11.2)$$

Selectivity and conversion in FTS depend on feedstream composition (H_2/CO ratio), pressure, temperature, catalyst and reactor type and size.[31,69] Table 11.4 provides an overview of the influence of various FTS operating parameters on product composition and syngas conversion.[61]

FTS technologies are classified into high-temperature FT (HTFT) and low-temperature FT (LTFT) technologies, based on the operating temperature. Typical temperatures for HTFT and LTFT are 300–350 °C and 170–270 °C, respectively.[35] Pressure is typically in the range of 20–30 bar and mainly affects productivity.[70] Table 11.5 summarizes the typical FTS process conditions for HTFT and LTFT.[35,56,70,71]

Most transition metals can hydrogenate CO. However, FTS is typically catalyzed by supported metals from the VIII group of transition metals. Iron, cobalt, nickel and ruthenium are considered to exhibit a sufficient reaction

Table 11.4 Influence of the operating parameters on the FT product selectivity.[61]

	Process parameter change (+ = increase, − = decrease, 0 = no change)		
Parameter	Temperature	Pressure	H_2/CO ratio
α-Value	−	+	−
Methane selectivity	+	−	+
Alkene selectivity	*a*	*a*	−
Oxygenate selectivity	*a*	+	−
Aromatics selectivity	+	*a*	−
Syngas conversion	+	+	*a*

[a]The change is the result of a more complex relationship.

Table 11.5 Comparison between HTFT and LTFT.[70,71]

Process parameter	HTFT	LTFT
Temperature	300–350 °C	170–270 °C[35]
Pressure	20–30 bar	20–30 bar
Catalyst	Fused Fe-Cat.	Precipitation Fe-Cat., supported Co-Cat.
Reactor types	Circulating fluidized-bed, stationary fluidized-bed	Tube bundle fixed-bed reactor, slurry-reactor
H_2/CO-ratio	0.5–2.5 (1.4)[56] (Fe-Cat.)	2.0–2.3 (Co-Cat.) 0.5–2.5 (Fe-Cat.)

rate and polymerization probability[61] However, only Fe- and Co-based catalysts are used industrially.[61] Since Co-catalysts have very low WGS activity compared to Fe catalysts, Co-based catalysts are often preferred for low temperatures, if the H_2/CO ratio at reactor entry is already adjusted to the synthesis.[61] An H_2/CO ratio of slightly above 2.1 is required for the Co-based catalyst FTS.[56]

Due to the highly exothermic FT reactions, insufficient heat removal inhibits the reaction equilibrium and may lead to localized overheating, which causes high carbon deposition and subsequent deactivation of the catalyst.[69] Consequently, FT reactor designs have also focused on heat removal and temperature control. Many types of reactors have been developed for FTS, the most common being fixed-bed and slurry bubble column reactors (SBCR).

SBCRs are preferred for LTFT up to 250 °C. They employ mainly Co-based catalysts and offer several advantages over fixed bed reactors;[72,73] *e.g.* they allow easy isothermal operation in the reactor and the possibility of continuous catalyst makeup and withdrawal. This enables constant production to be maintained in the case of catalyst deactivation or partial poisoning.[72] Compared with multitubular fixed-bed reactors, slurry reactors have a lower pressure drop over the reactor, resulting in lower cost.[69] Slurry reactors also offer higher capacity and higher conversion per pass.[59] The advantages of LTFT with the Co-catalyst inside an SBCR are also evident in terms of thermal efficiency, C turnover, saturated steam production and FTS product yield.[74]

11.3.4.2 Product Distribution and Selectivity towards Sustainable Aviation Fuel

The FT reaction product consists of a complex multicomponent mixture of hydrocarbons.[61] Typically, the majority of the products are *n*-alkanes, with the remaining part being composed mainly of 1-alkenes and neglectable, small amounts of oxygenates.[26,75] A common approach to describe the product formation is by using the chain growth probability factor α of the catalytic system.[61,76] The idealized chain length distribution of the FT

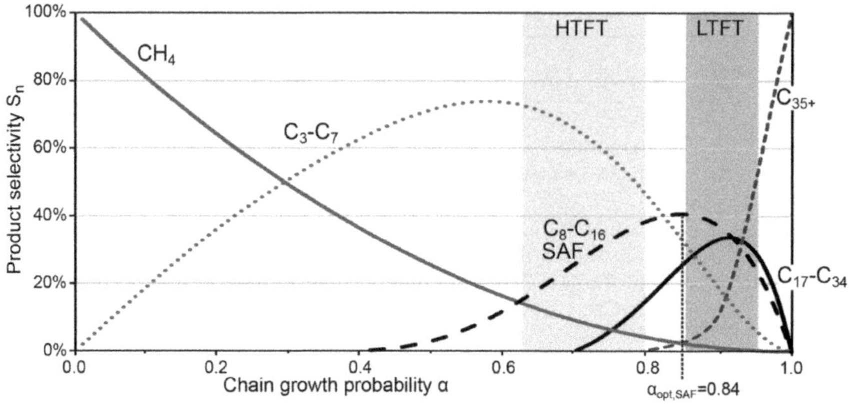

Figure 11.3 Dependency of the product selectivity on the chain growth probability α according to an ideal Anderson–Schulz–Flory distribution for different carbon lumps.

products follows the Anderson–Schulz–Flory (ASF) distribution. The α-value represents the probability of adding $n-1$ carbons and $(1-\alpha)$ is the probability of terminating the chain growth toward either alkanes or alkenes.[77–79] In the first approximation, with a constant α value, the reaction is not selective with regard to chain length, and the products follow a distribution that is only described by the chain growth probability α.[74,79] The selectivities for the product classes methane, light naphtha (C_2–C_7), jet fuel (C_8–C_{16}), heavy hydrocarbons (C_{17}–C_{34}) and FT waxes (C_{34+}) are shown in Figure 11.3. It shows that the longer-chain products increase with increasing α.

The selection of an FTS process depends on the desired target product. The ASF distribution can be used to estimate the amount of desired fraction to be produced by FTS. For an ideal chain growth model, jet fuel in the hydrocarbon range C_8–C_{16} is favored at $\alpha_{opt}=0.84$, with the maximum selectivity for jet fuel being about 41% (see Figure 11.3). With the chain growth probability α for HTFT being in the order of 0.65–0.8 (see Table 11.5), 50 wt% of the product is in the range of C_1–C_4 and less than 20 wt% of diesel and above.[74] HTFT also produces low-molecular-weight olefins (such as ethylene and propylene),[71] less alkenes and more olefins, aromatics and oxygenates than the LTFT[72,73] LTFT mainly yields high-molecular-weight waxes.[71] In LTFT α-values of close to 0.84 can be reached to produce as much C_8–C_{16} as possible while producing a minimum of methane. Thus LTFT is generally beneficial for producing jet fuel.

11.3.5 Separation, Refining and Upgrading

After the production of FT syncrude, further fractionation, refining and upgrading through hydrotreating, hydrocracking and isomerization is required. First, the syncrude produced *via* FTS is typically divided into fractions, according to the chain length. The unconverted syngas is separated from the FT products and recycled into the FTS, while the products are separated using their specific boiling points, in order to feed them to the respective refining operations. In addition, oligomerization of short hydrocarbons for HTFT or hydrocracking of long-chain hydrocarbons for LTFT can be used to increase the jet fraction to 50–60% in the final product.[35]

At this stage of conditioning, the products differ from usual oil cuts and thus do not meet the required SAF specifications; therefore, further upgrading is required.[35,80] In addition, even the straight-run jet-range fractions require hydroprocessing. Alkenes and oxygenates from HTFT are hydrogenated to saturate double bonds and remove oxygen. For liquid syncrude produced *via* LTFT, hydroprocessing is mainly used to saturate alkenic double bonds.[35]

In practice, refining and upgrading FT products can be carried out in existing petroleum refineries. This integration of FT syncrude refining into existing facilities, through coprocessing or other means, might also have a positive impact on SAF expansion.[81]

11.4 Simulation Model Design and Scope

In this study, a holistic modeling approach was used to simulate the entire integrated process chain. The process models were developed using the Aspen Plus® process simulator. A BtL-FT model was designed to ensure a consistent and transparent process. First, a reference simulation model was described, which represents a once-through BtL-FT plant to produce SAF. Material flow and heat integration were not included at this stage, and natural gas was used as a supplementary fuel to heat the pretreatment phase.

Based on this reference model design, an advanced BtL-FT process model was developed. Variation of process parameters, heat integration and material recycling were used to maximize SAF yield and selectivity. With biomass as the sole carbon source in the BtL-FT processes, the carbon efficiency (η_C) is closely connected to the SAF yield. Carbon efficiency is an important performance indicator for BtL-FT processes, since it defines the amount of initial material being transformed into valuable product:

$$\eta_C = \frac{\dot{m}_{C,\text{prod}}}{\dot{m}_{C,\text{feed}}} = \frac{\dot{m}_{C,\text{fuel}}}{\dot{m}_{C,\text{biomass}}} \tag{11.3}$$

where $\dot{m}_{C,\text{feed}}$ is the carbon mass flow rate fed to the process; $\dot{m}_{C,\text{biomass}}$ is the carbon mass flow rate in the biomass feed; $\dot{m}_{C,\text{prod}}$ is the carbon mass flow rate of the main product stream; $\dot{m}_{C,\text{fuel}}$ is the mass flow rate of carbon in the fuel that is produced.

In order to evaluate the energy utilization of the raw materials used, the energy yield (EY) based on the LHV indicates how much of the initial energy is retained in the product:

$$\text{EY} = \frac{\dot{E}_{\text{prod}}}{\dot{E}_{\text{feed}}} = \frac{\dot{m}_{\text{fuel}} \cdot \text{LHV}_{\text{fuel}}}{\dot{m}_{\text{biomass}} \cdot \text{LHV}_{\text{biomass}}} \tag{11.4}$$

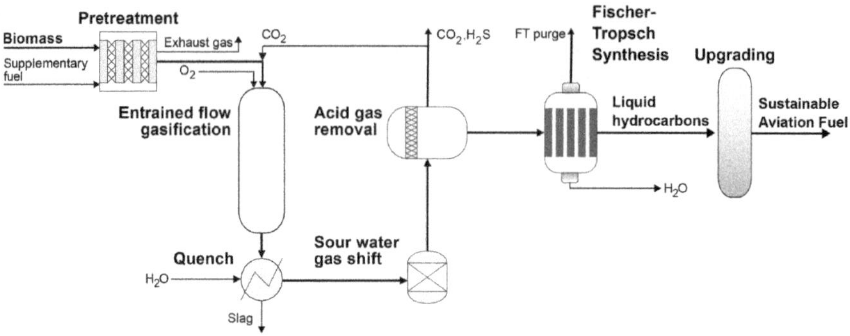

Figure 11.4 Simplified schematic BtL-FT process flow sheet of the once-through reference simulation model for the production of sustainable aviation fuel, including pretreatment (drying, torrefaction, grinding), oxygen-blown entrained flow gasification, water quench, sour water-gas shift, acid gas removal, Fischer–Tropsch synthesis and upgrading.

where $\dot{E}_{\text{feed}}$ is the enthalpy flow rate fed to the process; $\dot{E}_{\text{prod}}$ is the enthalpy flow rate of the main product; $\dot{m}_{\text{biomass}}$ is the mass flow rate of the biomass feed; $\text{LHV}_{\text{biomass}}$ is the weight-specific LHV of the biomass feed; $\dot{m}_{\text{fuel}}$ is the mass flow rate of the produced fuel; LHV_{fuel} is the weight-specific LHV of the produced fuel.

The overall net energy efficiency η_{E} also considers the auxiliary energy requirements ($\dot{E}_{\text{aux}}$), which are either electricity or sensible/latent heat carriers:

$$\eta_{\text{E}} = \frac{\dot{E}_{\text{prod}}}{\dot{E}_{\text{input}}} = \frac{\dot{m}_{\text{fuel}} \cdot \text{LHV}_{\text{fuel}}}{\dot{E}_{\text{feed}} + \dot{E}_{\text{aux}}} \tag{11.5}$$

where $\dot{E}_{\text{input}}$ is the overall energy flow fed to the process.

Exclusively lignocellulosic biomass residues were used as feedstock for this process model. The respective biomass input remains unchanged at 200 MW_{th}[‡] for all process models. Due to the complexity of the upgrading process and the benefits of using existing petroleum refineries for this purpose, this work was limited to splitting the product stream into the respective fractions: syngas, methane, naphtha (C_2–C_7), jet fuel (C_8–C_{16}), gas oil (C_{17}–C_{22}) and heavier waxes (C_{23+}). The individual fractions were used to evaluate the performance characteristics of the overall process. Figure 11.4 shows the simplified process flow sheet of the simulated once-through BtL-FT process.

[‡]Dry biomass input of 40 t per h^{1}; thermal energy flow based on LHV.

11.4.1 The BtL-FT Once-through Reference Simulation Model

In the reference model, hot exhaust gas from the combustion of torrefaction gas and supplementary fuel was used for drying. To maintain a temperature of 150 °C for the stream entering the dryer, a portion of the gas leaving the drying section is separated from the exhaust gas stream and recycled. Thus hot exhaust air (at 150 °C) and wet biomass (nonconventional component, $\varphi = 35\%$) are fed to the dryer (RStoic) to form a stream of dried biomass ($\varphi = 10\%$) and moist gas at 80 °C.

The input stream to the torrefaction reactor (RStoic) is the solid product from the dryer. In the torrefaction reactor, biomass is treated at 250 °C. Solid mass yield, gas phase composition and heating value of the solid product are calculated depending on the torrefaction temperature. The elaborated models used in this work are based on the data of Prins *et al.* and Nocquet *et al.*[82,83] Heat losses through the dryer and torrefaction reactor body are 0.5% of the higher heating value (HHV) of the respective feed and 1% of the HHV of the gaseous feed for the flare.[41] The drying and torrefaction steps are validated based on the simulation results of Peduzzi *et al.*[46]

Since a particle size of less than 300 μm is essential for EFG,[84] a crusher block (modeled on the Hardgrove grindability index) was used to model the required grinding step. The respective values are determined as a function of temperature for torrefaction conditions between 250 and 290 °C, based on the data reported by Manouchehrinejad.[85]

The pyrolysis step in gasification, where the solid feed is decomposed into its basic constituents, was modeled using an isobaric RYield reactor. To consider the complex reaction pathway of nitrogenous components, an additional RStoic reactor was employed to model the formation of HCN and NH_3, based on data reported by Aho *et al.*[86] An equilibrium reactor based on the Gibbs free energy minimization (RGibbs) was used to model the main conversion of biomass into syngas. CO_2 was used as the carrier agent for pneumatic transport of the solid fuel. Part of the CO_2 separated from the syngas in the AGR section was recycled to be used for that purpose. To account for a carbon conversion of 99%, which is typical for entrained flow processes,[48] 1% of the conventional solid carbon is specified as the inert component. HCN and NH_3 are specified as inerts as well, as their formation is modeled beforehand.

In the reference simulation, the oxygen feed was controlled to maintain gasification temperatures of 1500 °C. To minimize the cost of the downstream equipment, the gasifier in the reference model was operated at 37 bar. Oxygen supply by means of ASU was assumed. The ASU was not explicitly modeled, but the oxygen was fed into the process at the appropriate purity level, and the corresponding energy cost of 0.5 kW h_{el} N $m_{O_2}^{-3}$ was applied.[47,64,87] Heat loss through the gasifier walls was considered, depending on the reactor size. A heat loss of 1.5% was assumed for the 200 MW_{th} biomass input.[47]

A full water quench was used in the simulation model. Hot syngas from gasification was quenched by direct evaporation of injected water. The slag leaves the gasifier at the bottom of the quench reactor. The developed gasification and quench model was validated using data on the commercially available PRENFLO® process.[88]

Since the water quench and wet scrubber has already removed all the halide compounds, no adsorption bed is required before the WGS reactor. The isobaric, adiabatic, WGS reactor model calculates the equilibrium composition for the specified reactions. The WGS reaction itself was modeled using a specified molar extent. The sour WGS model was validated using data from Boissonnet *et al.*[89]

The AGR models of Field *et al.*, Kaptaki *et al.* and Adams *et al.* were used as the basis of the Selexol® process design.[64,65,90] The Selexol® process consists of a set of absorbers and steam strippers used for H_2S removal, followed by another set of absorbers and air strippers for CO_2 removal. For further reduction of H_2S to the FTS requirement, adsorption in a ZnO bed was employed.[60]

The FT reaction (RYield) was modeled isothermally at 220 °C and 25 bar, with a feed H_2/CO ratio of 2.1. A conventional LTFT process, using Co-based catalysts in a SBCR, was used for the production of SAF in this work. The heat loss was specified as 0.5% of the enthalpy flow entering the reactor, based on the HHV. After the FT reaction, the product stream is separated into a gaseous phase, a liquid organic product and a liquid watery phase.

Since FT kinetics are a function of product distribution at the reactor outlet, the complex mathematical description of the FT kinetics requires iterative solving of the reaction kinetics. Therefore, the existing unit models in Aspen Plus® cannot be used, and the FT reactor model was implemented using a Python routine and object-oriented FT reactor models *via* a FORTRAN subroutine. The models contain the FT reactor attributes and the kinetic models used to determine the product distribution based on the operating conditions. An alpha correlation developed by Ostadi *et al.*[91] was refitted to the experimental data of Todic *et al.*,[92] and used to model the FT product distribution in this work.[62] The kinetic model used was validated based on the data reported by Todic *et al.*[93–95]

11.4.2 Advanced and Highly Integrated BtL-FT Simulation Model

Heat and material integration, as well as an advanced light end hydrocarbons recovery process with FT light gas recycling, was integrated into the BtL-FT process to maximize SAF yield and selectivity, as well as carbon and energy efficiency. Further process optimization at a component and system level completed the advanced BtL-FT simulation model. Figure 11.5 shows a simplified process flow sheet of the advanced and highly integrated BtL-FT process.

11.4.2.1 Material and Heat Integration

Instead of supplementary fuel, exhaust heat from gasification was used in the advanced BtL-FT process to supply the pretreatment section. In that way,

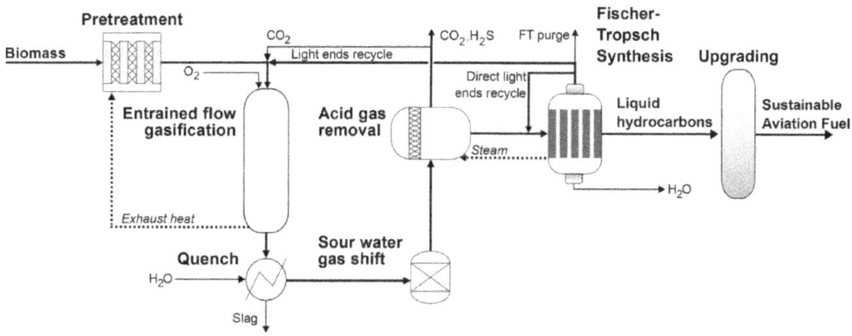

Figure 11.5 Simplified schematic BtL-FT process flow sheet of the advanced and highly integrated simulation model for the production of SAF, including pretreatment (drying, torrefaction, grinding), oxygen-blown entrained flow gasification, water quench, sour water-gas shift, acid gas removal, Fischer–Tropsch synthesis and upgrading.

the required heat for drying and torrefaction is mainly supplied *via* the syngas cooler after gasification. Utilizing exhaust heat for the biomass pretreatment allows carbon conversion of virtually 100% in the pretreatment section, with torrefaction gases being fed to the subsequent gasifier.

In order to make use of the methane-rich gas mixture that leaves the FTS, it can be recycled to the gasification section. At gasification conditions, the light end hydrocarbons are cracked to syngas to be refed to the FT reactor. In general, gasification is used to convert solid fuels to gaseous mixtures of high value. The gaseous recycle streams from torrefaction and FTS – consisting of light hydrocarbons – enter the gasifier at a section where most of the pulverized solid fuel is already gasified, and the temperature levels are reduced compared to the inlet hot spot. For this purpose, a reformer block was used, which also resembles a gas quench configuration, with colder gases used to quench the gasification products[48] The reformer model employed matched the specifications of the gasifier but was operated at 1050 °C. The heat required for the reformer was provided *via* heat coupling to the gasifier. The resulting gas mixtures from the gasifier and reformer were fed to a syngas cooler.[48]

If the torrefaction and all FT light gases were fed to the gasifier directly, cold gas efficiency on a LHV basis of the gasification process would result in 64.25% with a O_2 demand of 0.73 t_{O_2} per $t_{biomass,dry}$. With the reformer used, cold gas efficiency increased to 68.9% with 0.68 t_{O_2} per $t_{biomass,dry}$ fed to the gasification section. Overall carbon conversion efficiency can be further increased to 34.0%, since the molar fraction of $H_2 + CO$ in the syngas after gasification increased from 67–70% and showed an H_2/CO ratio of 0.92 instead of 0.71.

Although the configuration performed well in terms of reforming methane and other light hydrocarbons, the unconverted H_2 and CO leaving the FT reactor were also recycled through the entire BtL-FT process from gasification *via* quench, WGS and AGR. This requires larger equipment, which is associated with higher costs, and also causes an increased equivalence ratio in the gasifier. This is due to the increased heat requirement for the reformer and results in a cold gas efficiency of 65.0%.

One way to avoid recycling unconverted reactants through the entire process is to refeed part of the product gas directly to the FT reactor. To improve the light gas recovery after the FT reaction, a flash cascade is employed. To ensure that neither CO_2 nor any trace component was concentrated due to the closed recycle loop, a separator was used to remove 50% of the CO_2 in the FT light ends product stream. Part of these light ends are then recycled directly to the inlet of the FTS reactor, while the remaining part is fed to the reformer block of the gasifier section.

11.4.2.2 Optimization of Process Parameters

To further increase the system's carbon and energy efficiency, as well as the SAF yield and selectivity, various adaptions can be made throughout the entire process. For the advanced BtL-FT process model, the pressure and temperature levels of the main conversion steps and the light end recycle ratio are optimized in terms of key performance indicators.

In the pretreatment section, the specified torrefaction temperature is the only degree of freedom. Since both the gaseous and the solid products are fed to the gasifier, solid mass and energy yield after torrefaction play only a minor role in the overall efficiency of the process. Lower temperatures are favorable for the overall process, as they require lower heat requirements. Although a higher torrefaction temperature results in a lower energy requirement for the grinding process, overall energy efficiency and product yield increase with a lower torrefaction temperature.

Operating conditions in EFG also influence the overall process performance. Although pressurized gasification offers the advantage of a lower compression being required downstream, a higher pressure can shift the steam–methane equilibrium reaction toward the educt side.[52] Therefore, there is an increase in CH_4 and CO_2 content in the syngas with increasing pressure, while the H_2 and CO yield drops.[48] Figure 11.6A shows that, with increasing pressure, overall product yield and energy efficiency decrease. As a trade-off between performance and a reduction in equipment size, gasification pressure of 30 bar was chosen for this study.

With an operating pressure of 30 bar, a high temperature above 1300 °C is required to produce syngas with a low methane content.[48] The minimal gasification temperature is also limited by the ash melting temperature of about 1260 °C, as mentioned in Section 11.3.2. However, gasifying at an elevated temperature increases oxygen consumption in the EFG process and reduces overall process efficiency. Figure 11.6A shows that with an increase in gasification temperature, overall product yield and energy efficiency decrease. The EFG process was simulated at 1400 °C in order to remain well above the ash melting temperature while allowing maximum performance.

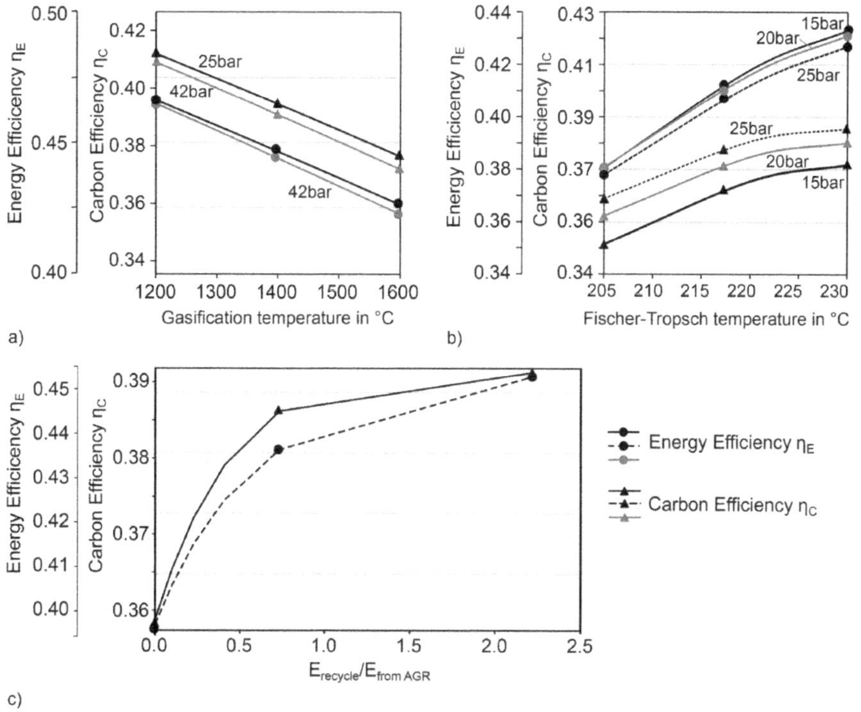

Figure 11.6 Influence of (a) gasification pressure and temperature ($T_{FT} = 230$ °C, $p_{FT} = 20$ bar); (b) Fischer–Tropsch pressure and temperature ($T_{Gasif} = 1500$ °C, $p_{Gasif} = 37$ bar); (c) amount of light ends being directly recycled to the FTS reactor ($T_{FT} = 230$ °C, $p_{FT} = 20$ bar, $T_{Gasif} = 1400$ °C, $p_{Gasif} = 30$ bar) on carbon and energy efficiency.

In terms of FTS conditions, a higher reaction temperature allows for a higher conversion rate and thus improves the overall product yield and energy efficiency. Higher pressure also results in a higher conversion rate and thus higher efficiency, but the recycle streams, in particular, require recompression to either gasification or FT reactor pressure. Thus overall energy efficiency decreases with increasing FT pressure, as shown in Figure 11.6B. Therefore, pressure of 20 bar was selected, and an FT temperature of 230 °C was chosen, which is the upper validity value for the FT model employed.

Another important operating factor is the H_2/CO ratio at the reactor inlet. In general, conversion increases with an increase in the H_2/CO ratio. However, this effect was limited to a ratio of about 2 in this simulation, which is considered the optimum inlet ratio. Additionally, the implemented macro-kinetic FT model was only valid up to a ratio of 2.1. For that reason, the sour WGS was set to reach an H_2/CO ratio of 2.0.

Table 11.6 summarizes the parameter influences in terms of torrefaction temperature, gasification conditions and FT reactor conditions on FT CO

Table 11.6 Influence of operating parameters on the FT CO conversion, carbon and energy efficiency, and SAF selectivity.

	Process parameter change (+ = increase, − = decrease, 0 = no change)					
	Torrefaction	Gasification		Fischer–Tropsch		
Parameter	Temp.	Temp.	Pres.	Temp.	Pres.	H_2/CO^a
FT CO conversion	0	0	0	+	+	$+^b$
Carbon efficiency	−	−	−	+	+	$+^b$
Energy efficiency	−	−	−	+	−	$+^b$
SAF selectivity	0	0	0	+	−	$+^b$

[a] H_2/CO ratio varied between 1.4 and 2.1.
[b] No notable increase between 2.0 and 2.1 H_2/CO.

conversion, product yield, jet fuel selectivity and energy efficiency. The results are in accordance with the literature (see Table 11.4).

As a consequence, the advanced BtL-FT process was conducted at a torrefaction temperature of 250 °C and a gasification temperature of 1400 °C in the gasifier and 1050 °C in the reformer, and at 30 bar in both. FTS took place at 230 °C and 20 bar, and the amount of water to be added to the quench section was minimized. The sour WGS reaction was aimed at achieving an H_2/CO ratio of about 2, after mixing fresh and recycled gas before the FT reaction. An inlet temperature of 250 °C was specified for the adiabatic WGS reactor.

The ratio of recycled light ends going directly into the FT reactor, and the amount going into the gasifier has an impact on overall system performance. The light ends direct recycle decisively influences the inlet conditions of the FTS. The composition of the gas stream entering the FT reactor was obtained from the mixture of clean gas from AGR and the direct recycle. The molar content of the reactants in the direct recycle was low, since it had already been reacted in the FT reactor. Because the FTS is sensitive to the partial pressure of H_2, CO and H_2O, the H_2/CO ratio at the reactor inlet is crucial. The conversion in the WGS reactor is adjusted so that the H_2/CO ratio after the direct light end recycle matches the required H_2/CO ratio for the synthesis.

Both the overall energy and the carbon efficiency increase with an increasing amount of light ends being recycled directly to FTS, as shown in Figure 11.6C. However, a higher recycle ratio also requires larger equipment, which means higher cost. Therefore, an enthalpy ratio of 1.1 of the fresh syngas and the direct light ends to FT recycle was selected to maximize the process performance. The rest of the light ends are recycled to the gasification section.

11.5 Simulation Results of the BtL-FT Process

The reference Aspen Plus® simulation model resembles the once-through BtL-FT pathway used to produce SAF from lignocellulosic biomass. The advanced process model is highly integrated, in order to achieve maximum performance. The results of the two simulation models are presented, compared and evaluated in the following sections.

11.5.1 Once-through BtL-FT Reference Model Results

A product yield of 0.108 t_{fuel} per $t_{biomass,dry}$ was obtained for the once-through BtL-FT reference model. All the hydrocarbons produced, which are liquid at SATP conditions (Standard Ambient Temperature and Pressure) of 25 °C and 1 atm, were considered valuable liquid products. The resulting product fraction corresponds to a carbon efficiency of 18.2%. With the chain growth probability α as the sole responsible parameter, a selectivity of 28.7% toward the hydrocarbon range C_8–C_{16} was obtained. Figure 11.7 shows the carbon flow from biomass feedstock to final product. In the once-through simulation, about 50% of the initial carbon was lost as CO_2, which was mainly produced in the gasifier and WGS reactions. The remaining carbon was lost in the FT purge containing mainly light ends and methane.

The product yield corresponds to an energy yield of 21.8%, which means that almost 22% of the chemical energy stored in the biomass feedstock is stored in the liquid fuel at the end of the process chain (on LHV basis). Figure 11.8 shows the corresponding energy flow diagram. Apart from the biomass input, electricity and supplementary fuel are also fed to the process. The resulting energy efficiency is 18.5%.

11.5.2 Advanced, Highly Integrated BtL-FT Model Results and Comparison

For the advanced and highly integrated BtL-FT process, the carbon efficiency and product yield can be improved when compared to the reference model. Using the gasifier's exhaust heat in the torrefaction process and feeding the torrefaction gases to the gasifier result in a 100% carbon efficiency for the pretreatment step. Since the FT light ends are no longer required as fuel for

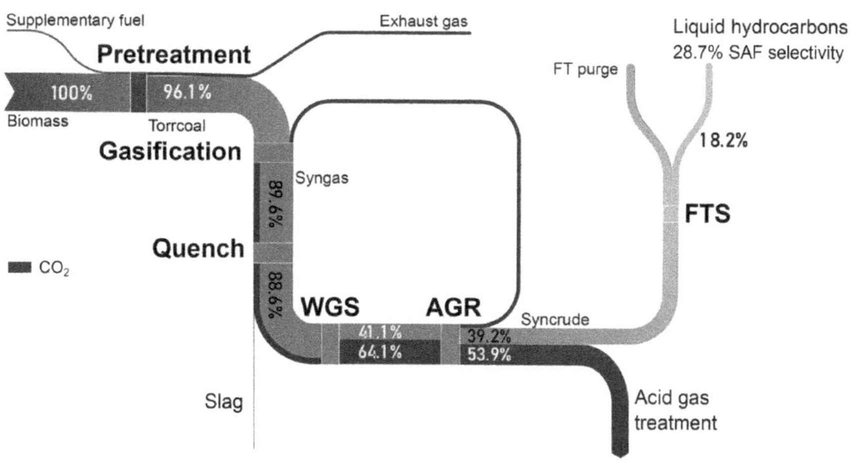

Figure 11.7 Simplified carbon flow simulation results for the once-through BtL-FT model, normalized to pure carbon biomass input. Minor flows of less than 1% of the carbon input are neglected.

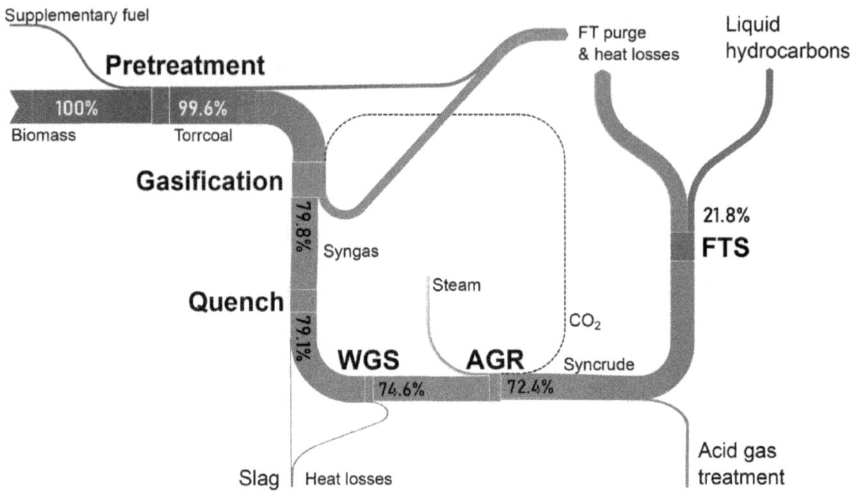

Figure 11.8 Simplified energy flow simulation results for the once-through BtL-FT model, normalized to pure energetic biomass input. Minor flows of less than 1% of the energy input are neglected.

the pretreatment section, the FT purge can be reduced, and the FT light ends can be recycled to gasification and FTS itself. Advanced light end recovery enables a high recycle ratio at a low purge rate, which conserves most of the carbon. Using the additional reformer and the optimized process conditions leads to a further improvement in terms of carbon efficiency and SAF yield. A carbon efficiency of 39.1% can be achieved, which corresponds to a product yield of 0.23 t_{fuel} per $t_{biomass,dry}$. This is more than double the amount of fuel produced compared to the reference simulation model: data for BtL-FT processes in the literature indicated a product yield of 0.026–0.258 t_{fuel} per $t_{biomass,dry}$.[33] The simulation results obtained are well within this range. Due to the process parameters being adjusted, the selectivity toward SAF increased to 32.5% for the advanced BtL-FT process. Figure 11.9 is a carbon flow diagram that shows where most of the unused carbon leaves the process as CO_2 after AGR.

The advanced BtL-FT simulation model shows an increase in electric parasitic load compared to the reference process. However, as product yield increases, so does energy efficiency. The ASU that supplies oxygen to the EFG is the biggest contributor to the electric parasitic load. The increase in electricity required for gasification in the advanced BtL-Ft process is mainly due to compression of the higher gas flows within the process. An energy efficiency of 40–53% is estimated in the literature for BtL-FT processes.[27] The highly integrated model achieves an energy efficiency of 44.4%, which is in accordance with the data in the literature. Figure 11.10 shows the energy flow diagram.

By modeling the advanced BtL-FT, it was demonstrated that using exhaust heat after gasification to heat the torrefaction unit in the pretreatment

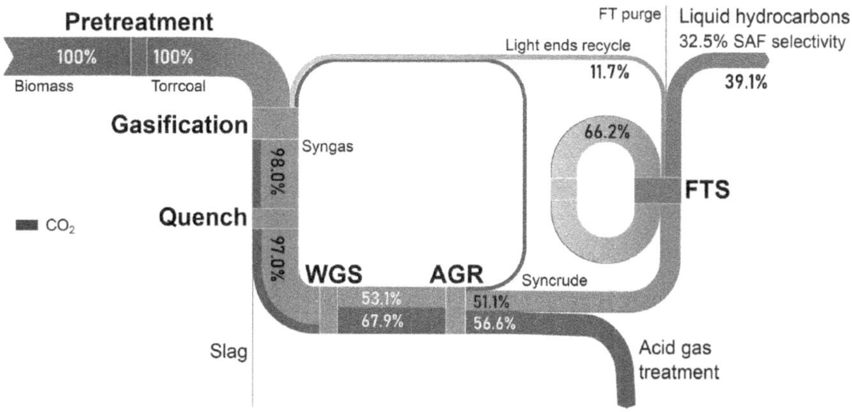

Figure 11.9 Simplified carbon flow simulation results for the advanced BtL-FT model, normalized to pure carbon biomass input. Minor flows of less than 1% of the carbon input are neglected.

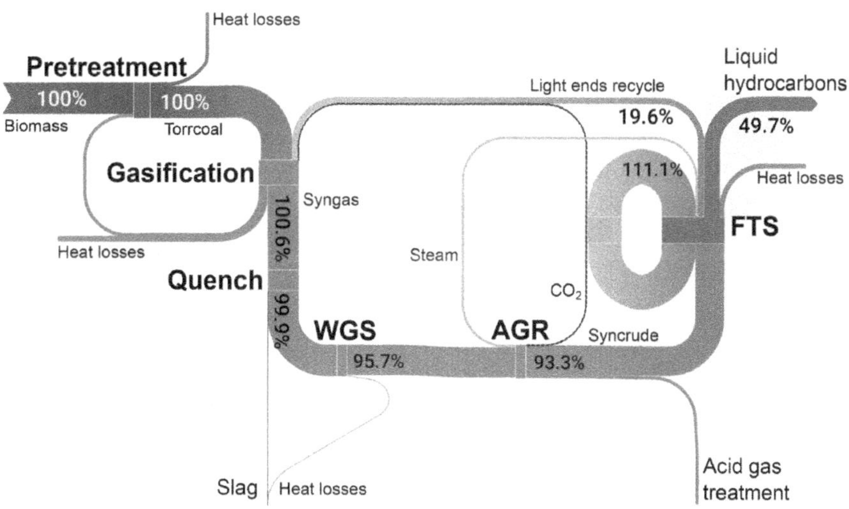

Figure 11.10 Simplified energy flow simulation results for the advanced BtL-FT model, normalized to pure energetic biomass input. Minor flows of less than 1% of the energy input are neglected.

section results in a higher overall process performance. Table 11.7 summarizes the most important overall key performance indicators. Product yield, carbon, energy efficiency and selectivity toward SAF show a significant increase with the highly integrated process.

11.6 Discussion and Outlook

The results of the three process models – reference, once-through, advanced and highly integrated BtL-FT – show the high potential for efficient

Table 11.7 Comparison of simulation results based on LHV for BtL-FT process variations.

Description	Unit	Simulation results	
		Once-through	Highly integrated
Product yield[a]	t_{fuel} per $t_{biomass,dry}$	0.108	0.231
Carbon efficiency	%	18.2	39.1
Energy efficiency	%	18.5	44.4
SAF selectivity	%	28.7	32.5

[a]Here, product corresponds to all produced hydrocarbons that are liquid at SATP conditions (standard ambient temperature and pressure).

production of SAF. It was shown that carbon efficiency of 40% and selectivity toward SAF of 32.5% are possible for a highly integrated BtL-FT pathway.

Regardless of selectivity considerations for aviation fuel, it is important to note that the carbon efficiency of the BtL-FT process is limited. Since carbon efficiency is the determining factor for product yield and thus for the performance of the process, a comparatively low system efficiency results even with highly advanced and integrated processes. Furthermore, in order to evaluate the sustainability of the produced fuel and compare the so-derived SAF with fossil-based CAF, an assessment of the process-related GHG emissions is required for the BtL-FT process.

One way to overcome the carbon limitation of the BtL-FT process, as well as a simplified assessment of the operational GHG emissions of the pathway, are discussed in the following section. In addition, remarks are made on the future development of the BtL-FT pathway.

11.6.1 Overcoming the Carbon Efficiency Limitation for BtL-FT Processes

The reason for the carbon efficiency limitation of the BtL-FT process is the basic composition of biomass compared to that of liquid hydrocarbons. For the given fuel analysis data, the raw biomass consists of about 32% carbon and 48% hydrogen, with the rest being mostly oxygen (molar basis). This is equivalent to a molar H-to-C ratio of 3/2. The final liquid fuel only comprises hydrogen and carbon with a molar ratio of 2/1. Thus in the syngas produced from the biomass, less hydrogen is present than what is needed for FTS.

Figure 11.11 shows the H, C and O molar fractions for the syngas-to-fuel process. The process steps and intermediate products of the advanced and highly integrated BtL-FT process are shown in the ternary diagram.

Torrefaction converts the raw biomass into a product with an increased carbon content. Oxygen-blown gasification then yields a gas mixture with a decreased carbon content. Recycling FT light ends and adding torrefaction gas result in an increase in hydrogen and oxygen content, with the raw syngas mainly consisting of hydrogen, water and carbon monoxide with less than 10% CO_2. The subsequent WGS, which is accompanied by water condensation and removal, is used for the shift toward H_2 and CO_2 to ensure an

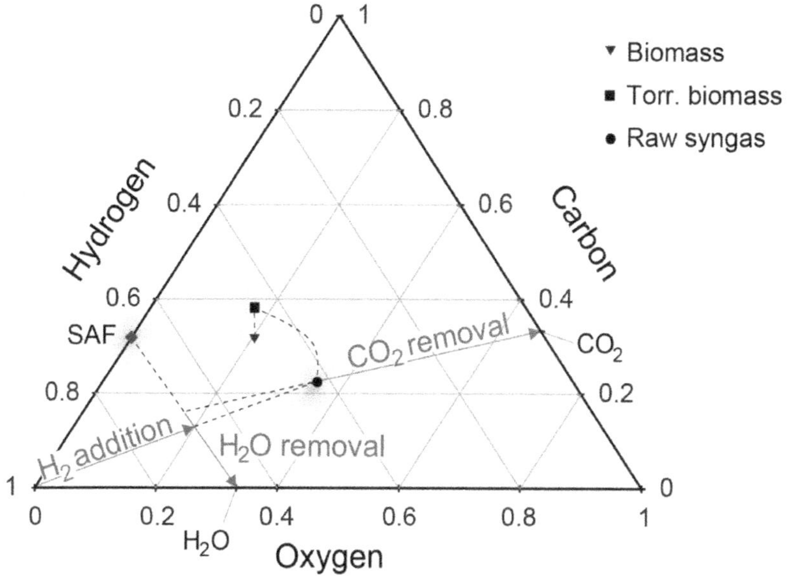

Figure 11.11 Ternary diagrams and H, C and O molar fractions for the intermediate steps of the advanced and highly integrated BtL-FT process.

H_2/CO ratio of about 2 at the inlet of the FT reactor. In the final step to produce liquid fuels, *i.e.* FTS, the formation and removal of H_2O is done to obtain the hydrocarbon product. In order to obtain the required molar composition of the syngas mixture, it is inevitable that carbon, in the form of CO_2, must be removed from the process, as shown in Figure 11.11.

To further enhance the BtL-FT process for SAFs, the product yield of the overall process needs to be maximized; thus the carbon efficiency limitation needs to be overcome. Shifting the low H_2 biogenic syngas while producing CO_2, which then needs to be removed from the process, should be avoided. As an alternative, an H/C ratio of 4, which is required for FTS, can be attained by adding hydrogen to the process, instead of removing carbon from the process. Figure 11.11 shows the potential with H_2 being added after gasification to obtain the necessary syngas composition for FTS and water removal. In the case of hydrogen enhancement, carbon efficiency is no longer limited.

One way of sustainably producing the required hydrogen for such a process is water electrolysis using renewable energy sources. When combining the concepts of BtL-FT and power-to-liquid (PtL), the process is considered a hybrid power-and-biomass-to-liquid (PBtL) approach. One of the advantages of using this type of combined system is complete utilization of the electrolysis products, as hydrogen integration in biomass-based processes allows the oxygen produced during electrolysis to be used for biomass gasification. Thus oxygen production *via* cryogenic air separation, which contributes 33–41% of the BtL-FT electricity requirement, becomes obsolete.

By integrating renewable power into the BtL-FT process, carbon efficiency can be increased to up to 97%.[96]

11.6.2 GHG Emissions

In the analysis of the GHG emissions of the presented process chains, an assessment of a complete life cycle is not within the scope of this work. Indirect emissions that are difficult to determine were not taken into account (such as plant construction, wear and maintenance). Thus biomass residues were regarded as climate neutral, as indirect emissions from agriculture and logistics were neglected. The operative process emissions are thus related to indirect emissions from the supply of electricity and possibly heat.

Natural gas, which is used in the reference model as a supplementary fuel in the pretreatment section, employs an emission factor of 56 $g\,MJ^{-1}$.[97] With no additional heating being required for the process, the remaining operative emissions produced within the process relate to the electricity input. Grid electricity emission factors were used to obtain a rough estimate of the respective emissions.

Figure 11.12 shows the resulting specific GHG emissions in mass CO_2-equivalents per product (MJ_{LHV}). The employed grid electricity emission factors range from 0–600 $g_{CO_2\text{-eq}}$ per kW h_{el} and correspond to a specific share of carbon-neutral electricity sources in the grid. The given performance target of 30.6 $g_{CO_2\text{-eq}}$ per MJ equals the required emission reduction potential of a minimum of 65% compared to CAF (87.5 $g_{CO_2\text{-eq}}$ per MJ). For example, based on Germany's estimated emission factor in 2019 (378.6 $g_{CO_2\text{-eq}}$ per kW h_{el} with about 40% of the electricity coming from renewable sources[98,99]), product-specific GHG emissions of

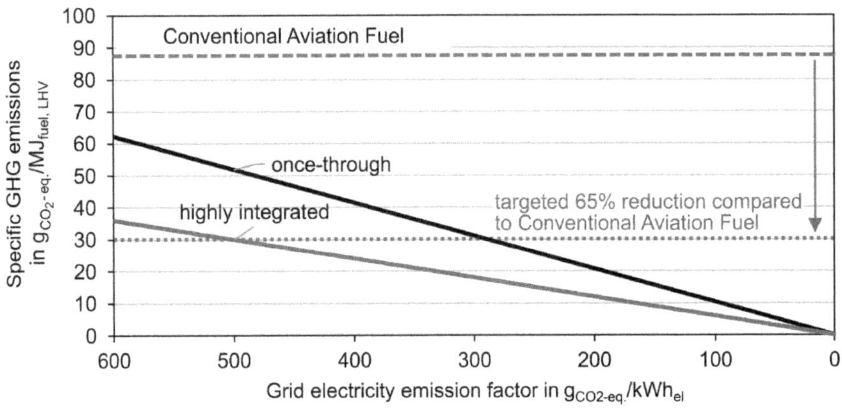

Figure 11.12 Comparison of simulation results for BtL-FT processes in terms of operative GHG emissions, depending on the grid electricity emission factor.

39.25 $g_{CO_2\text{-eq}}$ per MJ were obtained with the once-through BtL-FT reference process. Compared to CAF, these process emissions correspond to a GHG emission reduction of 55%.

Though absolute emissions increase in the advanced model, in comparison with the reference once-through BtL-FT process, due to an increase in electric parasitic load, the specific emissions per MJ liquid fuel decrease. This is due to the increasing energy yield, which allows for more product to be produced per unit of biomass. EASA estimates an emission coefficient of between 5.3 $g_{CO_2\text{-eq}}$ per MJ and 28.5 $g_{CO_2\text{-eq}}$ per MJ[18] for the BtL-FT pathway towards SAFs. The highly integrated BtL-FT process achieves GHG emissions of 22.75 $g_{CO_2\text{-eq}}$ per MJ for the German electricity mix. Thus the advanced process is well within the range suggested in the literature and also meets the performance target of an emission reduction of at least 65%.

11.6.3 Outlook and Future Developments

The results of the BtL-FT process modeling show a maximum SAF selectivity of 32.5% for a highly integrated BtL-FT pathway. The theoretical maximum selectivity toward SAF for an FT process that follows an ideal ASF distribution is about 41%. Further catalyst research could enhance SAF selectivity in BtL-FT up to this limit. Furthermore, the development of advanced catalytic systems such as bifunctional catalysts shows great potential to further increase FT selectivity toward the jet fraction.[81,100,101] For example, bifunctional catalysts with Y-type mesoporous zeolites can be used, in combination with cobalt nanoparticles to tailor the type of fuel produced, by engineering changes in the porosity and acidic properties of the zeolites.[102] Since these bifunctional catalysts are not subject to the limitations of traditional ASF distribution, selectivity of up to 72% could be achieved for the jet fuel fraction.[102] In an integrated selective FTS process used to produce SAF with a high iso-paraffinic content, downstream hydrorefining of waxes becomes obsolete.[102]

SAF produced *via* the BtL-FT pathway is similar in performance to CAF. It has excellent fuel properties and allows existing aircraft to be operated without any technical modifications (drop-in fuel). However, SAF still has to be blended with fuel from fossil sources. Since the blending requirement of a maximum of 50% SAF is mainly due to the required aromatic content in the final blend, the use of synthetic aromatics could eventually lead to the production of a fully synthetic SAF with aromatic compounds specifically designed to minimize the production of emissions that contribute to particulate matter.[103]

The ASTM D7566 Annex A4 provides approval for a fully synthetic fuel based on the addition of aromatics (FT-SPK/A), and this seems to be predestined to meet all CAF requirements. According to the ASTM D7566 specification, the SPK/A must be derived from FT-SPK combined with synthesized aromatics.[6] Thus FT-SPK/A could enable higher blend rates than are permitted when aromatics are not included. Alternative conversion steps are needed to produce aromatics, while similar feedstock to those used to produce FT-SPK are required. Therefore, FT is supplemented by the alkylation of light aromatics (primarily benzene) for FT-SPK/A production.[30] Recognition of such a SAF, which would not have to be blended with CAF, is still pending while it is under consideration by ASTM officials.[104]

11.7 Conclusion

BtL processes that use the syngas-to-aviation-fuels pathway and FTS allow for the sustainable use of biomass residues to produce aviation fuels with a high emission reduction potential. Since aviation requires high-energy-density fuels, the use of SAF is the most promising strategy for the reduction of GHG emissions in the aviation industry in the near future.

In this work, selected process routes were modeled – including torrefaction pretreatment of lignocellulosic biomass, oxygen-blown EFG, gas conditioning and a detailed macrokinetic model of FTS – which yield liquid hydrocarbon fuels for further SAF upgrading. Different path configurations, including material and heat integration, were modeled to compare the different process options available. A holistic approach was taken to maximize product yield, carbon, energy efficiency and SAF selectivity by varying the process conditions. It was shown that carbon efficiency of 40% and selectivity toward SAF of 32.5% is possible with a highly integrated BtL-FT pathway. This process achieves an energy efficiency of more than 44%, resulting in a GHG emission reduction of 74% compared to CAF.

In conclusion, the BtL-FT pathway to produce SAF is one of the most sophisticated technologies with large-scale facilities already built, and it enables the highest GHG emission savings of all the pathways considered. However, this conversion pathway results in rather low carbon conversion and energy efficiency, which has limited the use of these biofuels to date. One way to increase carbon and energy efficiency is by adding hydrogen *via* water electrolysis using renewable energy sources. One advantage of this type of combined PBtL process is complete utilization of the electrolysis products, as oxygen produced as a by-product during electrolysis can be used for biomass gasification. By focusing on increased productivity, optimal reaction conditions, a high level of energy, material integration and massive scale-up, this process design could help to further reduce the cost of SAF production.

In order to meet the emission reduction targets of ICAO and IATA, making carbon-neutral growth from 2020 economically viable and reducing risks for SAF producers, politically driven financial incentives are needed to recognize and counterbalance the environmental benefits of SAF. The possibility for aircraft operators to receive carbon credits for using SAFs and fossil-based "lower carbon aviation fuels" under ICAO's CORSIA is a first step towards defossilization of the airline industry. The BtL-FT route will play a crucial role in the production of large quantities of SAF that will soon be required.

Abbreviations

AGR	Acid gas removal
ASF	Anderson–Shulz–Flory
ASU	Air separation unit
ATM	Air traffic management
BtL	Biomass-to-liquid
CAF	Conventional aviation fuel
CNG2020	Carbon-neutral growth from 2020
CORSIA	Carbon offsetting and reduction scheme for international aviation
EASA	European Union Aviation Safety Agency
EFG	Entrained flow gasification
FRL	Fuel readiness level
FT	Fischer–Tropsch
FTS	Fischer–Tropsch synthesis
GHG	Greenhouse gas
HEFA	Hydroprocessed esters and fatty acids
HHV	Higher heating value
HTFT	High-temperature FT
IATA	International Air Transport Association
ICAO	International Civil Aviation Organization
LHV	Lower heating value
LTFT	Low-temperature FT
MSP	Minimum selling price
PBtL	Power-and-biomass-to-liquid
PtL	Power-to-liquid
SAF	Sustainable aviation fuel
SBCR	Slurry bubble column reactors
WGS	Water-gas shift

Acknowledgements

This study was carried out within the framework of the projects E2Fuels (project no. 03EIV011G) and VERENA (project no.: 03EE5044B) sponsored by the Federal Ministry for Economic Affairs and Energy (Germany). The financial support is gratefully acknowledged. In addition, the authors acknowledge the cooperation within the Network *TUM.Hydrogen and PtX.*

References

1. C. Gutiérrez-Antonio, F. I. Gómez-Castro, J. A. Lira-Flores and S. de Hernández, A review on the production processes of renewable jet fuel, *Renewable Sustainable Energy Rev.*, 2017, **79**, 709–729. Available from: http://www.sciencedirect.com/science/article/pii/S1364032117307463.
2. *Ullmann's Encyclopedia of Industrial Chemistry: Aviation Fuels*, 2019, ed. G. J. Bishop and B. Elvers.

3. W. G. Dukek, Jet Fuels, in *Kirk-Othmer Encyclopedia of Chemical Technology*, Wiley, 2011, pp. 1–31.
4. ASTM, ASTM D1655 – 19a: Specification for Aviation Turbine Fuels, D02 Committee, West Conshohocken, PA.
5. IATA, *IATA Guidance Material for Sustainable Aviation Fuel Management*, 2nd edn, 2015.
6. ICAO, *Status of Technical Certification of Aviation Alternative Fuels, Conference on Aviation and Alternative Fuels*, 2017.
7. B. M. Bomani and R. C. Hendricks Biofuels for Green Aviation, in *Encyclopedia of aerospace engineering*, ed. R. Blockley and W. Shyy, Wiley Interscience, Hoboken, NJ, 2010, pp. 1–15.
8. ASTM, ASTM D7566 – 20c: Standard Specification for Aviation Turbine Fuel Containing Synthesized Hydrocarbons, D02 Committee, West Conshohocken, PA.
9. IATA, IATA Guidance Material for Biojet Fuel Management.
10. J. I. Hileman, D. S. Ortiz, J. T. Bartis, H. M. Wong, P. E. Donohoo and M. A. Weiss *et al.*, *Near-term feasibility of alternative jet fuels*, Santa Monica, CA, 2009.
11. M. Kousoulidou and L. Lonza, Biofuels in aviation: Fuel demand and CO_2 emissions evolution in Europe toward 2030, *Transp. Res. Part D: Transp. Environ.*, 2016, **46**, 166–181.
12. Air BP Ltd, *Air BP Handbook of Products*, 2000.
13. UN data, Kerosene-type Jet Fuel: Record view, 2020 [accessed 2020 Apr 19]. Available from: http://data.un.org/Data.aspx?d=EDATA&f=cmID:JF;trID:01&c=2,5,6,7,8&s=_crEngNameOrderBy:asc,_enID:asc,yr:desc&v=1.
14. U.S. Energy Information Administration, International Energy Outlook 2019 [Table L8. Total commercial jet fuel use], 2019.
15. K. Ploetner, M. Urban, A. Habersetzer, A. Roth and G. Tay, *Fulfilling long-term emission reduction goals in aviation by alternative fuel options: An evolutionary approach, Preparation of Papers for AIAA Technical Conferences*, 2018.
16. International Air Transport Association, Economic Performance of the Airline Industry: 2019 Mid-year report, IATA, 2019 [accessed 2021 Jul 26]. Available from: https://www.iata.org/en/iata-repository/publications/economic-reports/airline-industry-economic-performance—june-2019—report/.
17. ICCT, CO_2 emissions from commercial aviation, 2018, Working Paper 2019-16, The International Council on Clean Transportation, September 2019.
18. EEA, *European Aviation Environmental Report 2019*, 2019.
19. ICAO. Aviation and Environment Outlook, In *2019 Environmental Report – Aviation and Environment: Destination Green – The Next Chapter*, 2019, ed. ICAO, ch. 1.
20. IATA, *IATA Annual Review*, International Air Transport Association, 2015.
21. In *2019 Environmental Report – Aviation and Environment: Destination Green – The Next Chapter*, 2019, ed. ICAO.

22. ICAO. Icao Global Environmental Trends – Present and Future Aircraft Noise and Emissions: Environmental Protection – General Provisions, *Aircraft Noise and Local Air Quality– Policy and Standardization; Assembly – 40th Sessio*, 2019.
23. ICAO, Committee on Aviation Environmental Protection Report: Tenth Meeting, CAEP/10, Montréal Doc 10069.
24. S. Farokhi, *Future Fuels and Energy Sources in Sustainable Aviation: Future Propulsion Systems and Energy Sources in Sustainable Aviation*, 2019.
25. F. Perry, *Sustainable Aviation Fuels Fact sheet*, 2020.
26. M. Hillestad, M. Ostadi, G. Alamo Serrano, E. Rytter, B. Austbø and J. G. Pharoah, *et al.*, Improving carbon efficiency and profitability of the biomass to liquid process with hydrogen from renewable power, *Fuel*, 2018, **234**, 1431–1451. Available from: http://www.sciencedirect.com/science/article/pii/S0016236118313632.
27. H. Wei, W. Liu, X. Chen, Q. Yang, J. Li and H. Chen, Renewable bio-jet fuel production for aviation: A review, *Fuel*, 2019, **254**, pp. 115599ff. Available from: http://www.sciencedirect.com/science/article/pii/S0016236119309433.
28. IATA, *Carbon offsetting for international aviation*, 2019.
29. IATA, What is Carbon Offsetting and Reduction Scheme for International Aviation? 2020 [accessed 2020 May 5]. Available from: https://www.iata.org/en/about/worldwide/asia_pacific/What-is-CORSIA/.
30. CAAFI, CAAFI Fuel Qualification: Commercial Aviation Alternative Fuels Initiative, 2020 [Accessed 2020 Apr 13]. Available from: http://www.caafi.org/focus_areas/fuel_qualification.html.
31. C. Hamelinck, A. Faaij, H. Den Uil and H. Boerrigter, Production of FT transportation fuels from biomass: Technical options, process analysis and optimisation, and development potential, *Energy*, 2004, **29**(11), 1743–1771.
32. K. Maniatis, M. Weitz and A. Zschocke, *2 million tons per year: A performing biofuels supply chain for EU aviation*, 2013.
33. H. Wang, B. Yang, Q. Zhang and W. Zhu, Catalytic routes for the conversion of lignocellulosic biomass to aviation fuel range hydrocarbons, *Renewable Sustainable Energy Rev.*, 2020, **120**, 109612. Available from: http://www.sciencedirect.com/science/article/pii/S1364032119308202.
34. G. W. Diederichs, M. Ali Mandegari, S. Farzad and J. F. Görgens, Techno-economic comparison of biojet fuel production from lignocellulose, vegetable oil and sugar cane juice, *Bioresour. Technol.*, 2016, **216**, 331–339. Available from: https://www.sciencedirect.com/science/article/pii/S0960852416307313.
35. A. De Klerk, Aviation Turbine Fuels Through the Fischer–Tropsch Process, in *Biofuels for aviation: Feedstocks, technology and implementation*, ed. C. J. Chuck, Academic Press, London, UK, San Diego, CA, 2016, ch. 10, pp. 241–259.

36. P. Schmidt, V. Batteiger, A. Roth, W. Weindorf and T. Raksha, Power-to-Liquids as renewable fuel option for aviation: A Review, *Chem. Ing. Tech.*, 2018, **90**(1–2), 127–140.
37. A. Fivga, L. G. Speranza, C. M. Branco, M. Ouadi and A. Hornung, A review on the current state of the art for the production of advanced liquid biofuels, *AIMS Energy*, 2019, 7(1), 46–76. Available from: https://www.scopus.com/inward/record.uri?eid=2-s2.0-85065606775&doi=10.3934%2fENERGY.2019.1.46&partnerID=40&md5=a63339a3fdb2ba7fac35af36b338077b.
38. EU-VRi, Alfa-Bird: Alternative Fuels and Biofuels for Aircraft Development: Synthesis – *The European Virtual Institute for Integrated Risk Management (EEIG)*, 2012.
39. G. E. Dorrington, Certification and Performance, in *Biofuels for aviation: Feedstocks, technology and implementation*, ed. C. J. Chuck, Academic Press, London, UK, San Diego, CA, 2016, ch. 3, pp. 35–44.
40. J. Koppejan and S. Madrali, Status overview of torrefaction technologies, 2012. Available from: https://www.semanticscholar.org/paper/Status-overview-of-torrefaction-technologies-Koppejan-Madrali/4e0bb08cb3a7cc2b44832bd787a1e50d6b7fe22c.
41. P. C. A. Bergman, A. R. Boersma, R. W. R. Zwart and J. H. A. Kiel, Torrefaction for biomass co-firing in existing coal-fired power stations. Energy Research Centre of the Netherlands, 2005. Available from: https://www.researchgate.net/publication/204978559_Torrefaction_for_Biomass_Co-Firing_in_Existing_Coal-Fired_Power_Stations.
42. J. S. Tumuluru, S. Sokhansanj, J. R. Hess, C. T. Wright and R. D. Boardman, A review on biomass torrefaction process and product properties for energy applications, *Ind. Biotechnol.*, 2011, 7(5), 384–401.
43. M. Manouchehrinejad and S. Mani, Process simulation of an integrated biomass torrefaction and pelletization (iBTP) plant to produce solid biofuels, *Energy Convers. Manage.: X*, 2019, **1**, 100008.
44. P. C. A. Bergman, A. R. Boersma, J. H. A. Kiel, M. J. Prins and F. J. J. G. Janssen, Torrefaction for entrained flow gasification of biomass, 2005 [Accessed 2020 Jul 2]. Available from: https://repository.tudelft.nl/view/tno/uuid%3A185d6bf8-ec13-42a8-b188-b5a042a2fa47.
45. M. Van der Stelt, H. Gerhauser, J. Kiel and K. J. Ptasinski, Biomass upgrading by torrefaction for the production of biofuels: A review, *Biomass Bioenergy*, 2011, 3748–3762.
46. E. Peduzzi, G. Boissonnet, G. Haarlemmer, C. Dupont and F. Maréchal, Torrefaction modelling for lignocellulosic biomass conversion processes, *Energy*, 2014, **70**, 58–67.
47. A. Tremel, D. Becherer, S. Fendt, M. Gaderer and H. Spliethoff, Performance of entrained flow and fluidised bed biomass gasifiers on different scales, *Energy Convers. Manage.*, 2013, **69**, 95–106.
48. C. Higman and M. Van der Burgt, *Gasification*, Amsterdam: Elsevier/GPP Gulf Professional Publ, 2nd edn, 2008, Available from:

http://search.ebscohost.com/login.aspx?direct=true&scope=site&db=nlebk&db=nlabk&AN=220264.

49. M. Stemmler and M. Müller, Theoretical evaluation of feedstock gasification using H2/C ratio and ROC as main input variables, *Ind. Eng. Chem. Res.*, 2010, **49**(19), 9230–9237.
50. M. Kremling, L. Briesemeister, M. Gaderer, S. Fendt and H. Spliethoff, Oxygen-blown entrained flow gasification of biomass: Impact of fuel parameters and oxygen stoichiometric ratio, *Energy Fuels*, 2017, **31**(4), 3949–3959.
51. F. Weiland, M. Nordwaeger, I. Olofsson, H. Wiinikka and A. Nordin, Entrained flow gasification of torrefied wood residues, *Fuel Process. Technol.*, 2014, **125**, 51–58.
52. F. Weiland, H. Wiinikka, H. Hedman, J. Wennebro, E. Pettersson and R. Gebart, Influence of process parameters on the performance of an oxygen blown entrained flow biomass gasifier, *Fuel*, 2015, **153**, 510–519.
53. A. Buttler, *Technoökonomische Bewertung von Polygenerationskraftwerken und Power-to-X-Speichern in einem nachhaltigen Energiesystem. 1. Auflage*, Verlag Dr. Hut, München, 2018.
54. P. Chiesa, S. Consonni and T. Kreutz, Co-production of hydrogen, electricity and CO from coal with commercially ready technology. Part A: Performance and emissions, *Int. J. Hydrogen Energy*, 2005, **30**(7), 747–767.
55. H. Meerman, *Perspectives on gasification systems to produce energy carriers and other chemicals with low CO_2 emissions: Techno-economic system analysis on current and advanced flexible thermo-chemical conversion of fossil fuels and biomass*, Dissertation, Universiteit Utrecht, 2012.
56. M. Claeys and E. Van Steen, Basic studies, in *Fischer-Tropsch technology. Reprinted*, ed. A. Steynberg and M. Dry, Amsterdam, 2006, Studies in Surface Science and Catalysis, Elsevier, vol. 152, pp. 601–80.
57. L. P. R. Pala, Q. Wang, G. Kolb and V. Hessel, Steam gasification of biomass with subsequent syngas adjustment using shift reaction for syngas production: An Aspen Plus model, *Renewable Energy*, 2017, **101**, 484–492.
58. P. J. Robinson and W. L. Luyben, Integrated gasification combined cycle dynamic model: H 2 S absorption/stripping, water – gas shift reactors, and CO_2 absorption/stripping, *Ind. Eng. Chem. Res.*, 2010, **49**(10), 4766–4781.
59. E. Rytter, E. Ochoa-Fernández and A. Fahmi, Biomass-to-Liquids by the Fischer-Tropsch Process, in *Catalytic Process Development for Renewable Materials*, ed. P. Imhof and J. C. Van der Waal, Wiley-VCH, 1. Aufl. s.l., 2013, pp. 265–308.
60. J.-C. Viguié, N. Ullrich, P. Porot, L. Bournay, M. Hecquet and J. Rousseau, BioTfueL project: Targeting the development of second-generation biodiesel and biojet fuels, *Oil Gas Sci. Technol.*, 2013, **68**(5), 935–946. Available from: https://www.scopus.com/inward/record.uri?eid=2-s2.0-84888151420&partnerID=40&md5=7a1b4befc4bac04bfbf3e7766fc2190f.

61. A. De Klerk, *Fischer-Tropsch refining*, [elektronische ressource], Wiley, Wiley-VCH, Hoboken, NJ, Weinheim, 1st edn, 2011.
62. M. Ostadi, E. Rytter and M. Hillestad, Boosting carbon efficiency of the biomass to liquid process with hydrogen from power: The effect of H2/CO ratio to the Fischer-Tropsch reactors on the production and power consumption, *Biomass Bioenergy*, 2019, 127. Available from: https://www.scopus.com/inward/record.uri?eid=2-s2.0-85067668213&doi=10.1016%2fj.biombioe.2019.105282&partnerID=40&md5=1723faf52ef6aaf79829932e8bbdbd7e.
63. D. J. Kubek, E. Polla and F. P. Wilcher, Purification and Recovery Options for Gasification, 2000 [accessed 2020 Aug 12].
64. R. P. Field and R. Brasington, Baseline flowsheet model for IGCC with carbon capture, *Ind. Eng. Chem. Res.*, 2011, **50**(19), 11306–11312.
65. Z. Kapetaki, P. Brandani, S. Brandani and H. Ahn, Process simulation of a dual-stage Selexol process for 95% carbon capture efficiency at an integrated gasification combined cycle power plant, *Int. J. Greenhouse Gas Control*, 2015, **39**, 17–26.
66. A. L. Kohl and R. B. Nielsen, *Gas purification*, Gulf Pub, Houston, Tex, 5th edn, 1997, Available from: https://books.google.de/books?id=rpzrIZW-OcEC.
67. F. Fischer and H. Tropsch, Über die direkte Synthese von Erdöl-Kohlenwasserstoffen bei gewöhnlichem Druck. (Erste Mitteilung), *Ber. Dtsch. Chem. Ges. A/B*, 1926, **59**(4), 830–831.
68. R.-U. Dietrich, F. G. Albrecht, S. Maier, D. H. König, S. Estelmann and S. Adelung, *et al.*, Cost calculations for three different approaches of biofuel production using biomass, electricity and CO_2, *Biomass Bioenergy*, 2018, **111**, 165–173. Available from: https://www.scopus.com/inward/record.uri?eid=2-s2.0-85024369169&doi=10.1016%2fj.biombioe.2017.07.006&partnerID=40&md5=6447c4a2a5c1309b50c33ba4853670ee.
69. H.-J. Lee, *Optimization of Fischer-Tropsch plant*, The University of Manchester, Manchester, UK, 2010.
70. P. M. Maitlis and A. De Klerk, *Greener Fischer-Tropsch processes for fuels and feedstocks. 1. Aufl*, Wiley-VCH, Weinheim, 2013, Available from: http://search.ebscohost.com/login.aspx?direct=true&scope=site&db=nlebk&db=nlabk&AN=531837.
71. S. S. Ail and S. Dasappa, Biomass to liquid transportation fuel via Fischer Tropsch synthesis – Technology review and current scenario, *Renewable Sustainable Energy Rev.*, 2016, **58**, 267–286.
72. S. Fedou, E. Caprani, D. Douziech and S. Boucher, Conversion of syngas to diesel [Axens], PTQ, 2008 [accessed 2020 Jun 29], Available from: https://www.axens.net/product/process-licensing/10101/gasel.html.
73. J. G. Speight, *Synthetic fuels handbook: Properties, process, and performance*, McGraw-Hill, New York, NY, 2008, (McGraw-Hill handbooks). Available from: https://www.accessengineeringlibrary.com/content/book/9780071490238/chapter/chapter8.

74. M. Trompelt, T. Weber and T. Kuchling, Modellierung der Fischer-Tropsch-Synthese mit Aspen Plus®, 2009.
75. D. Leckel, Diesel production from Fischer – Tropsch: The past, the present, and new concepts, *Energy Fuels*, 2009, **23**(5), 2342–2358.
76. M. Hillestad, Modeling the Fischer–Tropsch product distribution and model implementation, *Chem. Prod. Process Model.*, 2015, **10**(3), 147–159.
77. D. Förtsch, K. Pabst and E. Groß-Hardt, The product distribution in Fischer–Tropsch synthesis: An extension of the ASF model to describe common deviations, *Chem. Eng. Sci.*, 2015, **138**, 333–346.
78. J. M. Fox and S. S. Tam, Correlation of slurry reactor Fischer-Tropsch yield data, *Top. Catal.*, 1995, **2**(1–4), 285–300.
79. E. Iglesia, S. C. Reyes, R. J. Madon and S. L. Soled, Selectivity control and catalyst design in the Fischer–Tropsch Synthesis: Sites, pellets, and reactors, in *Advances in Catalysis*, ed. Elsevier, Elsevier, 1993, pp. 221–302.
80. N. Marchal-George, J. Cosyns, V. Coupard, E. Caprani, D. Douziech, A. Dandeu *et al.*, inventors. Process for Producing Middle Distillates by Hydroismerizing and Hydrocrackinga Heavy Fraction from a Fischer-Tropsch Effluent, PCT/FR2008/000753.
81. S. Van Dyk and J. Saddler, Progress in the commercialisation of biojet fuels [accessed 2021 Jun 29], Available from: https://www.ieabioenergy.com/wp-content/uploads/2021/06/Task-39-Progress-in-the-commercialisation-of-biojet-fuels-FINAL-May-2021.pdf?utm_campaign=IEA%20Bioenergy%20Press%20Release%2029%20June&utm_medium=email&utm_source=EOACLK.
82. M. J. Prins, K. J. Ptasinski and F. J. Janssen, More efficient biomass gasification via torrefaction, *Energy*, 2006, **31**(15), 3458–3470. Available from: https://research.tue.nl/en/publications/more-efficient-biomass-gasification-via-torrefaction-2.
83. T. Nocquet, C. Dupont, J.-M. Commandre, M. Grateau, S. Thiery and S. Salvador, Volatile species release during torrefaction of wood and its macromolecular constituents: Part 1 – Experimental study, *Energy*, 2014, **72**, 180–187.
84. M. Kaltschmitt, Energie aus Biomasse: Grundlagen, *Techniken und Verfahren. 2., neu bearb. und erw. Aufl.*, Springer, Berlin, 2009, Available from: http://site.ebrary.com/lib/alltitles/docDetail.action?docID=10310211.
85. M. Manouchehrinejad, I. Van Giesen and S. Mani, Grindability of torrefied wood chips and wood pellets, *Fuel Process. Technol.*, 2018, **182**, 45–55.
86. M. Aho, J. Hamalainen and J. Tummavuori, Importance of solid fuel properties to nitrogen oxide formation through HCN and NH_3 in small particle combustion, *Combust. Flame*, 1993, **95**(1–2), 22–30.
87. IKTS, Dezentrale Sauerstoffproduktion [accessed 2020 Oct 3], 2020.
88. Thyssen Krupp Uhde GmbH, *Gasification Technologies* [accessed 2020 Aug 12], 2017.

89. G. Boissonnet, M. Hecquet, P. Avenier, L. Bournay, D. Chiche, J.-P. Heraud *et al.*, inventors. Improved Method for Converting a Feedstock Containing Biomass for the Production of Hydrocarbons by Means of Fischer-Tropsch Synthesis: Fascicule De Brevet Europeen, EP 2 914 701 B1.
90. T. A. Adams, Y. K. Salkuyeh and J. Nease Processes and simulations for solvent-based CO_2 capture and syngas cleanup, in *Reactor and Process Design in Sustainable Energy Technology*, ed. F. Shi, Elsevier, 1. Aufl. Amsterdam, 2014, ch. 6, pp. 163–231, Available from: http://www.sciencedirect.com/science/article/pii/B9780444595669000065.
91. M. Ostadi, E. Rytter and M. Hillestad, Evaluation of kinetic models for Fischer–Tropsch cobalt catalysts in a plug flow reactor, *Chem. Eng. Res. Des.*, 2016, **114**, 236–246.
92. B. Todic, W. Ma, G. Jacobs, B. H. Davis and D. B. Bukur, Effect of process conditions on the product distribution of Fischer–Tropsch synthesis over a Re-promoted cobalt-alumina catalyst using a stirred tank slurry reactor, *J. Catal.*, 2014, **311**, 325–338.
93. B. Todic, T. Bhatelia, G. F. Froment, W. Ma, G. Jacobs and B. H. Davis, *et al.*, Kinetic Model of Fischer–Tropsch Synthesis in a slurry reactor on Co–Re/Al_2O_3 catalyst, *Ind. Eng. Chem. Res.*, 2013, **52**(2), 669–679.
94. B. Todic, W. Ma, G. Jacobs, B. H. Davis and D. B. Bukur, CO-insertion mechanism based kinetic model of the Fischer–Tropsch synthesis reaction over Re-promoted Co catalyst, *Catal. Today*, 2014, **228**, 32–39.
95. B. Todic, W. Ma, G. Jacobs, B. H. Davis and D. B. Bukur, Corrigendum to: CO-insertion mechanism based kinetic model of the Fischer–Tropsch synthesis reaction over Re-promoted Co catalyst, *Catal. Today*, 2015, **242**, 386.
96. M. Dossow, V. Dieterich, A. Hanel, H. Spliethoff and S. Fendt, Improving carbon efficiency for an advanced Biomass-to-Liquid process using hydrogen and oxygen from electrolysis, *Renewable Sustainable Energy Rev.*, 2021, pp. 111670ff.
97. FfE, Basisdaten von Energieträgern [accessed 2020 Oct 3], 2010.
98. I. Forsyth, Country specific electricity grid greenhouse gas emission factors, 2020 Grid Electricity Emissions Factors v1.3 – July 2020. www.carbonfootprint.com; 2020 [accessed 2020 Dec 14], Available from: https://www.carbonfootprint.com/docs/2020_07_emissions_factors_sources_for_2020_electricity_v1_3.pdf.
99. AIB, 2019 European Residual Mix Factors: Version 1 (29 May 2020), 2020 [accessed 2020 Dec 15], Available from: https://www.aib-net.org/facts/european-residual-mix.
100. H. Jahangiri, J. Bennett, P. Mahjoubi, K. Wilson and S. Gu, A review of advanced catalyst development for Fischer–Tropsch synthesis of hydrocarbons from biomass derived syn-gas, *Catal. Sci. Technol.*, 2014, **4**(8), 2210–2229.

101. M. Yang, L. Zhu, Y. Zhuo, J. Liang and S. Wang, Selective Fischer–Tropsch synthesis for jet fuel production over Y3 + modified Co/H-β catalysts, *Sustainable Energy Fuels*, 2020, **4**(7), 3528–3536.
102. J. Li, Y. He, L. Tan, P. Zhang, X. Peng and A. Oruganti, *et al.*, Integrated tuneable synthesis of liquid fuels via Fischer–Tropsch technology, *Nat. Catal.*, 2018, **1**(10), 787–793. Available from: https://www.nature.com/articles/s41929-018-0144-z.
103. J. I. Hileman and R. W. Stratton, Alternative jet fuel feasibility, *Transp. Policy*, 2014, **34**, 52–62.
104. F + L Magazine, Are specifications for 100% sustainable aviation fuel on the way? [accessed 2021 Jun 29], 2021.
105. H. Boerrigter, *Green Diesel from Biomass* via *Fischer-Tropsch synthesis: New insights in gas cleaning and process design*, 2002, Available from: https://www.academia.edu/31971944/Green_Diesel_from_Biomass_via_Fischer_Tropsch_synthesis_New_Insights_in_Gas_Cleaning_and_Process_Design.
106. ICAO, *Trends and scenario on alternative fuels – Working Paper, Conference on Aviation and Alternative Fuels*, 2017, Available from: https://www.icao.int/Meetings/CAAF2/Documents/CAAF.2.WP. 006.4.en.pdf.
107. D. Thrän and J. Ponitka Government Policy on Delivering Biofuels for the Aviation Sector, in *Biofuels for aviation: Feedstocks, technology and implementation*, ed. C. J. Chuck, Academic Press, London, UK, San Diego, CA, 2016, ch. 13, pp. 295–313.
108. IATA, Jet Fuel Price Monitor [accessed 2020 Aug 7], 2020. Available from: https://www.iata.org/en/publications/economics/fuel-monitor/.

CHAPTER 12

Biomass to Liquid-syngas to Olefins

SIXU LIU,[a,b] DANFENG WU,[a,b] BINYONG LIN,[a,b] JINLIN LI*[c] AND HAIFENG XIONG*[a,b]

[a] State Key Laboratory of Physical Chemistry of Solid Surfaces, College of Chemistry and Chemical Engineering, Xiamen University, 361005 Xiamen, China; [b] Innovation Laboratory for Sciences and Technologies of Energy Materials of Fujian Province (IKKEM), Xiamen 361000, China; [c] Key Laboratory of Catalysis and Energy Materials Chemistry of Ministry of Education & Hubei Key Laboratory of Catalysis and Materials Science, South-Central Minzu University, Wuhan 430074, China
*Emails: jinlinli@aliyun.com; haifengxiong@xmu.edu.cn

12.1 Introduction

The lower olefins are ethylene, propylene and butylene. They are the key building blocks used in the chemical industry for the production of plastics and fine chemicals.[1,2] Traditionally, these components were mainly produced by cracking naphtha or by fluid catalytic cracking (FCC) of gas oil and light alkanes from petroleum reserves.[3] In recent years, the production of syngas (a mixture of H_2 and CO) to lower olefins with high selectivity has received increasing attention.[4–6] Although the syngas conversion has been studied for decades, the direct conversion of syngas to lower olefins *via* Fischer–Tropsch synthesis (FTS) without any intermediate steps was only developed recently. It is usually referred to as the Fischer–Tropsch-to-olefins (FTO) process.[7] One of the advantages of FTO is that the source syngas can

Catalysis Series No. 44
Chemicals and Fuels from Biomass *via* Fischer–Tropsch Synthesis: A Route to Sustainability
Edited by Joshua Gorimbo, Xinying Liu, Yali Yao and Diane Hildebrandt

Published by the Royal Society of Chemistry, www.rsc.org

be produced from natural gas, coal or biomass. Therefore, FTO enhances the flexibility of olefin production, which is highly desirable. However, one limitation of the FTO process is that the product selectivity typically follows the Anderson–Schulz–Flory (ASF) model, with a maximum C_2–C_4 hydrocarbon weight fraction of ~56% and an undesired methane fraction of ~29%, due to the surface polymerization mechanism.[2] Recent investigations have focused on preparing smart catalysts that are used to catalyze syngas to lower olefins with high selectivity.

Recent developments in catalyst design strategies for selective production of lower olefins in syngas conversion are reviewed in this chapter.

12.2 Strategies for the Conversion of Syngas to Olefins

Direct and indirect processes for the conversion of syngas to lower olefins have been developed to break the selectivity distribution posed by the ASF distribution restriction in FTS. Much progress has been made in catalyst design for improving the selectivity of low olefins, and these catalysts can be divided into two main types: one is the modified FTS catalysts, and the other is the metal oxide coupling with zeolites.

12.2.1 Modified Fischer–Tropsch Synthesis Catalysts

FTS is a heterogeneous catalytic process in which syngas (H_2 and CO) is transformed into valued hydrocarbons (paraffins and olefins) and oxygenates. The products are diverse, and the reaction mechanism is complex. Many different mechanisms have been proposed over the past few decades, but no consensus has been reached on the mechanism of various metal catalysts. However, it is generally accepted that FTS proceeds through the following key elementary steps that take place on the catalyst surface: (1) adsorption and dissociation of syngas; (2) formation of CH_x ($x = 0$–3) species on the catalyst surface; (3) C–C bond formation through coupling of CH_x species, which leads to chain growth and surface C_nH_m intermediates ($n \geq 2$), CH_x species hydrogenation to CH_4, or coupling of CH_x species to C_nH_m intermediates ($n \geq 2$); (4) formation of olefin or paraffin compounds as products through C_nH_m intermediates dehydrogenation or hydrogenation.[8,9] Co, Fe and Ru are three typical catalysts used for FTS, which have different abilities for CO dissociation, hydrogenation and chain growth.

In FTS, ~200 kinds of products can be produced. These products can be divided into the following main types: alkanes, olefins and oxygenates. The selectivity regulation is therefore an important challenge in FTS because chain growth *via* C–C coupling is difficult to control on catalyst surfaces, which leads to the broad distribution of Fischer–Tropsch (FT) products. This is known as ASF distribution.

Based on the ASF distribution model, product selectivity can be described using the chain growth probability α. Different factors have an influence on the parameter α, such as reaction conditions, catalysts and promoters.[10] ASF

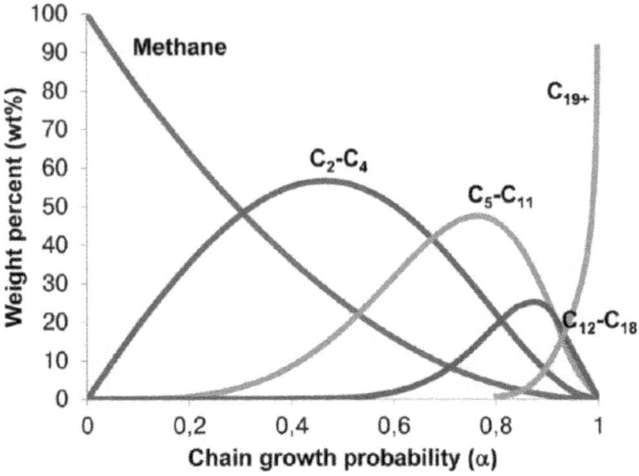

Figure 12.1 Anderson–Schulz–Flory (ASF) model for product distribution in FTS.[1] Reproduced from ref. 1 with permission from American Chemical Society, Copyright 2013.

distribution as a function of α is illustrated in Figure 12.1. It shows that the maximum theoretical selectivity of C_2–C_4 and methane is 56% and 29%, respectively. It is clear that in order to maximize the selectivity of lower olefins, methane production should be reduced, and the CO_2 emitted must be suppressed. According to the ASF distribution (see Figure 12.1), α value between 0.4 and 0.5 leads to the maximum selectivity toward C_2–C_4 olefins. One of the most efficient ways to shift product selectivity to a lower α value is to increase the reaction temperature. However, as shown in the ASF product distribution, the decrease in chain growth probability caused by an increase in temperature leads to higher undesired methane selectivity.[1] As for the ASF model, C_{5+} hydrocarbon selectivity increases as methane selectivity decreases. Therefore, the challenge is how to reduce the selectivity of both methane and C_{5+} at the same time.

There are two main types of modified FTS catalysts used in FTO: modified Fe- and Co-based catalysts. The active phase of Fe catalysts in FTO is iron carbide, where a smaller particle size leads to increased CO conversion and methane selectivity. Meanwhile, Co with a hexagonal close-packed (HCP) phase and Co_2C nanoprisms with the (101) and (020) facets exposed are preferred for FTO.[11,12]

12.2.1.1 Fe-Based Catalysts

Fe-Based catalysts have been extensively studied for FTO. Compared to other FT catalysts (*i.e.* Co-based and Ru-based catalysts), Fe catalysts with iron carbide species as the active phase have a relatively weaker ability for olefin hydrogenation and a higher selectivity for lower olefins. The ASF distribution model indicates that the FTO selectivity of C_2–C_4 hydrocarbons (including paraffins and olefins) is limited to 56%. De Jong *et al.*[7] reported an optimized Fe-based FTO catalyst, which was composed of iron carbide nanoparticles dispersed on α-Al_2O_3 and carbon nanofiber supports, with sodium promoter and the presence of sulfur. The C_2–C_4 olefins selectivity reached up to 60% at a low CO conversion over the Fe catalyst. The selectivity of the lower olefin was close to the upper limit indicated by the ASF distribution model.

Alkali metal prompters have been widely used to improve the selectivity of C_2–C_4 olefins of iron-based catalysts by improving chain growth probability and the olefin/paraffin ratio. For example, the effect of K on the catalytic performance of the reduced-graphene-oxide- (rGO-) supported iron catalysts was studied in the FTO process.[13] With an increase in K content in the Fe/rGO catalysts, the FTY (iron-time-yield to hydrocarbons) first increases, showing a maximum FTY of 646 $\mu mol_{CO}\, g_{Fe}^{-1}\, s^{-1}$ over the FeK1/rGO catalyst (K loading = 1 wt%); it then decreases as the K content continues to increase.[13] The lower-olefin selectivity increases steadily when the amount of K promoter increases. A FeK2/rGO catalyst (K loading = 2 wt%) exhibited the best catalytic performance: the selectivity to C_2–C_4 olefins and the olefin/paraffin (O/P) ratio in the C_2–C_4 hydrocarbons reached 68% and 11%, respectively.[13] The reason why FTY showed volcanic evolution with the increased amount of K promoter is that the appropriate amount of K could promote the formation of active phase iron carbide, while excess K increased the particle size of the iron carbide and made the catalyst more prone to carbon deposition on the surface of the iron carbide. The monotonic increase in selectivity to C_2–C_4 olefins is due to the presence of K, which improves the chain growth probability and surface CO/H_2 ratio. This is beneficial to inhibiting the undesired CH_4 and the secondary hydrogenation of C_2–C_4 olefins.

Na- and Zn-promoted iron catalysts performed well in terms of the production of C_2–C_{12} olefins in CO hydrogenation, and the selectivity of the C_2–C_{12} olefins reached 78%.[14] Na^+ plays an important role in improving the selectivity of C_2–C_{12} olefins by inhibiting the formation of CH_4 and C_2–C_{12} paraffins. Moreover, C_3H_6 was used as a model molecule to demonstrate the positive role of Na^+ in inhibiting olefin hydrogenation. Adding Na^+ to the ZnO–Fe_5C_2 catalyst significantly inhibited the hydrogenation of C_3H_6 to C_3H_8. The explanation for this is that Na^+ increased the basicity of the catalyst surface, which suppressed the hydrogenation ability of Fe_5C_2 and promoted olefin desorption.

The lower-olefin selectivity can be enhanced by using structural promoters. In this regard, the effects of structural promoters such as Zn, Al, Si and Ti in iron-based catalysts have been reported for the conversion of syngas to C_2–C_4 olefins.[15] All the catalysts were synthesized using the co-precipitation method, with sodium carbonate being used as a precipitation agent. The FeZn (Fe/Zn molar ratio = 1) catalyst showed excellent selectivity to C_2–C_4 olefins, and the water-gas-shift (WGS) reaction activity was inhibited by adding the Zn promoter. With the FeAl catalyst, selectivity to C_2–

C_4 olefins decreased, and no distinct change was observed for the selectivity to CH_4 and C_{5+} products. However, CO conversion for both the FeSi catalyst and the FeTi catalyst decreased from 90 to 43%. According to the XRD measurements, FeZn existed as $ZnFe_2O_4$ and ZnO after calcination. It was concluded that the presence of α-Fe during the reaction was beneficial for C_2–C_4 olefin desorption and increased the selectivity of C_2–C_4 olefins. Alkali-metal- (K- and Na-) promoted FeZn catalysts have also been studied for the conversion of syngas to C_2–C_4 olefins, and the addition of 1.86 wt% of Na^+ increased the CO conversion and selectivity to C_2–C_4 olefins. The catalyst was stable in a 100 h run, and the selectivity of C_2–C_4 olefins was 53% at a conversion of 95%.

The effect of the structure promoter Mn and the alkali promoter K on Fe/CNT catalysts was also investigated recently.[16] The amount of Mn and K was changed by modifying the concentration of $KMnO_4$. The results showed that, with an increase in Mn and K content, the activity of the catalyst decreased, and the selectivity of the C_2–C_4 olefin increased.

Core–shell structure catalysts were designed by Tao and coworkers to inhibit the formation of C_{6+} products and hence to lead to greater lower-olefin selectivity (see Figure 12.2).[17] The Fe_3C/C catalyst was first synthesized using a melting method. A layer of silica was then modified on the surface of the catalyst. Finally, a layer of SAPO-34 zeolite was deposited on the surface of the silica to obtain the Fe/C@Si-SAPO-34 catalyst. Compared to the Fe_3C/C catalyst, the Fe/C@Si-SAPO-34 catalyst exhibited a different product distribution in the FTO reaction. Moreover, owing to the confinement effect of SAPO-34, the formation of C_{6+} hydrocarbons was completely inhibited, and the main products were C_2–C_4 hydrocarbons. The SAPO-34 zeolite shell is beneficial for inhibiting the C_{6+} hydrocarbons and rapid mass transfer of light hydrocarbons, and it was suggested that the microporous structure and

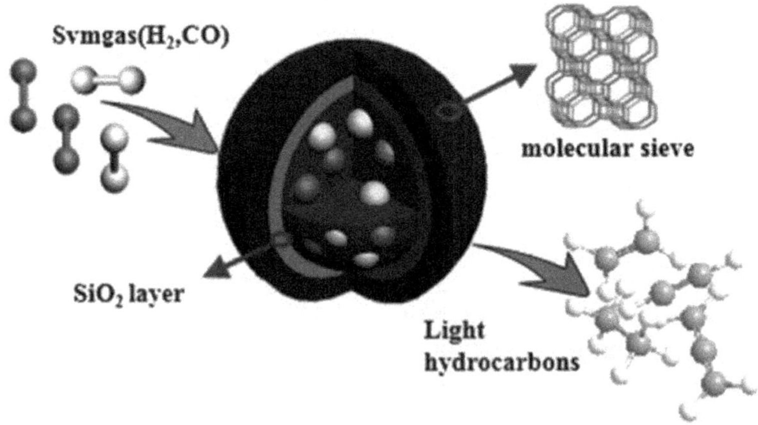

Figure 12.2 Schematic representation of Fischer–Tropsch synthesis to lower olefins on a core–shell structured catalyst.[17] Reproduced from ref. 17 with permission from Elsevier, Copyright 2017.

unique acidity of the SAPO-34 shell play an important role in increasing the selectivity to C_2–C_4 hydrocarbons and reducing the selectivity to C_{6+} hydrocarbons. The effect of the promoter alkali metal (K) was also studied.[17] The results showed that the K-promoted core–shell catalyst (Fe/C@Si-SAPO-34-K) showed high selectivity for the formation of C_2–C_4 olefins and low selectivity for the formation of methane in the FTO reaction. The C_2–C_4 olefin selectivity and the O/P values for the C_2–C_4 hydrocarbons were 40.3% and 3.0, respectively.

12.2.1.2 Co-Based Catalysts

Cobalt-based catalysts show high paraffin selectivity in syngas conversion. Adding Mn to a cobalt catalyst can promote catalyst activity but has a negligible effect on product selectivity at atmospheric pressure.[18] In contrast, when using TiO_2 as a support for loading cobalt, Mn can promote selectivity toward long-chain linear alcohols and olefins at high pressure.[19] However, a recent study demonstrated that the CoMn catalyst could form Co_2C nanoprisms with specifically exposed facets during the reaction process (see Figure 12.3a–f). It exhibited high C_2–C_4 olefin selectivity (~60.8%) but a relatively low CO conversion of ~30%.[20] Therefore, the CoMn catalysts are active for lower-olefin production in FTO.

The present authors[21] further found that cobalt-manganese (CoMn) oxides obtained through one-pot synthesis can be controllably reduced to

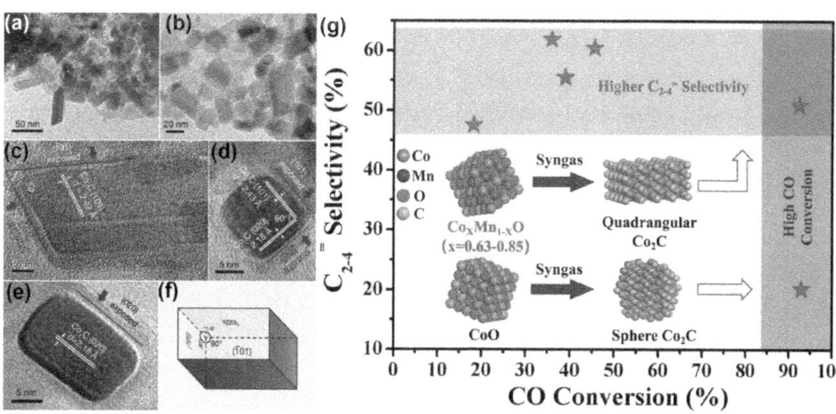

Figure 12.3 TEM images of the CoMn catalysts under steady state.[2] (a and b) Low-resolution TEM images. (c–e) High-resolution images of Co_2C nanoprisms with exposed facets of (101), (−101) and (020). (d) Distance (length) of the lattice fringes. (f) Co_2C nanoprism has a parallel piped shape, with four rectangular faces and two rhomboid faces. (g) Role of intermediate $Co_xMn_{1-x}O$ ($x = 0.6$–0.85) nanocrystals in the formation of active species for the direct production of lower olefins from syngas.[20] Reproduced from ref. 2 with permission from Springer Nature, Copyright 2016 and from ref. 20, with permission from the Royal Society of Chemistry.

$Co_xMn_{1-x}O$ nanocrystals for direct production of C_2–C_4 olefins with high selectivity (50.9 mol%) from syngas at a high CO conversion of 92.4%. The low olefin selectivity was higher than that predicted by the ASF distribution (see Figure 12.3g). The formation of the $Co_xMn_{1-x}O$ phase is important to generate an active Co_2C quadrangular nanoprism phase with exposure of the (101) and (020) facets formed *in situ* during the reaction, as this plays a pivotal role in obtaining a high yield of the lower olefin at a high CO conversion. During the durability test, no deactivation was observed during a 100 h run at high-conversion conditions. Furthermore, the activity was one order of magnitude higher than that of a traditional cobalt catalyst.

The relationship between the Co_2C nanoprism and the intermediate was also studied using CoMn catalysts prepared by coprecipitation. The catalysts were pretreated using different reduction agents (H_2 and CO) at different temperatures.[21] Co_2C nanoprisms were formed during reaction after the catalysts were reduced by 10% CO-300 °C, 10% H_2-300 °C and 10% H_2-250 °C; Co_2C nanospheres were found with the catalysts reduced by CO-300 °C and 10% H_2-400 °C. However, with the catalyst reduced by 10% H_2-350 °C, both the Co_2C nanoprisms and nanospheres were present during the reaction. From the XRD and HRTEM characterization results, it was found that the formation of a $Co_xMn_{1-x}O$ phase *via* reduction is crucial for the formation of Co_2C nanoprisms during reaction.[21]

Kinetics studies showed that Co_2C nanocrystals have higher activation energy and higher sensitivity to hydrogen than Co_2C nanocrystals. Density functional theory (DFT) calculations have clearly elucidated the relationship between Co_2C nanostructures and the performance of syngas in terms of low olefins.

In order to investigate the role of Na in a CoMn catalyst for FTO, Li *et al.*[22] designed CoMn catalysts promoted by different amounts of Na. With the Na-free sample, it was found that Co^0 was the active phase of the spent catalyst, which was characterized by XRD. Similar to the traditional Co-based catalyst used for syngas conversion: the olefin/paraffin (O/P) ratio of the Na-free CoMn catalyst was found to be low; the selectivity to low olefins was only 15.4% (see Figure 12.4). However, Co_2C quadrilateral nanoprisms with exposed (101) and (020) facets were formed when the Na content increased. Co_2C nanoprisms showed high selectivity for lower olefins (54.2%) and low selectivity for methane (5.9%) under reaction conditions of 0.1 MPa and 250 °C. The O/P ratio for C_2–C_4 hydrocarbons reached 23.9. Therefore, Co_2C nanoprisms exposed (101) and (020) facets were considered to be the active phase for highly efficient production of lower olefins in FTO. The Na promoter played an important role in promoting the formation of Co_2C nanoprisms. Furthermore, Na was considered to provide electrons to cobalt, which enhanced CO adsorption and CO dissociation; this was beneficial for the formation of the Co_2C nanoprisms phase. Other alkali metals promoted CoMn catalysts were also investigated for use in the FTO process. It was found that only the K promoter, which is similar to the Na promoter, could catalyze the FTO reaction over CoMn catalyst.[22]

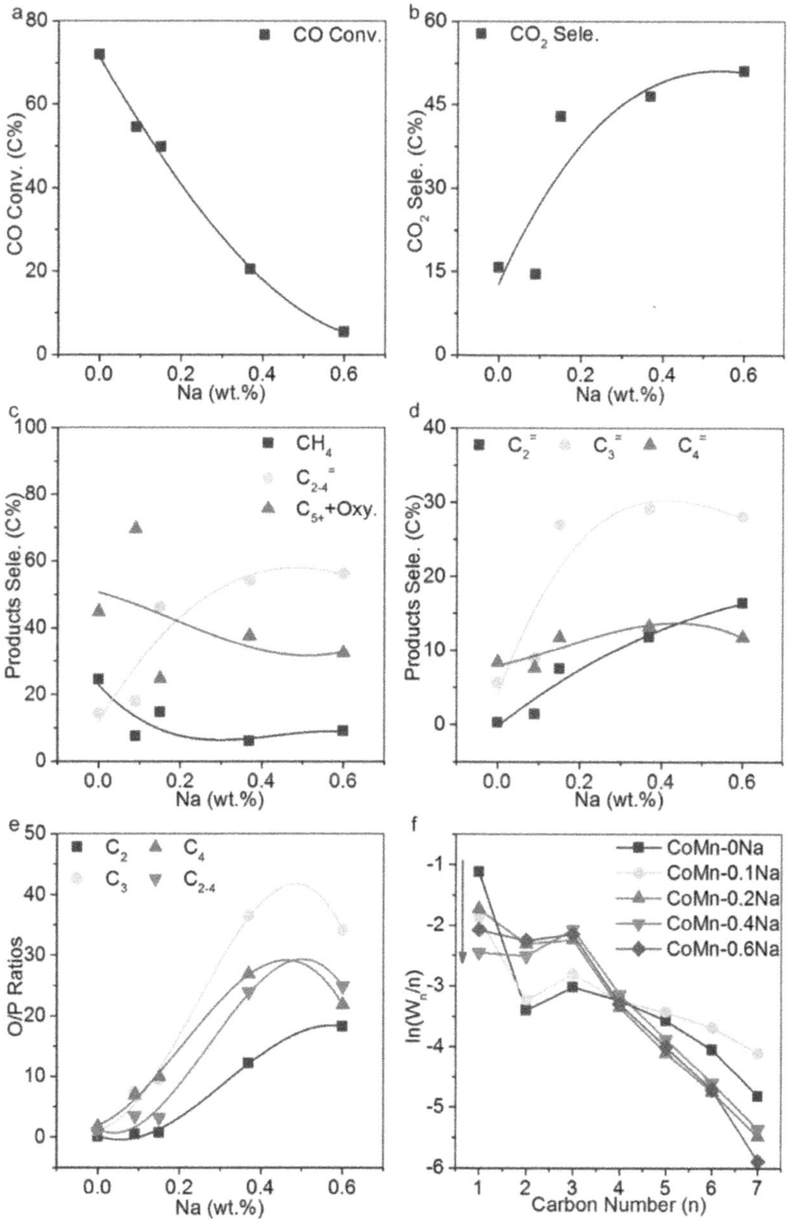

Figure 12.4 Catalytic performance of the CoMn catalysts with different Na concentrations at 250 °C, 1 bar, 3000 ml $(\mathrm{h\,g_{cat}})^{-1}$, and $H_2/CO = 2$. (a) CO conversion; (b) CO_2 selectivity; (c) product selectivity; (d) lower olefin selectivity; (e) O/P ratios of C_2–C_4; (f) product distribution.[22] Reproduced from ref. 22 with permission from American Chemical Society, Copyright 2017.

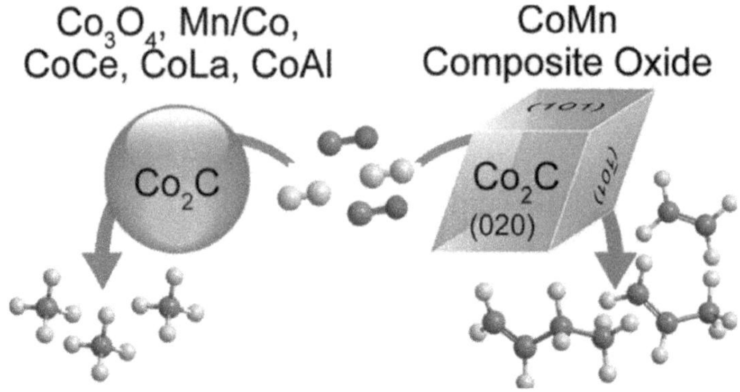

Figure 12.5 Schematic representation of CoX (X = Ce, La, Al) oxide for FTO.[23] Reproduced from ref. 23 with permission from American Chemical Society, Copyright 2017.

With the CoMn catalyst system, the role of Mn in FTO is still unknown. Li *et al.* designed several catalysts to investigate the effect of the Mn promoter on the morphology of the Co_2C active phase in the FTO process.[24] The catalysts promoted with Ce, La and Al were prepared using the coprecipitation method (see Figure 12.5). After being reduced by H_2, the main phase of the CoMn catalyst was transformed to $Co_xMn_{1-x}O$, and during the reaction process Co_2C nanoprisms exposed (101) and (020) facets were formed. The Co_2C nanoprisms exhibited significant FTO catalytic performance, showing selectivity for C_2–C_4 olefins of 54.2% and a lower ethane selectivity of 5.9%. However, a CoMn catalyst synthesized using the impregnation method exhibited higher methane selectivity and lower C_2–C_4 olefin selectivity, which led to the formation of Co_2C nanospheres after reaction.

Similar to the CoMn catalyst prepared by impregnation, a CoLa catalyst was synthesized by coprecipitation and displayed similar catalytic performance and the same formation of Co_2C nanospheres. Furthermore, the CoCe catalyst showed a high selectivity (~50%) to lower olefins without the formation of Co_2C nanoprisms. For the Al-promoted Co catalyst, the $Co_xAl_{1-x}O$ phase formed after reduction, and the phase was unchanged during the reaction process. In contrast to this conclusion on the CoCe catalyst, Li *et al.*[25] found that increasing the Ce content promoted the transformation of the Co catalyst from Co_2C nanospheres to Co_2C nanoprisms, which led to increased selectivity for lower olefins in FTO.

12.2.2 Metal Oxide Coupling with Zeolites

As already indicated, the previous studies for direct conversion of syngas to olefins were mainly focused on Fe- and Co-based catalysts. However, the limitation of the ASF distribution hindered the development of the process, which shows that the maximum selectivity of C_2–C_4 and methane calculated are 56% and 29%, respectively. Recently, a promising strategy for the conversion of syngas to lower olefins was developed, by combining metal oxides with zeolites (OX-ZEO). These are referred to as "hybrid catalysts" or "relay catalysts".[9] In this process, syngas is first activated by metal oxides, followed by C–C propagation on zeolites to form lower olefins. The combination of metal oxides with zeolites was successfully used to produce gasoline from syngas using Zn–Cr oxide and ZSM-5 as the hybrid catalysts.[26] The hybrid catalyst was also applied in the conversion of syngas to C_2–C_4 paraffins (LPG) with core–shell CuZnAl/H-β zeolite catalysts.[27]

Recently, syngas conversion to lower olefins *via* OX-ZEO catalysts has seen significant progress, and different mechanisms involving the intermediates (ketene and methanol) route have been proposed.[28]

12.2.2.1 Ketene as Intermediate

In 2016, Bao *et al.*[28] reported that a combination of $ZnCrO_x$ oxide and MSAPO (mesoporous SAPO) zeolite converted syngas directly to lower olefins. This new strategy for syngas conversion was referred to as the OX-ZEO route. The catalytic performance of the route showed that the selectivity of both CH_4 and C_{5+} are very low, while the selectivity of C_2–C_4 hydrocarbons was up to ~94% (see Figure 12.6). More interestingly, the selectivity of C_2–C_4 olefins reached 80% at a CO conversion of 17%, and the catalyst system was stable during 110 h of reaction.

In order to investigate the activation mechanism of syngas on the $ZnCrO_x$ surface, Bao *et al.* used highly sensitive synchrotron vacuum ultraviolet ionization mass spectrometry (SVUV-PIMS) to detect the presence of ketene (CH_2CO) in the gas phase, following the interaction of syngas with $ZnCrO_x$.

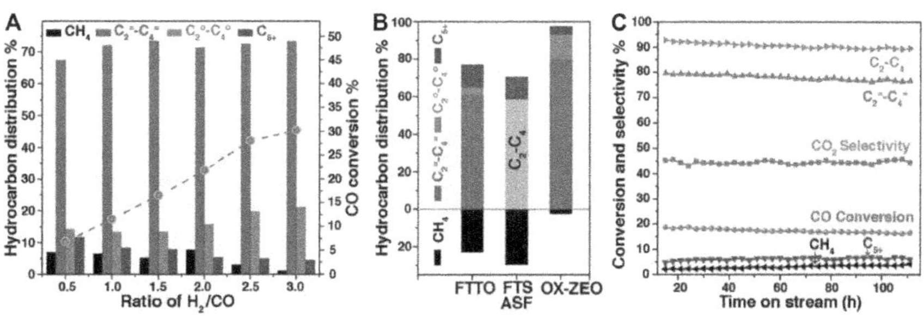

Figure 12.6 (A) CO conversion and product distribution at different H_2/CO ratios in syngas over a catalyst with a mass ratio of $ZnCrO_x$/MSAPO = 1.4 at a space velocity of 4800 ml $h^{-1} g_{cat}^{-1}$. (B) Hydrocarbon distribution in OX-ZEO in comparison to that reported for FTO (4) and that in FTS predicted by the ASF model at a chain growth probability of 0.46, with the yellow bar representing selectivity of C_2–C_4 hydrocarbons. (C) Stability test of a composite catalyst with $ZnCrO_x$/MSAPO ratio = 0.9 at 6828 ml $h^{-1} g_{cat}^{-1}$ and H_2/CO of 2.5.[28] Reproduced from ref. 28 with permission from the American Association for the Advancement of Science, Copyright 2016.

In addition, a model catalytic reaction was carried out, and it was found that the conversion of ketene on MSPAO produced C_2H_4 and C_3H_6.[28] Theoretical calculation results revealed that the formation of olefins from ketene in MSPAO is possible. Therefore, it was proposed that ketene originally formed from syngas on $ZnCrO_x$ and subsequently converted to C_2–C_4 olefins in MSAPO.

The same group[29] also found that MnO_x can replace $ZnCrO_x$ to combine with SAPO-34 as a bifunctional catalyst to convert syngas to C_2–C_4 olefins. Ketene also acted as an intermediate during the FTO reaction over the catalyst system. With the MnO_x/SAPO-34 catalyst, it was revealed that the spent catalyst was crystallized MnO, regardless of the manganese oxide (MnO_2, Mn_2O_3 or MnO) used in the fresh bifunctional catalyst system. This implies a reduction of Mn^{4+} or Mn^{3+} to Mn^{2+} during the reaction. In addition, the surface O/Mn ratio was characterized by means of ambient-pressure XPS, and an O/Mn ratio of 0.73 was detected under the reaction conditions (atmosphere and 300 °C) that is less than the optimal ratio of 1 (MnO). Therefore, the amount of O was less than Mn, which indicated that the Mn was further reduced during the reactions. Therefore, the partially reduced MnO surface has many oxygen vacancy sites, which are active sites for CO activation.

The ZnO/SAPO-34 catalyst can catalyze syngas to lower olefins.[30] Previous work performed found that ZnO in combination with SAPO-34 could not convert syngas to C_2–C_4 olefins. However, when the ZnO particle size decreased from 79 to 33 nm, the CO conversion increased significantly in FTO. Furthermore, CO conversion changed slightly with the further decrease in the ZnO particle size. However, the selectivity of C_2–C_4 olefins increased and that of C_2–C_4 paraffins decreased stably with a decrease in ZnO particle size (see Figure 12.7). The space time yield (STY) of lower olefins increased significantly. The apparent turnover frequency (TOF) of lower olefin formation based on surface zinc sites increased slightly with a decrease in ZnO particle size. Furthermore, the selectivity of C_2–C_4 paraffins increased with an increase in ZnO particle size.

Bifunctional catalyst $ZnCrO_x$ combined with mordenite (MOR) was also synthesized for FTO.[31] The results revealed that product selectivity was effectively regulated toward ethylene, by effectively shielding the 12-membered ring (12MR) of MOR. The combination of the partially reduced $ZnCrO_x$ and the 8MR MOR exhibited extremely high selectivity for ethylene (73%) at a CO conversion of 26% (see Figure 12.8). However, based on the ASF distribution model, the maximum value of ethylene selectivity was 30%. Compared with other results in the reaction of syngas to lower olefin, the selectivity value of 73% to ethylene was the highest. Selective site blocking experiments showed that this highly selective pathway was achieved through ketene as an intermediate at a catalytic site in the 8MR side pockets of MOR, while the 12MR was completely nonselective for ethylene. This study provides experimental evidence for a new type of syngas chemistry that ketene acts as the key intermediate.[31]

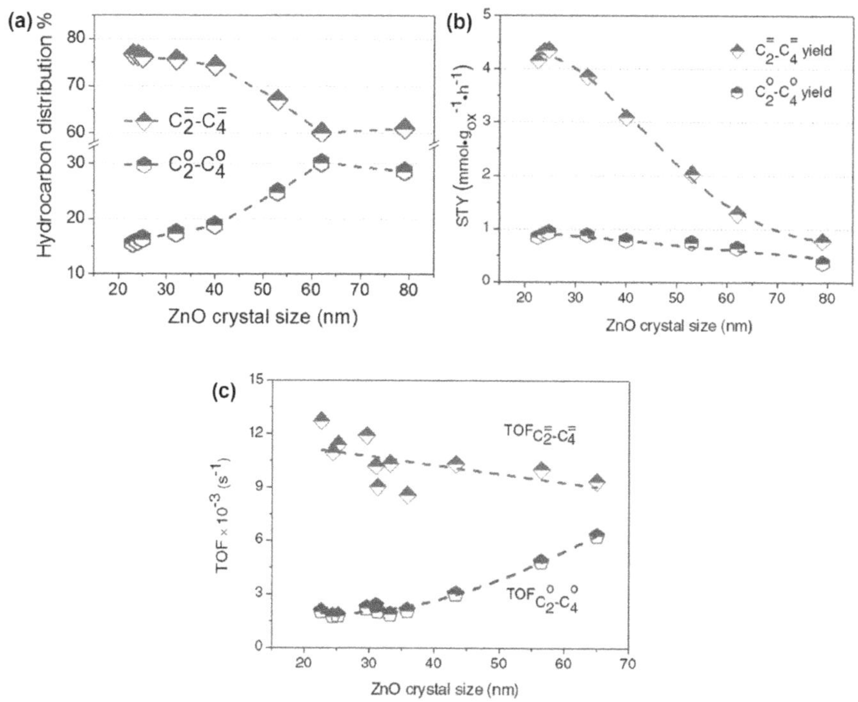

Figure 12.7 Effect of the size of ZnO particles on the catalytic behavior of ZnO/SAPO-34 catalysts.[30] Reproduced from ref. 30 with permission from American Chemical Society, Copyright 2018.

12.2.2.2 *Methanol as the Intermediate*

Cu–Zn–Al mixed oxide is an efficient catalyst for syngas conversion to methanol, and SAPO-34 is an excellent methanol-to-olefin (MTO) catalyst.[32,33] Therefore, designing a hybrid catalyst that integrates syngas conversion to methanol/dimethyl ether (DME) and CH_3OH/DME to olefins (MTO) could achieve direct conversion of syngas into C_2–C_4 olefins. However, there are still challenges to overcome. First, the MTO reaction usually occurs at ≧400 °C, which is due to the kinetic requirements of high activation energy formed by the C–C bond.[5] However, in terms of thermodynamics, it is not feasible to synthesize methanol when the temperature is ≧400 °C. Thermodynamic analysis shows that the MTO reaction can be driven by the methanol intermediate to convert the syngas to C_2–C_4 olefins, according to le Chatelier's principle. Second, it is not easy to find a high-temperature methanol synthesis catalyst because most current methanol synthesis catalysts work at ~250 °C, *e.g.* Cu–Zn–Al mixed oxide catalyst. High temperatures lead to the formation of methane rather than CH_3OH as the primary product. Third, it is difficult to match the two components that require different reaction temperatures and to avoid the hydrogenation of olefins to alkanes. In fact, earlier studies by Fujimoto *et al.*[34] reported that

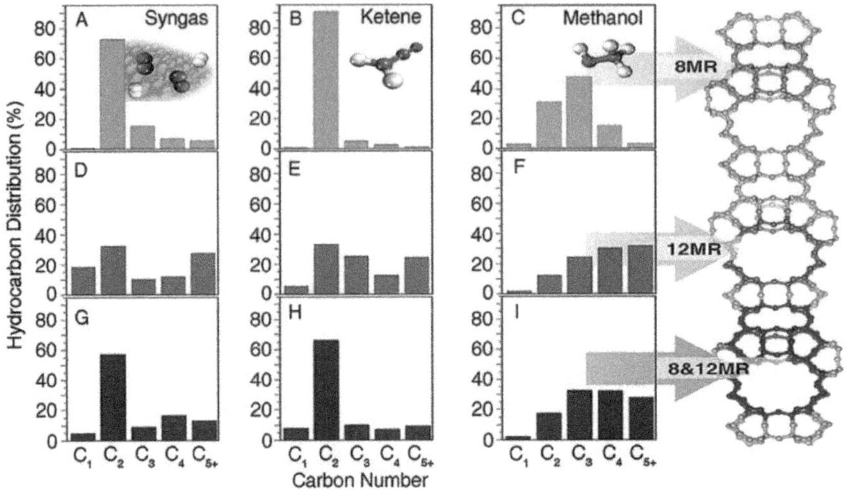

Figure 12.8 Hydrocarbon distributions in the conversion of syngas, ketene and methanol over different sites of MOR zeolites at 375 °C.[31] Reproduced from ref. 31 with permission from John Wiley & Sons, Copyright © 2018 Wiley-VCH Verlag GmbH & Co. KGaA, Weinheim.

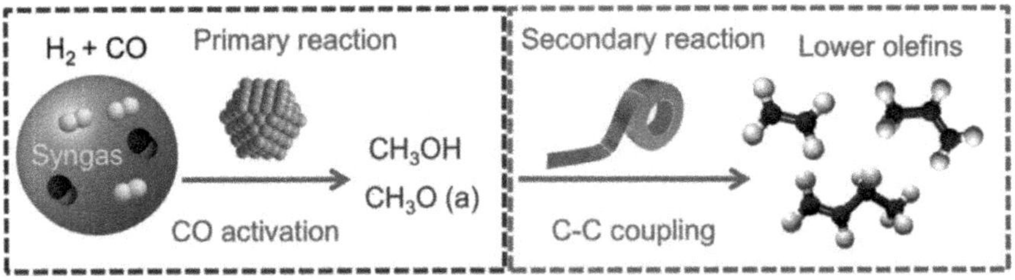

Figure 12.9 Schematic illustration of direct conversion of syngas into lower olefins by integrating methanol/DME synthesis and methanol/DME to lower olefins.[35] Reproduced from ref. 35 with permission from John Wiley & Sons, Copyright © 2016 Wiley-VCH Verlag GmbH & Co. KGaA, Weinheim.

combining the syngas conversion catalyst (Pd/SiO_2 or Cu/ZnO) with methanol synthesis catalysts to generate hydrocarbons (Y or ZSM-5 zeolites); however, during the syngas conversion reaction, the main products were LPG instead of olefins.

Wang and coworkers[35] successfully designed a ZnO–ZrO_2/SAPO-34 bifunctional catalyst that efficiently converted syngas to lower olefins *via* methanol/dimethyl ether intermediates (see Figure 12.9). Under the reaction conditions 400 °C and 1.0 MPa (H_2/CO = 2), the CO conversion was 11%, and the olefin selectivity was 74%. The durability of the catalyst was tested during 100 h of reaction, and no obvious deactivation was observed. At 400 °C, owing to thermodynamic limitation, the CO conversion was very low for ZnO–ZrO_2 alone, but methanol and dimethyl ether were generated.

However, the combination of $ZnO–ZrO_2$ and SAPO-34 not only transformed the main product of CH_3OH/DME into $C_2–C_4$ olefins but also improved the CO conversion as expected, due to the thermodynamic driving force.

Due to the CHA topological structure of SAPO-34, its special channel structure has the ability of shape to selectively generate $C_2–C_4$ olefins.[36] The effects of the Zn/Zr ratio and Brønsted acid site density of the zeolite on the reaction performance were also investigated. The $C_2–C_4$ olefin selectivity of the $ZnO–ZrO_2$/SSZ-13 catalyst with optimal Zn/Zr ratio and Brønsted acid site density reached 87% at a CO conversion of 10%. When the reaction pressure was changed to 3.0 MPa, the selectivity of $C_2–C_4$ olefins reached 77% on the $ZnO–ZrO_2$/SSZ-13 catalyst, with a CO conversion of 29%. However, the CO_2 selectivity of both the $ZnO–ZrO_2$/SAPO-34 and the $ZnO–ZrO_2$/SSZ-13 catalyst was as high as 40–45% at 400 °C. It is obvious that both the $ZnO–ZrO_2$/SAPO-34 and the $ZnO–ZrO_2$/SSZ-13 catalyst could provide about 80% of lower olefins, which is much higher than the calculated value based on the ASF distribution.

The effects of the Zn/Zr molar ratio of $ZnO–ZrO_2$, the Brønsted acid site density of zeolites, the $ZnO–ZrO_2$/zeolite mass ratio and the proximity of metal oxide to SSZ-13 zeolite on the catalytic performance were also studied.[36] The results revealed that the CO conversion increased with an increase in the Zn/Zr ratio. They showed a maximum CO conversion of ~28% at the Zn/Zr ratio of 1/1. CO conversion began to decrease as the Zn/Zr ratio increased further. Lower olefin selectivity decreased steadily when the ratio of Zn/Zr increased. The combination of ZrO_2 and SAPO-34 or SSZ-13 reached 80% selectivity of $C_2–C_4$ olefins but showed a lower CO conversion (see Figure 12.10a).[36] The density of the Brønsted acid sites in SAPO-34 or SSZ-13 was also considered as an important parameter. When the density of the Brønsted acid site was low, CH_3OH and DME were the main products. With an increase in the Brønsted acid site density, $C_2–C_4$ olefins were the main products (see Figure 12.10b). This provides strong evidence that the C–C bond is formed at the Brønsted acid site of SAPO-34 and SSZ-13. However, the extremely high density of the Brønsted acid site was unfavorable for lower olefin formation, owing to overhydrogenation of olefins to alkanes. Furthermore, as the density of the Brønsted acid increased, the conversion of CO accelerated significantly. This could be attributed to the conversion of products from intermediates to $C_2–C_4$ olefins, which could drive the syngas conversion thermodynamically.

It was also demonstrated that spatial proximity of the active sites plays an important role in product selectivity for a multifunctional catalyst.[37] Therefore, different distances between $ZnO–ZrO_2$ and SSZ-13 catalysts have been investigated.[36] It was revealed that both the lower-olefin selectivity and the CO conversion increased when $ZnO–ZrO_2$ and SSZ-13 were closer (see Figure 12.10c).[36]

Other metal oxide composites, such as $ZnO–Al_2O_3$, $ZnO–ZrO_2–Al_2O_3$ and $In_2O_3–ZrO_2$ coupled with zeolites, have also been studied as catalysts for direct conversion of syngas to lower olefins.[38–40] These catalysts exhibit

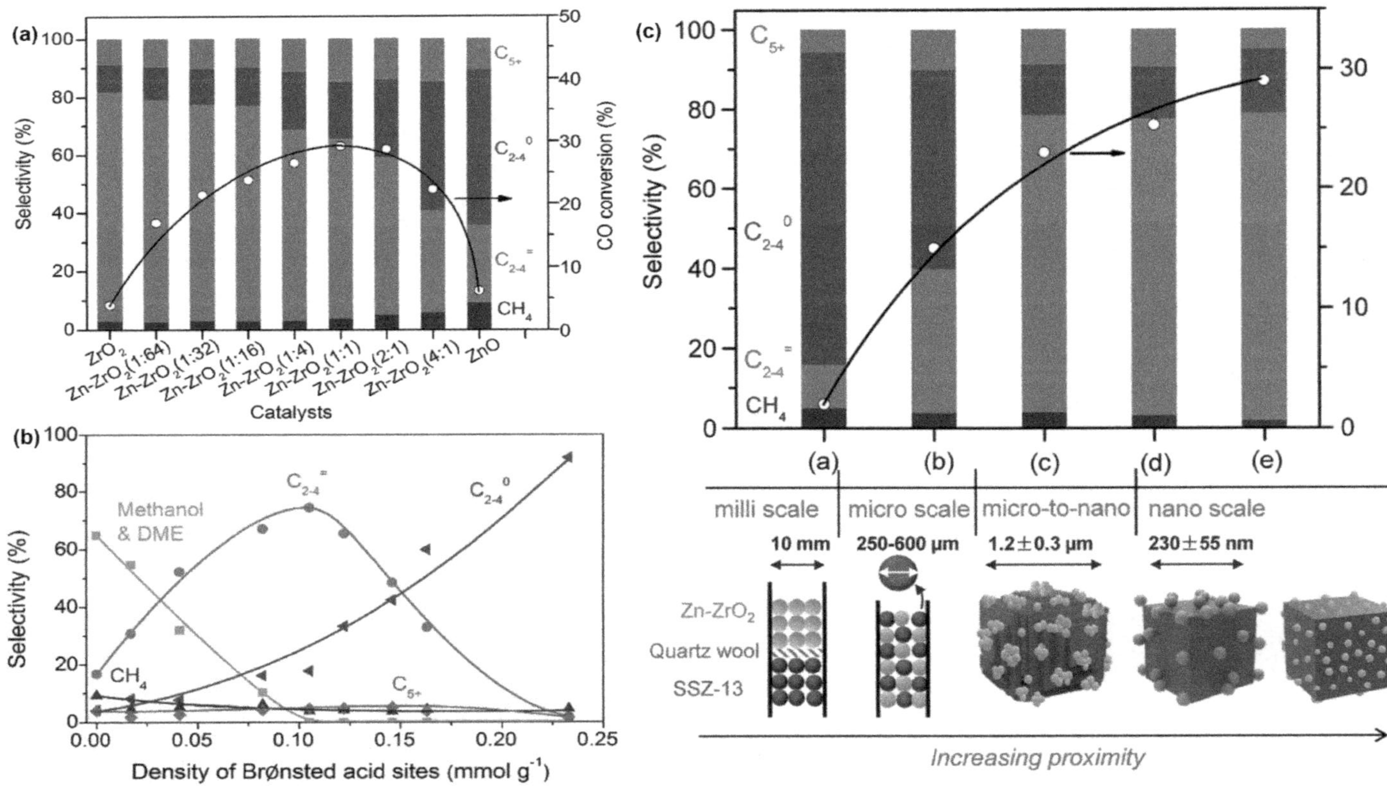

Figure 12.10 Effect of Zn/Zr molar ratio, density of Brønsted acid site and proximity between $ZnO–ZrO_2$ and zeolite on the catalytic behavior of $ZnO–ZrO_2$/SSZ-13 catalysts.[36] Reproduced from ref. 36 with permission from the Royal Society of Chemistry.

significant catalytic performance for FTO and lower olefin selectivity (70–80%). In addition, the combination of MoS_2 and SAPO-34 as a bifunctional catalyst was also investigated.[41] The K-promoted MoS_2/SAPO-34 catalyst was active for FTO, and the selectivity of C_2–C_4 olefins reached 64.7% at a CO conversion of 9.4%.

12.3 Conclusions and Challenges with Syngas to Olefins

Lower olefins (C_2–C_4) are important building blocks in the chemical industry in terms of the production of plastics and chemicals. Lower olefins can be produced *via* dehydrogenation at elevated temperatures. Alternatively, lower olefins can be achieved through the process of syngas to lower olefins at lower temperatures. This chapter discussed two types of catalysts used for the efficient production of low olefins from syngas: modified FTS catalysts and metal oxide coupling with zeolites.

All the OX-ZEO catalysts showed significant catalytic performance at a temperature >380 °C and pressure ≥ 1.0 MPa, although the selectivity of C_2–C_4 olefins was much higher than the calculation using the ASF distribution model. However, CO_2 selectivity is always >40% in all cases. In a recent work, Wang *et al.*[43] found that a combination of $Zn_xCe_{2-y}Zr_yO_4$ oxide and SAPO-34 zeolite as a bifunctional catalyst exhibited significantly low CO_2 selectivity (<6%) in FTO under mild reaction conditions.[42] The selectivity to C_2–C_4 olefins was ~83% at reaction conditions of 300 °C and 0.1 MPa.

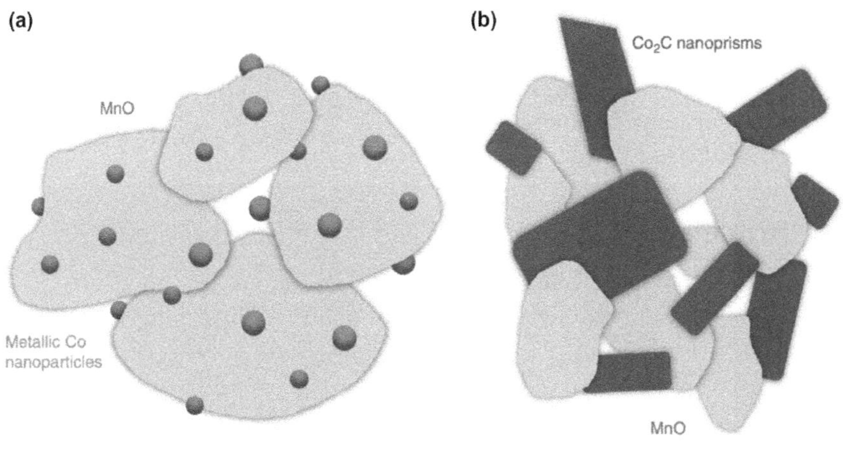

Figure 12.11 Structure of the active catalysts based on TEM and XRD analysis. (a) Co_1Mn_3–Na_2S consists of metallic Co nanoparticles of ~10 nm dispersed on MnO support. (b) Co_3Mn_1–Na_2O consists of Co_2C nanoprisms of ~10–50 nm. Whereas the former catalyst restricts WGS and thus CO_2 formation, the latter leads to large amounts of CO_2.[44] Reproduced from ref. 44 with permission from Springer Nature, Copyright 2019.

However, the CO conversion was ~7% at this low temperature. Theoretical calculations revealed that doping Ce in the Zn–Zr oxide significantly increased the free energy barrier of the WGS reaction and therefore restrained the CO_2 formation under mild reaction conditions. Adding CO_2 to syngas to avoid the WGS reaction was also considered as a method to suppress CO_2 emission with a bifunctional catalyst.

Another issue with nanoprism Co_2C catalysts is producing a high yield of CO_2, since the Co_2C phase is active for WGS. In order to solve this problem, De Jong and coworkers designed a Co/Mn/Na/S catalyst (see Figure 12.11), which exhibited negligible WGS activity and restrained CH_4 formation during the FTO reaction.[44] The catalyst displayed selectivity to lower olefins of 54% under the reaction conditions 10 bar and 240 °C. The spent catalyst phase was analyzed, and the active phase was 10 nm Co nanoparticles with hcp Co metal phase supported on MnO. Theoretical calculations showed that Na and S play a synergistic role in the regulation of selectivity for valuable hydrocarbons. In addition, compared with the Na_2O-promoted CoMn catalyst, the Na- and S-promoted Co_1Mn_3 catalysts were less active for the WGS reaction.

Abbreviations

ASF	Anderson–Schulz–Flory
DFT	Density functional theory
FCC	Fluid catalytic cracking
FT	Fischer–Tropsch
FTO	Fischer–Tropsch to olefins
FTS	Fischer–Tropsch synthesis
FTY	Iron-time-yield
O/P	Olefin/paraffin
rGO	Reduced graphene oxide
TOF	Turnover frequency
WGS	Water-gas shift

References

1. H. M. Torres Galvis and K. P. de Jong, *ACS Catal.*, 2013, **3**, 2130.
2. L. Zhong, F. Yu, Y. An, Y. Zhao, Y. Sun, Z. Li, T. Lin, Y. Lin, X. Qi, Y. Dai, L. Gu, J. Hu, S. Jin, Q. Shen and H. Wang, *Nature*, 2016, **538**, 84.
3. J. J. Sattler, J. Ruiz-Martinez, E. Santillan-Jimenez and B. M. Weckhuysen, *Chem. Rev.*, 2014, **114**, 10613.
4. U. Olsbye, S. Svelle, M. Bjørgen, P. Beato, T. V. Janssens, F. Joensen, S. Bordiga and K. P. Lillerud, *Angew. Chem., Int. Ed.*, 2012, **51**, 5810.
5. P. Tian, Y. Wei, M. Ye and Z. Liu, *ACS Catal.*, 2015, **5**, 1922.
6. X. Zhou, J. Ji, D. Wang, X. Duan, G. Qian, D. Chen and X. Zhou, *Chem. Commun.*, 2015, **51**, 8853.
7. H. M. Torres Galvis, J. H. Bitter, C. B. Khare, M. Ruitenbeek, A. I. Dugulan and K. P. de Jong, *Science*, 2012, **335**, 835.

8. R. A. van Santen, A. J. Markvoort, I. A. Filot, M. M. Ghouri and E. J. Hensen, *Phys. Chem. Chem. Phys.*, 2013, **15**, 17038.
9. W. Zhou, K. Cheng, J. Kang, C. Zhou, V. Subramanian, Q. Zhang and Y. Wang, *Chem. Soc. Rev.*, 2019, **48**, 3193.
10. S. Abello and D. Montane, *ChemSusChem*, 2011, **4**, 1538.
11. J. X. Liu, H. Y. Su, D. P. Sun, B. Y. Zhang and W. X. Li, *J. Am. Chem. Soc.*, 2013, **135**, 16284.
12. D. I. Enache, B. Rebours, M. Roy-Auberger and R. Revel, *J. Catal.*, 2002, **205**, 346.
13. Y. Cheng, J. Lin, K. Xu, H. Wang, X. Yao, Y. Pei, S. Yan, M. Qiao and B. Zong, *ACS Catal.*, 2015, **6**, 389.
14. Z. Zhang, H. Yin, G. Yu, S. He, J. Kang, Z. Liu, K. Cheng, Q. Zhang and Y. Wang, *J. Catal.*, 2021, **395**, 350.
15. M. Zhao, C. Yan, S. Jinchang and Z. Qianwen, *Catal. Today*, 2018, **316**, 142.
16. B. Chen, X. Zhang, W. Chen, D. Wang, N. Song, G. Qian, X. Duan, J. Yang, D. Chen, W. Yuan and X. Zhou, *Ind. Eng. Chem. Res.*, 2018, **57**, 11554.
17. T. Qiu, L. Wang, S. Lv, B. Sun, Y. Zhang, Z. Liu, W. Yang and J. Li, *Fuel*, 2017, **203**, 811.
18. A. Dinse, M. Aigner, M. Ulbrich, G. R. Johnson and A. T. Bell, *J. Catal.*, 2012, **288**, 104.
19. J. Paterson, M. Peacock, R. Purves, R. Partington, K. Sullivan, G. Sunley and J. Wilson, *ChemCatChem*, 2018, **10**, 5154.
20. S. Liu, B. Sun, Y. Zhang, J. Li, D. E. Resasco, L. Nie and L. Wang, *Chem. Commun.*, 2019, **55**, 6595.
21. Y. An, Y. Zhao, F. Yu, T. Lin, Y. Lu, S. Li, Z. Li, Y. Dai, X. Wang, H. Wang, L. Zhong and Y. Sun, *J. Catal.*, 2018, **366**, 289.
22. Z. Li, L. Zhong, F. Yu, Y. An, Y. Dai, Y. Yang, T. Lin, S. Li, H. Wang, P. Gao, Y. Sun and M. He, *ACS Catal.*, 2017, **7**, 3622.
23. Z. Li, T. Lin, F. Yu, Y. An, Y. Dai, S. Li, L. Zhong, H. Wang, P. Gao, Y. Sun and M. He, *ACS Catal.*, 2017, 8023.
24. Z. Li, T. Lin, F. Yu, Y. An, Y. Dai, S. Li, L. Zhong, H. Wang, P. Gao and Y. Sun, *ACS Catal.*, 2017, 7, 8023.
25. Z. Li, D. Yu, L. Yang, J. Cen, K. Xiao, N. Yao and X. Li, *ACS Catal.*, 2021, **11**, 2746.
26. X. Yang, X. Su, B. Liang, Y. Zhang, H. Duan, J. Ma, Y. Huang and T. Zhang, *Catal. Sci. Technol.*, 2018, **8**, 4338.
27. P. Lu, J. Sun, D. Shen, R. Yang, C. Xing, C. Lu, N. Tsubaki and S. Shan, *Appl. Energy*, 2018, **209**, 1.
28. F. Jiao, J. Li, X. Pan, J. Xiao, H. Li, H. Ma, M. Wei, Y. Pan, Z. Zhou and M. Li, *Science*, 2016, **351**, 1065.
29. Y. Zhu, X. Pan, F. Jiao, J. Li, J. Yang, M. Ding, Y. Han, Z. Liu and X. Bao, *ACS Catal.*, 2017, **7**, 2800.
30. N. Li, F. Jiao, X. Pan, Y. Ding, J. Feng and X. Bao, *ACS Catal.*, 2018, **9**, 960.
31. F. Jiao, X. Pan, K. Gong, Y. Chen, G. Li and X. Bao, *Angew. Chem., Int. Ed.*, 2018, **57**, 4692.

32. U. Olsbye, S. Svelle, M. Bjorgen, P. Beato, T. V. Janssens, F. Joensen, S. Bordiga and K. P. Lillerud, *Angew. Chem., Int. Ed.*, 2012, **51**, 5810.
33. U. Olsbye, S. Svelle, M. Bjørgen, P. Beato, T. V. W. Janssens, F. Joensen, S. Bordiga and K. P. Lillerud, *Angew. Chem., Int. Ed.*, 2012, **124**, 5910.
34. Q. Zhang, X. Li, K. Asami, S. Asaoka and K. Fujimoto, *Fuel Process. Technol.*, 2004, **85**, 1139.
35. K. Cheng, B. Gu, X. Liu, J. Kang, Q. Zhang and Y. Wang, *Angew. Chem., Int. Ed.*, 2016, **128**, 4803.
36. X. Liu, W. Zhou, Y. Yang, K. Cheng, J. Kang, L. Zhang, G. Zhang, X. Min, Q. Zhang and Y. Wang, *Chem. Sci.*, 2018, **9**, 4708.
37. J. Zecevic, G. Vanbutsele, K. P. de Jong and J. A. Martens, *Nature*, 2015, **528**, 245.
38. G. Raveendra, C. Li, Y. Cheng, F. Meng and Z. Li, *New J. Chem.*, 2018, **42**, 4419.
39. G. Raveendra, C. Li, B. Liu, Y. Cheng, F. Meng and Z. Li, *Catal. Sci. Technol.*, 2018, **8**, 3527.
40. J. Su, D. Wang, Y. Wang, H. Zhou, C. Liu, S. Liu, C. Wang, W. Yang, Z. Xie and M. He, *ChemCatChem*, 2018, **10**, 1536.
41. H. Zhou, S. Liu, J. Su, C. Liu, L. Zhang, W. Jiao and Y. Wang, *Ind. Eng. Chem. Res.*, 2018, **57**, 6815.
42. S. Wang, P. Wang, D. Shi, S. He, L. Zhang, W. Yan, Z. Qin, J. Li, M. Dong, J. Wang, U. Olsbye and W. Fan, *ACS Catal.*, 2020, **10**, 2046.
43. X. Liu, M. Wang, C. Zhou, W. Zhou, K. Cheng, J. Kang, Q. Zhang, W. Deng and Y. Wang, *Chem. Commun.*, 2018, **54**, 140.
44. J. Xie, P. P. Paalanen, T. W. van Deelen, B. M. Weckhuysen, M. J. Louwerse and K. P. de Jong, *Nat. Commun.*, 2019, **10**, 167.

CHAPTER 13

Application of Fischer–Tropsch Synthesis and Hydroformylation in Syngas Conversion to Oxygenates

JIANLI CHANG,[a] YALI YAO,*[a] XINYING LIU,*[a]
JOSHUA GORIMBO[a] AND DIANE HILDEBRANDT[b]

[a] Institute for Development of Energy for African Sustainability, University of South Africa, Private Bag X6, Roodepoort 1710, South Africa; [b] Future Energy African Energy Leadership Centre, Wits Business School, and Molecular Sciences Institute, School of Chemistry, University of the Witwatersrand, Johannesburg 2050, South Africa
*Emails: Yaoy@unisa.ac.za; Liux@unisa.ac.za

13.1 Introduction

Oxygenates have a broad-spectrum application in chemical industrials.[1] Methanol (C_1), ethanol (C_2), high-molecular-weight alcohols (C_{2+}) and aldehydes (C=O) are frequently used oxygenates in upgrading fuels and chemicals.[2]

Methanol is used to produce chemicals such as formaldehyde, methyl *tert*-butyl ether (MTBE), *tert*-amyl methyl ether (TAME), acetic acid and dimethyl ether (DME).[2] Ethanol is mainly used as a blending stock for petrol and is also employed in synthesizing 1,3-butadiene, isobutene, butanol and acetaldehyde.[2] Formaldehyde is a crucial chemical in producing thermosetting polymers; resins, explosives, adhesives and dyes.[2] Butyraldehyde is typically

Catalysis Series No. 44
Chemicals and Fuels from Biomass *via* Fischer–Tropsch Synthesis: A Route to Sustainability
Edited by Joshua Gorimbo, Xinying Liu, Yali Yao and Diane Hildebrandt

Published by the Royal Society of Chemistry, www.rsc.org

used to make *n*-butanol and 2-ethylhexanol, which are both used in the production of plasticizer.[2] Aldehydes are essential as both final products and as intermediates in the case of synthesizing bulk chemicals such as esters and amines.[3]

Oxygenates are synthesized directly from syngas (H_2/CO) derived from fossil fuels, commonly natural gas and coal.[1,4–7] However, due to the limited reserves and environmental impact of fossil fuels, the development of oxygenates from biomass is increasingly important in the requirement for cleaner and renewable sources.[1,8,9] This alternative production of oxygenates is achieved through Fischer–Tropsch synthesis (FTS).[10] The conversion of biomass to oxygenates *via* FTS starts with the gasification of biomass with air and steam to generate syngas, which is a combination of CO, CO_2, H_2 and H_2O.[5,9] After gas cleanup and conditioning, the syngas can be reacted over a metal catalyst to alcohol.[11,12] Additionally, aldehyde and alcohol can be produced by hydroformylation of a syngas and olefin stream feed.[13–16] This chapter focuses on the technology in syngas conversion to oxygenate *via* FTS and on the application to a case study.

13.2 Catalysts for Syngas to Oxygenates Conversion

13.2.1 Catalysts for Syngas to Alcohols Conversion *via* FTS

The catalysts employed in producing mixed alcohols include single-metal species, bimetallic catalysts and even triple-metal catalysts such as Mo_2S, Mo_2C, Cu–Co, Rh–Fe, Pt–Fe, Cu–Zn–Al and Zn–Cr–K.[11,17,18] The main catalysts used for the syngas to ethanol reaction are Rh based, Cu based, Mo based and modified FTS catalysts.[10]

13.2.1.1 Rh-Based Catalysts

The products of Rh-catalyzed CO hydrogenation, among others, can be methane, alcohols and other oxygenates. The product distribution is governed by the support and promoter used, as well as by the reaction conditions.[10] The following is an example to illustrate the results for syngas-to-ethanol conversion using Rh-based catalysts.[19]

The three catalyst samples, namely Rh/$Ce_{1-x}Zr_xO_2$ ($x = 0$–1); Rh/MgO and Rh/SiO_2, were all synthesized with a 2% Rh loading.[19] The catalytic experiments were conducted in a fixed-bed flow reactor at a pressure of 2.4 MPa, temperature 275 °C, with an H_2/CO ratio of 2/1 and a W/F of 10 g h mol^{-1}.[19] Comparatively, under identical reaction conditions, the Rh/CeO_2 catalyst yielded the highest CO conversion of 23.7%, followed by Rh/ZrO_2 (18.2%), then Rh/MgO (10.8%) and lastly Rh/SiO_2 (10.1%).[19] Furthermore, the Rh/$Ce_{0.8}Zr_{0.2}O_2$ catalyst achieved a CO conversion of 27.3%, significantly better than that achieved by the Rh/CeO_2 catalyst (23.7%).[19] The Rh/$Ce_{0.8}Zr_{0.2}O_2$ catalyst was comparatively more selective toward the ethanol (35.2%) than other catalysts.[19]

13.2.1.2 Cu-Based Catalysts

$Cu/ZnO/Al_2O_3$ and $Cu/ZnO/Cr_2O_3$ promoted with one or more alkali metals are the main components used for Cu-based catalysts.[20,21] The selectivity toward ethanol and/or higher alcohols is governed by a plethora of factors, mainly type and amount of promoter(s), the composition of the feed, pressure, temperature and flow rate.[21] A multifaceted catalyst with a composition Co/Mn and Cu/Zn/Al/Zr oxide markedly increased the selectivity to oxygenates to 58.1 wt% with 92.0 wt% of the produced oxygenates being $C_{2+}OH$. The total selectivity toward valuable olefins and oxygenates reached 80.6 wt% at a CO conversion 29.0%. The undesired C1 products can be suppressed by adding appropriate promoters at correct locations on the catalyst surface. The synergetic effects of the different active species enhance the reaction selectivity toward higher oxygenate production.[22]

13.2.1.3 Modified FTS Catalysts

Cobalt is a very active metal for FTS, while copper tends to promote the formation of alcohols.[23] Thus bimetallic Cu–Co-based catalysts could be promising for the synthesis of alcohols from syngas.[23,24] The investigation of direct ethanol synthesis from syngas was conducted in two reactors, a slurry reactor and a fixed-bed reactor, using a $Cu–Co/Al_2O_3$ catalyst. The reaction in a slurry reactor was more selective toward ethanol and yielded 70 wt% in total alcohols.

13.2.1.4 Modified Mo-based Catalysts

Alkali-promoted Mo-based catalysts are more selective to CO hydrogenation to alcohols.[10,23] The effect of promoting the catalyst with alkali metals on MoS_2 for the synthesis of alcohol was observed to increase in the order of $Li < Na < Cs < Rb < K$, indicating that moderate primary promotion is desired.[10] Nickel-promoted MoS_2 catalysts also exhibited excellent performance for the production of ethanol from syngas.[25] The three catalysts – namely nickel-promoted MoS_2, unsupported catalysts and a laponite-supported catalyst for alcohol synthesis – were synthesized following the microemulsion (ME) and hydrothermal (HT) methods. The unsupported Ni/MoS_2 catalyst synthesized using the ME method gave a comparatively higher CO conversion (33 mol%) and proved to be more selective toward alcohol (61 mol%) than catalysts prepared following the HT method (see Table 13.1 for hydrogenation activity).

13.2.2 Catalysts for Syngas Conversion to Alcohols *via* Hydroformylation

The metals in the neighborhood of rhodium in the periodic table are active for hydroformylation, and a generally accepted order of activity is $Rh \gg Co > Ir, Ru > Os > Pt > Pd \gg Fe > Ni$.[3,26,27] The hydroformylation process is the largest industrially homogeneously catalyzed process with high selectivity (well above 95%) to oxygenates under optimal reaction parameters

Table 13.1 Summary of the reaction conditions and the product selectivity of the heterogeneous catalysts for syngas co-feeding C_2H_4 to oxygenates.

	Reaction condition			Products selectivity (%)				
Catalyst	Feed ratio (C_2H_4: CO: H_2)	T (°C)	P (MPa)	C_nH_{2n+2}	R–CHO[a]	R–OH[b]	OXY[c]	Ref.
Rh/GNF-R	1:1:1	240	—	2.40	2.31	—	—	39
Rh/GNF-P	1:1:1	240	—	8.91	6.15	—	—	
Rh/GNF-H	1:1:1	240	—	6.35	2.53	—	—	
Rh/SiO_2	1:1:1	240	—	21.38	1.42	—	—	
Rh/GNF-R	1:1:1	240	—	4.36	3.30	—	—	
Rh/GNF-P	1:1:1	240	—	10.54	6.49	—	—	
Rh/GNF-H	1:1:1	240	—	5.57	3.00	—	—	
Rh/SiO_2	1:1:1	240	—	16.06	2.24	—	—	
Rh/GNF-R	1:1:1	240	—	2.25	2.25	—	—	
Rh/GNF-P	1:1:1	240	—	3.47	3.42	—	—	
Rh/GNF-H	1:1:1	240	—	2.67	1.81	—	—	
Rh/SiO_2	1:1:1	240	—	18.25	1.44	—	—	
Rh/GNF-R	1:1:1	240	—	4.95	2.48	—	—	
Rh/GNF-P	1:1:1	240	—	10.65	4.32	—	—	
Rh/GNF-H	1:1:1	240	—	7.40	2.71	—	—	
Rh/SiO_2	1:1:1	240	—	18.78	1.82	—	—	
1%Rh/SiO_2	3:4:2	180	2	—	—	—	39	40
5%Rh/SiO_2	3:4:2	180	2	—	—	—	26	
5%Rh/SiO_2	3:4:2	180	2	—	—	—	25	
9%Rh/SiO_2	3:4:2	180	2	—	—	—	6.4	
10%Rh/SiO_2	3:4:2	180	2	—	—	—	5.5	
23%Rh/SiO_2	3:4:2	180	2	—	—	—	28	
$Rh_4(CO)_{12}/SiO_2$	1:1:1	180	2	—	—	—	23	
$Rh_4(CO)_{12}/MgCl_2/SiO_2$	1:1:1	180	2	—	—	—	20	
$Rh_4(CO)_{12}/CaCl_2/SiO_2$	1:1:1	180	2	—	—	—	25	
$Rh_4(CO)_{12}/SrCl_2/SiO_2$	1:1:1	180	2	—	—	—	21	
$Rh_4(CO)_{12}/ZnCl_2/SiO_2$	1:1:1	180	2	—	—	—	32	
$Rh_4(CO)_{12}/ScCl_3/SiO_2$	1:1:1	180	2	—	—	—	33	
$Rh_4(CO)_{12}/TiCl_4/SiO_2$	1:1:1	180	2	—	—	—	19	
$Rh_4(CO)_{12}/ZrCl_4/SiO_2$	1:1:1	180	2	—	—	—	17	
$Rh_4(CO)_{12}/VCl_3/SiO_2$	1:1:1	180	2	—	—	—	24	
$Rh_4(CO)_{12}/CrCl_3/SiO_2$	1:1:1	180	2	—	—	—	24	
$Rh_4(CO)_{12}/Mo(CO)_6/SiO_2$	1:1:1	180	2	—	—	—	23	
$Rh_4(CO)_{12}/MnCl_2/SiO_2$	1:1:1	180	2	—	—	—	19	
$Rh_4(CO)_{12}/FeCl_3/SiO_2$	1:1:1	180	2	—	—	—	18	
$Rh_4(CO)_{12}/CuCl_2/SiO_2$	1:1:1	180	2	—	—	—	22	
1%Rh/SiO_2	1:1:1	150	0.1	—	—	—	52	
2%Rh/SiO_2	1:1:1	150	0.1	—	—	—	43	
2%Rh/SiO_2	1:1:1	150	0.1	—	—	—	33	
Pd/SiO_2	1:1:1	160	0.3	80.7	—	—	—	42
Pd/SiO_2	1:1:1	180	0.3	87.6	—	—	—	
Pd/SiO_2	1:1:1	200	0.3	91.3	—	—	—	
2Pd3Zn/SiO_2	1:1:1	160	0.3	5.4	—	—	—	
2Pd3Zn/SiO_2	1:1:1	180	0.3	7.6	—	—	—	
2Pd3Zn/SiO_2	1:1:1	200	0.3	8.8	—	—	—	
3Pd1Zn/SiO_2	1:1:1	160	0.3	49.1	—	—	—	

Table 13.1 (*Continued*)

Catalyst	Reaction condition: Feed ratio (C_2H_4: CO : H_2)	*T* (°C)	*P* (MPa)	Products selectivity (%): C_nH_{2n+2}	R–CHO[a]	R–OH[b]	OXY[c]	Ref.
3Pd1Zn/SiO_2	1:1:1	180	0.3	51.2	—	—	—	
3Pd1Zn/SiO_2	1:1:1	200	0.3	59.3	—	—	—	
Rh-SILP catalysts	1:1:1	—	—	—	—	—	—	43
$Co(NO_3)_2·6H_2O$	1:1:1	150	0.1	—	—	—	—	44
$CoCl_2·6H_2O$	1:1:1	150	0.1	—	—	—	—	
$Co_2(CO)_8$	1:1:1	150	0.1	—	—	—	—	
$Ru(NO_3)_3$	1:1:1	150	0.1	100	—	—	—	
$RuCl_3·nH_2O$	1:1:1	150	0.1	100	—	—	—	
$Ru_3(CO)_{12}$	1:1:1	150	0.1	100	—	—	—	
$HRuCo_3(CO)_{12}$	1:1:1	150	0.1	19	10	71	—	
$Ru(NO_3)_3 + Co(NO_3)_2·6H_2O$	1:1:1	150	0.1	27	9	64	—	
$RuCl_3·nH_2O + CoCl_2·6H_2O$	1:1:1	150	0.1	31	12	57	—	
$Ru(NO_3)_3 + Co_2(CO)_8$	1:1:1	150	0.1	29	7	64	—	
$Ru(NO_3)_3 + Co_2(CO)_8$	1:1:1	150	0.1	19	9	72	—	
$RuCl_3·nH_2O + Co_2(CO)_8$	1:1:1	150	0.1	32	15	53	—	
$Ru_3(CO)_{12} + Co_2(CO)_8$	1:1:1	150	0.1	17	14	69	—	
Co/AC	1:1:1	120	3	8.6	56.9	34.5	—	45
Co/AC	1:1:1	120	3	4.3	40.6	55.1	—	
Co/CMK-3	1:1:1	120	3	22.9	76.6	0.5	—	
Co/CB	1:1:1	120	3	63.8	36.2	0	—	
Co/SiO_2	1:1:1	120	3	63.5	36.2	0.3	—	
Co/CNTs	1:1:1	120	3	59.9	40	0.1	—	
Co/B-zeolite	1:1:1	120	3	31	67.2	1.8	—	
Rh/POL–PPh_3	1:1:1	120	1	—	—	—	96.1	46
Rh/POL–PPh_3	1:1:1	120	1	—	—	—	96.4	
$Rh(CO)_2(acac)/SiO_2$	1:1:1	120	1	—	—	—	99.9	
$HRh(CO)(PPh_3)_3/SiO_2$	1:1:1	120	1	—	—	—	99.5	
Rh/POL–PPh_3	1:1:1	120	1	—	—	—	95.4	
Rh/POL–PPh_3	1:1:1	120	1	—	—	—	94.6	
Rh/SiO_2	1:1:1	115	0.1	62	38	0	—	47
	1:1:1	115	0.1	34	65	<1	—	
V-Rh/SiO_2	1:1:1	115	0.1	36	7	57	—	
	1:1:1	115	0.1	31	8	61	—	
Mo–Rh/SiO_2	1:1:1	115	0.1	55	4	41	—	
	1:1:1	115	0.1	68	1	31	—	
Zn–Rh/SiO_2	1:1:1	115	0.1	18	50	32	—	
	1:1:1	115	0.1	45	33	22	—	
Fe–Rh/SiO_2	1:1:1	115	0.1	44	35	21	—	
	1:1:1	115	0.1	60	23	17	—	
Co–Rh/SiO_2	1:1:1	115	0.1	42	49	9	—	
Zr–Rh/SiO_2	1:1:1	115	0.1	36	36	28	—	
Nb–Rh/SiO_2	1:1:1	115	0.1	54	36	10	—	
5%Rh/Al_2O_3	1:1:1	205	2	—	28	7	—	35
1%Co/Al_2O_3	1:1:1	205	2	—	18	6	—	
0.5%Rh–0.5%Co/Al_2O_3	1:1:1	205	2	—	19.6	9.7	—	

Table 13.1 (*Continued*)

Catalyst	Reaction condition: Feed ratio (C_2H_4 : CO : H_2)	*T* (°C)	*P* (MPa)	Products selectivity (%): C_nH_{2n+2}	R–CHO[a]	R–OH[b]	OXY[c]	Ref.
Rh/SiO_2 Mo–Rh/SiO_2	1:1:1	115	—	62	38	0	—	48
Mo–Rh/SiO_2	1:1:1	115	—	50	47	3	—	
RhCl(PPh_3)$_3$	1:1:1	185	0.1	—	0.3	—	—	49
RhCl(PPh_3)$_3$	1:1:1	185	0.3	—	0.76	—	—	
RhCl(CO)(PPh_3)$_2$	1:1:1	185	0.1	—	0.16	—	—	
RhCl(CO)(PPh_3)$_2$	1:1:1	185	0.3	—	0.38	—	—	
Decomposed RhCl(PPh_3)$_3$	1:1:1	185	0.3	—	0.44	—	—	
7-Rh/C(C)	1:1:1	110	0.5	24	58	—	—	50
7-Rh/C©	1:1:1	175	—	45	50	—	—	
6-Rh/SiO_2	1:1:1	110	—	35	61	—	—	
6-Rh/SiO_2	1:1:1	175	—	63	35	—	—	
2-Rh–PPh_3/SiO_2	1:1:1	110	—	5.3	90	—	—	
2-Rh–PPh_3/SiO_2	1:1:1	175	—	0	0	—	—	
8-Fibrecat	1:1:1	75	—	2.4	96	—	—	
8-Fibrecat	1:1:1	110	—	6.5	88	—	—	
8-Fibrecat	1:1:1	135	—	4.8	93	—	—	
3-Fibrecat	1:1:1	100	—	4.1	96	—	—	
3-Fibrecat	1:1:1	110	—	3.2	96	—	—	
3-FibrecatTM (Acet)	1:1:1	100	—	4.5	95	—	—	
4-FibrecatTM (Acet)	1:1:1	100	—	4.5	95	—	—	
6-FibrecatTM (Acet)	1:1:1	100	—	6.7	93	—	—	

[a]R–CHO refers to the selectivity of aldehyde products.
[b]R–OH refers to the selectivity of the alcohol products.
[c]OXY refers to the selectivity of the oxygenate products.

and process conditions.[27] The aldehydes can be manufactured stoichiometrically by adding carbon monoxide and hydrogen to the olefin in a single step with 100% atom efficiency.[28] However, the homogeneous process could not avoid its inherent pitfalls such as catalyst loss, difficult catalyst separation and recycling, high cost of rhodium-based catalyst and the contamination of the aldehydes products by metal species.[26] It is for these reasons that much effort has been made to develop a heterogeneously catalyzed hydroformylation process using solid catalysts.[28–30] One type of solid catalyst immobilizes the homogeneous metal complex catalysts on various organic polymers and inorganic materials by physical adsorption; hydrogen bonding or chemical anchoring such as SiO_2, Al_2O_3, MgO, ZnO clays; activated carbon and zeolites.[30] The other type of catalyst that is being investigated consists of supported metal catalysts prepared by decomposition or reducing a metal precursor (metal salt, oxide or complex) on a support (SiO_2, Al_2O_3, carbon, *etc.*).[31] As rhodium and cobalt are widely used in industrial processes, mono- or bimetallic cobalt and rhodium were studied in nanoparticles, clusters, and complexes.[3,14,26,32–38] Table 13.1 lists some heterogeneous catalysts reported in the literature.

13.2.3 Graphene-Supported Catalysts in FTS and Hydroformylation

Since its discovery in an experiment in 2004 by Novoselov *et al.*, graphene has become a well studied and used material.[39] Many researchers have investigated and reported on the attractive properties of this material and its use in different fields *e.g.* heterogeneous catalysis, electrochemical catalysis, as an adsorbent and sensors, *etc.*[40,41] Graphene has several unique properties such as an exceptional thermal conductivity of 5000 W m^{-1} K^{-1}, a high electron mobility of 250 000 $cm^2 V^{-1} s^{-1}$ at room temperature, attractive mechanical properties with Young's modulus of 1 TPa and a very high specific surface area of approximately 2600 $m^2 g^{-1}$. These unique properties make graphene an attractive option as catalyst support.[42]

Graphene-supported iron and cobalt catalysts have been used in FTS. One of the catalysts investigated for FTS, consisting of nanoparticles of iron oxide deposited on pyrolytic graphene oxide, has been used for controlling the selectivity of iron-based catalyst by, for instance, selective tuning, particularly for the production of C_{5+} hydrocarbons.[43] Qiao *et al.*[44] demonstrated a one-pot hydrothermal hydrolysis reduction pathway for making a Fe/rGO nanohybrid.[44] The γ-Fe_2O_3 nanoparticles of sub-3 nm in size that were produced were highly dispersed on the reduced graphene oxide (RGO). The catalyst showed excellent catalytic stability and thermal stability as the nanoparticles on the support remained highly dispersed even after a 16 h reduction at 450 °C followed by a 120 h of FTS reaction at 270 °C.[44] On the other hand, when cobalt was supported on graphene, the dispersion of cobalt nanoparticles increased, resulting in an observed decrease in the average particle size.[44] Comparatively, graphene-supported cobalt catalysts perform better (in terms of activity and stability) than the cobalt supported on carbon nanotubes under the same loading and reaction conditions.[45] Graphene-supported worm-like ruthenium was used as a catalyst for a photocatalytic FTS at a temperature of 150 °C, a H_2 pressure of 2.0 MPa and a CO pressure of 1.0 MPa under visible light irradiation; a catalytic activity of 14.4 $mol_{CO}\,mol_{Ru}{}^{-1}\,h^{-1}$ was achieved.[46]

In recent years, graphene has been used as a support for hydroformylation catalysts due to its remarkable properties.[47–50] Tan *et al.*[51] reported that an organic ligand of PPh_3 was used to reduce graphene oxide and fabricate a PPh_3 functionalized Rh/RGO catalyst simultaneously for 1-olefin hydroformylation. This catalyst exhibited excellent catalytic activity. In a study by Loni *et al.*,[49] methylated graphene oxide was used as a support that resulted in the formation of rhodium nanoparticles in a size range of 2–3 nm on the catalyst surface. The prepared catalyst exhibited a high conversion as well as regioselectivity in the hydroformylation of various olefins in the "green" medium of supercritical CO_2.[49] An RGO-supported rhodium nanoparticles (Rh/RGO) catalyst was prepared *via* a one-pot liquid-phase reduction method and applied in the hydroformylation 1-hexene for the first time.[52] However, to the best of knowledge, graphene-supported cobalt-based catalysts or bimetallic catalysts have not been tested for the hydroformylation of any olefins. In the following section, a case study by the authors for the production of oxygenates *via* FTS and hydroformylation using RGO-supported cobalt and cobalt–rhodium bimetallic catalysts is discussed.

13.3 Case Study for Oxygenate Production from Syngas *via* FTS and Hydroformylation

13.3.1 Catalyst Preparation

Two reduced graphene-oxide-supported catalysts were used in this case study. One of the catalysts was 20% cobalt supported on a reduced graphene oxide (CG catalyst), and the other was a 5%Rh–20%cobalt supported on a reduced graphene oxide (RCG catalyst). More detailed information on the catalyst preparation procedure has been reported in our published work.[53]

13.3.2 Reactor System

A fixed-bed reactor (FBR) was used for the simultaneous heterogeneous FTS ethene hydroformylation (EH) reactions to produce aldehydes and alcohols. As depicted in Figure 13.1, the feed gas flowed downward through the catalyst bed in a profile approximating that of plug flow.

Identical FBRs, as shown in Figure 13.1, were used in all the experiments. The reactors were comprised of stainless steel tubes with a tube length of 380 mm and an internal diameter of 8 mm. The heating jacket with thermocouple was used to adjust and maintain the operating temperature.

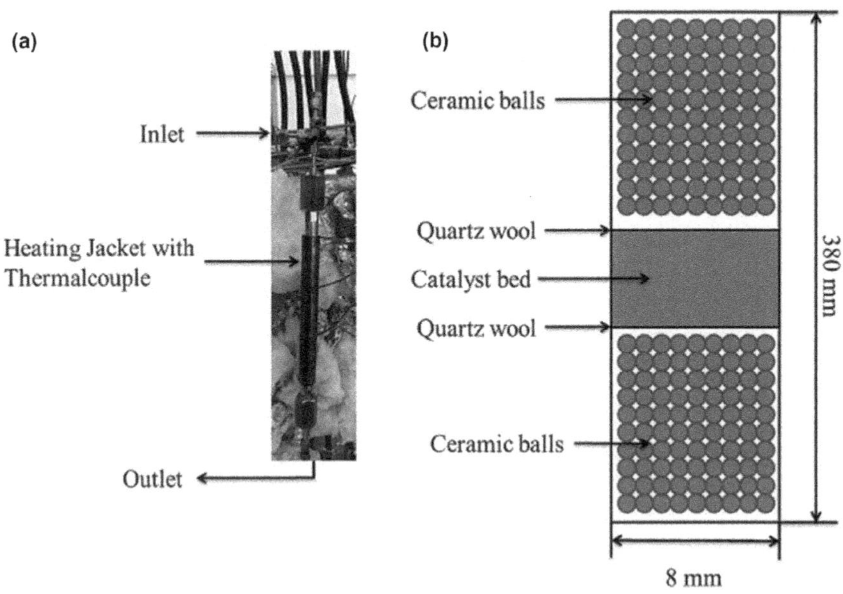

Figure 13.1 FBR: (a) digital portrait of the FBR and (b) sketch portrait of catalyst loading.

A layer of ceramic balls at the top of the catalyst bed was used for preheating the feed gas, while the layer of ceramic balls at the bottom of the reactor was used to support the catalyst bed. The plug of quartz wool at the top and the bottom of the catalyst bed served to keep the catalyst in place.

13.3.3 Product Analysis

The analytical equipment measured the concentrations of propionaldehyde and propanol, hydrocarbons, the components of the feed gas and the internal standard gas N_2. The tail gas was analyzed every 4 h using an online Agliant 7890B GC. An FID detector was used to analyze propionaldehyde, propanol and short-chain hydrocarbons, while two TCD detectors were used to analyze for N_2, H_2, CO, CO_2.

The tail gases from the FBR were sent to the GC sampling loops through heated and insulated lines (180 °C). The gases entered the GC through three multiple sampling valves maintained at a temperature of 150 °C. The temperature of the TCD and FID detectors was kept at 250 °C. The scheme for sampling flow for the online GC is presented in Figure 13.2. Information related to the columns, carrier gases and the oven temperature is summarized in Table 13.2.

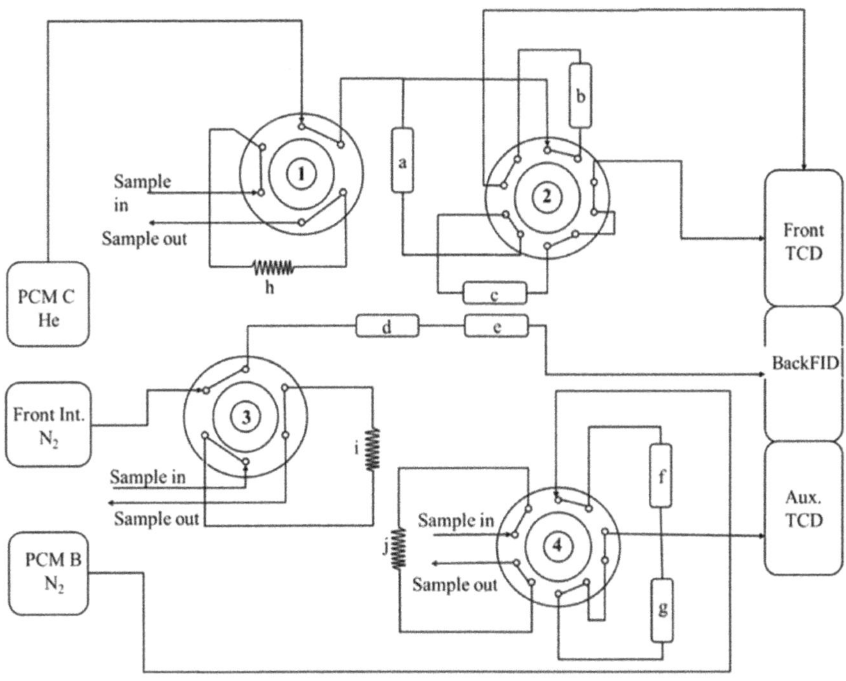

Figure 13.2 Sampling flow scheme for the online GC: (a) delay part; (b and f) mole sieve 13× column; (c) Heyasep Q column; (d) splitter; (e) LTM system with a CP-Sil 5 CB column; (g) plot Q column; (h–j) sampling loops.

Table 13.2 Parameters of an online GC.

Online GC	Agilent 7890B
Back detector	FID2 B, $T=250$ °C
Column 1	CP-SIL 5 CB (25 m×150 μm×2 μm)
Sample valve temperature	150 °C
Flame gas	Air with a flow rate of 40 ml(NTP) min^{-1} and UHP H_2 with a flow rate of 400 ml(NTP) min^{-1}
Carrier gas	UHP Ar with a flow rate of 25 ml(NTP) min^{-1}
Oven temperature program	35 °C for 3 min then 10 °C min^{-1} to 250 °C for 60 min
Product analysis	Hydrocarbons, propionaldehyde and propanol
Front detector	TCD1 A, $T=250$ °C
Column 2	HAYESEP Q 100/120 mesh, 1 m, 1 mm ID
Column 3	13× mole sieve 80/100 mesh, 1 m, 1 mm ID
Sample valve temperature	150 °C
Carrier gas	UHP He with a flow rate of 20 ml(NTP) min^{-1}
Oven temperature program	50 °C for 5 min, then 10 °C min^{-1} to 80 °C for 0 min, then 25 °C min^{-1} to 200 °C for 25 min
Product analysis	N_2, CO, CH_4, CO_2
Aux detector	TCD3 C, $T=250$ °C
Column 3	13× mole sieve 80/100 mesh, 1 m, 1 mm ID
Column 4	Plot Q
Sample valve temperature	150 °C
Carrier gas	UHP Ar with a flow rate of 20 ml(NTP) min^{-1}
Oven temperature program	50 °C for 5 min, then 10 °C min^{-1} to 80 °C for 0 min, then 25 °C min^{-1} to 200 °C for 25 min
Product analysis	H_2

13.3.4 Experimental Setup

The EH and FTS experiments were conducted in tubular fixed-bed reactors using different catalysts. The equipment for reaction and product analysis is described next (Figure 13.3).

The feed gas space velocity was controlled and monitored using mass flow controllers (Brooks Instrument 5850). The catalyst was loaded into the reactor, and the top and bottom were filled with ceramic balls (the structure of the catalyst bed was presented in detail in Section 13.3.2). The feed gas was preheated by the ceramic balls packed on the top of the catalyst bed. The products and unreacted reactant gases then passed out at the bottom of the reactor tube. For the EH reaction, the tail gases passed through a tail gas line to a back-pressure regulator (Swagelok), which was kept at a temperature of 180 °C with heating tape, while during the FTS reaction, a tail gas line carried the tail gases to a hot trap (which maintained the 150 °C temperature and reactor pressure), followed by a cold trap (kept at room temperature and reactor pressure). After the junction where the two tail gas lines joined, there was a back-pressure regulator to adjust the pressure in the reactor. As the tail gases passed through the back-pressure regulator which reduced the pressure to atmospheric level. Thereafter, the tail gas was sent to the sampling loop (maintained at a temperature of 150 °C) of an online GC (Agilent

7890B), while the remainder of the reactor tail gas was sent to the vent. To prevent condensation of products, all the tail gas lines were heated and kept at 180 °C. Half a gram of catalyst was loaded into each reactor and reduced with H_2 at 330 °C, 1 bar and 30 mL min^{-1} for 20 h. Thereafter the reactions were conducted under several different conditions.

13.3.5 Results and Discussion

After activating the catalysts with H_2, the performance of the CG and RCG catalysts was evaluated at different reaction conditions. The optimal reaction conditions for both the FT and the hydroformylation reactions are compared next.

13.3.5.1 FTS Reaction over the CG Catalyst

Table 13.3 lists the FTS reaction results of the CG catalyst at different GHSV. The high selectivity to C_{1-5} alcohols demonstrated that the CG catalyst successfully catalyzed the conversion of syngas to oxygenates.

GHSV has a significant influence on the performance of the CG catalyst during the conversion of syngas to oxygenates. A relatively high GHSV increases the selectivity to alcohols and short hydrocarbons while decreasing the selectivity to CO_2, CH_4, and long-chain products.

13.3.5.2 EH Reaction over the CG Catalyst

To investigate the EH performance of the CG catalyst, ethene was co-fed with syngas, at a 1 : 1 ratio of ethene to CO in the feed. Table 13.4 lists the EH results of the CG catalyst. Although the conversion of CO, C_2H_4 and H_2 increases with temperature, the maximum selectivity of C_3 oxygenates was achieved at a lowest temperature of 200 °C. These results thus show that C_3 oxygenates, propanol and propionaldehyde can be produced by co-feeding ethene to syngas over the CG catalyst.

13.3.5.3 EH Reaction over the RCG Catalyst

To improve the performance of the CG catalyst for EH, rhodium-promoted CG catalyst (RCG) was prepared and tested for the EH reaction. Table 13.5 summarizes the EH results for the RCG catalyst. By comparing the results for feed gases $CO/C_2H_4/H_2 = 1/1/1$ and 2/2/1 given in Table 13.5, it can be seen that the selectivity of C_3 oxygenates increased slightly with decreasing H_2 ratio in the feed gas. However, as can be seen by comparing the results for feed gases $CO/C_2H_4/H_2 = 1/1/1$ and 2/1/1 in Table 13.5, increasing the partial pressure of CO increased the selectivity of the C_3 oxygenates fairly significantly. After that, an optimal ratio of the feed gas, $CO/C_2H_4/H_2 = 4/2/1$, was tested, and a maximum selectivity to C_3 oxygenates of 45.76% was obtained.

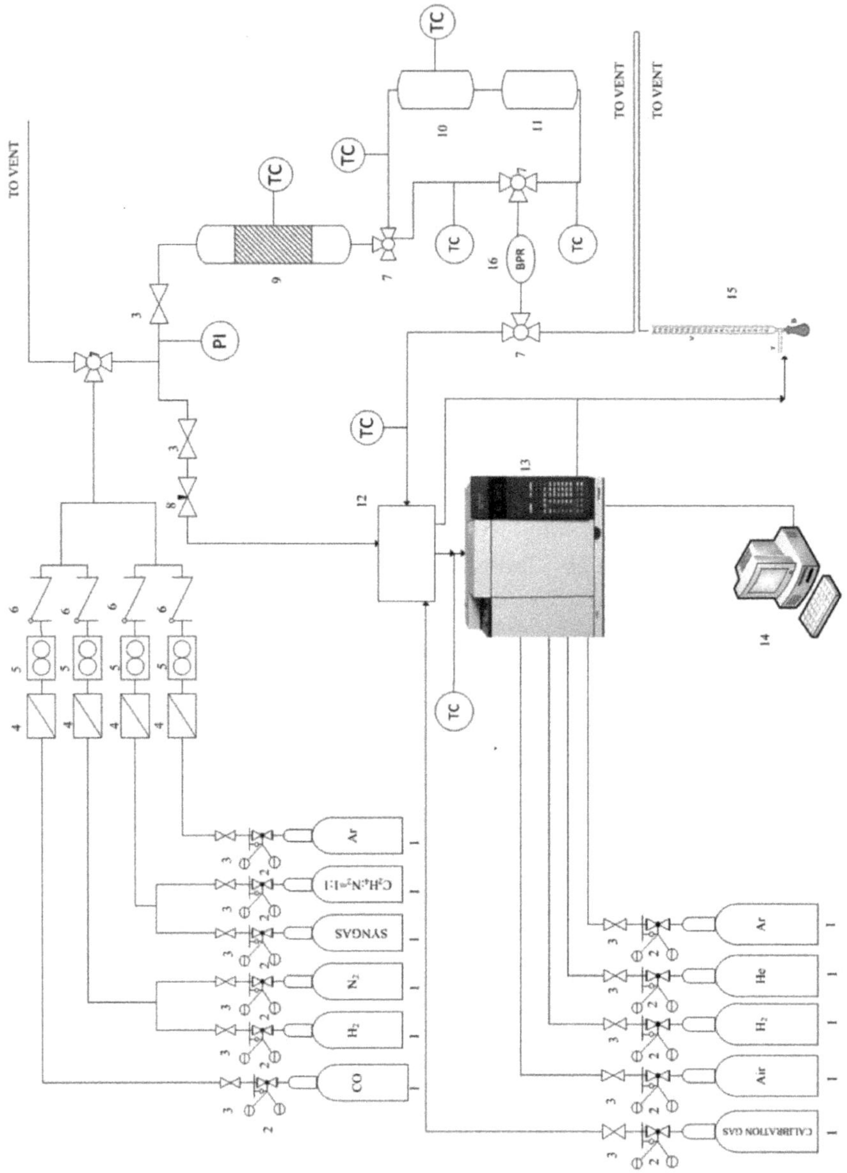

Figure 13.3 Experimental setup: (1) gas cylinders; (2) pressure regulators; (3) shutoff valves; (4) one-way valves; (5) mass flow controllers; (6) needle valves; (7) three-way valves; (8) needle valve; (9) fixed-bed reactor; (10) hot condensable product trap; (11) cold condensable product trap; (12) multivalve box; (13) online GC; (14) computer for data collection; (15) bubble meter; (16) back pressure regulator.

Table 13.3 FTS reaction results over a CG catalyst. Reaction conditions: 20 bar, 250 °C and a syngas of $CO:H_2=1:2$.

		Selectivity (%)				
$GHSV^{-1}$	CO conv. (%)	CO_2	CH_4	C_{2-4}	C_{1-5} alcohols	Others[a]
600	11.47	6.90	14.79	28.67	27.56	22.08
300	17.70	8.80	16.90	24.44	23.36	26.50

[a]The other products are higher alcohols and hydrocarbons.

Table 13.4 Catalytic results of CG for ethene hydroformylation. Reaction conditions: 20 bar, $CO:C_2H_4:H_2=1:1:2$ and $GHSV=2400\ h^{-1}$.

	Conversion (%)			Selectivity based on converted C_2H_4 (%)				Selectivity based on converted C (%)			
T (°C)	CO	C_2H_4	H_2	C_2H_6	C_2H_5CHO	C_3H_7OH	C_3O^a	C_2H_6	C_2H_5CHO	C_3H_7OH	C_3O^a
200	12.79	31.72	22.69	56.07	16.79	12.71	29.49	47.79	21.46	16.24	37.70
210	17.35	51.15	39.83	69.45	9.77	11.35	21.12	60.61	12.79	14.85	27.65
250	22.53	95.31	74.35	75.85	0.71	6.17	6.88	68.03	0.95	8.30	9.25

[a]Total selectivity of ($C_2H_5CHO+C_3H_7OH$).

Table 13.5 Catalytic results of RCG for ethene hydroformylation. Reaction conditions: 20 bar, 250 °C and $GHSV=1200\ h^{-1}$.

	Conversion (%)			Selectivity based on converted C_2H_4 (%)				Selectivity based on converted C (%)			
$CO:C_2H_4:H_2$	CO	C_2H_4	H_2	C_2H_6	C_2H_5CHO	C_3H_7OH	C_3O^a	C_2H_6	C_2H_5CHO	C_3H_7OH	C_3O^a
1:1:1	8.19	16.94	23.80	57.07	5.26	23.39	28.65	45.44	6.28	27.93	34.22
2:2:1	8.31	17.21	44.02	49.95	11.87	21.98	33.85	39.94	14.23	26.36	40.59
2:1:1	7.32	31.90	39.32	43.33	11.90	23.86	35.76	35.01	14.42	28.92	43.34
4:2:1	7.16	17.71	25.87	47.77	23.32	28.92	52.24	27.90	20.43	25.34	45.76

[a]Total selectivity of ($C_2H_5CHO+C_3H_7OH$).

13.4 Conclusions

The performance of catalysts for the production of oxygenates *via* the FTS route have been reviewed. The process, equipment and experimental setup have been demonstrated at laboratory scale. New Co and Rh/Co catalysts supported by RGO were prepared and tested for both the FTS and EH reactions. It was shown that mixed alcohols can be produced *via* the FTS process, while propionaldehyde can be produced from syngas *via* the EH route. A higher flow rate, lower temperature and higher partial pressure of CO enhance the conversion of syngas to oxygenates. The experimental results suggest that RGO is a potential support for the heterogeneous production of oxygenates from syngas *via* FTS.

References

1. D. Damma and P. G. Smirniotis, *Catal. Sci. Technol.*, 2021, **11**, 5412–5431.
2. R. G. Grim, A. T. To, C. A. Farberow, J. E. Hensley, D. A. Ruddy and J. A. Schaidle, *ACS Catal.*, 2019, **9**, 4145–4172.

3. R. Franke, D. Selent and A. Borner, *Chem. Rev.*, 2012, **112**, 5675–5732.
4. Y. Li, Z. Zhao, W. Lu, M. Jiang, C. Li, M. Zhao, L. Gong, S. Wang, L. Guo, Y. Lyu, L. Yan, H. Zhu and Y. Ding, *ACS Catal.*, 2021, **11**, 14791–14802.
5. J. J. Spivey and A. Egbebi, *R. Soc. Chem.*, 2007, **36**, 1514–1528.
6. M. R. Takehiko Matsuzaki, T.-A. Hanaoka, K. Takeuchi, H. Arakawa, Y. Sugi, K. Wei and T. Dong, *Catal. Today*, 1997, **36**, 311–324.
7. T.-W. Kim, F. Kleitz, J. Won Jun, H.-J. Chae and C.-U. Kim, *J. Ind. Eng. Chem.*, 2017, **51**, 196–205.
8. S. C. Antonio Molino, V. Larocca and D. Musmarra, *Energies*, 2018, **11**, 1–31.
9. A. Kumar, D. D. Jones and M. A. Hanna, *Energies*, 2009, **2**, 556–581.
10. J. J. Spivey and A. Egbebi, *Chem. Soc. Rev.*, 2007, **36**, 1514.
11. K. Fang, D. Li, M. Lin, M. Xiang, W. Wei and Y. Sun, *Catal. Today*, 2009, **147**, 133–138.
12. H. T. Luk, C. Mondelli, C. Ferre, J. A. Stewart and J. Pe, *Chem. Soc. Rev.*, 2016, **46**, 1358–1426.
13. I. Edition, T. E. Kron, O. G. Karchevskaya, S. N. Rush, D. V Marochkin, G. A. Korneeva, M. N. Mikhailov, I. M. Bedrina, A. E. Bessudnov and D. Centre, *Russ. Chem. Bull.*, 2019, **68**, 1662–1668.
14. J. Y. Kim, J. H. Park, O. S. Jung, Y. K. Chung and K. H. Park, *Catal. Lett.*, 2009, **128**, 483–486.
15. C. Hydroformylation, S. Olefins, S. Hanf, L. A. Rupflin and R. Gläser, *Catal.*, 2020, 510.
16. M. Lenarda, L. Storaro and R. Ganzerla, *J. Mol. Catal. A: Chem.*, 1996, **111**, 203–237.
17. A. Fukuoka, T. Kimura, N. Kosugi, H. Kuroda, Y. Minai, Y. Sakai, T. Tominaga and M. Ichikawa, *J. Catal.*, 1990, **126**, 434–450.
18. K. Xiao, Z. Bao, X. Qi, X. Wang, L. Zhong, K. Fang, M. Lin and Y. Sun, *Chin. J. Catal.*, 2013, **34**, 116–129.
19. Y. Liu, K. Murata, M. Inaba, I. Takahara and K. Okabe, *Catal. Today*, 2011, **164**, 308–314.
20. Y. Liu, Z. Zuo, C. Li, X. Deng and W. Huang, *Appl. Surf. Sci.*, 2015, **356**, 124–127.
21. M. Gupta, M. L. Smith and J. J. Spivey, *ACS Catal.*, 2011, **1**, 641–656.
22. T. Lin, X. Qi, X. Wang, L. Xia, C. Wang, F. Yu, H. Wang, S. Li, L. Zhong and Y. Sun, *Angew. Chem., Int. Ed.*, 2019, **58**, 4627–4631.
23. J. He and W. Zhang, *J. Zhejiang Univ., Sci., A*, 2008, **9**, 714–719.
24. Q. Yang, A. Cao, N. Kang, H. Ning, J. Wang, Z.-T. Liu and Y. Liu, *Ind. Eng. Chem. Res.*, 2017, **56**, 2889–2898.
25. M. Konarova, F. Tang, J. Chen, G. Wang, V. Rudolph and J. Beltramini, *ChemCatChem*, 2014, **6**, 2394–2402.
26. K. Wiese and D. Obst, *Catalytic Carbonylation Reactions*, 2006, vol. 18, pp. 1–33.
27. H.-W. Bohnen and B. Cornils, *Adv. Catal.*, 2002, **17**, 1–64.
28. C. Li, W. Wang, L. Yan and Y. Ding, *Front. Chem. Sci. Eng.*, 2018, **12**, 113–123.

29. X. Wang, *J. Flow Chem.*, 2015, **5**, 125–132.
30. Â. C. B. Neves, M. J. F. Calvete, T. M. V. D. Pinho, E. Melo and M. M. Pereira, *Eur. J. Org. Chem.*, 2012, 6309–6320.
31. V. A. Likholobov and B. L. Moroz, *Handbook of Heterogeneous Catalysis*, 2008.
32. S. I. Reut, G. L. Kamalov and G. I. Golodets, *React. Kinet. Catal. Lett.*, 1991, **44**, 191–195.
33. L. Wang, W. Zhang, S. Wang, Z. Gao, Z. Luo, X. Wang, R. Zeng, A. Li, H. Li, M. Wang, X. Zheng, J. Zhu, W. Zhang, C. Ma, R. Si and J. Zeng, *Nat. Commun.*, 2016, 7, 1–8.
34. X. Qiu, N. Tsubaki, S. Sun and K. Fujimoto, *Catal. Commun.*, 2001, **2**(2), 75–80.
35. N. Navidi, J. W. Thybaut and G. B. Marin, *CHISA 2012–20th International Congress of Chemical and Process Engineering and PRES 2012–15th Conference PRES*, 2012, **469**, 357–366.
36. L. Huang, Y. Xu, G. Piao and A. Liu, *Catal. Lett.*, 1994, **23**, 87–95.
37. M. Marchionna and G. Longoni, *J. Mol. Catal.*, 1986, **35**, 107–118.
38. L. Huang, Y. Xu, G. Piao and A. Liu, *Catal. Lett.*, 1994, **23**, 87–95.
39. K. S. Novoselov, A. K. Geim, S. V Morozov, D. Jiang, Y. Zhang, S. V Dubonos, I. V Grigorieva and A. A. Firsov, *Science*, 2004, **306**, 666–669.
40. J. Xu, R. S. Ruoff and H. Zhu, *Chem. Soc. Rev.*, 2017, **46**, 4417–4449.
41. Y. Cheng, Y. Fan, Y. Pei and M. Qiao, *Catal. Sci. Technol.*, 2015, **5**, 3903–3916.
42. V. Singh, D. Joung, L. Zhai and S. Das, *Prog. Mater. Sci.*, 2011, **56**, 1178–1271.
43. H. Zhao, Q. Zhu, Y. Gao, P. Zhai and D. Ma, *Appl. Catal., A*, 2013, **456**, 233–239.
44. B. Sun, Z. Jiang, D. Fang, K. Xu, Y. Pei, S. Yan, M. Qiao, K. Fan and B. Zong, *ChemCatChem*, 2013, **5**, 714–719.
45. A. Tavasoli, K. Sadagiani, F. Khorashe, A. A. Seifkordi, A. A. Rohani and A. Nakhaeipour, *Fuel Process. Technol.*, 2008, **89**, 491–498.
46. X. Guo, Z. Jiao, G. Jin and X.-Y. Guo, *ACS Catal.*, 2015, **5**, 3836–3840.
47. V. Singh, D. Joung, L. Zhai, S. Das, S. I. Khondaker and S. Seal, *Prog. Mater. Sci.*, 2011, **56**, 1178–1271.
48. M. Tan, Y. Ishikuro, Y. Hosoi, N. Yamane, P. Ai, P. Zhang, G. Yang, M. Wu, R. Yang and N. Tsubaki, *Chem. Eng. J.*, 2017, **330**, 863–869.
49. Y. V. Ioni, S. E. Lyubimov, V. A. Davankov and S. P. Gubin, *Russ. Chem. Bull.*, 2014, **63**, 2243–2249.
50. M. Tan, G. Yang, T. Wang, T. Vitidsant, J. Li, Q. Wei, P. Ai, M. Wu, J. Zheng and N. Tsubaki, *Catal. Sci. Technol.*, 2016, **6**, 1162–1172.
51. M. Tan, Y. Ishikuro, Y. Hosoi, N. Yamane, P. Ai, P. Zhang, G. Yang, M. Wu, R. Yang and N. Tsubaki, *Chem. Eng. J.*, 2017, **330**, 863–869.
52. M. Tan, G. Yang, T. Wang, T. Vitidsant, J. Li, Q. Wei, P. Ai, M. Wu, J. Zheng and N. Tsubaki, *Catal. Sci. Technol.*, 2016, **6**, 1162–1172.
53. J. Chang, Y. Zhang, X. Lu, Y. Yao, X. Liu and D. Hildebrandt, *Appl. Catal., A*, 2021, **614**, 118050.

CHAPTER 14

Biomass to Liquid Fuel via *Fischer–Tropsch (BTL-FT) Synthesis: Process Description and Economic Analysis*

CHIKE GEORGE OKOYE-CHINE,*[a,b] JOSHUA GORIMBO,[b] MAHLULI MOYO,[b] YALI YAO,[b] XINYING LIU,[b] DIANE HILDEBRANDT[b] AND JAMES ALISTAIR FOX[b]

[a] Department of Chemical & Life Science Engineering, Virginia Commonwealth University (VCU), 737 North 5th Street, Richmond, VA 23219, USA; [b] Institute for the Development of Energy for African Sustainability (IDEAS), University of South Africa (UNISA), cnr Christiaan De Wet and Pioneer Street, Florida, Johannesburg 1710, South Africa
*Email: george.okoyechine@gmail.com

14.1 Introduction

Fischer–Tropsch (FT) synthesis is the catalytic transformation of synthesis gas (a mixture of CO and H_2) derived from biomass, coal or natural gas into value-added hydrocarbons, which can be upgraded to fuel (*i.e.* gasoline, diesel and kerosene) and chemicals.[1–4] Presently, research on BTL-FT focuses on the possible ways in which biomass can be utilized to best advantage, given the intense concern about the devastating effects on the environment from using fossil fuel.[5,6] Hydrocarbon fuel derived from the FT synthesis process is described as clean fuel because of the very low sulfur

Catalysis Series No. 44
Chemicals and Fuels from Biomass *via* Fischer–Tropsch Synthesis: A Route to Sustainability
Edited by Joshua Gorimbo, Xinying Liu, Yali Yao and Diane Hildebrandt

Published by the Royal Society of Chemistry, www.rsc.org

content compared to hydrocarbons obtained from fossil fuel.[7] Ru-, Ni-, Fe- and Co-based catalysts are the commonly used catalysts for FT synthesis. However, Fe- and Co-based catalysts are used in commercial FT synthesis processes because they are active, readily available and cost-effective.[8] Fe-based catalysts seem to be the most suitable catalysts for BTL-FT production due to their high intrinsic activity for the water-gas-shift (WGS) reaction, which helps to adjust the low H_2/CO ratio that is typical with biomass-derived syngas.[9]

For more than a century, fossil fuels have been a major contributor to the global energy supply, in turn contributing significantly to global economic and social development.[10] However, there is concern that the overreliance on fossil fuels has resulted in elevated CO_2 emissions, which have in turn led to global warming (*i.e.* a steady rise in global temperature) and climate change.[11,12] This has necessitated research into sustainable and environmentally friendly energy vectors for industrial and domestic use.[13] The emission of greenhouse gases by the transportation industry could be reduced by 28–69% by converting to biomass-based fuels produced *via* BTL processes using straw, forest wood or short-rotation woody crops as the feedstock.[14] Thus, in a bid to reduce the current high dependency on fossil fuels, biomass could be converted into fuels that could replace current fossil-based transportation fuels.

According to the statistical analysis done by the U.S. Energy Information Administration (EIA), in 2020, biomass totaled about 4.9% of total U.S. primary energy consumption, which is equivalent to about 1.33 billion megawatts (MW). About 0.62 billion MW were produced from wood and wood-derived biomass, about 0.59 billion MW were obtained from biofuel (mainly ethanol), and 0.13 billion MW were derived from biomass in municipal waste.[15]

The different biofuels can be categorized into three types based on the source of biomass: first-generation biofuel, second-generation biofuel and third-generation biofuel. First-generation biofuel is produced from food sources, such as sugarcane and corn starch, which is not sustainable because of the potential challenge that production would put on food supplies. The economic aspects of first-generation biofuel depend on the feedstock type and the area where the feedstock has been cultivated. The price of food would be affected due to increased production of energy crops that potentially compete with food crops for land use.[16] Statistics show that approximately 700 million people suffer globally from a lack of healthy food; therefore, ensuring continuous access to food in order to meet this basic human need is very important.[17] Second-generation biofuel is derived from non-food-based biomass sources such as perennial energy crops and agricultural/municipal waste. Third-generation biofuel is obtained from microalgae.

Biomass-to-liquid (BTL) technology is considered a good approach to obtain environmentally friendly transportation fuel and reduce the high dependency on fossil-based transportation fuel.[18,19] BTL technology is a multistep process that involves the gasification of carbonaceous material

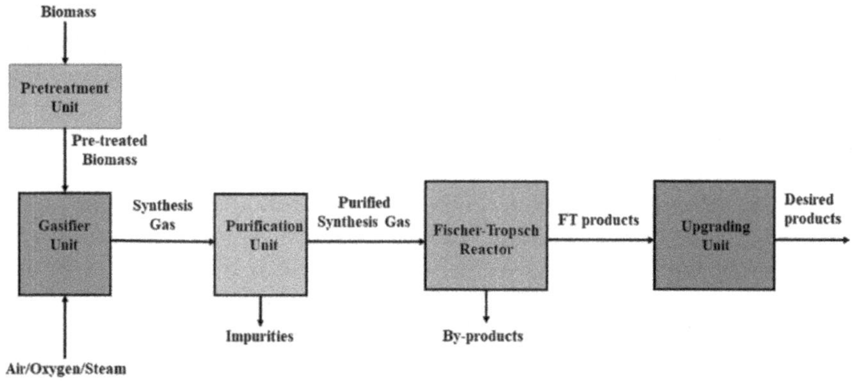

Figure 14.1 Schematic of BTL-FT.

(biomass) into synthesis gas, followed by conversion of the biomass-derived synthesis gas into synthetic fuel and chemicals (see Figure 14.1). The benefit of BTL is the ability to use virtually any type of biomass with little pre-treatment other than moisture control because the feedstock undergoes gasification in the first stage of the process. The synthesis gas produced then undergoes cleanup and conditioning to remove tar, particulates and gaseous contaminants in order to obtain a clean synthesis and to adjust the ratio of hydrogen and carbon monoxide to the desired ratio. The synthesis gas is catalytically converted into hydrocarbons *via* FT synthesis and upgraded to the required synthetic fuel or chemicals. Large-scale gas-to-liquid *via* FT synthesis (GTL-FT) and coal-to-liquid *via* FT synthesis (CTL-FT) processes have been commercialized for decades.

However, this is not the case with BTL-FT technology, with only a few BTL-FT plants having been built to date at the pilot and demonstration scale.[20,21] One significant challenge in scaling up the BTL-FT process is the economic aspect, which is a major motivation for developing novel processes that can use existing unit operations in a more effective process design.[22] This chapter provides an up-to-date overview of the economic feasibility of the BTL-FT process, by reviewing the methodologies used and the assessments of the various bioproducts, especially liquid fuel used in transportation.

14.2 Economic Analysis

14.2.1 Economic Estimates for Biofuel Technologies

The adequate supply and low cost of biomass resources are vital for the production of BTL biofuel and chemicals. One of the ways to evaluate the growth and development of BTL technology is *via* economic estimation of the capital investment and production costs.[23] This helps to uncover economically favorable BTL technology, as a technoeconomics assessment of BTL-FT technology is essential for its development and commercialization.

Swanson *et al.*[24] did a study on the coproduction of liquid transportation fuels and electricity *via* FT synthesis when using technology that will be available commercially within the next 5–8 years. The capital and production costs of two lignocellulosic BTL production plants (based on the thermochemical pathway of gasification) were compared by feeding 2000 tons per day (dry basis) of corn stover into two biorefineries. The first biorefinery setup was a low-temperature (870 °C) fluidized-bed gasifier, while the second was a high-temperature (1300 °C) entrained flow (EF) gasifier. In both instances, the gasifier process was followed by FT synthesis and hydroprocessing to produce naphtha and distillate liquid fractions.

Table 14.1 shows the economic analysis of the two biorefinery setups. It indicates that the capital investment for the high temperature setup and low temperature setup is approximately \$600 and \$500 million, respectively. Meanwhile, the fuel product values of the high-temperature and low-temperature setups are approximately \$4.30 and 4.80 per gallon (\$1.10 and 1.30 per liter) of gasoline gallon equivalent (GGE), respectively. This shows that despite the higher capital investment for the high temperature biorefinery setup, the fuel product value is lower. This was attributed to the increased fuel revenue from the higher fuel yield.[24] Furthermore, the lower fuel yield of the low-temperature setup influenced the fuel product value negatively, despite the lower total capital investment.

Figure 14.2 shows the average annual operating cost for the two setups over the lifetime of the refineries. The average return on investment (ROI), cost of feedstock and capital depreciation are the main expenses incurred.

Trippe *et al.*[25] investigated the gasification process step, which converts a biomass-derived intermediate called "slurry" into synthesis gas, followed by the production of fuel and chemicals. The technoeconomic evaluation was carried out for different process configurations of the gasification facility using the Aspen Plus process simulation software combined with investment and cost estimation based on literature data and vendor quotes. The varying process parameters used were operating pressure, gasification agent, syngas composition and feedstock composition.

It was observed that it is possible to produce synthesis gas at a cost of about \$25.99 Cent per Nm^3 based only on biomass feedstock. Production costs can be cut by almost half when co-gasification of 10% biomass-derived

Table 14.1 Economic analysis of the biorefinery setups. Reproduced from ref. 24 with permission from Elsevier, Copyright 2010.

Biorefinery setup	Total capital investment (\$ MM)	Total installed cost (\$ MM)	Fuel yield (GGE per ton)	Annual fuel output (MMGGE per year)	Fuel product value (\$ per GGE)
High temperature	606	309	61.0	41.7	4.27
Low temperature	498	254	47.2	32.3	4.83

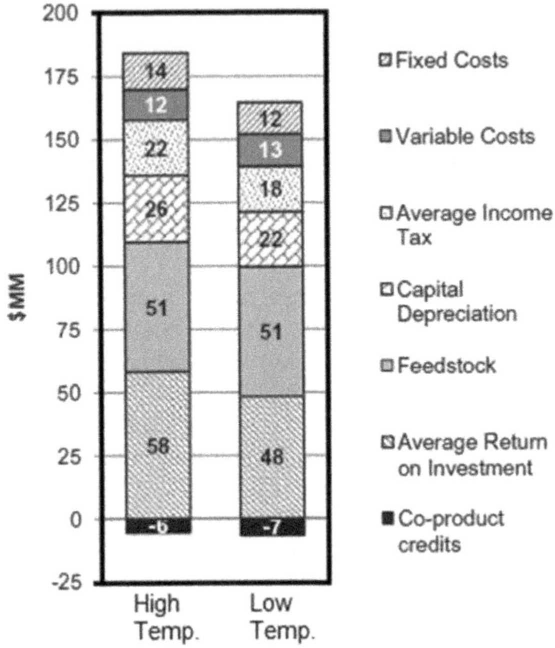

Figure 14.2 Annual operating costs ($ million per year) for the two biorefinery setups. Reproduced from ref. 24 with permission from Elsevier, Copyright 2010.

slurry and 90% coal are considered. Furthermore, a higher operating pressure used in the gasification process was identified as a major influencer of production cost reduction. The use of excess heat for electricity generation using either an Organic Rankine Cycle or Kalina process would allow about 4–7% savings in production costs. In addition, the sale of excess nitrogen produced as a by-product by the air separation unit would lead to savings in production costs of about 4–8%.[25] Furthermore, Kempegowda *et al.*[26] showed that the torrefaction of woodchips could help to reduce the overall cost of BTL production.

The production cost of BTL-FT biofuel and chemicals from a pressurized gasifier-based plant is higher than for an atmospheric gasifier-based plant.[27] This is primarily because the capital cost of pressurized equipment is considerably higher than that of atmospheric equipment. One reason for the high cost of pressurized equipment is the high cost of the feeding process due to the multifaceted feeding segment of the equipment. Nonetheless, despite the higher capital cost of the pressurized gasifier, it is more efficient than the atmospheric system, especially at high capacity.[27]

Dimitriou *et al.*[20] investigated the technoeconomic assessment of six BTL systems producing liquid fuels based on the use of a circulating fluidized-bed (CFB) and EF gasification of wood biomass. Thereafter, three fuel synthesis technologies were used for the production of liquid fuels. The fuel synthesis methods included FT synthesis, methanol synthesis followed by methanol-to-gasoline (MTG) and Topsoe integrated gasoline (TIGAS) synthesis. Table 14.2 shows that the overall energy efficiency of the assessed BTL systems ranges from 37.9 to 47.6% LHV, while Table 14.3 indicates that the total production cost ranges from \$20.20 to \$28.71 per GJ of produced liquid fuels.

FT synthesis appears to be the most promising fuel synthesis technology for the commercial production of liquid fuels *via* biomass gasification, considering that higher efficiency and lower production costs were achieved compared to MTG and TIGAS. This could be attributed to the extra synthesis steps required in MTG and TIGAS, which adds to the overall costs and leads to a lower mass yield and thus to lower overall energy conversion efficiency.

The BTL system with the highest fuel energy efficiency and the lowest production cost incorporates CFB gasification and FT synthesis. Even though a BTL system combined with CFB gasification technology has a higher capital cost than the corresponding BTL system that incorporates EF gasification technology, the CFB gasifier exhibits higher carbon conversion due to the inclusion of a tar cracker, which has a compensating effect that leads to a lower production cost. The resultant production cost for liquid fuel obtained from the BTL system that incorporates CFB gasification technology is 7.7%, 17.6% and 42.7% higher than the current market fuel price for FT synthesis, TIGAS and MTG, respectively.[20] Furthermore, Dimitriou *et al.*[20] did a sensitivity analysis showing that fuel production cost is mostly affected by a change in the fuel product output (*i.e.* conversion efficiency), capital investment and biomass feedstock cost. The BTL system that incorporates CFB gasification and FT synthesis was the most economical and would require a subsidy of about \$13.56 per ton of dried wood to match conventional fuel prices.

14.2.2 Subsidy Estimates to Ensure Biofuel Market Competitiveness

This chapter has suggested that, in order for biomass-derived liquid fuel to be economically competitive against existing fossil-derived liquid fuel, some form of subsidy or incentive is required for biomass-derived liquid fuel.

In this section, the data already provided in this chapter are used to provide a class 5 (screening and feasibility) estimate for the subsidy amount required to ensure the competitiveness of biomass-derived liquid fuel. This estimate is calculated by comparing the competitiveness of a biomass-to-liquid (BTL) fuel process to that of a similar coal-to-liquid (CTL) fuel process.

Table 14.2 provides data on the processing rates of various BTL processes. From that data, consider a BTL process that processes 84 000 kg h^{-1} of dry biomass and produces 16 500 kg h^{-1} of liquid fuel. Next, assume that there is also a CTL process that is identical to the previous BTL process in terms of processing capacity. Since this is only intended to be a Class 5 estimate, it is also assumed that the two processes have similar operating and maintenance costs.

Table 14.2 Mass and energy balances. Reproduced from ref. 20 with permission from Elsevier, Copyright 2018.

	Biomass				Liquid fuels				Yield and efficiency	
BTL system	Mass flow (wet), kg h^{-1}	Mass flow (dry), kg h^{-1}	LHV,[a] kJ kg^{-1}	Power input,[a] MW	Mass flow, kg h^{-1}	LHV, kJ kg^{-1}	Fuel energy output, GJ h^{-1}	Power output, MW	Fuel energy efficiency, %	Fuel mass yield, %
EF-FT	120 000	84 000	19 897	459.6	16 958	43 917	744.84	206.9	45	20.4
EF-MTG	120 000	84 000	19 897	459.6	14 810	42 308	626.76	174.1	37.9	17.8
EF-TIGAS	120 000	84 000	19 897	459.6	15 831	42 338	670.32	186.2	40.5	19
CFB-FT	120 000	84 000	19 897	459.6	17 930	43 917	787.32	218.7	47.6	21.6
CFB-MTG	120 000	84 000	19 897	459.6	16 980	42 307	718.2	199.5	43.4	20.4
CFB-TIGAS	120 000	84 000	19 897	459.6	17 541	42 312	742.32	206.2	44.9	21

[a]Values are given on a dry ash-free basis.

The market value for energy-crop biomass and coal as of the end of 2021/beginning of 2022 was found to be 113 US$ per ton of biomass and 149 US$ per ton for coal. Once the value of the feed materials and the production rates are determined, the revenue of the two processes can be calculated as the difference between the sale of the fuel and the cost of the feed.

Figure 14.3 shows the revenue generated by selling the liquid fuel produced, at the value of the *x*-axis fuel price, after the cost of the feed has been subtracted. It shows that if the fuel price drops below a certain level, neither the BTL nor the CTL process can be operated profitably, as the value of the feed material is greater than the value of the fuel produced. The BTL process appears to be more profitable than the CTL process, regardless of the fuel price. This difference is entirely due to the higher value of coal.

This result seems to contradict the findings of this chapter. However, such a simplistic calculation does not tell the whole story. Recall from Chapter 2, that biomass has one-tenth the energy density of coal. This means that the BTL process has to be larger than a similar CTL process, which comes with a higher capital cost.

Table 14.3 Production costs for liquid fuels of evaluated BTL systems. Reproduced from ref. 20 with permission from Elsevier, Copyright 2018.

BTL system	Capital $ per GJ	O&M $ per GJ	Biomass $ per GJ	Total $ per GJ
EF-FT	9.07	4.70	7.07	20.84
EF-MTG	13.37	6.93	8.42	28.72
EF-TIG	9.82	5.09	7.86	22.77
CFB-FT	8.90	4.61	6.69	20.20
CFB-MTG	12.41	6.43	7.33	26.17
CFB-TIG	9.54	4.94	7.11	21.59

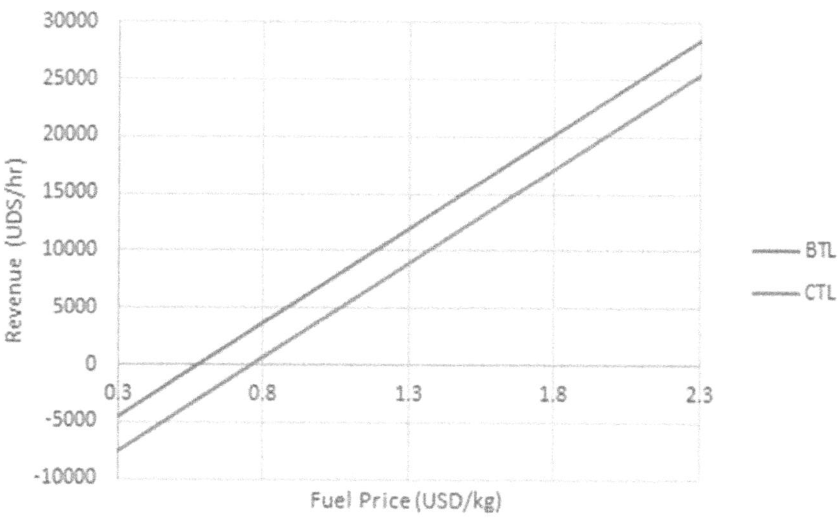

Figure 14.3 Revenue generated from similar capacity BTL and CTL processes.

Without any form of communition, a BTL process would need to be some 7.5 times larger than a CTL process to match the production capacity of the CTL process. However, both the BTL process and the CTL process make use of communition facilities. For BTL, this has the advantage of increasing the density of the feed material, and biomass communition technology often increases the density of biomass from 250 kg m^{-3} to 700 kg m^{-3}.[28] Taking this increase into account, the BTL process needs to be 2.8 times larger than a similar CTL process.

To estimate how much more capital will be needed to build a BTL process that is 2.8 times larger than the CTL process, the "sixth-tenth rule" of process costing is used:

$$\frac{C_{\text{BTL}}}{C_{\text{CTL}}}=\left(\frac{S_{\text{BTL}}}{S_{\text{CTL}}}\right)^{0.6}=\left(\frac{2.8S_{\text{CTL}}}{S_{\text{CTL}}}\right)^{0.6}=1.82 \tag{14.1}$$

where:

- C_{BTL} is the capital cost of the BTL process.
- S_{BTL} is the size of the BTL process.
- C_{CTL} is the capital cost of the CTL process.
- S_{BTL} is the size of the CTL process.

In order to construct a BTL process that is 2.8 times larger than its CTL counterpart, 1.8 times as much capital is required. A return can be calculated to determine whether the additional investment is worthwhile, and, while there is insufficient information to calculate a detailed ROI, it is possible to determine a return on capital (ROC).

$$R=\frac{I}{C}=\frac{\text{Income}-\text{Expenses}}{C} \tag{14.2}$$

where:

- R is the ROC.
- C is the capital invested.
- I is the revenue generated by the process.
- Income is the money generated from selling the liquid fuel produced by the process. It can also include other sources of income, such as a subsidy on biomass-derived fuels.
- Expenses are the money spent to purchase the raw material for the process.

This calculation contains numerous assumptions (similar operating costs, similar maintenance costs, no interest), which greatly simplify the calculation but also decrease its accuracy. However, the assumptions should be acceptable for the purposes of a comparison.

The ROC can be written for the BTL and CTL process separately, and the following equation can be derived:

$$\frac{R_{\mathrm{BTL}}}{R_{\mathrm{CTL}}} = \frac{I_{\mathrm{BTL}}}{1.82 I_{\mathrm{CTL}}} = \frac{\mathrm{Income}_{\mathrm{BTL}} + \mathrm{Subsidy} - \mathrm{Expenses}_{\mathrm{BTL}}}{1.82\left(\mathrm{Income}_{\mathrm{CTL}} - \mathrm{Expenses}_{\mathrm{CTL}}\right)} \tag{14.3}$$

where:

- R_{BTL} is the ROC for the BTL process;
- R_{CTL} is the ROC for the CTL process.
- I_{BTL} is the revenue of the BTL process.
- I_{CTL} is the revenue of the CTL process.
- $\mathrm{Income}_{\mathrm{BTL}}$ is the income generated by selling the liquid fuel produced by the BTL process.
- $\mathrm{Income}_{\mathrm{CTL}}$ is the income generated by selling the liquid fuel produced by the CTL process.
- Subsidy is the subsidy paid to make BTL competitive with CTL.
- $\mathrm{Expenses}_{\mathrm{BTL}}$ is the cost of purchasing biomass for the BTL process.
- $\mathrm{Expenses}_{\mathrm{CTL}}$ is the cost of purchasing coal for the CTL process.

In order for the BTL process to be considered competitive with the CTL process, the two processes should have a similar ROC. In other words, the ratio of the ROC should equal 1. At this point, the two processes have the same returns and can be considered equivalent in an economic sense. But how much of a subsidy is required to ensure the BTL process is competitive with the CTL process? This depends on the market value of the liquid fuel being produced.

Figure 14.4 shows that the subsidy required for biomass-derived fuel increases as the market value of the liquid fuel increases. The CTL process is more profitable when the fuel price is higher. When the fuel price falls below 1 US$ per kg, the BTL process becomes more profitable and starts to outperform the CTL process, and it is the coal-derived fuels that require a subsidy to remain competitive. If the fuel price falls too far, as shown in Figure 14.3, neither process is economically viable.

The market value of diesel fuel, at the time of writing this chapter, was about 1.2 US$ per kg. If the liquid fuel from the BTL process is sold as diesel, a subsidy of about 0.2 US$ per kg would be required for the BTL process to achieve a similar return to a similarly sized CTL process. If the market value of liquid fuel increases dramatically in the future, the required subsidy could become excessive and potentially reach half the value of the fuel itself.

This example has treated the subsidy payment as a direct income per kg of product produced. However, subsidies are not always dealt with in this way, as other models include tax breaks or direct funding of the capital cost. Choosing the optimal mode for improving the economic competitiveness of biomass-derived fuels is well beyond the scope of this chapter, as the point of this exercise was to show that biomass-derived fuel still requires some form of assistance to compete with fossil fuel refineries.

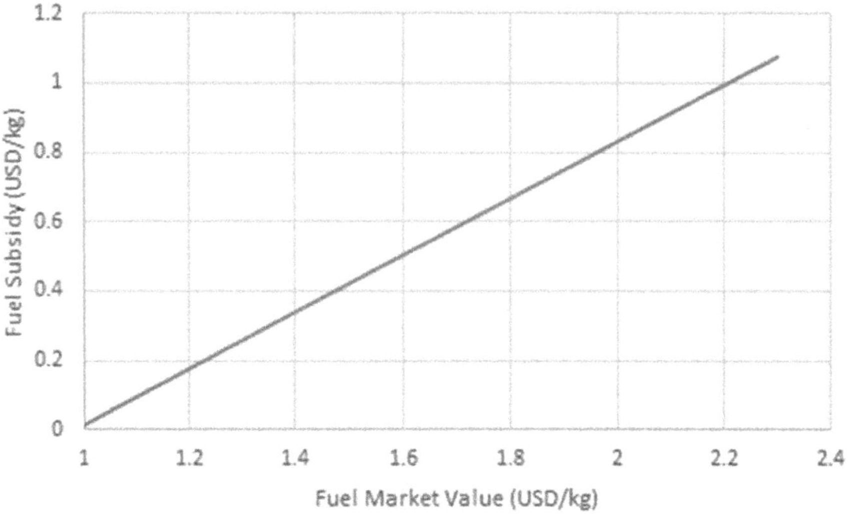

Figure 14.4 Fuel subsidy required to make BTL competitive with CTL, presented as a function of the market value of fuel.

14.3 Environmental Impact Assessment

The transportation industry is considered one of the main sources of environmental impact. As a result, increasing efforts are being made to develop alternative energy sources to address environmental concerns toward economic profitability and social benefits. This section focuses on the environmental impact evaluation of liquid biofuel production *via* biomass gasification and FT synthesis. The environmental impact assessment is a key aspect to be considered in evaluating the successful development of BTL biofuel production. From a life-cycle perspective, the use of biosyngas as a feedstock for FT synthesis does not guarantee a beneficial environmental performance. Life-cycle assessment (LCA) is a distinct method used for the evaluation of both the current and potential environmental impact associated with the production and use of a product.[29]

Bressanin *et al.*[30] did an environmental impact study on advanced liquid biofuel production *via* gasification of biomass and FT synthesis integrated into a conventional sugarcane distillery. It was reported that the BTL biofuel derived can be decreased by more than 76% in comparison to fossil fuel, depending on the methodological approach, feedstock and capital costs. The environmental impact assessment encompassed the whole biomass production system: preplanting operations, land for plantation, soil preparation, mechanized planting, plant cultivation, mechanized harvesting, and transporting of the biomass to the industrial site. Biomass material such as eucalyptus could be cultivated in marginal areas unsuitable for agriculture, which can help to minimize any extra pressure in terms of the need for productive land for biofuel production.[30] The emission of greenhouse gas from the whole process of BTL biofuel and chemical production and their use were considered during the environmental impact assessment. The results obtained by Bressanin *et al.*[30] show that integrating the biorefinery provided an 85–95% reduction in greenhouse gas emission compared to the fossil equivalent, which is higher than that expected from a typical sugarcane biorefinery.

An environmental impact analysis of a BTL process for the coproduction of green diesel and electricity was done by Im-orb *et al.*[31] Rice straw was used as the biomass feedstock, and the modeling was done using the Aspen Custom Modeler (ACM). A comparison of the performance of three BTL processes, *i.e.* with and without a tar removal unit based on steam reforming and autothermal (ATO) reforming, was carried out. The BTL process with steam reforming showed the highest amount of electricity and green diesel, followed by ATO reforming and then without any reforming. However, the BTL process without any reforming had the lowest energy consumption and the least environmental impact.

Iribarren *et al.*[29] carried out a LCA of the coproduction of fuel and electricity from a gasification-based biosyngas feedstock *via* FT synthesis coupled with a combined-cycle process. The environmental performance of the system was evaluated *via* process simulation in terms of abiotic depletion (ADP), global warming potential (GWP), ozone layer depletion (ODP), photochemical oxidant formation (POFP), land competition (LC), acidification (AP) and eutrophication (EP). In addition, the cumulative non-renewable energy demand (CED) of the FT system was measured and used in the estimation of the life-cycle energy balance of the FT system, which was positive. Biosyngas generation was identified as the main contributor of environmental impact.

The products of the FT system were electricity, diesel, gasoline and surplus hydrogen from a gasification-based biosyngas feedstock. The environmental profile of these FT products was calculated and compared with that of fossil diesel, rapeseed biodiesel, soybean biodiesel, fossil gasoline, corn bioethanol, steam-methane reforming hydrogen and the EU electrical grid. The environmental suitability of the FT bioproducts was observed to be dependent on the impact categories examined. In particular, considerably high greenhouse gas savings were observed for each of the FT bioproducts; however, FT bioproducts presented highly unfavorable results for LC because of high land requirements for biomass production (see Figure 14.5a and b). Generally, the evaluated FT products are favorable alternatives to the present energy products from an environmental life cycle perspective.

14.4 Conclusion

The development of BTL-FT fuel and chemical production has been gathering great interest in recent times due to the urgent search for sustainable and environmentally friendly energy sources that ensure industrial and societal benefits. The technoeconomics assessment of BTL-FT production of

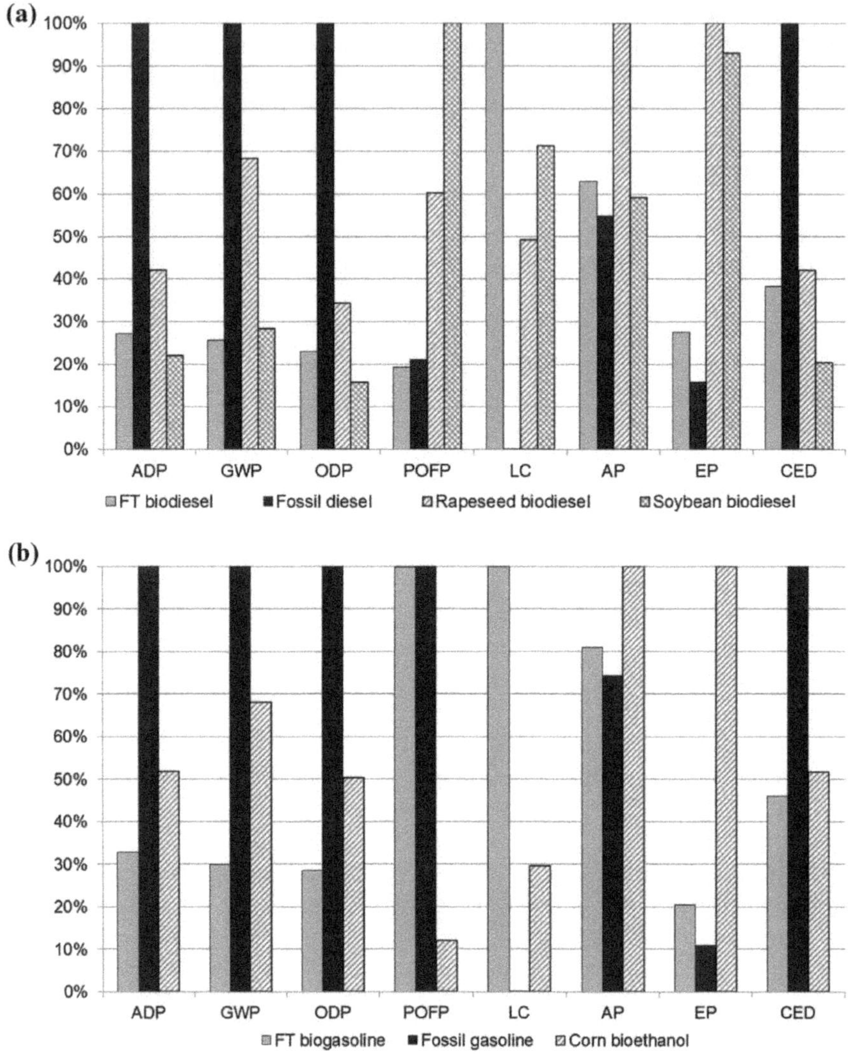

Figure 14.5 Relative comparison of the environmental impacts of (a) different diesel fuels and (b) different gasoline fuels. Reproduced from ref. 29 with permission from Elsevier, Copyright 2013.

biofuel has shown that the estimated production costs of the BTL-FT process do not compete with conventional fossil fuel refineries, since the biofuel production cost is approximately 25% higher than the current market price. At the current market price, the BTL process would require a subsidy of approximately 0.2 US$ per kg of fuel produced, in order to compete with similar coal-based technologies. Large-scale BTL-FT biofuel production may be feasible in the future through subsidies, lower biomass prices, growth of biomass commodity markets, a high petroleum price and favorable, stricter climate legislation on transportation fuels. The financial support needed for BTL competitiveness can become significant, but BTL technologies are unlikely to gain any traction without such support.

Furthermore, the gasification unit is central to the technical operation of liquid fuel production technology, but biomass gasification currently struggles in terms of economic viability. If biomass gasification can be made more economical, the viability of BTL will also increase. This is an avenue for future research and development.

While biomass-derived fuel is a more environmentally sustainable solution than fossil-derived fuel, clearly many technical and economic hurdles need to be overcome before BTL can be turned into a viable technology.

Abbreviations

ADP	Abiotic depletion
AP	Acidification
ACM	Aspen Custom Modeler
ATO	Autothermal
BTL	Biomass-to-liquid
BTL-FT	Biomass to liquid fuel *via* Fischer–Tropsch
CFB	Circulating fluidized bed
CTL	Coal-to-liquid
CTL-FT	Coal-to-liquid *via* FT synthesis
CED	Cumulative nonrenewable energy demand
EF	Entrained flow
EIA	Energy Information Administration
EP	Eutrophication
FT	Fischer–Tropsch
GGE	Gasoline equivalent
GWP	Global warming potential
LC	Land competition
LCA	Life-cycle assessment
MW	Megawatts
MTG	Methanol-to-gasoline
ODP	Ozone layer depletion
POFP	Photochemical oxidant formation
ROC	Return on capital
ROI	Return on investment
TIGAS	Topsoe integrated gasoline
WGS	Water-gas shift

References

1. C. Niu, M. Xia and C. Chen, *et al.*, Effect of process conditions on the product distribution of Fischer–Tropsch synthesis over an industrial

cobalt-based catalyst using a fixed-bed reactor, *Appl. Catal., A*, 2020, **601**(27), 117630.

2. C. G. Okoye-Chine, C. O. L. Mbuya, T. S. Ntelane, M. Moyo and D. Hildebrandt, The effect of silanol groups on the metal-support interactions in silica-supported cobalt Fischer–Tropsch catalysts. A temperature programmed surface reaction, *J. Catal.*, 2020, **381**, 121–129.
3. A. C. Ghogia, B. F. Machado and S. Cayez, *et al.*, Beyond confinement effects in Fischer–Tropsch Co/CNT catalysts, *J. Catal.*, 2021, **397**, 156–171.
4. C. G. Okoye-Chine, M. Moyo and D. Hildebrandt, The effect of hydrophobicity on SiO_2–supported Co catalysts in Fischer–Tropsch synthesis, *Fuel*, 2021, **296**, 120667.
5. M. Rothaemel, K. Firing Hanssen, E. A. Blekkan, D. Schanke and A. Holmen, The effect of water on cobalt Fischer–Tropsch catalysts studied by steady-state isotopic transient kinetic analysis (SSITKA), *Catal. Today*, 1997, **38**, 79–84.
6. A. Martínez, C. López, F. Márquez and I. Díaz, Fischer–Tropsch synthesis of hydrocarbons over mesoporous Co/SBA-15 catalysts: The influence of metal loading, cobalt precursor, and promoters, *J. Catal.*, 2003, **220**(2), 486–499.
7. A. Y. Khodakov, Fischer–Tropsch synthesis: Relations between structure of cobalt catalysts and their catalytic performance, *Catal. Today*, 2009, **144**(3–4), 251–257.
8. C. G. Okoye-chine, M. Moyo, X. Liu and D. Hildebrandt, A critical review of the impact of water on cobalt-based catalysts in Fischer–Tropsch synthesis, *Fuel Process. Technol.*, 2019, **192**, 105–129.
9. G. Jacobs, J. A. Chaney, P. M. Patterson, T. K. Das and B. H. Davis, Fischer–Tropsch synthesis: Study of the promotion of Re on the reduction property of Co/Al_2O_3 catalysts by in situ EXAFS/XANES of Co K and Re L_{III} edges and XPS, *Appl. Catal., A*, 2004, **264**(2), 203–212.
10. L. Yildiz, Fossil Fuels, in *Comprehensive Energy Systems*, 2018, **1**, 521–567.
11. R. Gao, C. Zhang and K. W. Jun, *et al.*, Green liquid fuel and synthetic natural gas production via CO2 hydrogenation combined with reverse water-gas-shift and Co-based Fischer–Tropsch synthesis, *J. CO2 Util.*, 2021, **51**, 101619.
12. F. J. Gutiérrez Ortiz, B. Alonso-Fariñas, F. J. Campanario and A. Kruse, Life cycle assessment of the Fischer–Tropsch biofuels production by supercritical water reforming of the bio-oil aqueous phase, *Energy*, 2020, **210**, 118648.
13. L. D. Gomez, C. G. Steele-King and S. J. McQueen-Mason, Sustainable liquid biofuels from biomass: The writing's on the walls, *New Phytol.*, 2008, **178**(3), 473–485.
14. N. Jungbluth, S. Büsser, R. Frischknecht and M. Tuchschmid, *Ökobilanz von Energieprodukten: Life Cycle Assessment of Biomass-to-Liquid Fuels*, 2008.
15. Biomass explained. U.S. Energy Information Administration (EIA).
16. M. Naqvi and J. Yan, First-Generation Biofuels. *andbook of Clean Energy Systems*2015 , pp. 1-18.

17. K. Kurowska, R. Marks-Bielska, S. Bielski, H. Kryszk and A. Jasinskas, Food security in the context of liquid biofuels production, *Energies*, 2020, **13**(23), 1–16.
18. S. S. Ail and S. Dasappa, Biomass to liquid transportation fuel via Fischer Tropsch synthesis - Technology review and current scenario, *Renewable Sustainable Energy Rev.*, 2016, **58**, 267–286.
19. N. Nkosi, E. Muzenda and J. Gorimbo, Developments in waste tyre thermochemical conversion processes: gasification , pyrolysis and liquefaction, *RSC Adv.*, 2021, **11**, 11844–11871.
20. I. Dimitriou, H. Goldingay and A. V. Bridgwater, Techno-economic and uncertainty analysis of biomass to liquid (BTL) systems for transport fuel production, *Renewable Sustainable Energy Rev.*, 2018, **88**, 160–175.
21. N. Dahmen and J. Sauer, Evaluation of techno-economic studies on the bioliq® process for synthetic fuels production from biomass, *Processes*, 2021, **9**(4), 1–23.
22. R. C. Baliban, J. A. Elia and C. A. Floudas, Biomass to liquid transportation fuels (BTL) systems: Process synthesis and global optimization framework, *Energy Environ. Sci.*, 2013, **6**(1), 267–287.
23. F. Trippe, M. Fröhling, F. Schultmann, R. Stahl and E. Henrich, Techno-economic analysis of fast pyrolysis as a process step within biomass-to-liquid fuel production, *Waste Biomass Valorization*, 2010, **1**(4), 415–430.
24. R. M. Swanson, A. Platon, J. A. Satrio and R. C. Brown, Techno-economic analysis of biomass-to-liquids production based on gasification, *Fuel*, 2010, **89**(Suppl. 1), S11–S19.
25. F. Trippe, M. Fröhling, F. Schultmann, R. Stahl and E. Henrich, Techno-economic assessment of gasification as a process step within biomass-to-liquid (BtL) fuel and chemicals production, *Fuel Process. Technol.*, 2011, **92**(11), 2169–2184.
26. R. S. Kempegowda, G. Del Alamo Serrano, B. M. Güell and K. Q. Tran, Techno-economic analysis of biomass to Fischer–Tropsch diesel production with and without CCS under Norwegian conditions, *Energy Procedia*, 2014, **61**(1876), 1248–1251.
27. M. Patel, X. Zhang and A. Kumar, Techno-economic and life cycle assessment on lignocellulosic biomass thermochemical conversion technologies: A review, *Renewable Sustainable Energy Rev.*, 2016, **53**, 1486–1499.
28. S. Clarke and F. Preto, Biomass Densification for Energy Production, Ministry of Agriculture, Food and Rural Affairs, 2011.
29. D. Iribarren, A. Susmozas and J. Dufour, Life-cycle assessment of Fischer–Tropsch products from biosyngas, *Renew Energy.*, 2013, **59**, 229–236.
30. J. M. Bressanin, B. C. Klein and M. F. Chagas, *et al.*, Techno-economic and environmental assessment of biomass gasification and Fischer–Tropsch synthesis integrated to sugarcane biorefineries, *Energies*, 2020, **13**, 4576.
31. K. Im-orb and A. Arpornwichanop, Techno-environmental analysis of the biomass gasification and Fischer–Tropsch integrated process for the co-production of bio-fuel and power, *Energy*, 2016, **112**, 121–132.

CHAPTER 15

Future Aspects of BTL-FTS Processes

JOSHUA GORIMBO,*[a] GRATITUDE CHARIS,[b,c] YUSHENG ZHANG,[a] YOHAN RICHARDSON[d] AND GWIRANAI DANHA[b]

[a] Institute for the Development of Energy for African Sustainability (IDEAS) Research Unit, University of South Africa (UNISA), Florida Campus, Private Bag X6, Johannesburg 1710, South Africa; [b] Botswana International University of Science and Technology, Plot 10071, Boseja Ward, Private Bag 16 Palapye, Botswana; [c] Manicaland University of Applied Sciences, Stair Guthrie Road, Mutare, Zimbabwe; [d] Laboratoire Energies Renouvelables et Efficacité Energétique (LabEREE), Institut International d'Ingénierie de l'Eau et de l'Environnement (2iE), 01 BP 594 Ouagadougou 01, Burkina Faso

*Email: joshuagorimbo@gmail.com

15.1 Introduction

Increased human intervention has shifted the world out of a relatively stable Holocene period into a new geological era, often termed the "Anthropocene".[1] The reality of this current epoch (the Anthropocene) is that, through various activities including industrialization, humans have significantly altered the chemical composition of the atmosphere. This alteration has given rise to a number of questions and challenges regarding climate change and global warming. It has greatly impacted human and natural systems, resulting in increased droughts, declining air quality, floods and

Catalysis Series No. 44
Chemicals and Fuels from Biomass *via* Fischer–Tropsch Synthesis: A Route to Sustainability
Edited by Joshua Gorimbo, Xinying Liu, Yali Yao and Diane Hildebrandt

Published by the Royal Society of Chemistry, www.rsc.org

the emergence of other types of extreme weather phenomena, such as a rise in the sea level and biodiversity loss.[2,3] These extreme weather events are causing risks to various populations, with direct and indirect health implications, including heat-related illnesses, vector and waterborne diseases, environmental poisoning, cardiovascular and respiratory diseases caused by bad air quality and mental health disturbances.[4,5]

Recent measurements of air (see Figure 15.1) trapped in ice core samples indicate the upward trajectory of carbon dioxide from a preindustrial CO_2 level of about 260–270 ppm[7] to 416.34 ppm as of March 2021.[6] An understanding of the history of variations in atmospheric CO_2 concentration is crucial for a number of reasons. In order to control future CO_2 levels, anthropogenic activities, especially fossil fuel combustion that transfers carbon from rocks and rock formations to the atmosphere, need to be reduced. Biomass waste fractions from agriculture, forestry and municipal solid waste (MSW) also release nitrogen oxides and methane into the atmosphere, which are strong greenhouse gases (GHGs). The International Energy Outlook 2019 report suggested that global CO_2 emissions from the energy sector would increase at a rate of 0.6% per annum from 2018 to 2050.[8]

A direct consequence of these human-induced GHG emissions is the apparent warming of the Earth's atmosphere over the last half century. For instance, the average rate of global annual temperature increase since 1981 (+0.18 °C per decade) (was more than double the average rate for the period 1880–1980 a (0.08 °C per decade).[9] By May 2021, the global surface temperature (GMST) was 0.81 °C above the 20th-century average of 14.8 °C.[10]

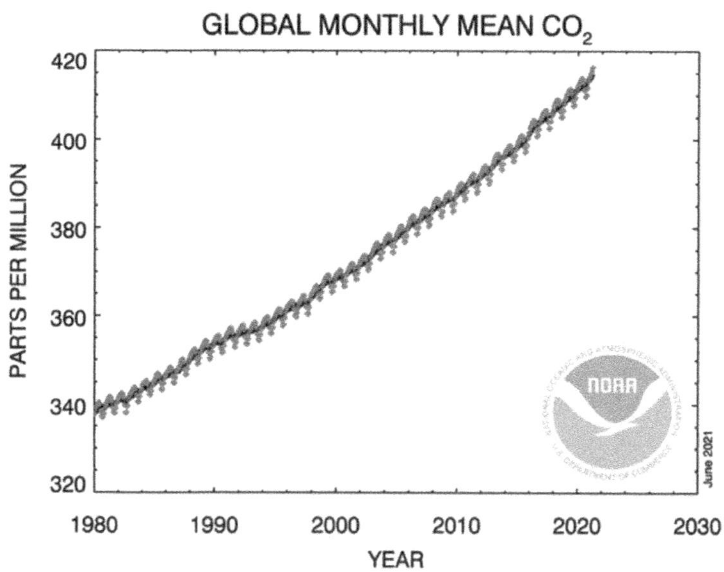

Figure 15.1 CO_2 from the atmosphere, measured at the National Oceanic and Atmospheric Administration at Mauna Loa, Hawaii. Reproduced from ref. 6 with permission from NOAA Global Monitoring Laboratory.

Global CO_2 emissions declined by 5.8% in 2020, or almost 2 Gt CO_2 due to the COVID-19 pandemic, although global energy-related CO_2 emissions are projected to increase significantly from 2021 onward.[11] The demand for coal alone is anticipated to increase by 60% more than the summative demand for the rest of renewables, resulting in a rise in emissions of ~5%, or 1500 Mt. This would reverse 80% of the drop in emissions seen in 2020 and would effectively result in a 1.2% (or 400 Mt) drop below 2019 emissions levels.[11]

Human activities in the form of industrialization have been estimated to contribute an approximate temperature rise of 1.0 °C of global warming above preindustrial levels. The Intergovernmental Panel on Climate Change (IPCC) forecasted that global warming was likely to jump up by 1.5 °C between 2030 and 2052 if atmospheric CO_2 continued to increase at current rates.[12] The "preindustrial period" refers to the multicentury period before the onset of large-scale industrial activity, that is, around 1850–1900.[12] The Paris Agreement, premised on the United Nations Framework Convention on Climate Change, aims to maintain the difference between the present global average temperature and the preindustrial levels at 1.5 °C above the latter to reduce the risk and impacts of climate change.[13] Reaching and maintaining global net zero in anthropogenic CO_2 emission demands a rapid total overhaul to our energy and industrial systems where innovations in pathways limiting global warming are well funded.

This chapter first demonstrates the significance of biomass-to-liquid Fischer–Tropsch synthesis BTL-FTS as a conversion route that potentially reduces the carbon footprint of a currently fossil-biased conversion industry dominated by processes such as coal to liquid (CTL) and gas to liquid (GTL). However, for future BTL-FTS processes to be technoeconomically competitive, their efficiency has to be significantly improved. This chapter explores the aspects of the BTL-FTS systems that would need attention: the upstream supply chain (SC) logistics for the biomass through the use of spatiotemporal tools and the mid- and downstream stages through the application of process intensification. It also shows the research trends in these fields that are indicative of future directions and scope.

15.2 Overall Distribution of Energy Resources

The main challenge of the Anthropocene era is that it requires the development of high-quality energy-efficiency strategies and research toward sustainable materials and renewable sources of energy. The negative impact of energy consumption is being felt globally, making it essential that systems are put in place to reduce the further buildup GHGs. In Figure 15.2, the breakdown of global primary energy consumption for 2019 is depicted, with Table 15.1 showing the nature of the energy sources. The majority of global energy comes from fossil fuels.

Fossil fuels (coal, oil and gas) are the dominant fuels used for power generation in most parts of the world. The United States, Germany, Japan,

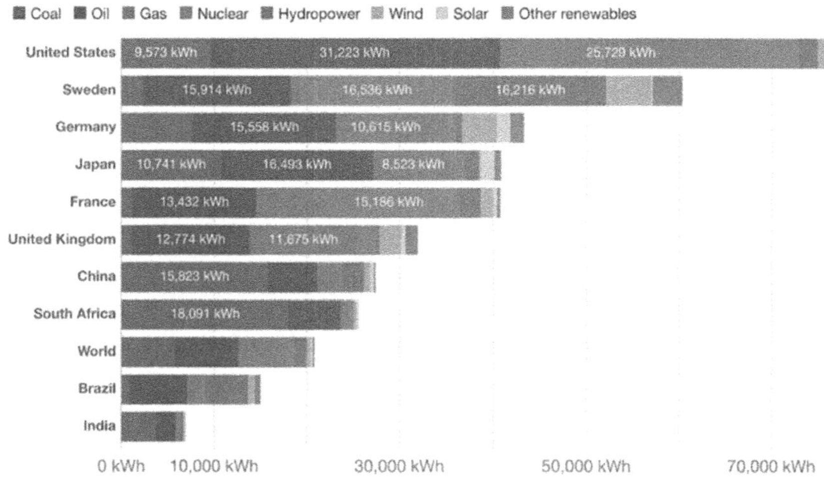

Figure 15.2 Primary energy consumption per capita in 2019. The global energy mix is still dominated by fossil fuels in most parts of the world. Reproduced from ref. 14, https://ourworldindata.org/energy, under the terms of the CC BY 4.0 license https://creativecommons.org/licenses/by/4.0/.

Table 15.1 Global primary energy consumption by source. Reproduced from ref. 14, https://ourworldindata.org/energy, under the terms of the CC BY 4.0 license https://creativecommons.org/licenses/by/4.0/.[a]

Source	Percentage contribution (%)	
Oil	33.1	84.4% of global energy comes from fossil fuels
Coal	27	
Gas	24.3	
Nuclear	4.3	15.6% of global energy comes from low-carbon sources
Hydropower	6.4	
Wind	2.2	
Solar	1.1	11.4% comes from renewables
Biofuels	0.7	
Other renewables	0.9	

[a]NB: Table does not include traditional biomass, which can be a source of energy in lower-income settings.

United Kingdom, China and South Africa get more than half of its power from fossil fuels with renewables in a narrow range.

The global trends in renewable energy are more skewed to more affluent countries. Data obtained from Scopus (see Figure 15.3) search engine indicates that research in renewable energy is on the rise, and Figures 15.4 and 15.5 indicate that the bulk of that research is in rich countries. The same trend is reflected in advanced BTL processes at the demonstration and commercialization stages, which include gasification + methanol synthesis; gasification + mixed alcohols synthesis; gasification + Fischer–Tropsch (BTL-FTS); and lignocellulosic ethanol *via* fermentation. These technologies

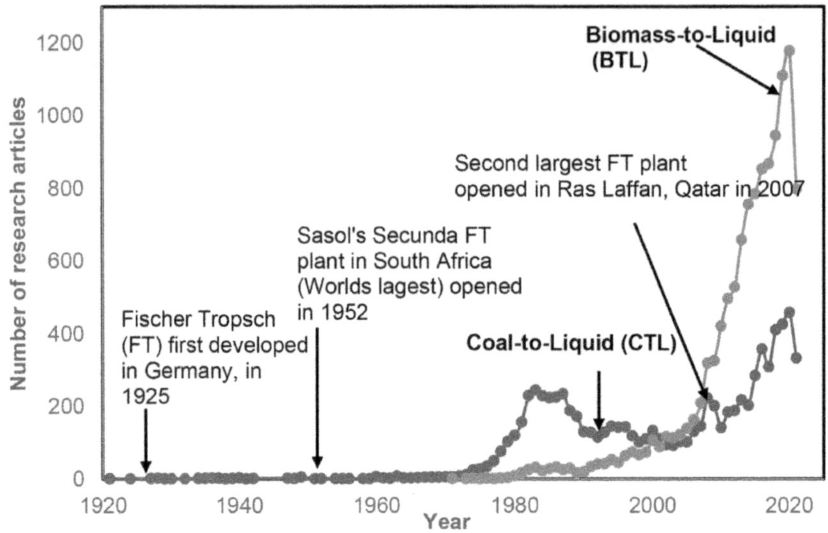

Figure 15.3 Timeline of the number of scientific journal publications per year on the topic "Coal to liquid" and "Biomass to Liquid" (based on Scopus search engine) with some key events annotated.

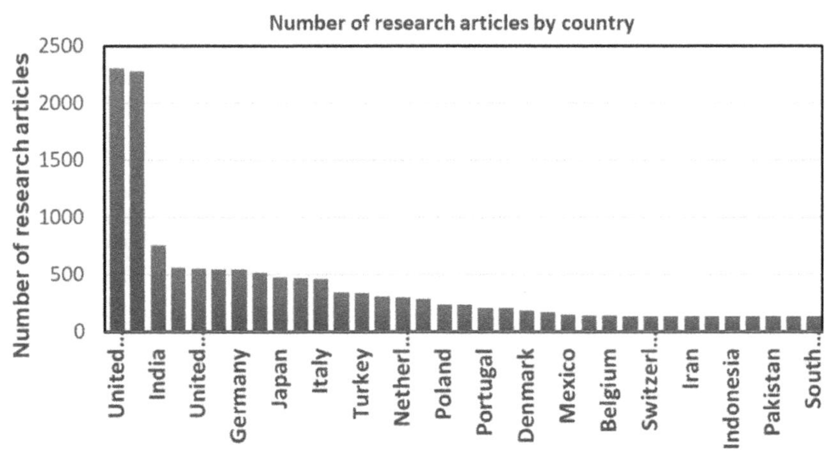

Figure 15.4 Number of research articles on biomass-to-liquid (BTL) by country from 1920 to 2021.

are particularly applicable to the conversion of lignocellulosic parts of biomass rather than edible seeds, which are being abandoned in favor of the former to avoid food security issues and land use change, as already discussed in other chapters. Following the inception of FT technology around 1925, there was an initially slow but gradual increase in the number of scientific research contributions in this field. However, as the gasification of lignocellulosic biomass became the most attractive thermochemical route

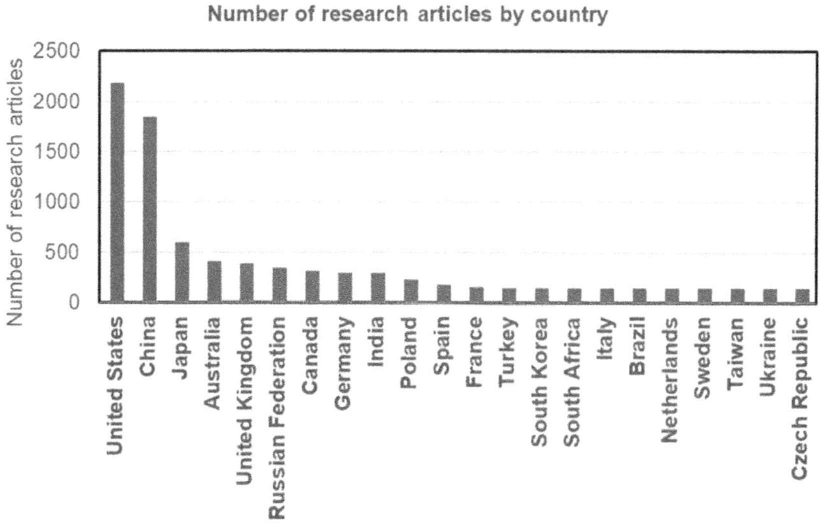

Figure 15.5 Number of research articles on coal-to-liquid (CTL) by country from 1920 to 2021.

of recovering useful chemicals (including fuels) from biomass, FT also began to gain popularity as a subsequent conversion route.

Comparatively, research on BTL is more rapidly increasing than CTL research (see Figure 15.3); countries are beginning to make systemic changes toward integrating renewable energy and averting climate change, addressing growing energy demands and depletion of conventional energy resources. The United States is the main user of fossil fuels (see Figure 15.2) and the leader in this field of CTL and BTL research, together with China (see Figures 15.4 and 15.5). Research is often directed by the views of public officials and world organizations such as the United Nations (UN), World Health Organization (WHO), *etc.*, which influence the distribution of research finances, the coordination of funding agencies and the perceptions of scientists. For example, officials in Africa (due to poverty levels) dedicate public research funds toward research programs that prioritize issues of national concern, such as food security. Such circumstances have affected the kind and quality of funding available from donors.

Climate change has taken years for its effects to manifest and to cause tremendous damage to the environment. Although climate change is a natural phenomenon, human activities have contributed significantly and heavy industrialization has contributed immensely to the problem. Developed states are the pioneers of industrialization and are therefore more responsible for climate change problems than less developed ones. As such, there is a strong notion that they must take more responsibility in dealing with the effects of climate change. Despite the research efforts of various developed countries into reducing emissions, most of these countries are still not producing positive results, as indicated by the continued rise in temperature. It is acknowledged that states generally have the sovereign right to exploit their resources to their benefit. But along with this privilege comes the responsibility to ensure that their conduct does not cause harm to the territory of another state. Such transboundary damage could give rise to the question of state responsibility and compensation for damage caused due to climate change.

15.3 Supply Chain Optimization for BTL-FTS Processes with Assistance of Spatiotemporal Tools

Despite the promising environmental and socioeconomic benefits, it is not deemed prudent to abruptly transition from fossil sources to renewables such as BTL-FTS without proper prior technoeconomic assessment and adjustments.[15] Most of the technologies, plant-specific economies and SC logistics are still maturing toward a sustainable state and must be optimized to reduce unit production cost and environmental impact. When considering any bioenergy project, it is always critical to note that, although biomass feedstock is normally cheap to obtain, the supply logistics cost can contribute 40–70% of the production cost.[15,16] This is due to the scattered nature of biomass, the low energy density and the seasonal or intermittent quantities, which make transport and other logistics costly per unit of energy compared to fossil fuels.[17] These logistic issues also affect the carbon neutrality of the projects, and they should be evaluated properly. The sustainability of biorefineries is therefore highly dependent on the cost of obtaining biomass, and this makes SC factors such as facility location, size and transportation key. As such, locating a strategic site and sizing a small distributed scale gasification FT-based biorefinery is important in the early stages of the project. Such a facility location, as well as the biorefinery size optimization problem, would have to factor in the spatial and temporal characteristics of the biomass resource. This approach also applies to the rather static nature of any alternative feedstock or energy source. Clearly, alternative energy or feedstock resources such as conventional fossils (natural gas, coal or crude oil), nuclear and solar are more consistent and predictable in terms of supply volume and location, compared to biomass. Stand-alone biorefineries and those that integrate alternative energy sources therefore need to factor in the SC issues regarding biomass feedstock. GIS has proved versatile in terms of biomass SC management, both as a single tool and also when integrated with mathematical or software models such as mixed integer linear programming (MILP), multicriteria decision analysis (MCDA), analytic hierarchy process, and fuzzy and mathematical algorithms.[18]

15.4 Trends, Gaps and Opportunities in the Adoption of GIS in Bioenergy Projects

Although other researchers have used tools such as MILP, fuzzy algorithms and heuristics for strategic level optimization, Scival metrics (https://www.

scival.com/trends) shows a gradual, though unsteady increase of GIS-based methods for such facility location problems in comparison to alternatives.

Three searches were done on scholarly output for the period 2011 to 2020. The first was done using more specific and compulsory keywords "GIS, bioenergy, facility location" (case 1), while the second was done using fewer compulsory keywords "GIS, spatial, bioenergy" (case 2) to ensure a more generic and inclusive scope. The first case was limited to the strategic facility location problem and returned 29 articles, while the second captured all strategic (*e.g.* facility location) and even tactical (*e.g.* resource allocation) problems that used GIS, returning 125 research studies for the period.

Figure 15.6 reveals that both case 1 and case 2 show a gradual increase in research output over the past 10 years, with a peak in 2015. There were four research outputs for case 1 and 15 research outputs for case 2 in 2020 compared to only two research outputs for case 1 and five research outputs for case 2 recorded in 2011. Figure 15.6 also shows that the average for the last five years for case 2 was higher at 13.4 compared to 11.6 for case 1. A survey of the 50 most prevalent keywords used in case 2, as reflected in Figure 15.7.

Evidently, the generic word "bioenergy" is the most prevalent. This is an all-inclusive word that covers all forms of conversion of biomass feedstocks. What is interesting is the seemingly greater interest in "energy crops" and the use of "marginal land" suitable for rotation coppices, compared to the interest in crop/agricultural residue. Forestry biomass is also mentioned more in the research studies than crop residue, suggesting this could be a more promising feedstock for future BTL-FTS processes.

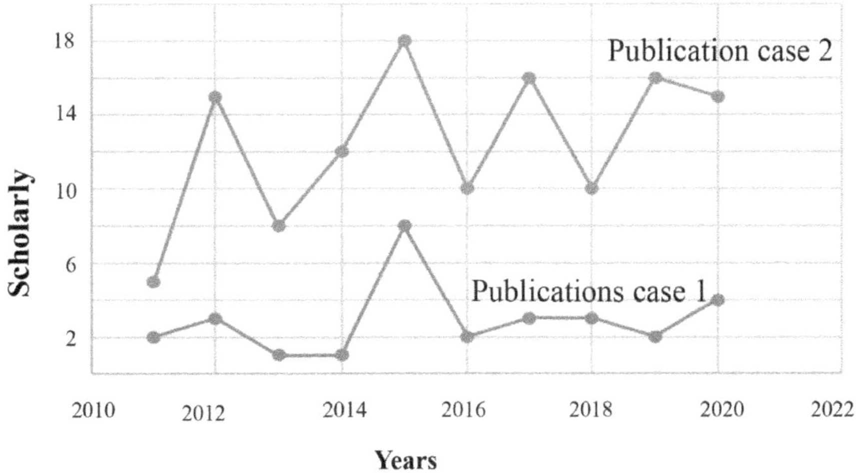

Figure 15.6 Research output on the application of GIS in bioenergy projects (2011–2020).

The third search used the compulsory keywords "biomass, bioenergy, GIS and Fischer–Tropsch". It was meant to identify both strategic and tactical level research studies that used GIS for FT conversion of biomass.

Figure 15.8 show that only two research studies on biomass second generation (2G) conversion through FT used GIS in the period 2011–2020. There are a number of research studies on the application of GIS to first-generation (1G) oil seed feedstocks such as rapeseed and soya bean to biodiesel[19,20] and also on biochemical conversion of 2G feedstock to ethanol.[21,22] However, in line with the trends shown in Figure 15.3, the increase in BTL research and the number of plants is most likely to be matched with an increase in SC research, particularly the mapping of biomass feedstock to be used in these

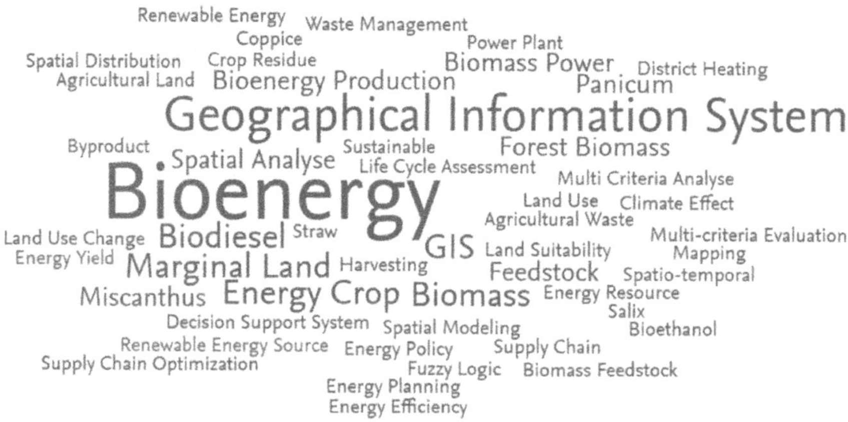

Figure 15.7 Keywords analysis for research output (case 2) (2011–2020).

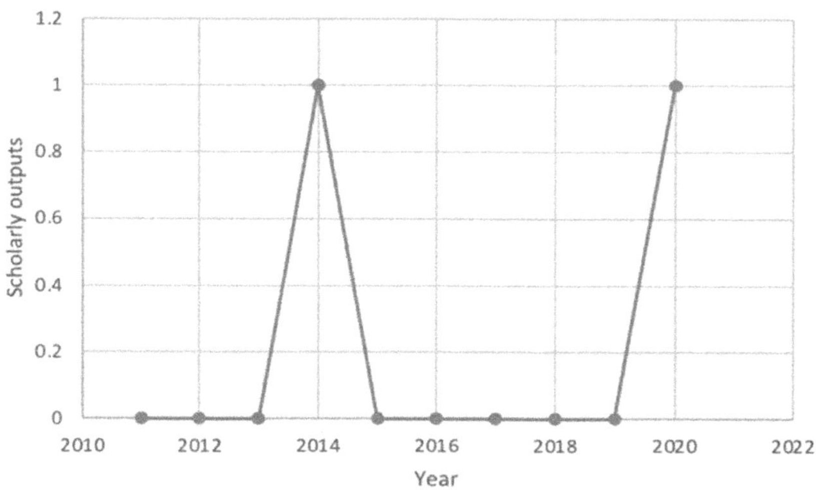

Figure 15.8 Research output on the application of GIS in FT-related bioenergy projects (2011–2020).

plants. A comparison of Figures 15.3 and 15.7 shows an initial bias toward the technical aspects of BTL research, while the SC aspect is yet to gain momentum. However, it is inevitable that these will become key elements of BTL-FTS processes, as it has been the trend with most thermochemical conversions for biomass. Researchers always start verifying the technical feasibility of the desired conversion for the selected feedstocks before they look at the pertinent economic factors such as SC logistics.[17]

15.5 Functionality of GIS-Embedded Tools and Relevance to Future BTL-FTS Processes

GIS is a modern and versatile tool that captures, stores, graphically displays, manipulates and analyzes spatial data.[23] It is able to integrate spatial data such as vector and raster resolutions of satellite imagery with nonspatial quantitative or qualitative data such as resource quantity, elevation profiles, topographic data, labor availability and accessibility to roads and logistic conditions. The data are then arranged into thematic layers represented by digital maps.[17] Multiple objectives are often associated with spatially oriented projects such as distributed scale gasification FT-based biorefineries. Examples include maximizing profitability, minimizing environmental impact and maximizing socioeconomic returns, such as employment and enterprise development. GIS is embedded with an MCDA algorithm, which is a key functionality that enables it to solve these objectives from a spatial standpoint.[24] However, this capability is limited computationally, necessitating outsourcing of extra computational flexibility from mathematical models such as MILP and support systems such as AHP or the factorial rating method. Analytical hierarchy process (AHP) and the factorial rating method can be used for ranking and weighing the suitability of potential facility sites using a set of environmental, socioeconomic and technical criteria. Factorial rating and AHP methods have various ways of showing such relative hierarchies of the criteria, *e.g.* a simple scale of: (1) - not very important; (2) - important; (3) - very important. In the factorial rating method, the facility locations are then scored using a scale of 1–10 or 0–1 for the various criteria, then the overall suitability index or score is a product of the two indicators. The AHP method uses a similar logic; however, it is a more sophisticated computer model that can deal with larger numbers of criteria and scores with a faster computational turnover period.

Table 15.2 shows some of the criteria that can be considered for the case of distributed scale gasification FT-based biorefineries and their relative importance. More criteria are included for cases in which energy crops and short-rotation coppices (2–5 years) or short rotation forestry (up to 10 years) are used as feedstock. This is a more flexible alternative feedstock, since the quantity and spatial distribution can be predetermined with a greater degree of freedom as opposed to depending on residues from fixed points. This alternative source of biomass feedstock seems to be more attractive for

Table 15.2 Criterion that can be considered for site selection of distributed-scale gasification FT-based biorefineries. Data taken from ref. 18 and 25.

Characteristics	Factor weight
Availability of raw materials (proximity to biomass and fossil source in case of integration)	3[c]
Availability of utilities	3[c]
Proximity to market	2[b]
Accessibility (road or railway)	2[b]
Availability of suitable land (topography, no land use change (LUC), costs, *etc.*)	2[b]
Environmental considerations and effluent disposal	1[a]
Additional factors for self-grown biomass (energy crops, short rotation coppices and short rotation forestry)	
Population density and land use profiles	3[c]
Rainfall patterns or average rainfall	2[b]
Altitude and topography	2[b]
Temperature	2[b]

[a]Not very important.
[b]Important.
[c]Very important.

BTL-FT refineries that will be integrated with fossil sources, since there is no guarantee that residue will be available in sufficient quantities within a feasible proximity to the fossil source.

For the additional factors, the availability of marginal land that cannot be used, especially for food crops, is more ideal. The energy crops and forestry coppices should not be very sensitive to the type of soil, rainfall, altitude, topography and temperature; hence the rating of 2 for these factors. However, this largely depends on the specific feedstock chosen.

The GIS then uses the AHP and its embedded MCDA capability to scan for optimal sites for the distributed scale gasification FT-based biorefineries from its spatial database.[24] At times, only economic objectives are considered initially, and a set of competitive sites are identified. After that, other sites are then eliminated using other social, environmental and legal criteria, depending on which ones are considered important enough. Essentially, a GIS model not only acts as a digital data bank for spatial characteristics (*e.g.* quantity) of waste or biomass but can also manipulate that data in reduced time and at reduced cost to give the best location and alternatives for a processing or storage facility.[26] Figure 15.9 exemplifies a stage-wise approach to a GIS optimal site selection or resource allocation problem. In a number of cases, the GIS is integrated with simulation or optimization tools and can either be embedded in the overall program or be a precursor used to predetermine the best candidate sites for subsequent SC optimization.

Although the problem of SC optimization has not been tackled in the right detail in terms of small distributed scale gasification FT-based biorefineries, it is similar to that of other bioenergy problems, especially when biomass

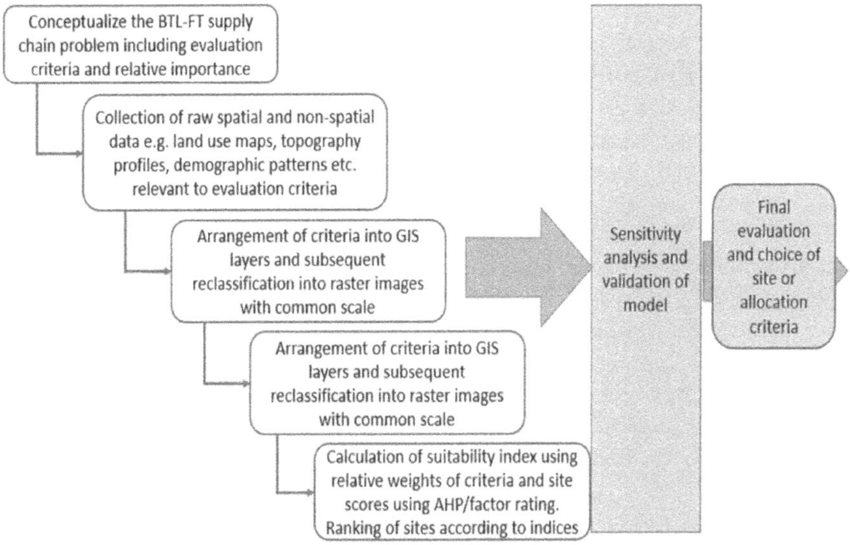

Figure 15.9 Stage-wise approach to a GIS optimal site selection or resource allocation problem.

residues are the feedstocks. There have to be adjustments where the gasification FT-based biorefineries are integrated with fossil sources, as one has to weigh the options of placing the biorefinery close to the biomass source or to the alternative fossil energy sources. In the latter case, the proprietor can decide to grow their own biomass at designated points that are predetermined using GIS tools. The optimum location of the facility would also need to factor in the environmental impact in order to ensure that there is no reversal of the gains to be made (by using biomass to curb the carbon footprint of fuel production) if the biomass has to be transported over long distances.

15.6 Technological Trends of BTL-FTS Processes Based on Process Intensification

The limited availability of local biomass feedstock has implications in terms of a reduced plant size for BTL compared to GTL or CTL. However, the technoeconomic feasibility of the FTS process is highly dependent on economies of scale. The future challenge for both researchers and industries is then to ensure that small-scale BTL plants are viable. Unless very large quantities of biomass can be provided at a single site, to benefit from economies of scale while avoiding high biomass transportation costs, it is currently well established that the future of BTL-FTS plants lies in improving the competitiveness of small-scale facilities (<100 000 t per a).[27–29] Although this approach goes against the principle of economies of scale, thereby compromising the profitability of these smaller facilities, the latter have the advantages of significantly lower capital expenditure (CAPEX), nimbler operation (easier modification, quicker commissioning and decommissioning), transportability of equipment (easier to transport, install, remove and implement in remote locations close to a biomass feed source), modularity (which cuts down on engineering cost and can lead to savings through mass production). Furthermore, it is possible, in such cases, to increase the plant size using small units in parallel while increasing the overall robustness of the facility.[27,28] Process intensification is considered a very promising pathway for efficient scaling down of both energy- and capital-intensive processes involved in BTL-FTS to enable the exploitation of small and remote biomass feedstock, thereby substantially reducing the production cost of BTL fuel.

The BTL process comprises several steps (see Figure 15.10): biomass gasification, syngas cleaning and conditioning, FT synthesis of syncrude and product upgrading through refining.

According to economic analyses,[30–34] in a typical BTL-FTS process, the biomass gasification and syngas cleaning steps contribute to about 50% or more of onsite CAPEX. FT synthesis CAPEX alone comprises 20% of the total cost, while the upgrading and refining section comprises about 10%. Consequently, one of the main goals of enhancing the onsite plant economics of a BTL-FTS process is to reduce the production cost of syngas *via* biomass gasification and syngas cleaning. These critical technological steps of the whole BTL-FTS process offer an opportunity for application of process intensification strategies. Process intensification (PI) has proved to be a promising approach in developing new processes that allow major abatements in unit size, operational cost and environmental emissions.[35,36]

PI has been a swarming engineering field since the 1970s, and several researchers from the chemical engineering community have contributed to establishing a comprehensive definition and the core principles of PI over the past 20 years.[35–45] PI consists of developing, designing and implementing breakthrough technologies and methods that drastically improve how chemicals are produced, while achieving a substantial reduction in installation size *vs.* the production capacity ratio, energy requirements,

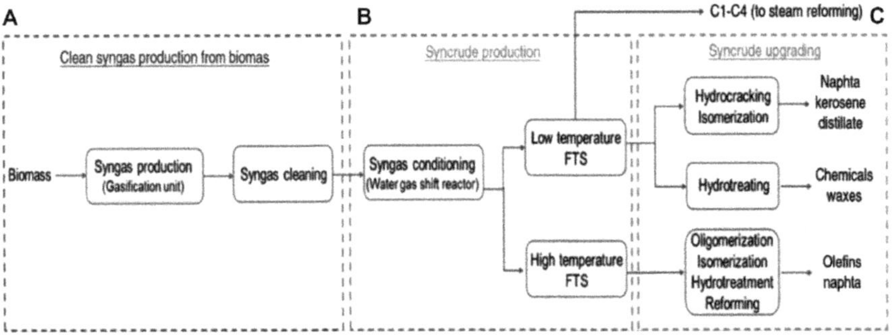

Figure 15.10 BTL conversion stages: (A) gasification to produce syngas; (B) production of syncrude; (C) upgrading and refining of syncrude.

waste generation. This results in more sustainable and less hazardous processes with a better economic balance.

Van Gerven *et al.*[39,42] postulated four theories for PI design, which can be summarized as follows:

Principle 1: Optimize process kinetics, as they are a defining factor in the optimization of conversion and selectivity.

Principle 2: Control the reaction conditions in order to produce products of a similar quality, which result in less waste and energy employed in downstream separation processes.

Principle 3: Employ best practice in terms of process and equipment design, which will result in optimum equipment size being used to process optimum amounts of feed material.

Principle 4: Integrate individual processes in order to come up with a combined process that has the benefit of improved efficiency.

PI can apply to various chemical process dimensions and scales: molecular, mesoscale and macroscale within the spatial (structural), thermodynamic (energy), functional (synergy) and temporal (time) scales.[39] The main approaches of PI in terms of BTL-FTS processes currently being considered target the most CAPEX-intensive sections of the process referred to earlier. In the following sections, some of the most promising PI techniques for both syngas production *via* biomass gasification and fuel production from Fischer–Tropsch synthesis are illustrated from the state-of-the-art research work that has been done.

15.7 Process Intensification Approaches in the Syngas Production System

The classical process of converting biomass into syngas comprises the gasification, gas cleanup and gas conditioning stages (see Figure 15.11). Syngas can also be derived, secondarily, from the reformation of biogas and bioethanol, products of methanization and the fermentation of various biomass forms.[46,47] Lignocellulosic biomass can also be converted into intermediate liquid or solid products, which are versatile, and economic feedstock products that can be further converted into syngas. This approach can entail flash pyrolysis and hydrothermal treatment to produce liquid bio-oil or syncrude and carbonization or torrefaction to produce solid char of various regimes. The resulting liquid intermediate (bio-oil) from flash pyrolysis can be used as a feedstock in both gasification and catalytic steam reforming technologies.[48] The intermediate pyrolysis oils are more energy dense and cleaner compared to the original feedstock, since most inorganic matter from biomass results in the by-product char.[49]

The four main stages are carried out in distinct unit operations sections, which results in a large plant size outlay along with a high capital and energy cost. Figure 15.12 shows the three main philosophies of PI currently being explored in the field of producing syngas from biomass.[50]

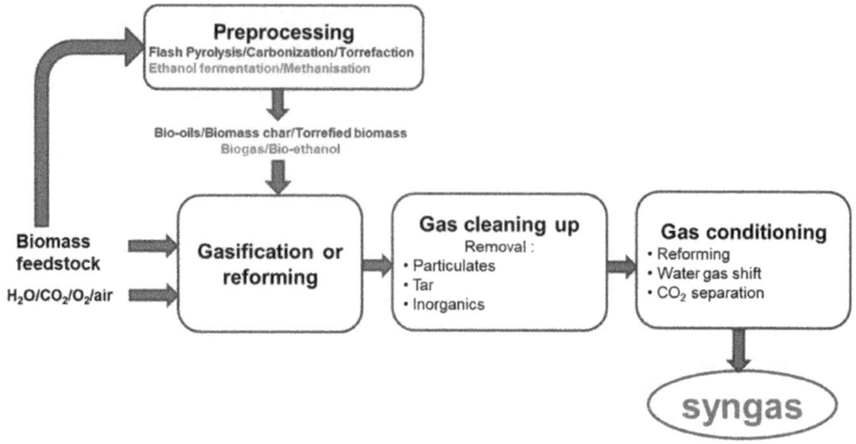

Figure 15.11 Conversion flowsheet of biomass into syngas or H_2 through gasification.

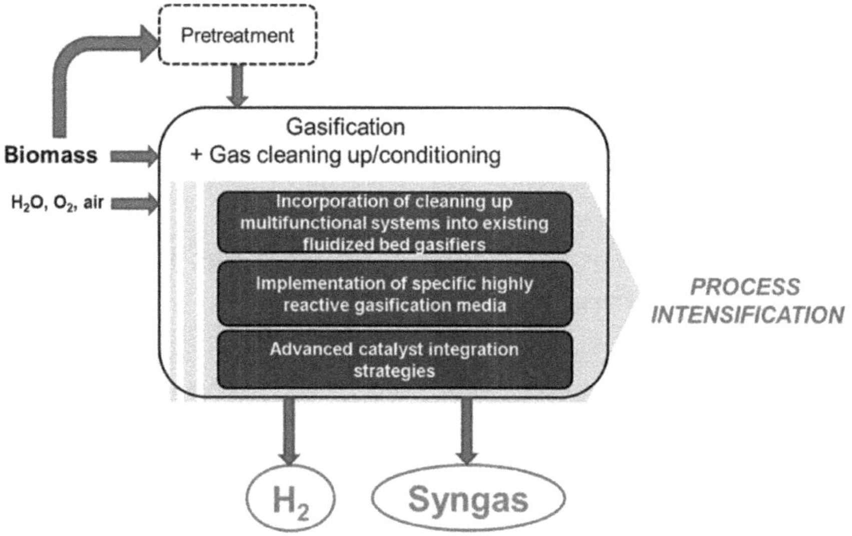

Figure 15.12 Three main approaches proposed in the literature targeting PI in producing syngas from biomass through gasification.

Integrating multifunctional systems, combining reforming or catalytic tar cracking and eliminating carbon dioxide/particles/other contaminants within the gasifier have been contemplated as a promising intensification strategy. Bubbling fluidized-bed (BFB) reactor technologies are ideal for integration with gasifiers that employ biomass as feed material *via* two critical components of a process: (1) the fluidized-bed (FB) media and (2) the freeboard section within the reactor. The use of catalysts as supplements in biomass FB reactors in catalytic tar reforming is a major study within this PI strategy, but there are intrinsic challenges.[50] Another application of this PI strategy is the dual fluidized-bed gasifier (DFBG), also known as the circulating fluidized-bed (CFB) technology, which enables integration of supplementary functionalities through the bed media. DFBG technology has been thoroughly described elsewhere.[51–55] DFBGs comprise two connected fluidized-bed gasifiers: an endothermal fluidized bed that employs steam to achieve gasification and an exothermal fluidized bed that provides the required heat for gasification through circulation of the bed medium. The mode of operation of a DFBG is such that the feed material is initially gasified in the fluidized-bed section, followed by combustion of residual char in the subsequent section. The low probability of generation of flue gas from biomass gasification in DFBGs results in a medium-heating-value gas being obtained. One of the most efficient DFBG designs in terms of fuel conversion and tar production couples the BFB and CFB for the combustion of char.[51] This DFBG process, depicted in Figure 15.13, is a successful approach that has been employed in the production of high-quality syngas.

One of the key elements of this technology related to PI relates to using the bed media to provide multiple functionalities to the DFBG, depending on the physicochemical properties. The bed media play the role of catalyst for tar cracking/reforming and are also a heat carrier from the CFB to the BFB. Besides, additional functionalities can be integrated through specific bed

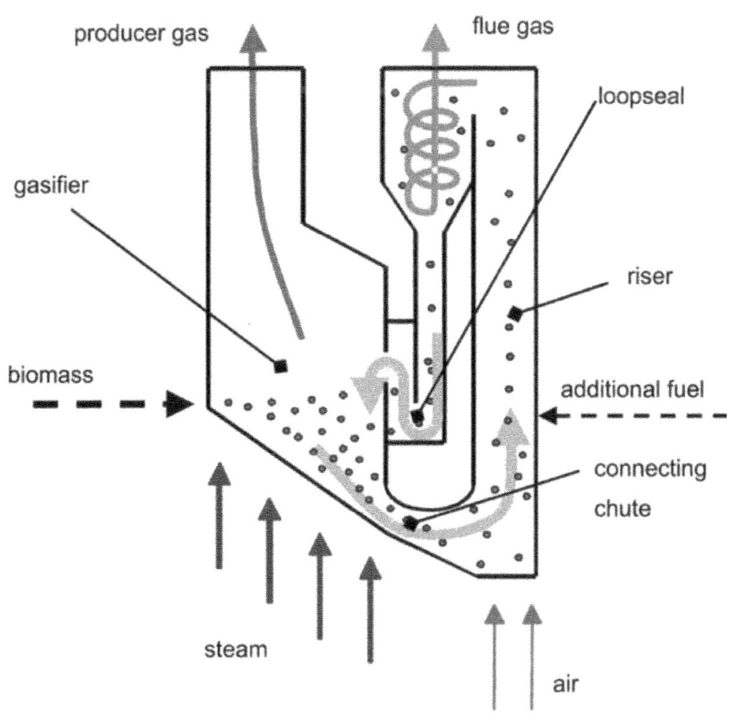

Figure 15.13 DFBG process.

media such as the oxygen transportation for the chemical loop combustion process, the prevention of agglomeration phenomena in fluidized beds, the reduction of undesirable compounds such as H_2S, HCl or alkali metals, and the sorption-enhanced gasification process that entails elimination of CO_2.[53–57] This latter DFBG process carries a dual purpose of fuel conversion and *in situ* CO_2 capture and sequestration, which is a key advantage for processes requiring a specific H_2/CO ratio, such as BTL-FTS.[57]

BFB gasifier experiments have shown that the majority of the thermochemical reactions take place within the bed and at zones above the bed, including a space called the freeboard zone.[58] In addition to the bed materials, the freeboard region has thereby been the subject of special interest in the design of PI. Rapagnà *et al.* successfully illustrated this PI strategy by integrating bed media with gas cleanup systems into the freeboard region of a BFB gasifier, as depicted in Figure 15.14.[59,60] In this concept, bed materials of the BFB gasifiers comprise tar reforming catalysts and adsorbents of selected trace elements, such as H_2S, HCl or alkali metals.

The catalytic reforming of tar and the hot gas filtration are then performed in a single stage, inside a very compact reactor vessel resulting in a genuine advance toward PI, notably through the reduction of the overall size of the biomass gasification unit and significant heat savings. Compared to a two-stage process consisting of a particle trap unit followed by catalytic tar elimination unit, using catalytic filters in the freeboard of a BFB gasifier is more energy efficient and has lower investment costs.[61] Interestingly, this concept could potentially be adapted to the freeboard region of the BFB

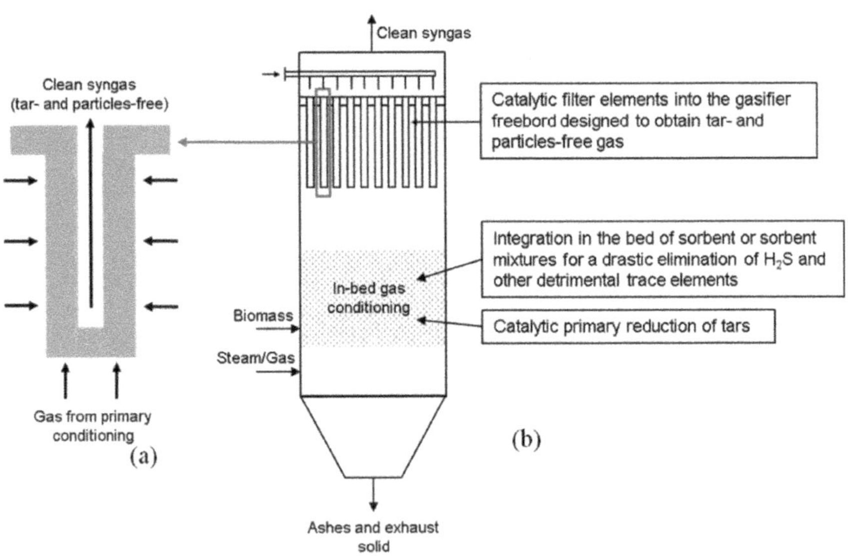

Figure 15.14 Gasification reactor with PI postulated by Rapagnà *et al.* Reproduced from ref. 59 with permission from American Chemical Society, Copyright 2009.

gasifier of the DFBG process to perform syngas conditioning through CO_2 absorption moving the water-gas-shift reaction equilibrium and to regenerate the various bed media. Such an association of PI strategies could lead to the process design of an optimum unit in which higher-purity hydrogen-rich syngas is produced.

Though many of these PI approaches were experimented on for post-conversion applications, they were predictably proven to be more difficult to implement inside the gasifier vessel. Design constraints and manufacturing complexity are increasingly important with multifunctionality integration inside the gasifier. Proper combinations of different multifunctionalities integration strategies still need to be studied in order to come up with an optimal gasification process design.

According to the aforementioned principles of PI, molten metal or supercritical water (SCW) may be employed as substitutes for steam, as they are also gasification media that are highly reactive. This method would be useful for the production of hydrogen or methane in the gasification of wet biomass (with a moisture content $\geq$30%) since there would be no need to dry it. This cuts energy expenditure and associated costs. Although SCW biomass gasification has been extensively studied,[62,63] further improvements are still necessary to overcome the bottlenecks related to process engineering, including developing the catalysts that perform efficiently under the severe SCW gasification atmosphere. The use of molten metal as a gasification medium has been much less studied than SCW but promises to be a cutting-edge technology that could result in notable PI. The specially chosen liquid metals (*e.g.* FeSn) then serve as heat and chemical carriers and catalysts for the reactions, as proposed and illustrated by Eatwell-Hall *et al.*[64] One of the specific features of this process is the production of both high-purity carbon monoxide and hydrogen in different streams from the reactor, which could be advantageously used to obtain a tailored H_2/CO ratio necessary for the further FTS process.

Other approaches of PI are based on catalyst integration strategies and gasification reactor miniaturization. In these approaches, several objectives are specifically targeted: (1) shortening the biomass residence time in the reactor, leading to a substantial reduction in the plant size/production capacity ratio; (2) reduction of reaction temperatures, leading to a reduction in energy consumption and simplification of reactor material requirements; and (3) direct production of clean syngas, simplifying downstream syngas processing. The gasification reactor miniaturization strategy has been illustrated by Dauenhauer *et al.*, where they reduced the short contact time with the catalyst (Rh–Ce based, supported on Al_2O_3 spheres) in the reactor.[65,66] This process is often referred to as reactive flash volatilization (RFV). The biomass feedstocks are introduced as micrometric particles and gasified in time scales <70 ms. Typically, when gasification of pyrolysis reaction kinetics are associated with high heat fluxes, they circumvent certain prohibitive steps such as tar and solid carbon formation.[67,68] These unwanted intermediates smear on active sites and the pores of the catalyst, hindering surface reactions in the process.[69] An illustration of the micro-structured catalytic reactor is shown in Figure 15.15a.

As shown in Figure 15.15b after continuously adding air and cellulose, the gases produced a few millimeters up the reactor profile begin to react catalytically, participating in water-gas-shift (WGS) and steam-reforming conversions, as they traverse the bed length. The gases at the exit are rid of all tar and mostly contain C_1 products with molar H_2/CO ratios of 1–1.3, which can be increased by steam addition or adjusting the steam-reforming and WGS reactions. Although the proof of concept was successfully provided for purified microcrystalline cellulose, the catalytic reactive flash volatilization of raw lignocellulosic feedstock is more challenging with respect to the catalyst stability. Indeed, the occurrence of inorganic species such as Na, K, Ca, Mg, Si, S and P inside real biomass particles remains an actual issue since those species are mostly volatilized at such process conditions and lead to catalyst poisoning.[70–73] This can be avoided by using moving catalyst beads instead of catalyst foam to continuously remove ash and its inherent organic species.[65] Indeed, RFV technology is a very promising PI method for producing high-quality syngas from biomass, but challenges arise from the high cost of the rodium–cerium catalyst and its stability with respect to volatile inorganics from real biomass feedstock.

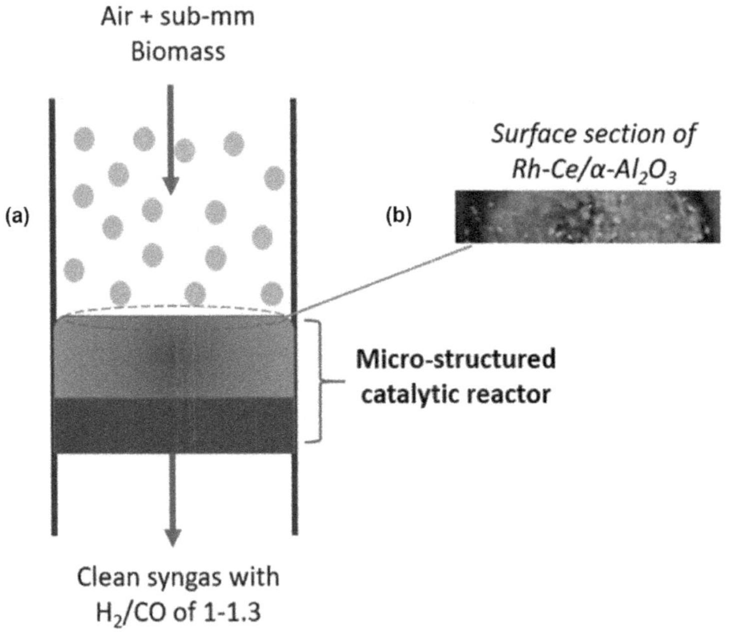

Figure 15.15 (a) Microstructured catalytic reactor used in the reactive flash volatilization of sub-millimeter-sized biomass particles. (b) Surface image of a rodium–cerium-based catalyst supported on α-Al_2O_3 monolith (45 ppi) during flash gasification of 200–400 mm cellulose particles in air, as it turns a brilliant orange color.[65]

If the integration of efficient catalysts into the gasification reactor is a promising PI strategy, a further catalyst integration strategy consists of integrating the catalyst inside the biomass fuel particles. This catalyst/biomass integration PI approach specially using transition metal catalysts is illustrated in Figure 15.16 and has been proposed instead of the high-cost noble metal-based catalytic microreactors using submillimeter-sized biomass feedstock.[74]

Based on similar technologies applied in other contexts such as the fuel-borne catalyst technology in automotive industry or the biomass conversion into biocrude, this approach is founded on the theory that the efficiency of heterogeneous catalysis for biomass gasification can be increased by a more intimate catalyst-to-biomass contact. The close particle contact between catalysts and biomass effects: (1) an alteration of initial biomass decomposition kinetics to enhance product selectivity and (2) a significant increase in catalysis efficiency and subsequently higher biomass conversion. The expected benefits of this strategy are (1) it is possible to operate at low temperatures with good productivity rates and (2) good-quality syngas is produced, making downstream gas treatment and upgrading less tedious for various applications. The distinct steps for this approach are shown in Figure 15.16. The approach was illustrated by Richardson *et al.*, who demonstrated the *in situ* formation of nickel metal Ni^0 nanoparticles *via* carbothermal reduction below 500 °C during biomass pyrolysis using a nickel

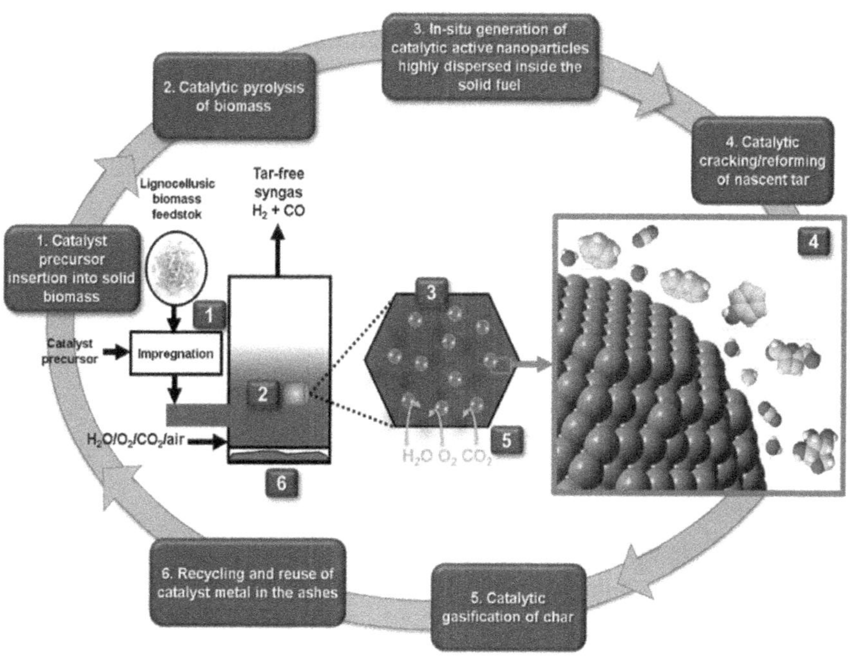

Figure 15.16 Integrated catalyst/biomass gasification approach. Adapted from ref. 74 with permission from American Chemical Society, Copyright 2013.

catalyst precursor.[75] These *in situ*-generated Ni^0 nanoparticles were assumed to furnish necessary active sites that would adsorb selected aromatic tar compounds and catalyze their conversion[74] and that of other gasification transformations.[76] Since char gasification is a rate-determining step, reducing the char's residence subsequently shortens the biomass residence time during gasification. It was demonstrated from the kinetics parameters of Ni-char gasification that char conversion times can be three times shorter than without a catalyst, enabling the use of a char gasification reaction thrice as small as ordinary. However, a major challenge of the integrated catalytic biomass gasification concept lies in implementing an efficient catalyst recycle and reuse philosophy. This has to be considered during the design of the proof of concept.

15.8 Process Intensification Approaches in the FTSs and Upgrading/Refining Processes

Two main PI approaches have been investigated to date in order to develop small-scale FTS applications: (1) combining multiple process tasks into the FTS reactor through membrane reactors; (2) miniaturization and thermal management mastering through microstructured catalysts and reactors.

FTS is a complex exothermic three-phase process. The goal of integrating different kinds of membranes into the FTS reactor is to drive the chemical equilibrium toward the intended product by feeding or removing selected reactants or products, respectively, while improving heat and mass transfer properties of the reactor. According to their functionality, membranes used into the FTS reactor can play different roles. Five concepts of the FTS membrane reactor have been reported so far:[29,77,78] (1) distributed reactant feed; (2) water extraction *in situ*; (3) forced-through membrane; (4) zeolite-impregnated catalysts; (5) dual-bed membrane reactor. The first four concepts are displayed in Figure 15.17.

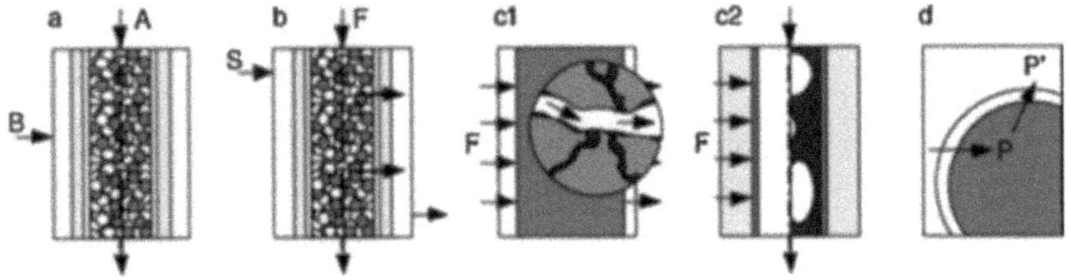

Figure 15.17 FTS membrane reactors in the literature: (a) distributed feed of B around the membrane and reaction with A; (b) water extraction *in situ* using a selective membrane (F, feed; S, sweep); (c1) plug-through contactor membrane (PCM) with wide transport pores, (c2) forced-through membrane, with product and heat extraction into liquid product; (d) zeolite encapsulated catalysts. Reproduced from ref. 77 with permission from Elsevier, Copyright 2005.

Syngas produced by biomass gasification is usually low in H_2 and not optimal for FTS. The H_2/CO ratio is typically close to 1, *i.e.* way below the optimum of 2, and therefore needs to be adjusted before the FTS reaction in the BTL plant. In this case, a membrane reactor based on H_2 distribution feeding is a promising way to ramp up CO conversion into selected products. In this concept, H_2 is administered by means of a cylindrical H_2-permselective membrane, while CO is delivered *via* the FTS catalyst packed bed within the membrane reactor configuration. Through this concept, the H_2/CO ratio is optimized, resulting in an improvement of activity and product selectivity of the FTS catalyst, especially if Co-based catalyts are used, while ensuring a better temperature control.[77] A demonstration of this concept has been performed by Liuzzi *et al.* with a Pd–Ag and Ru-based supports and/or catalysts.[79]

In situ removal of H_2O, the main by-product of the FTS reaction, through an extractive hydrophilic membrane is of particular interest for at least three reasons: (1) to reduce H_2O-promoted iron- and cobalt-based catalyst deactivation; (2) to shift the water-gas reaction equilibrium in favor of CO_2 conversion to hydrocarbons; (3) overall, to enhance selective production within the reactor. The removal of H_2O from a complex syngas + hydrocarbons mixture within the FT profile can be done by integrating a highly H_2O-permselective membrane into the reactor compartment. However, this is still a challenge in terms of choice of membranes and adequate membrane preparation techniques. Indeed, although several membrane candidates have been investigated to date, there has not yet been a breakthrough around the *in situ* removal of H_2O during the FTS reaction.[29]

In the concept of forced-through flow membrane contactor, a pressure differential is applied across a porous membrane having a catalyst embedded in its porous structure. The syngas is forced quickly across the membrane, resulting in an intensive mixing of the reactants inside the pores. Two variant concepts of forced-through flow membrane contactors were tabled: (1) a ceramic honeycomb monolith whose mesopore channels have layered coats of a microporous ceramic membrane and a FTS catalyst;[80] (2) a tubular membrane with a high Co/Al_2O_3 catalyst loading and good thermal conductivity.[81]

In the zeolite-encapsulated catalyst concept, the FTS catalysts can be integrated with acidic zeolites to change the FT product distribution immediately after product formation. Recently, a catalyst design comprising an oxide–zeolite (OXZEO) was reported to facilitate the direct conversion of syngas to mixed light olefins (~80% selectivity) and to ethylene (83% selectivity).[82]

The low-temperature FTS is often performed over supported cobalt-based catalysts at temperatures below 240 °C and 20–30 bar. This process is characterized by a strong exothermicity ($\Delta H_R^0 \approx 165$ kJ mol_{CO}^{-1}) and is highly sensitive to temperature variations, making process temperature control highly critical, especially for compact fixed-bed reactors with PI. Indeed, an effective extraction of the reaction heat is necessary to (1) abate the creation of hot spots in the catalyst bed, (2) to avoid a temperature gradient inside the catalyst bed that is detrimental to catalyst selectivity, and (3) to curtail safety risks that could emanate from thermal runaway.[83] Several structured reactors with PI based on superconductive structures have been reported for the compact and modular FTS process such as microchannel reactors, closed cross-flow reactors, microfibrous entrapped catalysts, honeycomb monoliths, foam monoliths reactors.[83] The microchannel reactor technology developed by Velocys is an example of PI. This microchannel FTS reactor system reduces the reactor volume required to produce a corresponding amount of product by an order of magnitude or more and is currently at the commercial stage.[84] The intensified heat and mass transfer capacity within the microchannel structure, coupled with the superactive cobalt-based catalyst, makes such productivities possible. The Velocys microchannel reactor concept is illustrated in Figure 15.18.[85] Each reactor module has thousands of reactor channels stuffed with FT catalyst, intertwined with coolant channels. In order to achieve commercially significant production volumes, multiple microreactors are arranged in parallel.

Open-cell architectures comprising very good heat conducting metals like Al or Cu are particularly attractive owing to their high porosity, low density, good mechanical strength and expansive surface area, thereby providing continuous thermally connected structures. These features afford a promising solution for enhanced heat transfer properties within the FT synthesis packed-bed reactor.[83] Fratalocchi *et al.* demonstrated experimentally the

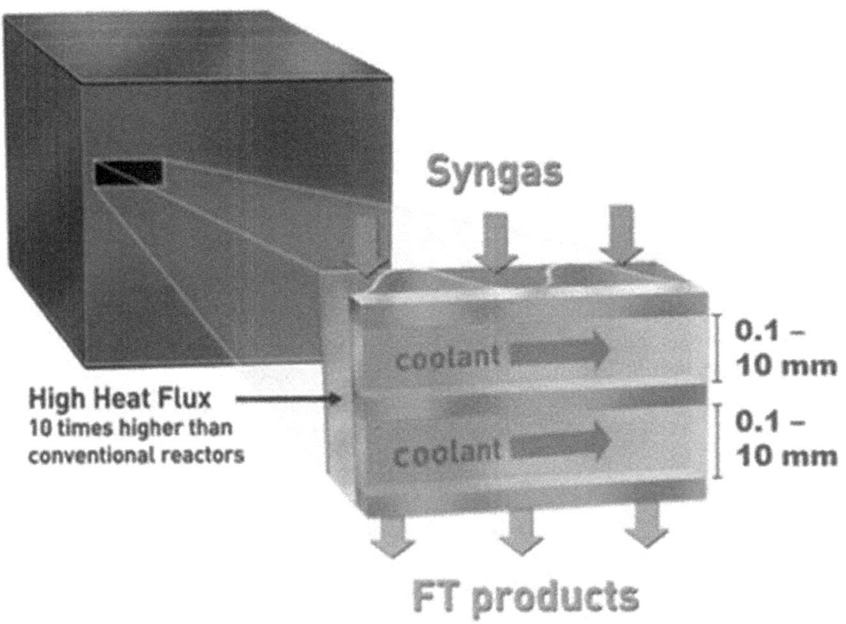

Figure 15.18 Velocys microchannel reactor concept. Adapted from ref. 85 with permission from Elsevier, Copyright 2010.

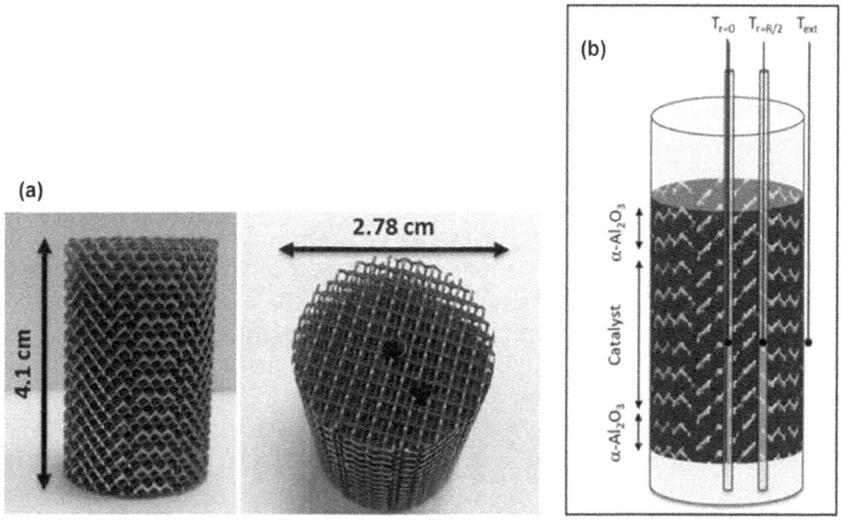

Figure 15.19 (a) Images of the Al POCS architecture and (b) open-section image of POCS within the reactor, impregnated with catalyst pellets and α-Al_2O_3.

ability of highly conductive open-cell aluminum foam packed with a Co/Pt/Al_2O_3 catalyst microspheres to contain a highly exothermic FTS reaction in a packed-bed reactor. This catalyst architecture demonstrated good qualities with duties exceeding 1300 kW m^{-3} and CO conversion >65%, coupled with excellent temperature containment.[86] Further improvements of open-cell foams of high conductivity have been recently provided through additive manufacturing by selective laser melting using $AlSi_7Mg_{0.6}$. Consequently, the design and printing of spatially arranged patterns of repeated unit cells with regular geometries such as cylindrical tubes, octahedrons or diamonds are possible.[87] Examples of 3D-printed periodic open cellular structures (POCS) packed with Co/Pt/Al_2O_3 catalyst pellets are shown in Figure 15.19. These proved to outperform the packed-foam reactor in terms of heat transfer with smaller radial temperature differences within the catalytic bed and nominal temperature gradients at the reactor wall. However, the thermal power released per unit volume was still remarkably larger with the POCS, whose regular geometry enhanced intrinsic thermal conductivity and improved the contacting scheme within the ordered structure that affects the wall heat transfer coefficient.

15.9 Concluding Remarks

There has been a gradual yet significant increase in interest regarding BTL-FTS systems, as lignocellulosic biomass has become a key resource for the production of chemicals and fuels in the future. The interest in using lignocellulosic biomass rather than fossils stems from its renewability, global abundance and general carbon neutrality. It also has an advantage over edible biomass resources in that it circumvents food security and land use change issues.

This chapter indicated that, for future BTL-FTS processes to be technoeconomically competitive, the efficiency of the SC logistics and conversion, upgrading and purification stages have to be improved. The upstream SC aspects can be tackled effectively using spatiotemporal tools, while the midstream and downstream stages can be improved through PI. When the conversion technology to be applied and its economic throughput/capacity have been established, it becomes imperative to decide on the location strategy of the small distributed-scale BTL FTS plants. GIS spatiotemporal tools have proved versatile in biomass SC management, as a single tool and also when integrated with mathematical or software models such as MILP, MCDA, the analytic hierarchy process, and fuzzy and mathematical algorithms. They have been used in strategic/optimal site location (centralized or decentralized), as well as transportation routes in a few cases. Trends have shown a gradual increase in the use of GIS tools in dealing with such problems.

Process intensification strategies have been extensively studied in both syngas production from biomass and syncrude production and upgrading, resulting in the emergence of promising technologies based on highly efficient small-scale and modular processes. These intensified technologies are not mature yet and have to be investigated in an integrated way enabling the intensification of the whole BTL-FTS process. The process intensification approach *via* miniaturization through the development of catalytic microreactors associated with high thermal management mastering of the processes appears to be one of the most promising. Heat transfer management is one of the most critical issues in both biomass gasification and FTS reaction when considering the scaling down of the processes. By reducing system sizes and enabling heat transfer through thermally conductive structures, the time scale of heat transfer is reduced by orders of magnitude compared to conventional reactors. Furthermore, the recent development of additive manufacturing in the field of catalytic reactors, associated with advances in both catalysis and process engineering applied to both syngas production and FTS processes, is expected to provide breakthrough leading-edge technologies, resulting in significant cuts to both capital and operational costs of BTL fuels at small scales and adaptation to the biomass feedstocks availability in most regions of the world.

Abbreviations

AHP	Analytical hierarchy process
BFB	Bubbling fluidized bed
BTL	Biomass to liquid
BTL-FTS	Biomass-to-liquid Fischer–Tropsch synthesis
CAPEX	Capital expenditure

CFB	Circulating fluidized-bed
CTL	Coal to liquid
DFBG	Dual fluidized-bed gasifier
FB	Fluidized bed
FTS	Fischer–Tropsch synthesis
GHG	Greenhouse gas
GIS	Geographical information system
GMST	Global surface temperature
GTL	Gas to liquids
IPCC	Intergovernmental panel on climate change
MCDA	Multicriteria decision analysis
MILP	Mixed integer linear programming
MSW	Municipal solid waste
PI	Process intensification
POCS	Periodic open cellular structures
RFV	Reactive flash volatilization
SC	Supply chain
SCW	Supercritical water
UN	United Nations
WGS	Water-gas shift
WHO	World Health Organization

References

1. Y. Malhi, *Annu. Rev. Environ. Resour.*, 2017, **42**, 77–104.
2. K. E. Trenberth and J. T. Fasullo, *J. Geophys. Res.: Atmos.*, 2012, **117**.
3. W. Takeuchi, A. Sekiyama and R. Imasu, *2013 IEEE International Geoscience and Remote Sensing Symposium – IGARSS 1805–1808*, 2013.
4. X. Wu, Y. Lu, S. Zhou, L. Chen and B. Xu, *Environ. Int.*, 2106, **86**, 14–23.
5. G. Luber and N. Prudent, *Trans. Am. Clin. Climatol. Assoc.*, 2009, **120**, 113–117.
6. NOAA Global Monitoring Laboratory, Earth System Research Laboratories: Trends in Atmospheric Carbon Dioxide, https://gml.noaa.gov/education/ (retrived 20 September 2021).
7. T. M. L. Wigley, *Clim. Change*, 1983, **5**, 315–320.
8. U.S. Energy Information Administration, International Energy Outlook 2019 With Projections to 2050, 2019.
9. NOAA National Centers for Environmental Information, State of the Climate: Global Climate Report for Annual, 2020.
10. NOAA National Centers for Environmental Information, State of the Climate: Global Climate Report for Annual, 2020.
11. International Energy Agency (IEA), Global Energy Review 2021 – Assessing the Effects of Economic Recoveries on Global Energy Demand and CO_2 Emissions in 2021.
12. V. Masson-Delmotte *et al.*, Global warming of 1.5 C. An IPCC Special Report on the Impacts of Global Warming of 1.5 C, 2018, 1.

13. Paris Agreement, *Report of the Conference of the Parties to the United Nations Framework Convention on Climate Change (21st Session, 2015: Paris)*, 2017, vol. 4.
14. H. Ritchie and M. Roser, Energy. Our World in Data, https://ourworld indata.org/energy (2020). [Online Resource], retrieved 21 September 2021.
15. IRENA, *Innovation Outlook Advanced Liquid Biofuels*, 2016, p. 132.
16. X. Ji and X. Long, A review of the ecological and socioeconomic effects of biofuel and energy policy recommendations, *Renewable and Sustainable Energy Reviews*, 2016, **61**, 41–52.
17. G. Charis, G. Danha and E. Muzenda, A review of the application of GIS in biomass and solid waste supply chain optimization: Gaps and opportunities for developing nations, *Detritus*, 2019, **6**, 96–106.
18. G. Charis, G. Danha and E. Muzenda, *Detritus*, 2018, **03**, 47–57.
19. F. Zhang, *et al.*, *Renewable Energy*, 2016, **85**, 740–748.
20. M. Dees, *et al.* Roadmap for regional end-users on how to collect, process, store and maintain biomass supply data. (2017), p. 78.
21. A. Zamboni, N. Shah and F. Bezzo, *et al.*, *Energy Fuels*, 2009, **23**, 5121–5133.
22. J. S. Koikai, *Utilizing GIS-Based Suitability Modeling to Assess the Physical Potential of Bioethanol Processing Plants in Western Kenya*, Saint Mary's University of Minnesota University Central Services Press, Winona, MN, 2008, vol. 10, pp. 1–8.
23. C. Chalkias and K. Lasaridi, in *International Waste Management*, ed. S. Kumar, InTech, 2011, vol. 1, pp. 417–434.
24. L. Quinta-Nova, P. Fernandez and N. Pedro, GIS-Based Suitability Model for Assessment of Forest Biomass Energy Potential in a Region of Portugal, *IOP Conf. Ser.: Earth Environ. Sci.*, 2017, **107**, 95.
25. H. Valin, *et al.*, The land use change impact of biofuels consumed in the EU Quantification of area and The land use change impact of biofuels consumed in the EU Quantification of area and greenhouse gas impacts, 2015.
26. I. Sufiyan, S. I. Dasuki and I. M. Kontagora, Design and Development of GIS Database for Informational Awareness on Waste Disposal in Keffi Nigeria, *IOSR Journal of Environmental Science, Toxicology and Food Technology*, 2015, **9**, 46–53.
27. Y.-W. Li and A. de Klerk, *Industrial Case Studies, in Greener Fischer–Tropsch Processes for Fuels and Feedstocks*, John Wiley & Sons, Ltd, 2013, pp. 107–129.
28. P. M. Maitlis and A. De Klerk, *Greener Fischer–Tropsch Processes for Fuels and Feedstocks*, John Wiley & Sons, Ltd, 2013, pp. 337–358.
29. D. Liuzzi, *et al.*, *Rev. Chem. Eng.*, 2020, 20190067.
30. M. J. A. Tijmensen, A. P. C. Faaij, C. N. Hamelinck and M. R. M. van Hardeveld, *Biomass Bioenergy*, 2002, **23**, 129–152.
31. C. N. Hamelinck and A. P. C. Faaij, *J. Power Sources*, 2002, **111**, 1–22.
32. P. L. Spath and D. C. Dayton, *Preliminary screening–technical and economic assessment of synthesis gas to fuels and chemicals with emphasis on the potential for biomass-derived syngas*, 2003.
33. M. Rafati, *et al.*, *Energy Convers. Manage.*, 2017, **133**, 153–166.

34. O. Emenike, S. Michailos, K. J. Hughes, D. Ingham and M. Pourkashanian, *Sustainable Energy Fuels*, 2021, **5**, 3382–3402.
35. C. Ramshaw, *Green Chem.*, 1999, **1**, G15–G17.
36. A. Stankiewicz and J. A. Moulijn, *Ind. Eng. Chem. Res.*, 2002, **41**, 1920–1924.
37. A. I. Stankiewicz and J. A. Moulijn, *et al.*, *Chem. Eng. Prog.*, 2000, **96**, 22–34.
38. J. A. Moulijn, A. Stankiewicz, J. Grievink and A. Górak, *Comput. Chem. Eng.*, 2008, **32**, 3–11.
39. T. van Gerven and A. Stankiewicz, *Ind. Eng. Chem. Res.*, 2009, **48**, 2465–2474.
40. S. Becht, R. Franke, A. Geißelmann and H. Hahn, *Chem. Eng. Process.: Process Intensif.*, 2009, **48**, 329–332.
41. F. J. Keil, *Rev. Chem. Eng.*, 2018, **34**, 135–200.
42. Y. Tian, S. E. Demirel, M. M. F. Hasan and E. N. Pistikopoulos, *Chem. Eng. Process.: Process Intensif.*, 2018, **133**, 160–210.
43. S. Sitter, Q. Chen and I. Grossmann, *Curr. Opin. Chem. Eng.*, 2019, **25**, 87–94.
44. Y. Tian and E. N. Pistikopoulos, *Curr. Opin. Chem. Eng.*, 2019, **25**, 101–107.
45. D. C. Boffito and D. Fernandez Rivas, *Can. J. Chem. Eng.*, 2020, **98**, 2489–2506.
46. N. Mota, C. Alvarez-Galvan, R. M. Navarro and J. L. G. Fierro, *Biofuels*, 2011, **2**, 325–343.
47. I. Rossetti, *et al.*, *Int. J. Hydrogen Energy*, 2016, **41**, 16878–16889.
48. Y. Zhang, T. R. Brown, G. Hu and R. C. Brown, *Biomass Bioenergy*, 2013, **51**, 99–108.
49. J. J. Chew and V. Doshi, *Renewable Sustainable Energy Rev.*, 2011, **15**, 4212–4222.
50. Y. Richardson, J. Blin and A. Julbe, *Prog. Energy Combust. Sci.*, 2012, **38**, 765–781.
51. G. Xu, T. Murakami, T. Suda, Y. Matsuzawa and H. Tani, *Ind. Eng. Chem. Res.*, 2006, **45**, 2281–2286.
52. J. Corella, J. Toledo and G. A. Molina, *Ind. Eng. Chem. Res.*, 2007, **46**, 6831–6839.
53. K. Göransson, U. Söderlind, J. He and W. Zhang, *Renewable Sustainable Energy Rev.*, 2011, **15**, 482–492.
54. J. Karl and T. Pröll, *Renewable Sustainable Energy Rev.*, 2018, **98**, 64–78.
55. N. Hanchate, S. Ramani, C. S. Mathpati and V. H. Dalvi, *J. Cleaner Prod.*, 2021, **280**, 123148.
56. M. Bartels, W. Lin, J. Nijenhuis, F. Kapteijn and J. R. van Ommen, *Prog. Energy Combust. Sci.*, 2008, **34**, 633–666.
57. A. M. Parvez, S. Hafner, M. Hornberger, M. Schmid and G. Scheffknecht, *Renewable Sustainable Energy Rev.*, 2021, **141**, 110756.
58. M. L. Mastellone and U. Arena, *AIChE J.*, 2008, **54**, 1656–1667.
59. S. Rapagnà, *et al.*, *Energy Fuels*, 2009, **23**, 3804–3809.
60. E. Savuto, A. Di Carlo, K. Gallucci, S. Stendardo and S. Rapagnà, *Chem. Eng. J.*, 2019, **377**, 120290.
61. M. Nacken, L. Ma, S. Heidenreich and G. V. Baron, *Appl. Catal., B*, 2009, **88**, 292–298.

62. Y. Guo, *et al.*, *Renewable Sustainable Energy Rev.*, 2010, **14**, 334–343.
63. C. S. Lee, A. V. Conradie and E. Lester, *Chem. Eng. J.*, 2021, **415**, 128837.
64. R. E. A. Eatwell-Hall, V. N. Sharifi and J. Swithenbank, *Int. J. Hydrogen Energy*, 2010, **35**, 13168–13178.
65. P. J. Dauenhauer, B. J. Dreyer, N. J. Degenstein and L. D. Schmidt, *Angew. Chem., Int. Ed.*, 2007, **46**, 5864–5867.
66. J. L. Colby, P. J. Dauenhauer and L. D. Schmidt, *Green Chem.*, 2008, **10**, 773–783.
67. J. Lédé, F. Blanchard and O. Boutin, *Fuel*, 2002, **81**, 1269–1279.
68. S. Wang, G. Dai, H. Yang and Z. Luo, *Prog. Energy Combust. Sci.*, 2017, **62**, 33–86.
69. A. V. Bridgwater, Renewable fuels and chemicals by thermal processing of biomass, *Chem. Eng. J.*, 2003, **91**, 87–102.
70. P. J. Dauenhauer, J. L. Colby, C. M. Balonek, W. J. Suszynski and L. D. Schmidt, *Green Chem.*, 2009, **11**, 1555–1561.
71. R. Chakrabarti, J. L. Colby and L. D. Schmidt, *Appl. Catal., B*, 2011, **107**, 88–94.
72. R. Chakrabarti, S. A. Tupy and L. D. Schmidt, *Energy Fuels*, 2011, **25**, 4763–4769.
73. R. Chakrabarti and L. D. Schmidt, *Energy Fuels*, 2015, **29**, 8102–8109.
74. Y. Richardson, J. Motuzas, A. Julbe, G. Volle and J. Blin, *J. Phys. Chem. C*, 2013, **117**, 23812–23831.
75. Y. Richardson, J. Blin, G. Volle, J. Motuzas and A. Julbe, *Appl. Catal., A*, 2010, **382**, 220–230.
76. Y. Richardson, S. T. Tanoh, A. Julbe and J. Blin, *Fuel*, 2015, **154**, 217–221.
77. M. P. Rohde, D. Unruh and G. Schaub, *Catal. Today*, 2005, **106**, 143–148.
78. R. Guettel, U. Kunz and T. Turek, *Chem. Eng. Technol.*, 2008, **31**, 746–754.
79. D. Liuzzi, *et al.*, *Catal. Today*, 2016, **268**, 37–45.
80. M. C. J. Bradford, M. Te and A. Pollack, *Appl. Catal., A*, 2005, **283**, 39–46.
81. A. A. Khassin, *et al.*, *Catal. Today*, 2005, **105**, 362–366.
82. X. Pan, F. Jiao, D. Miao and X. Bao, *Chem. Rev.*, 2021, **121**, 6588–6609.
83. F. Kapteijn and J. A. Moulijn, *Catal. Today*, 2022, **383**, 5–14.
84. S. LeViness, S. R. Deshmukh, L. A. Richard and H. J. Robota, *Top. Catal.*, 2014, **57**, 518–525.
85. J. J. Lerou, A. L. Tonkovich, L. Silva, S. Perry and J. McDaniel, Microchannel reactor architecture enables greener processes, *Chem. Eng. Sci.*, 2010, **65**(1), 380–385.
86. L. Fratalocchi, C. G. Visconti, G. Groppi, L. Lietti and E. Tronconi, *Chem. Eng. J.*, 2018, **349**, 829–837.
87. L. Fratalocchi, G. Groppi, C. G. Visconti, L. Lietti and E. Tronconi, *Chem. Eng. J.*, 2020, **386**, 123988.

Subject Index